Why Race Matters

Race Differences and What They Mean

MICHAEL LEVIN

Human Evolution, Behavior, and Intelligence
Seymour Itzkoff, Series Editor

Westport, Connecticut
London

Library of Congress Cataloging-in-Publication Data

Levin, Michael E.
  Why race matters : race differences and what they mean / Michael
Levin.
     p.   cm.—(Human evolution, behavior, and intelligence, ISSN
1063–2158)
  Includes bibliographical references and index.
  ISBN 0–275–95789–6 (alk. paper)
  1. Race. 2. Nature and nurture. 3. United States—Race
relations. 4. Blacks—Intelligence levels. I. Title. II. Series.
GN269.L49   1997
305.8—dc20   96–36361

British Library Cataloguing in Publication Data is available.

Library of Congress Catalog Card Number: 96–36361
ISBN: 0–275–95789–6
ISSN: 1063–2158

First published in 1997

Praeger Publishers, 88 Post Road West, Westport, CT 06881
An imprint of Greenwood Publishing Group, Inc.

Printed in the United States of America

The paper used in this book complies with the
Permanent Paper Standard issued by the National
Information Standards Organization (Z39.48–1984).

10 9 8 7 6 5 4 3 2 1

To Meg, Mark, and Eric

I'll teach you differences
                                        —*King Lear* I, iv.

Contents

Illustrations

FIGURES

TABLES

Acknowledgments

A number of individuals scrupulously reviewed earlier drafts of this work, making literally thousands of stylistic and substantive suggestions, and saving me from many errors. For encouragement, advice and assistance, I wish to thank my wife Margarita, Joseph Fulda, Walter Block, Ed Gordon, Jared Taylor, Wallace Matson, Max Hocutt, Louis Pojman, J. P. Rushton, Seymour Itzkoff, Arthur Jensen, Linda Gottfredson, Shelby Hunt, C. D. Ankney, Glayde Whitney, David Harder, Ethan Akin, Neven Sesardic, Richard Herrnstein, Charles Murray, David Stove, Richard Lynn, and Thomas Bouchard. Criticisms offered by David Weissman helped me clarify my own thinking. Lynn Zelem and Pattie Steele did an invaluable job of editing. My computer-literate sons Mark and Eric helped with the tables and graphs, and rescued me from countless computer panics and crises.

This book was made possible by a grant from the Pioneer Fund, under the auspices of the City University of New York Research Foundation. I deeply appreciate their help.

1

"So What?"

When I began writing in 1992, I foresaw battling strong taboos against discussing race differences in intelligence and motivation. Since that time, the topic has been made more accessible by a number of books—Richard Herrnstein and Charles Murray's *The Bell Curve*, J. P. Rushton's *Race, Evolution and Behavior*, Daniel Seligman's *A Question of Intelligence*, Dinesh D'Sousza's *The End of Racism*, Jared Taylor's *Paved with Good Intentions*, Seymour Itzkoff's *The Decline of Intelligence in America*, and Stanley Burnham's *America's Bimodal Crisis*. Herrnstein and Murray, Rushton, Seligman, and Burnham discuss race differences in intelligence, while Taylor, although he offers no positive account of the current unsatisfactory state of race relations, clearly rejects racism as an explanation. (D'Sousza blames cultural relativism.) Indeed, where I argue that race matters because it is connected with intelligence and motivation, *The Bell Curve* might have been called *Why Intelligence Matters*, explaining the connection of intelligence to life outcomes; this book completes the syllogism. At the same time, the reception of these books has been profoundly mixed: some respectful attention, some disregard (especially of Taylor, Seligman and Burnham), much demonizing. Race differences are still vehemently denied in most forums.[1] Hearing them affirmed is not quite the shock it once was, but much remains to be said.

This book seeks to sift what is sensible from what is confused in thinking about race, particularly but not exclusively as it concerns the United States. This winnowing demands the presentation of a good deal of factual material, in some cases meant to advance the empirical debate, but my overall concerns are analytic. Errors about race are due less to misinformation than to muddle. The astronomer Mark Chartrand once remarked that astrologers are very good at making 30-second assertions that are false, but require 30-minute lectures to explain why they are. Much that is said about race is little better than astrology; what follows are the half-hour corrections.

It is strange that race differences should ever have been taboo, since human groups obviously do vary, for instance in skin color and facial features.

No doubt differences in these merely "physical" qualities are easier to acknowledge than qualities of mind, which are felt to matter more. But behavior straddles the mind/body distinction; thought, the immediate control of behavior, is controlled in turn by the patently physical brain. And, since groups can differ in bone structure, they can presumably differ in brain structure, brain function, hormone levels, and other determinants of mental activity. Indeed, it is recognized that human groups not only look but act differently. Asians are comparatively restrained; rhythm is a more salient feature of black than European music. These differences are usually attributed to "society," but to seek to explain them is to admit they exist. If breeds of dog may differ in intelligence and temperament, there seems no reason evolution could not have differentiated human groups along similar lines.

As the accumulating evidence has made group differences harder to deny, one is apt to be told that even if they exist they do not matter. Psychologist Robert Sternberg dismisses black/white differences in intelligence with a curt "So what?" (Sternberg 1985: 244). Pursuing the issue, he says, "can only give comfort to those who would like nothing better than to hear the explicit message that blacks will have a greater handicap in the educational, occupational, and military assignments that are most highly correlated with measures of general intelligence." Noam Chomsky writes:

the relation, if any, between race and intelligence has little scientific importance. . . . [A] correlation between race and mean IQ (were this shown to exist) entails no social consequences except in a racist society in which each individual is assigned a racial category and dealt with, not as an individual in his own right, but as a representative of this category. . . . [W]e are left with little, if any, plausible justification for an interest in the relation between mean IQ and race. (1976: 295–297)

Christopher Jencks is less combative and more open-minded about biologically based race differences, but he too waves aside their "political" importance:

A correlation between genotype and school performance has no clear political implications. Knowing that migraine headaches are often inherited does not tell us whether they are treatable. . . . Just as with test scores, I conclude that genes do influence criminal behavior. But just as with test scores, I also conclude that this fact is politically neutral. (1992: 13–14; also see 98–100)

This book seeks to answer the "So what?" question. It argues that race differences, far from being neutral, undermine almost everything that has been said about race for the past 60 years, and the many policies based on this conventional wisdom. Much is now known about racial variation, but it remains to put this knowledge in a broad philosophical perspective. That is what I have attempted to do.

The importance of race differences is evident at an intuitive level. Whether fear of black crime is legitimate depends on whether blacks do in fact commit more crime, and, if so, why. Whether black poverty and academic failure are the fault of whites, and therefore impose compensatory obligations on whites—perhaps to be discharged by racial preferences—depends on why blacks fail. Received opinion not only ascribes black difficulties to oppression, black attainment is so far below white that, if blacks are as talented as whites, oppression seems the only explanation. And for decades just this conclusion has been drawn. To see how closely racial justice depends on assumptions about race differences, just consider Andrew Hacker's move from the "posit" that "within each race there will be a similar distribution of talents" (1992: 24) to the inference five pages later that "white America orchestrates [competition] to keep black Americans so far behind the starting line." The bearing of genetic race differences on questions of justice has not always gone unnoticed. As recently as the early 1970s critics of hereditarianism conceded that some arguments for quotas

do rely on assumptions concerning the causes of phenotypic differences. Compensatory arguments assume that some proportion of the phenotypic differences between groups is due to past unjust treatment. . . . To the extent that arguments are advanced for proportionate-to-population quotas which rely on assumptions about the distribution of genotypic abilities, it becomes relevant to assess the validity of such assumptions. (Block and Dworkin 1976a: 512; this essay first appeared in 1974)

One suspects that the growing severity of taboos since that time has made it harder to mention race differences even for purposes of repudiation.

The topic of racial variation is admittedly disturbing, and in an ideal world might be passed over in silence, but accusations against whites have made such discretion impossible. The right of the accused to present his case includes the right to raise issues that distress his accuser. A plaintiff demanding damages for a broken leg cannot ask at the same time that his leg not be talked about, nor take offense when the defendant presents evidence that the injury was congenital. By claiming harm he opens the question of why his leg is game. Claiming racial harm has opened the topic of race differences.

More than issues of justice warrant concentration on black/white differences. Yee, Fairchild, Weizmann, and Wyatt (1993) ask rhetorically why less attention is paid to white/Asian differences, but, granting this discrepancy, the reason is clear. Most white/Asians comparisons flatter Asians, while the direction in which black behavior differs from white is by common consent problematic. The black illegitimacy rate, always many times that of whites (Hacker 1992: 80), has risen sharply in the last 30 years to nearly 70% (Rector 1992a: 19; Hacker 1992: 80; also see Jaynes and Williams 1989: 518).[2] 42% of black children live in poverty; 56% of black households are headed by women, fewer than half of whom are self-supporting (Hacker 1992: 80, 86; Rector 1992a: 19; also see Jaynes and Williams 1989: 518). Blacks commit over 50% of the murders in the United States and over 60% of the robberies (Hacker 1992: 181). Comparable disparities exist at all levels: a black college student is 4.2 times more likely than a white to be enrolled in a remedial course (Knowlton 1995).[3] Editorial writers (e.g. Hyland 1992) list "crime, drugs, education [and] a squeeze on the middle class" as contemporary "urban crises": literalizing journalistic euphemy, "drugs" refers mainly to the chaos and violence attending narcotics in black slums (see Jaynes and Williams 1989: 464), "education" mainly to the low levels of achievement and order in schools with large black enrollments, and "urban crises" to all these phenomena together with the physical decay of cities with sizable black populations. A disproportionately white middle class is "squeezed" by taxation for services such as welfare, emergency medical treatment in public hospitals, and public schools, used disproportionately by blacks. On average, 20% of tuition at major colleges is used as aid for poorer, predominantly black, students (Goldin 1995).

Overtly race-neutral problems often have a tacit racial dimension. One with incalculable consequences for health, safety, and the habitability of cities is the growing cohort of urban derelicts. Although this is seldom stated openly, the median "homeless person" is a single black male; according to one survey, 91% of adult homeless in New York are black or Hispanic, according to another survey 94% (Hamill 1993). Explanations abound for the emergence of this cohort, but one neglected factor is black illegitimacy. The number of black males who have difficulty supporting themselves in cities is large to begin with, for urban life, unlike rural, requires money, which in turn requires a steady income. In the country a man with limited skills may own land on which to grow food, and get by by bartering odd jobs for other goods; in a city his skills must be marketable. White incompetents are generally cared for, directly or indirectly, by their families, whereas illegitimate urban blacks usually have no families. A 20-year-old male unwilling or unable to sell his labor cannot be helped by an unmarried, unemployed 35-year-old mother with other children and possibly grandchildren. He comes to live in public areas,[4] surviving by begging and scavenging. Supporting the hypothesis that loose black pair-bonding and a mismatch between black abilities and self-sufficiency in cities contribute to vagrancy is the fact that the homelessness phenomenon emerged within two generations of the large black migration from the rural South to the urban North (described in Lemann [1991]), and within one decade of the jump in black illegitimacy in the 1960s. "Homelessness" is a further reason for interest in black abilities and sexual behavior.[5]

The basic argument for studying race differences is that racial outcomes are currently viewed though a lens of guilt, and it is important to know whether this lens is distorting. One result of racial guilt feelings, already alluded to, is the idea that blacks deserve compensatory preferences in employment and education. Another is use of a "disparate impact" criterion for bias, according to which a standard or practice discriminates against blacks if blacks do not do as well as whites with respect to it. Since whites usually do outperform blacks, seemingly rational practices are besieged throughout society. One example is the cancellation of the presumption of innocence by the 1991 Civil Rights Restoration Act,[6] under which an employer accused of bias for hiring disproportionately few blacks bears the burden of proving that he did not discriminate. This use of numbers to create a presumption against the defendant is reasonable (if constitutionally objectionable; see Epstein 1992) only if there are in fact proportionately as many able blacks as whites in the workforce.

A striking application of disparate impact is the inference from the disproportionate rate of executions of blacks[7] and killers of whites to bias in capital punishment. Such bias as exists could of course be ended by executing more whites and (the almost always black) killers of blacks rather than by ending capital punishment, the remedy usually proposed, but in any case these statistical discrepancies show bias only if whites commit proportionally as many capital crimes as blacks and are no more frequently the victims of capital homicides. In fact, neither condition holds. To begin with, as already noted, blacks commit violent crimes at many times the rate whites do, and, while half of all black crimes victimize whites, less than 3% of white crimes victimize blacks. Now suppose in addition that blacks are on average less empathetic than whites, hence more inclined to regard other people merely as resources or obstacles. Blacks will then be more inclined to kill to overcome resistance or silence witnesses during felonies, the sorts of murders for which the death penalty is often reserved. Suppose, too, that blacks react more violently to frustration. As impulsive, anger-driven murders in personal disputes are usually *not* capitalized, relatively more blacks will be *victims* of noncapital homicides in personal disputes, which tend to involve members of the same race. Should proportionally more whites be killed in interracial robbery-murders than in white-on-white personal confrontations, which, we will see, is evidently the case, the two rate discrepancies would be effects of race differences in temperament rather than bias.[8] Indeed, even proponents of the bias-in-sentencing thesis admit that sentence discrepancies shrink when circumstances are controlled for. Baldus, Woodworth, and Pulaski (1992) nonetheless maintain that killers of whites are executed 6% more frequently when type of homicide is held constant; others claim that the discrepancy vanishes altogether when further controls are imposed (Katz 1989; Rothman and Powers 1994). The larger point is that rejecting the biosocial explanation out of hand, and demanding the end of capital punishment until its application is proportionate, commits one to urging the suspension of all punishment, since blacks are convicted of all FBI-indexed crimes far more frequently than whites. Should a higher black crime rate be admitted but ascribed to "racism," the logic of compensation used to justify racial preference to combat racism entitles blacks to reduced sentences.

In fact, this seeming reductio ad absurdum comes ever closer to being proposed. Bell (1983) suggests imprisonment quotas. One federal court overturned an anti-crack law on the grounds that crack is used disproportionately by blacks (London 1991); another has given black defendants lighter sentences than mandated by sentencing guidelines because of "institutional racism within the criminal justice system" (AP 1993). The Black and Puerto Rican Caucus of the New York State Legislature opposes anti-scofflaw measures lest they "target minorities" (Pierson 1993), and the American Civil Liberties Union opposes a crackdown on public urination as "unfair to poor minorities" (Rutenberg 1993). Mauer and Huling (1995: 27) urge "Legislative Racial/Ethnic Impact Statements" to determine whether criminalizing an act "would affect minorities disproportionately." At this writing a case pending in New York State (see Vacco and Donziger 1995) argues that, since the rate of imprisonment for voting-age blacks and Hispanics is 12 times that of whites, denying convicted felons the right to vote illegally "dilutes" black voter strength under the Voting Rights Act.

Fear of disparate impact inhibits seemingly sensible educational measures, such as strengthening academic standards, grouping children by ability, and asking that college athletes pass their courses. Congress declined to withhold loans to students at colleges with high default rates when it was noticed that this rule "would have hurt inner-city students" (AP 1994a). The Cincinnati school system "counsels" teachers who discipline black students in disproportionate numbers. A decision by Mississippi to raise admission standards for state universities up to white levels has been challenged as "a new strategy for locking blacks out of higher education" (Applebome 1995, A14). The press reported similar problems when the University of Texas sought to raise admission requirements:

[O]pponents say the proposal will limit access to the university, making it more difficult for women and minorities to enroll. Fifty-eight percent of all black students admitted between 1987 and 1990 would not have met the pending new standards, according to a study conducted by the university's research department. . . . "This puts me in a bad position," said Dr. [Ryan] Amacher, who took over as the university's president just six months ago and has said he opposes the new admissions policy. "Everyone is for better quality, but this institution has given people a first chance, a second chance and even a third chance." (Celis 1992)

Preemptive abandonment of standards requires practice in doublethink. Probably many educators, like Dr. Amacher, expect blacks to fare poorly under any objective screening procedure, yet they can neither admit to suspecting a black intellectual deficit, nor declare with any plausibility that, after decades of efforts at fairness, standards remain biased. With no coherent position left, discourse lapses into formulas about a "tradition" of black students being served poorly by education.

Finally, although this would be difficult to document, whites appear increasingly averse to norms intended to apply to society as a whole, or just to themselves, lest they be "racist" by implication. Whites when interviewed will refuse to condemn black youths who won't take low-paying "dead end" jobs, a hitherto unacceptable reason for idleness. There is a tendency in discussions of acquired immunodeficiency syndrome and illegitimacy to insist that teenage girls are going to have sex, and that talk of abstinence is foolish, intolerant, and repressive—a puzzling tendency, given that no previous Caucasoid society accorded teenage girls sexual liberty, or failed to condemn illegitimacy. I suspect that a major contributor to this sea-change has been fear of offending blacks, whose behavior often violates white sexual norms. In fact, mentioning the greater casualness of black sex invites retaliatory diagnoses of prurience and sexual problems:

Compounding the ordinary insecurities most men have in this sphere, white men face the mythic fear that black men may outrival them in virility and competence. . . . Aggravating this unease is a further foreboding: that white women may wonder whether black men could provide greater sexual satisfaction than they now get from their white mates. Notice, also, how white men glance a second time when they see a racially mixed couple: what, they seem to ask, does *she* see in—or do with—him? . . . [F]antasies persist that black men and women are less burdened by inhibitions, and can delight in primal pleasures beyond the capacity of whites. (The erotic abandon displayed in black dancing has no white counterpart.) Nor is it surprising that much of the commentary concerning women on welfare adds the charge that they share their beds with successions of men. Underwriting indolence is bad enough. That taxes also subsidize sexual excess stirs anger and, just possibly, envy. (Hacker 1992: 62)

Note that Hacker does not deny that black women on welfare are promiscuous. Nor does he reflect that the "erotic abandon" of black dance, along with statistics he cites elsewhere about black illegitimacy and early onset of intercourse, suggest that a stronger black sex drive is more than "fantasy."

The reader who accepts genetic race differences as fact may move immediately to chapter 6, where I begin to discuss their implications in earnest. Readers willing to allow that races might differ can also begin there, construing the subsequent discussion in a hypothetical spirit. They may take one as asking what follows *if* races differ genetically. But many readers will be disinclined to follow even that far, for conventional wisdom holds the evidence for race differences too meager to take seriously; worrying about them is like thinking about flying saucers or creationism. Nor can mere citation of previous hereditarian work be expected to impress the dubious.[9] I have therefore presented a self-contained statement of the empirical evidence in chapters 2–4, hoping to combine rigor with clarity, so that even the reader initially unfamiliar with the topic may judge the argument for himself at each step. A few stretches involve some mathematics, but summaries are always provided; more technical matters are consigned to footnotes or appendices. Although much of this material exists elsewhere in the literature, I try where possible to shed conceptual light on it. For this reason, some issues that may strike nonspecialists as abstruse or digressive are treated for the sake of completeness. Much of chapter 5, for instance, deals with whether a biological orientation is "determinist" or "reductionist," and the relation of genes to personal identity.

Chapter 2 summarizes key biological and statistical concepts; chapter 3, the evidence that whites are on average more intelligent and future-regarding than blacks. The familiar charge that standardized tests are biased is rejected in favor of a "realistic" view of intelligence and the intelligence quotient (IQ). A phenomenon may be considered real, I say, when it manifests itself in a variety of ways, and the multiple correlates of IQ show that it is no mere artifact or mathematical fiction. What IQ tests manifest coincides with what is ordinarily called "intelligence."

A mistake I constantly warn against is that of thinking individual or group differences must be genetic to be real. This is not so: a man with an IQ of 120 has more of what IQ measures than a man with an IQ of 100, whatever may have caused the difference, just as a 200-lb man is heavier than a 150-lb man, whether because of a genetically more sluggish metabolism or a fortuitous injury that keeps him from exercising. That intelligence is a phenotype—to use the language of genetics—allows the reality of race differences in this trait to be investigated apart from the question of genes. Nor does it follow, if race differences are real but wholly environmental in origin, that the crucial environmental factor is oppression. "Environmental factor" in this context usually is equated with oppression, however, because at one time blacks were mistreated, and, as noted, among environmental factors only intentional mistreatment seems able to explain the systematic depression of black attainment. That is why pressing on to the involvement of genes is imperative Chapter 4 presents the evidence that the race differences described in chapter 3 are significantly genetic in origin, or, again in the language of genetics, that genotypic variance explains much of the between-race phenotypic variance This position—not that genetic factors explain all of the black/white difference but that they explain a significant proportion—is "hereditarianism." "Environmentalism" attributes all race differences to environmental factors.

That there are race differences in learning ability has been the majority opinion of experts for some time. In 1968 it was possible to speculate in the quiet of the pages of the *Psychological Bulletin* that "As a group Negroes show up as deficient in abstract abilities. Possibly Negro genes limit such abilities" (Dreger and Miller 1968: 26). When over 600 scientists and educator were polled anonymously in the mid-1980s, 84% agreed that intelligence test do measure intelligence, as either a single general ability or a cluster of more specialized abilities, and 53% agreed that genes contribute to the race difference in IQ (Rothman and Snyderman 1988: 71, 285). Only 17% reported believing the difference to be entirely environmental.

Environmentalism has appeared to be the consensus in recent decades because of the sanctions directed against its critics.[10] In reality, its case is hollow. Of the two most widely cited environmentalist works, Lewontin, Rose and Kamin's *Not in Our Genes* (1984) is a farrago, while Stephen Jay Gould's *The Mismeasure of Man* (1981) is mainly a review of the early history and prehistory of mental testing, with attacks on the motives of hereditarians. A most 20% of *Mismeasure* engages issues of substance. Kamin's *The Science and Politics of IQ* (1974) makes a greater effort to deal with data, particularly twin studies, but it is entirely negative, pointing to possible flaws in studies that support a role for genes in behavior without offering any datum that environmental variation alone can explain. Lewontin in some other writings Hirsch, Block and Dworkin, and most recently Block (1995), Goldberger and Manski (1995), and Fischer et al. (1996) are more serious (although hardly free of ad hominems), but they, too, rest their objections to hereditarianism less on data than on speculation about what might be the case—or had better be the case if various moral ideals are to be realized. Often hereditarianism is rejected not on evidential but moral grounds, because it is bad, or "racist." That is why I take pains in chapter 5 to explain why talk of race differences in intelligence is not racist. Racism is by definition bad; if there are good reasons to discuss and accept, race differences, such discussions are not bad, hence not racist.

An honorable exception to these strictures is Flynn (1980). Despite some gratuitous introductory remarks about "racism," he keeps to the data, scorns bad environmentalist arguments, and treats his chief target, Arthur Jensen with respect. Many of the arguments Flynn develops in his (1980), and subsequent (1984) and (1987b) are treated below. However, Flynn (1980) has significant internal weaknesses: it takes at face value the early pronouncement of the Milwaukee Project (50, 160, 186, 205; see chapter 4), its treatment of twin and adoption studies is likewise dated (again see chapter 4), and it ignores cross-cultural comparisons. Flynn (1980) is also little cited in the environmentalist literature, perhaps because it concedes that blacks are les

telligent than whites—their "capacities," Flynn says, have been "damaged"
11)—and is concerned only to argue that this discrepancy is non-genetic in
rigin. Unlike most environmentalists, Flynn does not challenge the fairness of
Q tests and willingly concedes intelligence a fairly high heritability. Many
her egalitarians reject everything about race differences, IQ tests, intelligence
nd heredity, with scant regard for consistency.

Speaking broadly, the characteristic environmentalist fallacy is to infer,
om the possibility that an environmental factor might causes a race
ifference—from the failure of hereditarians to prove their case beyond all
oubt—that an environmental factor is actually involved. To be sure, there are
nvironmental variables that, when controlled for, reduce variation between
ces. The IQ gap shrinks by about 25% when children's socioeconomic status
held constant, for instance, more so when education is held constant as well;
ischer et al. (1996) present further examples. However, if status and education
re themselves partly determined by genes, they are not *nongenetic* factors.
Fischer et al. acknowledge this difficulty, but their sole response—a single
entence long—is that "Research should continue.") Environmentalists are
obliged to explain the race gap via variables uncorrelated with genes, and this
ey have not done. (So-called genotype/environment correlation is discussed
rther in chapter 4.)

Race differences need not be established with Cartesian certainty to carry
mportant implications. As I will have occasion to point out, many policies
st not merely on the denial of genetic race differences, but on the premise that
ese differences are known for certain not to exist. Such policies are
ndermined if it is merely more likely than not, or even somewhat likely, that
enetic race differences are real. Indeed, far from hereditarianism being a theory
or cranks, it is the idea that human evolution stopped at skin color, never
aching inward to the mind—that society is an Unmoved Mover of human
riation—which more resembles creationism. If part I convinces the reader
ho cannot agree that the races do differ that this proposition is at least too
edible to dismiss out of hand, it will have done its job.

Part II is a deep breath, a prolegomenon necessary to forestall
isunderstandings before the practical issues of Part III are taken up. The
ormative concepts used casually in chapter 5 are scrutinized, and four tenets
lvanced.

(1) There is no cosmic scale by which differences in intelligence and
otivation make one race superior to another. A gap separates facts from
dgments of value. The point seems self-evident to me, but I defend it at some
ngth for one unfortunate reason: the claim that blacks are less intelligent than
hites is regularly misdescribed as belief in black "inferiority," and
ereditarians are said to hold whites "higher" than blacks.[11] Indeed, critics
em so ready to think that anyone who says whites are more intelligent than
acks is saying whites are better, or that he believes it even if he does not say
, or that his words imply it even if he does not believe it, that perfunctory
sclaimers would be useless. So chapter 6 dwells on the normative neutrality
race differences.

(2) The issue of race and values ramifies further. There appear to be group
fferences in morality itself, continuous with and emerging from

differences in temperament and intelligence. A full understanding of rac
differences thus requires some account of the function of moral values and thei
possible evolutionary divergence, and an account is developed in chapter 6.

(3) Nor is this the end of the issue, since neutrality about values will be fel
to be inadequate in the present context. The "So what?" question plainl
demands a normative answer that addresses moral and social issues. I argue ii
chapter 7 that a satisfying answer can be given even in the absence of objectivt
values. A value-skeptic may appraise a system of norms for interna
consistency, and investigate the factual premises that guide its application. Hi
approach is not that of the moralist but of the jurist, who applies received lav
on the basis of the facts, and, at the level of judicial review, uses logic an
analogy to check laws and particular judgments for mutual coherence. M
concerns, then, are the factual premises by which widely accepted norms ar
applied to race, and some unacknowledged implications of the reader's ow
value system.

(4) I ultimately stray a bit from the path of empiricist orthodoxy. T
describe how the races differ is not to pass judgment, true, but leaving it ɛ
that is disingenuous. While the traits in which the races differ do not ii
themselves make one race more valuable than another, they are traits tha
people do value, do care about. It is useless to pretend, as value-neutral socia
scientists sometimes try to, that intelligence means no more to anyone thaɪ
blood type. Most individuals, particularly Caucasoids, judge others to
significant extent by their intelligence and self-restraint. The last half o
chapter 7 argues that, by the norms that *in fact* guide the judgment of th
typical reader of this book about matters other than race, whites are on averag
better people than blacks. However shocking this statement may sound,
believe it is a truth that must be dealt with.

Well, what do (or would) genetic race differences imply? Chapters 8–10 ge
down to important cases, deploying accepted norms of justice, self-defenst
reciprocity, and individual liberty in connection with racial preferences, crimt
civil rights, and the use of race in judging individuals.

Chapter 8 maintains that whites owe blacks no compensation in any form
because the limitations of blacks are by-products of genetic differences, no
injuries done by whites. It insists that virtually all arguments for preference aɪ
at bottom compensatory, which is why criticisms of affirmative action that dt
not engage the causes of race differences invariably miss the point.

Chapter 9 maintains that fear of black crime is rational, that ordinar
standards permit private race-conscious avoidance of black crime, and tha
race-conscious measures by the state in the exercise of its police function aɪ
also acceptable in principle. The evidence of a genetic component in blac
criminal deviance raises in a sharp new form the old issue of free will an
responsibility; I defend the "compatibilist" view that freedom is consistent wit
determinism, but add that, because of their lower mean intelligence and greatt
impulsiveness, blacks are less autonomous than whites. Blacks, on averag
less deterrable, are less deserving of punishment, but at the same time race
conscious guidelines for punishment-like deterrent sanctions may be warranted

Race-consciousness as such is finally taken up in chapter 10, which argue
that attention to race is consistent with "treating people as individuals.

orting people is inevitable, I say: what matters is sorting them correctly. ome classification by race is warranted, for example, by the correlation of ace with crime. Other classifications, such as those assumed by compensatory acial preference, are unwarranted. But racial categorizing is not objectionable er se, nor need it violate the principle, embodied in the 14th Amendment, that eople should be treated as "equals." This principle means only that ifferential treatment must be justified by some relevant trait, and, for many rivate and perhaps some public purposes, race is relevant. While races are roups, membership in a race is a trait of individuals.

I conclude that anti-discrimination laws violate versions of the golden rule nd freedom of association widely accepted in non-racial contexts. Others Nozick 1975;[12] Narveson 1987; Friedman 1989; W. Block 1976, 1992; Levin 984a; Epstein 1992; Fulda 1993; Bolick 1996) defend a similar position; I dd specifically that, by ordinary standards, the desire of whites to refuse oluntary association with blacks is often reasonable. Similarly, as the general ase against coercive transfer of wealth has been made elsewhere (see Nozick 975; Murray 1984; Friedman 1989; Fulda 1993; W. Block 1976), I consider pecifically whether, modulo ordinary standards of prudence, black time orizons make it wise to offer blacks the same "safety net" that whites offer ach other.

It is well to scout at the outset two common errors about group differences, ach the overestimation of a statistical cliché. One cliché runs that group ifferences permit overlap—that, for example, not all blacks are less intelligent han all whites. This is of course so; mean differences allow exceptions, and ecree nothing about specific individuals. However, mean group differences do upport conclusions about the aggregate characteristics of groups, and the *ikely* characteristics of individuals. That women are on average a half-foot horter than men allows some women to be taller than most men, but it rules ut an all-female NBA team, and makes it rational to bet that the next woman ou meet will be shorter than the next man.

A second oversold cliché is that variation within groups typically far xceeds the difference between their means. Again this is indisputable; the gap etween the brightest and dullest white is much greater than that between the verage white and the average black. But once again this does not vitiate etween-group differences. Within-sex variation in height is about 4', an order f magnitude greater than the 6" difference between the average man and the verage woman, yet this 6" difference still explains why tall women are xtremely rare. Likewise, that the 15-point mean difference in IQ between lacks and whites is less than the 100-point spread between the brightest and ullest whites still permits the 15-point gap to carry important consequences, uch as the relative scarcity of black scientists. For the same reason, the fact hat there is more genetic diversity among Africans than between Africans and on-Africans does not render a comparison of African to non-African means 'senseless" (Gould 1995: 13). There is much more diversity of height among logs than among giraffes, but it is quite sensible to observe that dogs are ypically shorter than giraffes.

My conclusions may sound cheerless, although a more positive word is 'sobering." To clarify just what attitude is appropriate, I ask you the reader to imagine that you are the president of the United States, keen to "do something about the race problem" but unsure how. In a week you are to give your recommendations to the nation. I have tried to write the book you would want to read; the Afterword is the speech you might be led to give.

I should say a word about what drew me to the topic of race, and the overall question of motives. In part my interest is a purely intellectual response to the challenge of clarifying such notions as "innate," "natural," and "genetic." Few problems are as intriguing as how we humans got the way we are, and what light human differences shed on evolution. Like many professional philosophers I am drawn to the conceptual foundations of these questions.

A more immediate concern has been affirmative action, which began just after my own academic career did. Since the early 1970s I have been aware of white male associates unable to get positions because of their race (and sex). From the first I sensed something amiss, but had trouble formulating what that something was. I gave some thought to compensatory justice (Levin 1980, 1981), and knew of Arthur Jensen's work on race differences, yet I did not put them together for many years. In 1982 I managed to review Jensen's *Straight Talk about Mental Tests* and Gould's *The Mismeasure of Man* (Levin 1982a) without fully connecting race differences in intelligence to the redress rationale for affirmative action.

It took me so long to see the obvious connection, and others still miss it, I believe, because of distraction by the red herrings that litter discussions of race. Repeated warnings against consequences race differences do not have obscure the consequences they do. Consider this characteristic volley from Lewontin:

But suppose the difference between the black and the white IQ distributions were completely genetic. What program for social action flows from that fact? Should all black children be given a different education from all white children, even the 11 percent who are better than the average white child? Should all black men be unskilled laborers and all black women clean other women's houses? (1970: 113)

A constant need to fend off attributions of idiocy—here, that all black women should be domestics—draws attention away from more serious reasons that race differences might be important. Having to contend with an army of straw men deflects one's concentration.

That opposition to hereditarianism is associated with the political Left may suggest that I stand on the Right. Understood as the view that democratic liberties are a facade for oppression, "leftism" does indeed strike me as absurd. Yet on matters of race, the orthodoxy of the Right is no more enlightened than that of the Left, and in some ways less coherent. The Left holds that blacks would do as well as whites but for racism, the Right that blacks would do as well as whites but for well-meaning government policies like welfare that sap black ambition. (The seductions of popular culture are sometimes added.) The Left's theory may be wrong, but it observes the forms of correct reasoning. It tries to deal with conflicting evidence, positing unconscious "structural" discrimination to explain black failure in the United States after the passage of civil rights measures, and internalization of the white man's image of blacks to explain the decline of postcolonial Africa. On the other hand, while

nservatives have made a strong case that welfare has accelerated black
ime, poverty, and illegitimacy, they ignore the failure of whites to respond as
acks do to welfare incentives available to both races, and explain black
ilure in the post–civil rights era as a legacy of slavery in language borrowed
om the Left (see, e.g., Jacoby 1992). Conservatives such as Thomas Sowell,
vare of the worldwide failure of black cultures to develop European/Asian
vels of technology, circularly attribute this failure to black culture. The
uism that a bad theory beats no theory may explain why the Right's account
race relations is seldom taken seriously.

Left and Right [13] tend to share four premises. One, of course, is that racial
fferences are caused by forces external to the races themselves—racism for the
eft, government intrusion for the Right. A second is impatience with
uantitative reasoning. Prior to inquiry, it is natural to ascribe the black/white
tainment gap to a number of causes, including genes, environmental
rrelates of genes, discrimination, and historical accidents. The substantive
sue is how much each factor contributes. Yet both sides cast the debate in all-
-nothing terms—it is all oppression or all welfare, with nothing between.
ome single factor or cluster of factors may well explain the entire race gap
ntelligence and temperament loom large in chapter 8), but the question
hether one factor dominates must be asked before it can be answered.
cidentally, the all-or-nothing mindset can coexist with extensive use of
escriptive statistics. Hacker (1992), for instance, presents pages of
uantitative data on race differences in residential patterns, crime, income,
productive trends and academic success, but goes on, without any causal
nalysis, to attribute these differences entirely to white racism. Fischer et al.
996) make more sophisticated use of regression analysis to isolate causes,
ut when in the crunch they must explain why blacks persistently fall short on
bjective tests, they offer only vague qualitative conjectures.

Third, a preoccupation with blame leads both the Left and the Right to
nflate questions of cause with questions of fault. The causal question is,
mply, "Why do the races differ?" The complex fault question is, "What
alice or folly created these differences?" For the Left it is malicious racism,
or the Right it is foolish welfare, with both sides ignoring the possibility that
uman action has nothing to do with it. The upshot is scolding and lecturing,
s the Left scolds whites for "racism" of which they may be innocent, the
ight (e.g., Mead 1990; Yates 1991; Magnet 1993; D'Souza 1995) lectures
veryone about a work ethic blacks may be unable to follow, and the Left
olds the Right right back for "blaming the victim."

Finally, both Left and Right see the failure of blacks to live like whites as a
roblem.[14] Certainly, blacks are less prosperous than whites. But this relative
ortfall does not imply that blacks are deprived in any absolute sense; a black
ith a TV set and flush toilet has treasures undreamed of by the Pharaohs.
hould the black/white discrepancy be an expression of more basic biological
ifferences, it is not clear why anything must be done about it. After all, it is
ot a problem that owls live in trees while gophers dwell underground, except
erhaps for an owl confined to a burrow. Consider the black/white difference in
nfant mortality, cited by Jaynes and Williams (1989: 401) as a paradigm
onsequence of preventable social conditions. Should this difference be a

biological phenomenon, as recent research suggests,[15] it is no more a soc: evil than is the susceptibility of Jews to Tay-Sachs. Humanitarian reaso would remain for trying to save more black babies, but the majority could . longer be held responsible for causing black infant mortality, or negligent permitting it. To be sure, biological differences can create problems—compa the underground owl. Behavior evolved in Africa may be maladaptive in urb societies created by Caucasians (but see chapter 6), making life in whi society a problem for blacks and the frequent conflict between black behavi and white norms a problem for everyone. But sheer differences in prosperity a not in themselves bad.

To be sure, blame-laying has its comforts. One can try to improve a ba situation caused by bad deeds by attacking the culprits, an option n available absent culprits. Racial conflict due to discrimination can be treate by suppressing discriminators; conflict not stemming from discrimination is ` that extent less tractable, hence, paradoxically, more alarming.

Talk of blame and evil leads inescapably to the Nazis, and whether t views expressed here resemble Hitler's. Strenuous denial would only strengthe the suspicion that there is something to deny, but one decisive point should I made. Nothing Hitler believed or did has any logical bearing on any issue substance. That a bad man believed races differ does not mean that all rac are the same. That his actions were caused by this belief does not make tho actions its corollaries. Physics has allowed the development of weapons, b that does not falsify physics, nor are the monstrous uses to which thes weapons have been put implicit in the discoveries that made them possible.

Invocation of Hitler is one of many ways in which discussions of race a warped into battles between good and evil. Another is the contrast betwee sympathy for the underdog, the decent motive driving many environmentalist with hereditarian "meanness." Still another is the counsel that genetic rac differences should be denied whether they exist or not because focusing nature breeds fatalism while focusing on nurture encourages social reform.

These versions of the debate are baseless: environmentalism appeals motives as suspect as "racism" and as likely to do harm. Blacks have practical interest in environmentalism inasmuch as it justifies racia preferences, and a psychological interest in it insofar as it explains awa failure. As for harm, telling blacks that their estate is due to oppression rathe than lack of ability can only stoke resentment. Furthermore, blamin crime—black or white—on oppression weakens society's will to restrain it, an hands criminals the rationalization that they are entitled to the possessions others that they covet. And, at the same time that bad motives abound fc denying race differences, those imputed to hereditarians are left unclea Hereditarians can hardly be preening themselves on the superiority of their ow race, since scholars (like most people) derive a sense of self-worth mainly fro personal, not group, achievement. I have already cited worthy reasons investigate race, among them scientific curiosity and a desire for justice; wh evidence is there that these concerns never inspire hereditarians? The classic legal question Cui bono? cuts two ways, and wounds environmentalists mo deeply.

Calling talk of race differences an excuse for perpetuating inequity begs th

uestion: black poverty and educational failure are inequities only if the races
o not differ. Should biological race differences exist, the arrangements
mplained of may be equitable after all. First deciding what is equitable and
en attacking extant social arrangements on those grounds puts the cart before
e horse. The proper course is to examine the facts first, then by their lights
sess the state of society. To this examination we now turn.

NOTES

1. According to the *New York Times'* editorial of Oct. 24, 1994, *The Bell Curve* "is
ced with tendentious interpretation. Once unlike-minded scholars have time to react,
ey will subject its findings to withering criticism. At its best, the Herrnstein-Murray
ory is an unconvincing reading of murky evidence. At its worst, it is perniciously and
urposely incendiary. . . . Mr. Murray has . . . not built a scientific case." A review in
e *Journal of Economic Literature* complains "HM and their publishers have done a
sservice by circumventing peer review. *The Bell Curve* was sprung full blown without
ternal scientific scrutiny. . . . [A] process of scientific review is now under way. But,
ven the [initial publicity], peer review of *The Bell Curve* is now an exercise in damage
ontrol rather than prevention" (Goldberger and Manski 1995: 776). Within 18 months
its publication, several volumes appeared devoted solely to refuting *The Bell Curve*,
e most responsible of which is Fischer et al. (1996). This outpouring testifies to the
mportance of *The Bell Curve*, but also the extraordinary hostility to hereditarianism
ithin the intellectual community.
2. The ratio of black/white illegitimacy rates decreased from 9.9 (16.1%/1.7%) in
950 to 4.3 (63.7%/14.9%) in 1988. (At this writing the white illegitimacy rate exceeds
e black rate of 1950.) This ratio will obviously tend to shrink as the black rate
pproaches 100%.
3. Blacks are 9.3% of the college population, and make up 30% of the students
rrolled in remedial courses; the corresponding figures for whites are 76.9% and 59%.
4. An increasingly common sight in New York City is public urination and
efecation.
5. The increase in black illegitimacy over the last half-century, while black genes
esumably remained constant, does not show that race differences in illegitimacy are
nvironmental rather than genetic in origin. The change may illustrate
enotype/environment interaction, with black genotypes responding more strongly to the
ncreased availability of welfare in the 1960s than they did to previous social
nvironments, while the white response across this change has been more stable.
6. This law merely codified a trend already present in court decisions made from the
970s to the late 1980s, and reversed the Supreme Court's partial retreat from the effects
andard announced in *Wards Cove Packing Co. v. Antonio* (1989).
7. Actually, although proportionately more blacks than whites are executed, black
urderers are executed at a lower rate than white murderers (Katz 1989). Generally
eaking, blacks accused of murder, robbery, rape, assault and burglary are all
nvicted at a lower rate than whites accused of these crimes; see Lerner 1996.
8. Consider a numerical example. Suppose 12 blacks commit 30 murders, 10 of
em victimizing whites, and 88 whites commit 25 murders, 2 of them victimizing
acks, and that, for both races, 10% of within-race murders but 50% of between-race
urders are capital crimes. Absent bias, 3 murderers of blacks and 7 murderers of
hites will face capital charges.
9. Moreover, Rushton and the relevant sections of Herrnstein and Murray ignore
ertain standard objections to "hereditarianism." Herrnstein and Murray stipulate with

little argument that there is such a thing as intelligence and that it is highly heritabl Gould (1994) rightly complains of this. Moreover, because they are exclusive concerned with intelligence, Murray and Herrnstein (although not Rushton) tend to tre blacks as distinguished simply by a lower mean IQ. Rushton for his part moves rath hastily from interindividual to intergroup heritability. (Given their purposes, tl decision of these authors to bypass certain issues is understandable.) Finally, neith Herrnstein and Murray, Rushton, nor Arthur Jensen explicitly concern themselves wi degrees of probability, but affirmative action is undermined, from an evidentiary poi of view, if it is merely more likely than not that genetic race differences expla outcome disparities.

10. When I mentioned race differences in intelligence in print, responses includ death threats, feces in the mail, arson, telephone harassment, and efforts by my colle, to revoke tenure, which necessitated a federal lawsuit. Other academics who ha publicly urged a role for genes in behavior, such as Jensen, Rushton, and E. O. Wilso report similar and often less pleasant experiences. Rushton has been investigated by tl Canadian government for dissemination of "hate literature," punishable by two yea imprisonment.

11. Gould allows himself great latitude in imputing belief in "racial ranking Despite Jensen's disavowals, Gould insists that Jensen accepts it—arguing that the id has "subtle power, even over those who would deny it explicitly" (1981: 159).

12. Nozick currently (1989: 291) believes freedom of association should be limite by the need to express social solidarity, and would apparently endorses a ban on spee "offensive" to blacks, Jews, homosexuals, Armenians, Asians, and American Indiar all groups "membership [in which] is part of their self-conception."

13. Herrnstein and Murray (1994) is again an exception.

14. The exception is Sowell, who insists that in every society different grou display different levels of achievement.

15. Schoendorf et al. (1992) found the mortality rate for infants born to colleg educated black women to be twice that for infants born to college-educated whi women, the same as the overall black/white difference. Deprivation is thus not relevant factor. The only relevant variable found by Schoendorf et al. was the rate prematurity among black births, three times the white rate.

Part I. The Empirical Argument

2

Preliminaries

is chapter presents basic biological and statistical ideas, with special ention to their bearing on race differences. This material is also intended to lp interested readers pursue the primary literature.

I. BASIC BIOLOGICAL CONCEPTS

Race

The concept of race is often said to lack scientific merit (see, e.g., Montagu 72, Yee et al. 1993, Hoffman 1994). Ironically, denial of the reality of race en prefaces a denunciation of race bias, with little explanation given of how ople can respond to a trait that no one possesses and no one understands. It ould be obvious as well that repudiating race forbids advocacy of racial eferences, although few critics of the race concept have faulted affirmative tion on this account. Whatever they may say, the parties to such disputes sume that the notion of race is reasonably clear.

It is true that human races no longer exist if "race" is taken to mean, as it metimes does in biology, a large, isolated breeding population. Yet, in dition to universal understanding of what foes of race prejudice oppose and ends of racial preference advocate, there is wide agreement on ascriptions of :e. One hundred randomly chosen individuals sorting passers-by on an urban eet would, without hesitation or collusion, almost always agree on who is ick, white, or Asian. Moreover, the race others would noncollusively ascribe an individual is almost always the race he unhesitatingly ascribes to mself. Such systematic agreement must rest on some objective sis—possibly misconstrued, but present and detectable.

The definition of race that captures ordinary usage, its usage in popular lemics (e.g., Hacker 1992: 7) and the usage of evolutionary biologists (e.g., Wilson 1978: 48–49) refers to birthplace of ancestors, although the precise

form to give this definition depends on the best theory of human origins. Let
assume the chronology currently favored by anthropologists and molecu
biologists (Stringer and Andrews 1988; Stringer 1990; Gibbons 1995; Horai
al. 1995; Aiello 1993, more cautiously; also see Cavalli-Sforza, Menozzi, a
Piazza 1993), according to which man evolved in Africa, branched off i
Europe 110,000+ years ago, and branched off from there into Asia abc
70,000 years later. The branches have interbred in historical time, Afric
isolation having ended two millennia ago. So, letting 25 years mark a sin
generation, a "Negroid" may be defined as anyone whose ancestors 40 to 44
generations removed were born in sub-Saharan Africa. "Mongoloid" a
"Caucasoid" are defined similarly, with Asia and Europe in place of Afri
Because comparisons of blood group frequencies in the white, African, a
conventionally identified American black populations indicate a wh
admixture of about 25% in the blacks in the American North and 10%
blacks in the American South,[1] an "American Negroid" can be defined
anyone 75% or more of whose ancestors 40 to 4400 generations removed w
born in sub-Saharan Africa. These definitions can be adapted to "polygeni
theories of human origins. If blacks, whites, and Asians evolved separate
over (say) the last million years, a Negroid is anyone (75% or more of) wh
ancestors 40 or more generations removed, with no upper bound, were born
Africa, and likewise for Mongoloids and Caucasoids.

Defining race by place of ancestry, although covering most humans, om
certain mixtures, such as Melanesians. Also, counting 75% African ancestry
Negroid will tend to understate any Negroid/Caucasoid genetic differenc
Still, the fact that it is recognized as appropriate for American blacks to c
themselves "African-Americans," or to call the dominant culture "Europea
shows that most people have the geographical conception of race in mind.

While race is conventionally equated with skin color, its familiar observa
criteria, which also include lip eversion, hair texture, facial bone structure, a
timbre of voice, do not define "race." Rather, these traits serve as continge
indicators of ancestry. They are "stereotypes" in Putnam's (1975; also s
Kripke 1973) technical sense that they fix the references of "Negroi
"Mongoloid," and "Caucasoid." Moreover, because people are grouped by
number of traits combined into gestalts, everyday ascriptions of race are hig
reliable. Indians, although dark-skinned, are seldom confused with Negroi
some Caucasoids may have thinner hair than a few Negroids, but
Caucasoid has thin curly hair, dark skin, everted lips, and a broad, flat no
The passers-by classified as black, white, and Asian in our thought experim
would almost certainly turn out to be of predominantly African, European, a
Asian ancestry, respectively.

It is sometimes objected that the ordinary indicia of race do not "rela
inherently to behavior and potentials" (Yee 1992: 110), but they do n
have to. They are observable correlates of geographical origin, used to identi
that less observable trait. Having certain facial features, say, picks c
individuals of Asian ancestry, just as being the first heavenly body visible
dusk picks out the planet Venus. The look of members of a given race bears t
same relation to the race's inherent properties that prominence in the even
sky bears to such inherent properties of Venus as its mass. Racial appearan

ssociates with further genetic differences, if such associations exist, because
e environments that differentiated appearance also differentiated gene pools.
characteristic appearance and (let us assume) level of genotypic intelligence
ould then be co-effects of a common cause, not causes one of the other—just
 the evening star does not have the mass it does because visibility at dusk
uses it to, but because visibility at dusk and that particular mass are co-
curring attributes of the single entity Venus.[2]

The geographical definition of race is operational; there is a routine
ocedure for applying it not subject to dispute by competent speakers of
nglish. Everyone understands "ancestor" and "Europe," so everyone
derstands "having European ancestors." It does not matter that the ancestries
 some individuals are unknown, and unguessable from the usual indicators.
Molybdenum is harder than chalk" is fully operationalized as "a sample of
olybdenum will scratch a sample of chalk," even though few people can
cognize samples of molybdenum.

When "race" is operationalized geographically, generalizations about races
quire clear empirical meaning. Ascribing a trait to the members of a racial
oup, whether to all or merely a disproportionate number, and whether the
ait is physical or mental, overt or hidden, behavioral or genetic, amounts to
cribing that trait to all or a disproportionate number of people whose
cestors were born in a certain part of the world. To say the mean intelligence
 whites exceeds that of blacks is to say that the mean intelligence of people
 European descent exceeds that of people of African descent. Every such
neralization may be false, but they are uniformly meaningful.

One might object (as does Diamond [1994]) that the race concept is
netheless arbitrary, it being possible to group people in many ways other
an ancestry—by the presence of a designated gene, say, or the ability to
gest milk, or fingerprint whorls. Still, given a classificatory criterion, it
comes a matter of empirical fact, not human choice, whether that criterion
rrelates with further, independently specified traits. There is no objectively
ight" way to classify land areas, which may be grouped by latitude, rainfall,
ight above sea level, or fauna, but once a principle of grouping is
opted—rainfall, say—the correlation of this variable with others, such as
op yield, does become a completely objective question. Nor does the greater
netic diversity of Negroids (defined geographically) compared to all other
pulations combined make talk of Caucasoid/Negroid differences "senseless"
ould 1995). Once again, there is more genetic diversity among dogs than
raffes, yet it makes perfect sense to say that giraffes are taller than dogs. It
mains possible that certain patterns hold across all populations of African
cestry.

In rather the reverse direction, it has been argued (by, e.g., Hoffman 1994)
at, since the races as defined by ancestry differ in genetic material by only
012%, their differences must be insignificant. This figure is misleading, to
gin with, for, after all, humans and chimpanzees share 98.5% of their genes
accone and Powell 1989; Gibbons 1990)—because humans and chimps
ree in having arms, legs, lungs, and other large gene-built structures. More
portant, subtle genetic differences can have large "non-linear" effects. The
rvous system of a virtuoso pianist differs very little in number of shared

genes from that of the average person. The consequences of an .0012⁽
difference in genetic material between ancestral lines is an empirical questic
that cannot be answered by a priori numerical considerations.

To anyone bent on denying come what may that race is a useful concept,
surrender the word "race." Such an individual may read what follows not as
discussion of *race* differences at all, but of *differences between descendants* ₀
Africans and Eurasians. Nothing is lost but a word.

Phenotype and Genotype

Central to genetics is the phenotype/genotype distinction. A *phenotype*
any trait of an organism, such as birth weight, adult height, lifespan, abilit
to recall telephone numbers, and, if there is such a thing, intelligenc
Phenotypes are the joint product of an organism's genes and the environmen
in which it develops. A gene, in turn, may be thought of for the moment as
stretch of chromosome short enough to retain its identity over mar
generations; the totality of an organism's genes is its *genotype.* (Tw
organisms with chemically identical chromosomes are said to have the sam
genotype.) The genes that determine a phenotype generally come in pai
matched by position on pairs of chromosomes. One of these "alleles" may i
more influential than, or *dominate,* the other. "The gene" for a trait often refe
to the chromosomal loci of its alleles. (I take these usages from Falcon
[1989: 111], although the totality of an organism's genes is sometimes call
its *genome,* and an allelic variant at particular loci that control a phenotyp
trait the genotype for that trait.)

A phenotype is said to "express" a gene; a phenotype expressing the joi
effect of many genes is "polygenic."[3] Conversely, a "pleiotropic" gene ma
express itself in more than one phenotype. Yet, while the gene-phenotype lir
is thus many-many, it remains convenient to speak of a "tallness gene," tl
"intelligence gene," and so on, as long as those phrases are understood
mean whatever bits of chromosome influence height or intelligence, n
necessarily single entities.

Environment always mediates the action of genes; a genotype may expre
itself differently in different environments, and different genotypes may expre
themselves differently in the same environment. Genes are sometimes said
"code for" enzymes directly, which enzymes build tissues, which tissues bui
organisms, whose phenotypic traits are thus coded for indirectly. However, tl
direct/indirect distinction, although natural, is a matter of degree, since ev
the smallest step of protein synthesis will occur only in certain chemic
environments. While a trait's intuitive "distance" from a gene may l
measured by the predictability of the trait given the gene's presence (s
Symons 1979: 43–44), this distance is never 0. Genes abstracted from a
environments would be inert, if sense could be made of a gene or anything el
situated in no environment.

Despite the interaction of genes and environment, it is an error to infer th
environmental manipulation can make any genotype express itself in any wa
and therefore that genes really don't matter. This non sequitur is particular
common in connection with race differences, and is discussed in chapters 4 a

True, no gene expresses itself without an environment, but a given gene's nge of expression may be quite limited, and exclude some phenotypes together. Good nutrition aids growth, but no diet can make a man 10 feet ll, and there may be genes from which no diet can coax heights greater than feet. Moreover, while the phenotypic expressions of each of a pair of genes ay be plastic, the difference between their expressions need not be. One diet ay elicit a height of 5' 10" from Mr. A's genes and another diet a height of 6' ; the same pair of diets may elicit heights of 5' 4" and 5' 8" from Mr. B's nes. The heights of both men vary, but in any one environment their height fference is a constant 6". Even restricting the favorable diet to B and posing the unfavorable one on A will leave B 2" shorter. A related error is at of regarding a phenotypic difference as somehow unreal if it expresses the ne gene in different environments. Should A be genetically identical to B it 2 inches taller because of better nutrition during childhood, A really is 2 ches taller than B. All that follows from A and B having the same gene for lness is that they would have been the same height had they been raised ke, not that they ("really") are. Similarly, if A has an IQ of 120 and B an of 100 for purely environmental reasons, A really does have more of atever IQ measures.

Selection

Organisms evolve via natural selection. Environments differentiate enotypes that enhance reproductive success, or "confer fitness," from those at do not; fitness-conferring phenotypes are transmitted by the genes these enotypes express. The unit of selection, the successful or unsuccessful producer, is thus the genotype.[4] By definition, then, a gene is fit, in an vironment, when its phenotypic expression facilitates its replication. Fitness elf—of genotypes or organisms—is measured by the number of reproducing pies of itself an entity produces. Fitness may be generalized to *inclusive* ness (Hamilton 1964), the number of relatives left in the next generation ighted by their genetic closeness. Thus, a man survived by two children, ch of whom shares half his genes, is less inclusively fit than a man survived five nephews, each of whom shares a quarter of his genes. One genotype is lusively more fit than another if the weighted sum of the partial copies of one after a single generation exceeds the weighted sum of the partial copies the other.

While the basic unit of selection is the genotype, the notion of "fitness" can extended to organisms and other entities. Groups can be assessed for fitness their likelihood of survival, and a group trait may be said to confer fitness other things being equal, groups with that trait outlast groups lacking it. As rule, the persistence of a trait cannot be explained by its contribution to lividual or group survival—contribution to inclusive genetic fitness alone is planatory—but even this rule can occasionally be relaxed.

2.2. STATISTICS

Descriptive Statistics

Readers familiar with elementary statistics may omit this sectio
nonmathematical readers may wish to consult the summary at the end.

The basic statistical concept is that of a variable taking a range of values
a population. Among men, for instance, the variable of height takes valu
between 36" and 96". If each value of a variable x is plotted against t
proportion of the population attaining that value, the resultant graph is call
a *distribution*. In Figure 2.1, each point on the x axis represents a height, a
the corresponding point on the y axis represents the proportion of men of t
height.[5] Height distributes "normally," in the familiar bell-curve pattern
effects of multiple independent causes. The *mean* of x, or $\bar{x}$, is its avera
value in the population. The mean height of American men is about 70", i
that a height of 76" deviates from the mean by 6 inches. The *standa
deviation* SD of a distribution (often denoted σ) is the typical distance of
value of x from $\bar{x}$. Because some deviations exceed the mean and others fa

Figure 2.1
The Normal Distribution for Height

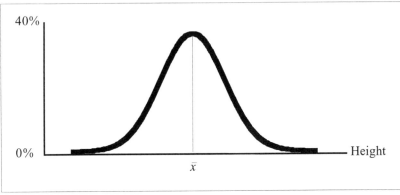

below it, canceling each other, SD is calculated by first squaring ea
deviation to make all numbers positive, averaging these squares, th
extracting the square root of this average. SD for height, for instance,
roughly 2". The position of a value in a distribution is often indicated by
distance from the mean in units of SD, denoted "z"; a height of 76", 3 S
above the mean, thus lies at $+3z$.

The square of a distribution's standard deviation, its *variance*, is often us
in describing connections between variables. Suppose the variance in heig
among men who consume the same number of calories per day is smaller th
that among men generally, suggesting a relation between appetite and heig
For each number n, find the ratio of the variance in height among men w
consume n calories per day to the overall variance in height; the average
these ratios measures the extent to which fixing calorie intake reduces variati

height. If that average is .25, for example, calorie consumption is said to *lain 25% of the variance* in height. This does not mean that men would be ⁄o shorter if they starved; it means that √25% = 50% of a man's deviation n the height mean is due to his deviation from the mean for caloric intake.

ıre 2.2
luction in Variance

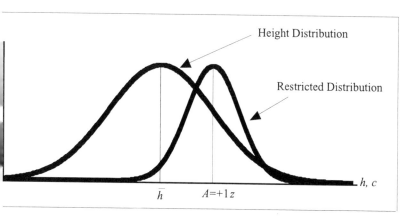

Height Distribution

Restricted Distribution

$\bar{h}$ $A=+1z$ h, c

he square root of the variance in one variable x explained by another y, tten r, is called their *coefficient of correlation*. An r of 1 expresses perfect ·elation between two variables, an r of −1 perfect inverse correlation, an r) that x and y are independent; in our example, r for height and calories sumed is √.25 = .5. In general, the r between x and y estimates how much ·e accurately you will predict the x-value of a member of a population from y-value than by guessing that he instantiates the mean value $\bar{x}$. To see ·, consider Figure 2.2 (similar to Figure 3.4 in the discussion of race bias in ɔter 3). Suppose height h and calorie intake c are both plotted along a ımon axis. Suppose too that Mr. A's calorie consumption lies 1 SD above calorie mean. The wide curve is the distribution of height among all men, narrow curve the distribution of height among men who eat like A. Since r ·, the SD of the height curve for men who eat alike, for instance like A, is ˙ the SD of the wide curve, so the mean height for men generally is two ·ow-curve standard deviations to the left of the narrow-curve mean. ording to the "z tables" found in statistics texts, 2.3% of a distribution lies ·e than 2 SD to the left of its mean. So the odds are 100/2.3 = 43 to 1 that · taller than average. Note that, since the variance in x explained by y is ned as a proportion of x's total variance, the sum of the variances lained by all variables associated with x is 1. This is not true of ·elation, because of the nonlinear relation between correlation and variance. hree points about correlation should be kept in mind. The first is not to be merized by the precept that "correlation does not imply causality." The ·ept itself is true as far as it goes. Food intake might correlate with height ıuse appetite controls growth, because growth controls appetite, or because

both are controlled by some underlying metabolic process. In the latter t
cases low caloric intake does not stunt growth, yet reducing variation in calc
intake still reduces variation in height—by screening out extreme bodies t
consume extreme numbers of calories or by screening out extremes of
underlying process that produces extreme heights. Correlation can
coincidental: Gould (1981: 242) cites a spurious ten-year correlation betwe
his age and the price of gasoline. However, in a wide range of cases correlat
does reveal causation of some sort. Changes in calorie intake patently cause
well as associate with changes in weight. Gould's example depends on be
defined over a minute population (of years), whereas a correlation holding o
an indefinitely large population is more probably nonaccidental, or w
inductive logicians call "projectible." A projectible correlation does not rev
the direction of causality or rule out underlying variables. Eating does
cause growth if growth causes eating or metabolism causes both. Still,
neither of the latter two cases would the correlation between eating and grov
be coincidental; it would indicate causal relations (see Meehl 1989: 542–54;

Second, while large correlations are usually more informative than sm
any significant r disconfirms a hypothesis predicting that two variables
unrelated. For instance, the idea that IQ tests measure white socializat
predicts a correlation of 0 between IQ and nonsocial factors like brain size,
a nonzero r between IQ and brain size suggests that IQ tests measure more.

The third caveat concerns the representation of correlation by *regress*
lines. Suppose height is graphed against calorie intake. The larger r is,
more collinear the points will be, as higher values of h match higher values
c. These data can be summarized by that line which *minimizes the sum of*
squares of the distance from each point to the line, as in Figure 2.3. T
equation for this line is height = $a \times$ calories + b. (In the example $b = 0$.)

Figure 2.3
Regression

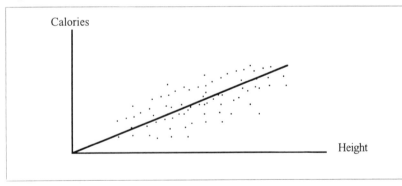

Regressing one variable on another yields an estimate of the value of
first from a known value of the second. If A consumes 2100 calories per d
his estimated height in inches, $\hat{h}(A)$, is $a \times 2100$. The larger r, the closer $\hat{h}$
will lie to A's real height, $h(A)$. As r approaches 1, the gap between $\hat{h}(A)$ a

(A) vanishes, so the ratio $|\bar{h} - \hat{h}(A)| \div |\bar{h} - h(A)|$ approaches 1. In fact, this ratio itself can be thought of as r: The more closely height and appetite associate, the more any deviation from the height mean is explained by deviation in calorie consumption. This situation is illustrated in Figure 2.4, which should be compared with Figure 3.2 in chapter 3.

Figure 2.4
Predicting Height from Calories

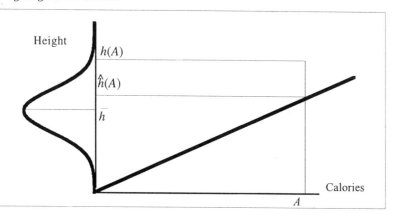

A variable x may vary less when both y and some further variable z are held constant. Variation in height among men with equal appetites decreases further when parents' height is controlled for. When both y and z explain some of the variance in x, the relation between all three can be expressed in a multiple regression equation of the form $x = by + cz + d$. Regression lines can also be curved.

This leads to the third caveat: whether the relation between two variables x and y is expressed as correlation (r) or proportion of variance explained (r^2) is a matter of indifference. Gould (1994) and others have complained that Herrnstein and Murray (1994) exaggerate the relation between IQ and correlates like poverty[6] by using r, which, when positive, always exceeds r^2. In fact, both numbers are simply alternative representations of the underlying datum, the difference between the average distance from the population mean of x when y varies and when y does not.

Summary: The standard deviation and its square, the variance, indicate the spread of a distribution. The correlation between two variables and the variance in one explained by the other measure the extent to which changes in the variables go together. Regression predicts the value of one variable from another it is "regressed on."

Testing Hypotheses

Scientific hypotheses are judged by their predictive accuracy. An hypothesis is confirmed by data that would be expected if and only if it were true, and disconfirmed by data that would not be expected were it true. Hypotheses too vague to yield predictions, as many environmental explanations of rac differences prove to be, are disregarded.

To be sure, many scientists favor a Popperian "falsificationist" methodology according to which hypotheses are never confirmed by data the predict, only, at best, shown to be "consistent" with those data. Scientists wh take more positive "maximum likelihood" and "Bayesian" approaches hee the principle that that hypothesis is most likely that makes the data mos likely. The more probable it is that the data would be observed were the hypothesis true, the more probable is the hypothesis itself.[7] On this approach if two competing hypotheses h and h' are initially equally probable, the available evidence e makes h more probable[8] than h' if e is more probable given h than given h'. The more the world looks the way a theory says i should look, the likelier the theory. Thus, the question to ask about race differences in phenotypic intelligence is whether the world looks more as i would be expected to look if the races were of equal intelligence or more as i would if they were not, and the question to ask when comparing environmenta to genetic explanations of a phenotypic race difference is whether the worle looks more as it would be expected to look if genetic variation caused tha difference, or more as it would be expected to look if environmental variatior caused it.

How improbable a hypothesis must be before it is too improbable to take seriously has no objective answer. A "confidence level" of .05 is customary. meaning that a hypothesis is rejected when the odds are less than 1 in 20 that the data would be observed were the hypothesis true, where "rejecting" a hypothesis means ignoring it when making plans.[9] Lower confidence levels are appropriate when rejection of truth is highly undesirable, higher ones when acceptance of error is highly undesirable.

Descriptive statistics provide data. However, statistical hypotheses predict only what will probably happen, not what definitely will, so cannot be compared to data directly, a circumstance that has prompted the developmen of measures of fit between the two. A description of one such measure, t, conveys an idea of the reasoning involved. Suppose it is conjectured that $\bar{x} = m$ in a certain population—for instance, that $\overline{IQ}$ for blacks = 100 = $\overline{IQ}$ for whites. A sample of size n is found to have a mean of m' and a standard deviation of s. How common are n-sized samples with mean m' and SD s when $\bar{x}$ is m? (How likely would a mean IQ of 85 and an SD of 12 be in a random sample of 10 blacks if $\overline{IQ}$ for blacks is 100?) We reason that, if the population mean is m, both $m - m'$, and therefore $(m - m')/(s /\sqrt{n})$, should be small. The latter term is t: the larger t is, the less likely it is that $\bar{x} = m$. In our example, t is 3.95. Statisticians have compiled "t tables" giving the probability of observed values of t for specified values of s, m, m', and n; the probability that $t \geq 3.95$ in our example turns out to be less than .005. The data strongly disconfirm the hypothesis that $\overline{IQ}$ for blacks = 100.

One must also bear in mind the idea of "second-order" evidence. Sometimes, the absence of evidence for a hypothesis is just that: no evidence has been observed, leaving our estimate of the hypothesis unchanged. Other times, however, absence of evidence for a hypothesis amounts to evidence against it, or, were the hypothesis true, some evidence favoring it would have been observed. We disbelieve in Santa Claus not only because his exploits defy known natural laws, but because, if there were a Santa Claus, he would have been spotted by now.[10] The very fact that there is no (first-order) evidence for his existence—that no-one has ever seen him—is evidence that he does not exist. Arguably, if the races are equal in intelligence there should be a good deal of evidence that they are; absence of such evidence is itself evidence that the races are not equal.

3. NULL HYPOTHESES

It is often computationally convenient to test hypotheses in a form that sets some parameter equal to 0. For instance, the hypothesis that race and IQ are unrelated amounts to supposing that r between race and IQ is 0. Hypotheses of this form are often called "null." The hypothesis that IQ is the same for blacks and whites, that is, that $\overline{IQ}$ for whites $-$ $\overline{IQ}$ for blacks $= 0$, is a null hypothesis.

It should be clear that computational convenience does not bestow credibility or epistemological priority. Hypotheses setting any parameter to any value can be assessed by significance tests; there is no reason to presume null hypotheses inherently more plausible than their competitors. The point is worth stressing because, when this presumption is made, the "null hypothesis" terminology creates bias against phenotypic and genotypic differences. That the difference between mean black and white IQ = 0, and that genetic variation explains 0% of any between-race variance in IQ, are both null in form, so, if null hypotheses are considered acceptable until refuted, the onus falls on their competitors. In fact, however, null hypotheses about race and IQ are not innocent until proven guilty. Any reasonable hypothesis about the size of race differences is as likely a priori as any other. Should it be insisted that null hypotheses do have special initial credibility, the null hypotheses about race may be taken to be (a) that the difference between the mean race difference in IQ and 1 SD is 0 and (b) that environmental variation explains 0% of any between-race variation in IQ. Hypotheses positing parameters of any size can be put in null form.[11]

NOTES

1. Reed (1969), Chakraborty et al. (1992); see Hendrick (1984: 75–81) for the statistical techniques involved.
2. On the relation of identifying traits to scientific definitions, see Kripke (1973), Putnam (1975), and Boyd (1988).
3. Plomin, Owen, and McGuffin (1994) suggest "QTL"—quantitative trait loci—for genes that partly determine a phenotype, but QTL yields no adjectival form for phenotypes, so I retain "polygenic."
4. I thus follow Dawkins (1986: 60). The "unit-of-selection" debate is still active (see Sterelny 1995), but the Dawkins view seems to be shared by a substantial majority of working biologists.
5. Strictly speaking, the curve in Figure 2.1 is a *density* function; the *distribution*

function for height gives the proportion of the population whose heights are, for a fixe x, less than or equal to x. However, I follow the widespread usage according to whic Figure 2.1 is a "distribution" because most of the statistical reasoning in this boo concerns normal "distributions" in this popular sense. In general, if $f(x)$ is a probabilit density function, its associated distribution function is $\int f(x)dx$. Clearly, the probabilit density at x is the first derivative of the distribution at x, so the probability of the valu x can be approximated by a "thin strip" around x in the distribution. This technique i used in the discussion of selection error bias in chapter 3.

6. The question is complicated by Herrnstein and Murray's use of "logistic regressions, whereby IQ is plotted against the logarithm of the probability of one valu of yes/no variables, such as being in poverty.

7. Let e be evidence, h a hypothesis, and $P(p|p')$ the probability of p given p', o $P(p\&p')/P(p')$. If in the latter expression p and p' are first replaced by h and e, and the by e and h, we get $P(h|e) = P(h\&e)/P(e)$ and $P(e|h) = P(h\&e)/P(h)$. Then $P(h|e)/P(e|h) = [P(h\&e)/P(e)]/[P(h\&e)/P(h)] = P(h)/P(e)$. Hence $P(h/e) = [P(h)/P(e)] \times P(e/h)$.

8. In the ratio $P(e|h)/P(e|h')$.

9. Dretske (1981: 197–209) judiciously discusses whether belief is definitionall connected to behavior.

10. More formally, let α be the sentence $(\exists e)[P(h|e) > P(h)]$. The principle of second order evidence is: $P(\alpha|h) > 0 \supset [(\sim(\exists e)[P(h|e) > P(h)] \supset P(h) < P(\sim h))]$.

11. Nowadays, "null hypothesis" often covers any hypothesis under test, regardles of mathematical form.

3

Race Differences in Intelligence and Temperament

1. THE BURDEN OF PROOF

natural starting point for discussing variations in intelligence and temperament is the psychometric literature. This literature does indeed reveal rge group differences, and I will soon turn to it, but to begin there creates a lse impression—that the scientific results are surprising, contrary to ppearances, or involve a special, technical sense of the word "intelligence." In ality, the races appear to differ apart from any tests. Black children do not erform in school as if they are as able and motivated as white or Asian hildren. Black adults do not succeed in science, art, commerce, or the rofessions as if they were as able and motivated as white or Asian adults. On ie whole, blacks encountered in everyday life, in the press, and on television ews broadcasts do not behave like whites or Asians. Moreover, evidence of ie equal intelligence of the races would presumably exist were the races in fact qual, and be prominently cited by the many social scientists who passionately elieve they are. Yet this does not happen. Authors like Gould and Kamin relessly criticize studies that show black intelligence to be lower than white, ut cite no black performances that indicate high mean intelligence. This is one f those cases in which absence of evidence for a hypothesis constitutes vidence against it. Everyday observation, together with the failure of galitarians to produce evidence that the races are equal, disconfirms racial arity.

To be sure, sufficiently many further assumptions may explain away rdinary observation and the psychometric data, but the possibility of xplaining away evidence against a hypothesis does not confirm it. In his *ialogues Concerning Natural Religion*, David Hume observed, a propos the ieological problem of evil, that if we already knew that God exists, we could e sure His existence is somehow consistent with the evil in the world. But as 'e are not given this in advance, we must base religious belief (or disbelief) on "known phenomena" (as Hume puts it); we must ask whether the suppositio of an omnipotent, benevolent deity predicts the order of things we actuall observe. In case it does not, we may (again in Hume's words) "pile conjectur atop hypotheses" to reconcile God with evil, but the whole structure is baseles. So too, if we knew in advance that the races are equally intelligent, we coul confidently attribute the appearance of lower black intelligence to distortin, factors such as bias. But lacking any prior assurance of racial parity appearances are all we have to go on. It is *possible* that various forms of bia rather than lower intelligence explain black academic failure. Withou independent confirmation that they do, however, this possibility plus failur does not *support* intellectual parity.[1] A race difference in performance may fa to show an underlying difference in ability, but it cannot show underlyin similarity.

To put matters bluntly, the question is not why anyone would believe th races are unequal in intelligence, but why anyone would believe them equa When someone asserts that black intellectual performance would equal that c whites were society free of bias, the proper response is "What makes you thin so?" The burden of proof, usually placed on those who deny the intellectu equality of the races, rests squarely on those who assert it.

3.2. STEREOTYPES

Everyday impressions might be dismissed as "stereotypes," but they are i fact a (modest) source of evidence. This evidence is not strictly essential t what follows, but "stereotypes" attract so much irrational scorn that som clarification is in order.

The received view is that ethnic generalizations serve psychological need often disreputable, such as—in the Sartre-Bettelheim theory—relief of guilt an enhancement of self-worth by projection of forbidden desires onto an outgrou On this view, the prevalence of a stereotype is unrelated to its truth. (Nieman and Secord's [1995] "ecological" theory is also adamantly anti-"cognitive." An alternative view (Goldberg 1977, 1992; Hagen 1979) is that stereotype extrapolate experience, so tend to be true or contain a germ of truth.

In favor of the second view is the fact that, like many incautiously worde statements, stereotypes become more plausible when interpreted charitabl Someone who says "Blacks are criminals" need not mean that all blacks ar criminals or all criminals are black, both plainly false, but tha disproportionately many blacks are criminals, which is true. Stereotypers ar either grossly stupid or misunderstood, and gross stupidity is less commo than misunderstanding.

Furthermore, the received view makes a mystery of flattering stereotype such as German efficiency, and neutral ones, such as the tendency of Italians t gesticulate while talking, neither suited to bolstering feelings of superiorit The mystery vanishes if Germans and Italians are ascribed these trai because they are observed to have them. Nor does the received view expla how particular groups are assigned particular traits. If the array of outgrou is a tabula rasa, with trait ascription independent of observation, which grou get which traits must be arbitrary. The conventional theory addresses th problem by calling stereotypes "self-fulfilling," made true by being accepted: thus, blacks are said to be good at basketball because everyone including blacks thinks they are, an expectation that leads black kids to practice a lot, for that reason excel, and reinforce the original belief. However, while positive feedback can perhaps help sustain a stereotype once it is accepted, the question remains of why it was accepted in the first place. Why were blacks ever thought to be good at basketball? This problem too vanishes if the ascription of stereotypes is initially controlled by perception of black ability at that sport.

Cross-culturally robust stereotypes are unlikely to manifest identical psychological needs. Generally speaking, the best explanation of a belief held by numerous noncolluding observers is that the belief is correct: should ten motorists report seeing a cow in a meadow, it would be a remarkable coincidence if all ten needed to see a cow. The hypothesis that they all had similar bovine visual experiences, which experiences were caused by something bovine, assumes the fewest coincidences. Likewise, the simplest explanation of robust stereotypes is that they report reliable observations.

Stereotypes about blacks in particular have shown great constancy. Pieterse (1992) records the agreement of all European cultures that blacks are unintelligent, brutal, and highly sexed, a view of blacks usually associated with pre–civil rights era America. Medieval Arab slave traders regarded blacks as rhythmic, highly sexed, unintelligent, and prone to "merriment" (Lewis 1990; also Rushton 1995b). Given their cultural achievements, Arabs would hardly have "needed" to feel superior to anyone. They may have been influenced by the servile status of the blacks they knew, but they did not regard their slaves from other groups as unintelligent. (Romans prized Greek and Jewish slaves for their intelligence.) The best explanation of this cross-cultural concordance is that the blacks observed by Europeans, Americans, and Arabs displayed the traits in question.

Gordon Allport's well-known *The Nature of Prejudice* (1954) devotes only a few unsystematic pages to the empirical truth of stereotypes. Helmreich (1982) explores the question more deeply, and reports that, of 75 generalizations about Jews, blacks, Italians, and other groups he considers, 25% are true and "about half have a factual basis." This is a higher proportion than is predicted by theories divorcing stereotypic belief from perception.[2] Moreover, Helmreich's criteria for accuracy are quite restrictive, for, like many people, he tends to view wrongfully caused traits as unreal, and therefore to discount stereotypes ascribing traits considered wrongfully caused. Thus, he rejects the stereotype of black oversensitivity on the grounds that oversensitivity is a response to oppression. However, if blacks are oversensitive for whatever reason, this stereotype is literally true. Blacks might not be to blame for oversensitivity due to oppression, but that is consistent with—in fact implies—that they *are* oversensitive.

The blameworthiness of a trait is often confused with its existence because the existence of a trait is confused with its innateness. Blameless consequences of an unfortunate environment are not innate, hence—assuming innate traits alone are real—not really present. And Helmreich clearly does think stereotypes like black oversensitivity are false, at bottom, because he takes their source to be nongenetic. Helmreich is not wrong to exclude environmentally caused traits in assessing stereotypes, since innateness and immutability[3] are elements in stereotypic thinking. Nonetheless, to the extent that stereotypes are construed a descriptions of group traits made without reference to their causes, more than Helmreich's "about half" of stereotypes will be wholly or partly true. A final point suggesting that stereotypes extrapolate experience is that they are deemed more acceptable when couched in positive language. *Tim* magazine criticized the National Football League's ban on "spiking" the ball after touchdowns for penalizing what it called "the black flair"; it is hard to imagine *Time* citing black "show-boating" or "extraversion," although these words differ from "flair" only in evaluative force. After decrying "the stereotype of the black man" as " 'big, black and hostile,' " and that of black baseball players in particular as "difficult," Joe Morgan goes on to make a case for more black managers:

Fully recognizing that individuals from any group are individuals, there is nevertheless an element in black life, in the everyday culture of the black community, that translates into the personalities of individuals. I can't find a name for it but I do know that whatever it is would make them actively different. . . . The NBA had a certain style before—and after—the arrival of black people in large numbers at every level of the league. The same would be true in baseball. (Morgan and Falkner 1993: 293–294)

Were racial generalizations suspect because they are thought to be false generalizations couched in favorable language would also be suspect, which they seldom are.

Stereotypes are strongly condemned in contemporary American society. Poll data are ambiguous, suggesting that whites, at least, are increasingly unwilling to assent publicly to any group generalization, but also that whites are very likely to be thought more intelligent and less rhythmic than black (Plous 1994). Stereotypes persist, so something must sustain them. An obvious candidate is observation.

3.3. RACE DIFFERENCES IN INTELLIGENCE: TEST DATA

The main scientific evidence of black/white differences in intelligence is black and white performance on standardized intelligence tests. Competent authorities agree that, as measured by the Wechsler Adult Intelligence Scale the Wechsler Intelligence Scale for Children, the Raven's Progressive Matrices and similar instruments, the mean IQ of whites exceeds that of American blacks by about one white standard deviation. When IQ is scaled so that th white mean is 100 and the SD is 15, the black mean is about 85 and the black SD slightly less than 15 (Sternberg 1994: 899–907).[4] This difference was first observed among Army recruits during World War I, and has remained fairly constant.

The most thorough survey of the literature through 1966 is Shuey's *Th Testing of Negro Intelligence* (1966), which reports 382 comparative studies involving 80 different tests administered to hundreds of thousands of black and white children, high school and college students, military personnel, civilian adults, deviates, and criminals. The average black score in these studies was bit below 85 and the average white IQ a bit above 100, with the difference in the means in the various studies ranging from 12 to 18 points. Numbers are somewhat misleading because of the variety of tests surveyed; the stable findings in the Shuey review were that the average black-white difference was always close to 1 SD and that the white variance almost always exceeded the black. Overlap, the proportion of black scores at or above the white comparison group mean, averaged 12% (501). W. E. B. Du Bois' reference to "the talented tenth" turns out to have been a remarkably accurate statistical intuition. In other words, 88% of blacks—and by definition 50% of whites—score below the white mean. This 50/88 ratio is not constant throughout the IQ distribution: as we shall see, the ratio falls when IQ rises.

A better known survey of American education that also appeared in 1966, the Coleman report (Coleman et al. 1966), reached identical conclusions: "[A]t the end of 12 years of high school . . . [t]he Negroes' averages tend to be about one standard deviation below those of the whites" (Coleman et al. 1966: 219; also see 221–251, Tables 3.11.1–3.11.31). The gap has persisted through renormings of IQ tests since Shuey and Coleman; the black mean on the Wechsler Intelligence Scale for Children-Revised is 86.4, and the white mean 102.2 (Jensen and Reynolds 1982: 425).

In what must be considered the definitive study of ability testing by a neutral scientific body, the National Academy of Science concluded:

Many studies have shown that members of some minority groups tend to score lower on a variety of commonly used ability tests than do members of the white majority in this country. The much publicized Coleman study provided comparisons of several racial and ethnic groups for a national sample of 3rd, 6th, 9th and 12th-grade students on tests of verbal and nonverbal ability, reading comprehension, mathematics achievement, and general information. The largest difference in group averages usually existed between blacks and whites on all tests and at all grade levels. In terms of the distribution of scores for whites, the average score for blacks was roughly one standard deviation below the average for whites. Differences of approximately this magnitude were found for all given tests at 6th, 9th and 12th grades. . . . The roughly one-standard-deviation difference in average test scores between black and white students in this country found by Coleman et al. is typical of results of other studies. (Garner and Wigdor 1982, vol. I: 71-72; also see vol. II: 365)

This conclusion was repeated in a subsequent NAS study (Hartigan and Wigdor 1989: 27). Table 3.1 presents additional studies.

The editors of *Profile of American Youth* (Office 1982) remark that "As in the civilian testing experience, there is unanimity of results in military testing: at each age level and under a variety of social and geographical conditions, blacks, on the average, regularly score below whites" (35). Scarr, who favors an environmental explanation of race differences in IQ, finds that "Mean differences in IQ scores between racial, ethnic and social-class groups are too well known to be restated at any length. . . . Briefly, there is often found an average difference of 10 to 20 points on IQ tests between black and white samples. . . . In general, Caucasian, American Indian, Eskimo and Oriental children are shown to have higher IQ scores and more rapid cognitive development than children of African or Australian aboriginal origin,

Table 3.1
Black-White IQ Differences

Source	Black IQ	Test
Baughman and Dahlstrom 1968	77	Primary Mental Abilities
Bodmer and Cavalli-Sforza 1970, citing Kennedy et al. 1963	80.7, 101.8[a]	Stanford-Binet
Brody 1992	1 SD	(None cited)
Broman 1987	90	Wechsler
Brooks-Gunn, Klebanov, and Duncan 1996	> 1 SD	SBIS, LM; WPPSI
Dreger and Miller 1968	1 SD	(Accept Shuey data)
Gottfredson 1986, from Hitchcock 1976	83.4, 101.8 [b]	Wechsler-C
Jensen 1973, 1980	15	WISC-R
Jensen 1981	85	(None cited)
Jensen and Figueroa 1975	85	WISC
Jensen and Inouye 1980	84	(Various)
Jensen and Reynolds 1982	1.14 SD [b,c]	WISC-R
Jensen and Johnson 1993	91.5	WISC
Jensen 1994	.72 SD [b]	Raven's Progressive Matrices
Kennedy et al. 1982	80.7, 101.8 [c]	Stanford-Binet
Klineberg 1935a, 1935b	72-88.5	Binet, National Intelligence Test
Lynn 1996	≈1 SD	Differential Ability Scale
Miele 1979	81	WISC
Montie and Fagan 1988	86	Stanford-Binet
Neisser et al. 1996	1 SD	(Various)
Peoples, Fagan, and Drotar 1995	1.15 SD [c]	Stanford-Binet, Fourth Edition
Scarr 1981	10-20 [b]	(None cited)
Profile of American Youth (Office 1982)	1.25 SD [b,d]	Armed Forces Qualification Test of the Armed Services Vocational Aptitude Battery

a. Sample from American South.
b. Mean difference.
c. Weighted average of black and white SDs.
d. Data base for Herrnstein-Murray (1994).

particularly after the first two years of life" (1981: 37). Bodmer and Cavalli Sforza, harsh critics of hereditarianism, cite without demurrer the Kennedy et al. study which finds an IQ gap of 21 points. Thoday, another critic, writes "The essential fact here is that United States 'Negroes' score on the average one standard deviation below the United States mean in IQ tests" (1976). Brody, something of a skeptic about the interpretation of IQ tests themselves, nonetheless concedes: "there is general agreement that on most tests of intelligence . . . the mean score of a representative sample of black Americans would be approximately one standard deviation below the mean score of a representative sample of white Americans" (1987: 507). So we may take a consensus to exist among behavioral scientists that whites typically outscore blacks by about 1 SD on what are conventionally called "intelligence tests."

Social and legal prohibitions have made the controlled testing of blacks increasingly difficult. At this writing, for instance, it is illegal in California to assess the IQ of black children for placement in special education classes (*Larry P. et al. v. Wilson Riles* 1979). There has been some decrease in the race gap on the Scholastic Achievement (neé Aptitude) Test (SAT) and American College Test through the early 1990s (Herrnstein and Murray 1994: 290–292; Zelnick 1996: 128–129), although the gap remains near 1 SD in all cases, and the possibility cannot be dismissed that test designers have eliminated some questions which differentiate the races, making the convergence partly artifactual. Vincent (1991) reports a marked closing of the race gap on several revised IQ tests for children born since 1970, but his figures may be misleading; when calibrated in terms of the standard deviations of the samples, in most cases the race gap is at or near the usual 1 SD. One of the two most recent studies known to me, that of Peoples, Fagan, and Drotar (1995), has found the IQs of black and white three-year-olds in Cleveland to differ by a bit more than 1 SD. This study used the most recent edition of the Stanford-Binet, which has been thoroughly screened for bias.[5] The other study, Brooks-Gunn, Klebanov, and Duncan (1996) also used the Stanford-Binet on black and white three-year-olds born in 1985 with low birth weight, finding an IQ gap of 20 (98 vs. 78); a follow-up using the Wechsler Preschool found a gap of 18 (103 vs. 85). In both studies the white SD exceeded the black.

I henceforth use "empirical egalitarianism," or sometimes simply "egalitarianism," to denote the hypothesis of racial equality in mean phenotypic intelligence, distinguishing it from "normative egalitarianism," the view that everyone deserves equal treatment, and "environmentalism," the view that any race difference in intelligence there may be is due wholly to environment.

3.4. CRITICISMS OF IQ

Informed egalitarians accept the race difference in IQ test scores but dispute its significance, usually for one of the following five reasons: 1) There is no such thing as intelligence—a claim based in turn on the multiplicity of human abilities, the supposed multiplicity of kinds of intelligence, or the supposed multiplicity of meanings given the word "intelligence" by different people and cultures. 2) There may be such a thing as intelligence, but so-called intelligence tests do not measure it. 3) Intelligence tests may measure the intelligence of whites, but not that of blacks. 4) Intelligence tests may measure intelligence for all races, but individual differences in intelligence are due entirely to environment. 5) Individual differences in intelligence may be due in part to genetic factors, but race differences are due entirely to environment.

The tendency of egalitarians to jump from one of these positions to another makes it hard to keep discussion on track, since each of them is incompatible with all the rest. If there is no such thing as intelligence, as 1) asserts, there is nothing for IQ tests to (fail to) measure, and no intelligence differences to explain environmentally or genetically. To put the point the other way around, there must *be* such a characteristic as intelligence for individuals to differ with respect to it, whatever the reason. Attributing a race difference to environmental factors is consistent with blaming it on whites, but not with denying it. To continue; if, as 2) asserts, IQ tests do not measure intelligence, they do not measure the intelligence of whites or blacks, nor do their results about individual or groups need explaining. If intelligence tests are biased, per 3), there is again no race difference in intelligence to explain either environmentally or genetically. If—per 4)—all variation in intelligence is due to environmental variation, genes play no role and the first clause of 5) is false.

These points of logic may be summed up in two simple precepts. First, you cannot say there is no such thing as intelligence *and* that everyone's intelligence is equal. If there is no such thing as intelligence, individuals cannot be compared with respect to it. Second, you cannot say that racism stunts the mental development black children *and* that black children are as intelligent as white. If the mental development of black children has been stunted by racism or by anything else, it follows that their intelligence has been limited.[6] Lower black intelligence might then be the fault of whites, but, once again, individuals and groups may possess traits for which they are not to blame.

That "environmentally caused" does not imply "unreal" is just an aspect of the phenotype/genotype distinction. Levels of intelligence are phenotypes, and whether individuals or groups share a phenotype is an issue quite separate from why they do (or do not). Phenotypic differences can be far more obvious than their causes. Should the black and white intelligence polygene turn out to be identical, and there is a mean difference in intelligence due entirely to environmental factors, all that would follow is that blacks would be on average as intelligent as whites if both were raised identically, and would have been on average as intelligent as whites had it not been for some environmental factors, perhaps including racism. It would not follow that blacks *are* as intelligent.

In addition to being an error in its own right, conflating phenotypes with genotypes makes phenotypic differences themselves sound more dubious than they are. Genotypic variance is a possible *explanation* of phenotypic variance, and no hypothesis is as certain as the data it is proposed to explain. On one favored view of explanation, these data must be entailed by the hypothesis, hence inherit any independent support the hypothesis enjoys; on the other hand, a purported explanatory hypothesis entailed by data just restates the data without explaining it. In the nature of the case, then, genotypic variance is always be less certain than phenotypic variance. Consequently, equating a difference in intelligence with a difference in genes makes the phenotypic ntelligence difference itself appear less certain than it actually is. Fischer et al. 1996) commit both blunders, persistently identifying belief in the existence of ntelligence with the belief that it is a "talent largely fixed at birth" that displays "*immutability*" (22). Rather than engage objections 1–5 directly while trying to make galitarians stay still, it is easier to consider them as they emerge from a review of IQ tests, and, in the next chapter, a discussion of genetic factors.

.5. DEFINING "INTELLIGENCE"

At the intuitive level, "intelligence" means "learning ability." Other uggested definitions include the ability to adapt to environmental changes, to olve problems, to abstract, to generalize, to detect redundancy, to iscriminate, to transform inputs, and, finally, susceptibility to conditioning, ut they are all roughly equivalent. Adaptation requires learning and eneralizing. We call *A* more intelligent than *B* when *A* catches on more uickly and thoroughly to which mushrooms are poisonous or which numbers re prime. In thus catching on, *A* solves the problem of discriminating oisonous from nonpoisonous mushrooms—an achievement equally well escribed as generalizing or abstracting from experience. This feat increases redundancy, for when *A* sees signs of poisonousness, he applies the eneralization he has learned and is unsurprised to find poisonousness itself. In et other terms, *A*'s responses to poisonous mushrooms become differentiated nore quickly than *B*'s. But an organism can be conditioned only to features it an discriminate; pigeons can be trained to peck at dots, but not at prime-numbered arrays of dots. Pigeons can't learn the prime/composite distinction.

Faced with these alternatives, I will conservatively follow Richard Weinberg, who characterizes intelligence as "the ability to learn from xperience and perform mental tasks expertly and effortlessly" (1989: 99).

Block and Dworkin insist (1976a: 416–424) that intelligence tests must rest n a "theory of intelligence," an indisputable point insofar as it means that ou cannot detect intelligence or anything else without some idea of hat you are looking for. But it is disputable indeed if it means that dentifying an ability such as intelligence requires knowledge of its functional r physical basis, or that psychologists holding rival theories of intelligence aay be investigating different phenomena. All theories of intelligence are neories of the same phenomenon, namely learning ability. It would be as silly s say that different cognitive theories yield divergent and possibly conflicting riteria for intelligence as to say that paleontologists who believe that birds volved from dinosaurs and those who believe otherwise use divergent criteria or "bird." However much paleontologists may disagree concerning where birds ame from, they agree on what creature they are talking about. Likewise, there s no disagreement among learning theorists as to the ability they are trying to neasure. Sternberg et al. (1981) do speak of and compare the "theories" of ntelligence held by laymen and experts, but what they are referring to (as they ometimes acknowledge) is the meaning given to the phoneme "intelligence" y laymen and experts. Since laymen and experts speak a common language,

the considerable agreement found by Sternberg et al. (1981) is hardl
surprising. The theories they survey are not of the sort Block and Dworkin sa
are necessary for intelligence testing. As Neisser et al. (1996) put it, "concept
of 'intelligence' " reflect efforts to organize a certain complex set c
phenomena, including the "ability to understand complex ideas, to adap
effectively to the environment, to learn from experience, to engage in variou
forms of reasoning, to overcome obstacles by taking thought" (77). Anyon
who uses "intelligence" to denote something else is simply changing th
subject.

Block and Dworkin and others press the possibility that current cognitiv
theories (or nontheories) are so deficient that, on learning more, "we ma
decide not to retain the term 'intelligence' at all" (Block and Dworkin 1976;
428). At one point they suggest as an analogy that, given how little physic
Galileo knew, "temperature" as he used it was meaningless and hi
thermometers measured nothing (420).[7] "Intelligence" has also been compare
to "phlogiston," banished from language by the discovery that no substance i
released during combustion (Gardner 1983: 297), to "witch," banished alon
with belief in supernatural powers, and to "sunrise," a metaphor that died wit
geocentrism. On reflection, however, such precedents show the reverse of wha
they are intended to. "Witch" remains a perfectly good English word, meaning
as it always has, "woman with supernatural powers." Indeed, the rejection o
supernatural powers ended accusations of witchcraft precisely because, b
perduring definition, these are the powers a witch must have. Likewise
"intelligence" will mean what it has always meant, namely "learning ability,
no matter what anyone discovers. Should human beings prove incapable o
learning—a most remote possibility—it would follow that no human bein
has ever been to any degree intelligent. Yet even then, although "intelligent
would have turned out not to describe anyone, it would, like "witch," remai
part of language.

But would not such a discovery still undermine "intelligence"? It hardl
matters that "witch" remains well-defined, now that we know there is no one
is true of. So too, Block and Dworkin seem to suggest, "intelligence" migl
fall into disuse through lack of reference. Obvious as it appears today tha
people are capable of learning (they might add), other apparently obviou
beliefs have proven false.[8]

Such broad appeals to human fallibility ignore a vital distinction. "Sunrise
and "witch" do not refer, and in retrospect never referred, because built int
each is an hypothesis which proved false. An object can rise only by increasin
its distance from a stationary horizon, as the Sun appears to but does no
since the Sun is actually motionless, it does not literally rise. For a woman t
be a witch she must consort with the devil, as some women were once though
to but no woman actually does. "Sunrise" differs in this respect from a wor
like "dawn," which atheoretically names the thickening of morning light
whatever its explanation, and names something as long as there are morning
Like "dawn," "intelligence" refers to a quasi-observable phenomenon, namel
learning, whatever explains it, and has a reference so long as this phenomeno
exists. Using the word "intelligence" assumes only that people learn an
perform mental tasks, a "theory," as noted, in little danger of refutation.

6. INTELLIGENCE TESTS AND INTELLIGENCE

"IQ tests," tests of learning ability, are families of tests of more specific ompetencies, such as vocabulary, reading comprehension, mathematical asoning, general information, and pattern completion. Some of their uestions are verbal, others are not, some call on background information and hers do not, but all test intelligence in the intuitive sense of asking the testee figure something out. Deciding which one of five shapes goes with four ven ones requires abstracting or "learning" the pattern common to the four; terpreting a proverb requires a grasp of the principle it expresses. A subject's Q" is essentially the ratio of the number of questions he answers correctly to e average number answered correctly by members of a reference group, ually a large random sample of children or adults asked the same questions eviously. (The relativity of IQ to reference group as a source of bias is taken) below.) This ratio is multiplied by 100 to yield a convenient number. Thus, meone who correctly answers exactly as many questions as the average ember of the reference group is by definition of average intelligence; the ratio ` his total to the reference group mean is 1, and his IQ is 100. If he manages ore correct answers than the reference group mean, his IQ exceeds 100; fewer rrect answers yields an IQ below 100. Scores tend to distribute normally, ith an SD of 15.

Hacking (1995), Fischer et al. (1996: 30–38), and others object that, cause psychometrists retain just those questions that on average half the ference group answers correctly, the normality of the IQ distribution is tificial. The standard and quite adequate reply is that many traits are known distribute normally and there is no evidence that whatever IQ measures does)t, so inducing normality is harmless. A reply more pertinent here is that, so ng as the adjustments needed to achieve normality are not themselves cially biased, the unadjusted shape of a score distribution is *irrelevant* to race fferences. Suppose 75% of a large reference group score below 40, 15% score tween 41 and 90, and 10% score above 91 on 100 reasonable questions, and at the white population as a whole performs likewise. Should 90% of blacks ore below 40, 6% score between 41 and 90, and 4% score above 91, the test, spite its skewness, reveals an overall black/white difference. The somewhat tificial normality of the IQ distribution does not in itself indicate bias.

Many people profess to doubt that a few dozen questions can test a general ental ability. Yet in other contexts it is accepted that general abilities can be sted by highly specific tasks: time over a 2-mile run, for instance, is an cellent indicator of overall physical fitness. The intuitive quality of IQ tests, parent to anyone who has ever taken one, should by no means be ignored, t this is not the main reason for saying that IQ measures intelligence—or, ore circumspectly, that IQ measures something, which turns out to be what is dinarily called "intelligence." More important is the positive mutual rrelation between performance on the various IQ subtests, and, decisively, ose between aggregate IQ score and a variety of external variables.

The correlations between scores on the various IQ subtests, and between ores on different IQ tests, gravely weaken the familiar argument that there is such thing as intelligence because "you can be smart at one thing and dumb

at another." Gilbert Ryle (1974: 54–55), an eminent English philosopher, offe
an uncompromising statement of this view:

The ingenious punster may be a silly car-driver; the boy who copes well wit
intelligence-tests (so called) might be a lame conversationalist, slow and unretentive i
learning foreign languages or easily outclassed in the school chess-tournaments. . .
Only occasionally is there even a weak inference from a person's possession of a hig
degree of one species of intelligence to his possession of a higher degree of another. . .
[A]t least some species of intelligence—say, solving crossword-puzzles—are, so far a
we know, only randomly correlated with some others. [9]

One can readily imagine Ryle denying that there is any such thing a
general athletic ability on the grounds that a good boxer might be a terrib
tennis player and that running speed correlates only randomly with catchin
grounders. But suppose good boxers do in fact play better than average tenni
run fast, and catch grounders uncommonly well. In that case we would begi
to suspect that there *is* such a thing as athletic ability, or a small number o
abilities, tapped by different sports. And, Ryle's confident pronouncemen
notwithstanding, able arithmetical reasoners do do unusually well on verba
analogies, are unusually adept at spotting patterns, and display larg
vocabularies. Those who "cope well" with IQ tests are seldom slow an
unretentive in foreign languages, and even Ryle stops short of claiming, wh
he has committed himself to, that a boy who copes ill "on intelligence-tests (s
called)" might prove "quick and retentive" in advanced mathematics. So it
natural to suspect some one ability, or small group of abilities, being tappe
by all IQ subtests. In psychometric jargon, a collection of correlations betwee
several variables is a "manifold," said to be "positive" when all th
correlations are. The positive manifold between IQ subtests, and betwee
various IQ tests, supports the existence of a general mental ability.

To grasp something of the mathematics involved in this inferenc
imagine tests of three physical abilities: Grip Pressure, Push-ups in On
Minute, and Running Speed with Full Pack, all positively correlated as i
Table 3.2. Doing so is not mathematically necessary, but one can construe th
manifold as arising from the correlation of each test with one underlyin
factor, which might be called "power." For suppose the "loadings" of G, I
and R on power are as in Table 3.3. One may think of 69% of (the deviatio
from the mean for) grip pressure as due to (the deviation from the mean fo
power, and 84% of (the deviation. from the mean for) push-ups as due
(deviation from the mean for) power. So if A and B exert the same gr
pressure, 69% of what they have in common is power, and power is 84%
doing push-ups. A and B thus share 84% of 69% = 58% of what it takes to c
push-ups, or, more simply, the common overlap of grip pressure with push-u
ability is $.69 \times .84 = .58$, the original G/P correlation. The other correlatio
in Table 3.2 fall out similarly. When the positive correlations between I
subtests are "factor analyzed" in this way, the common factor extracted, fir
identified by Charles Spearman, is usually denoted "g," or "general ability
The higher the g-loading of a test, the more it is taken to measure intelligenc
The g-loadings of some IQ tests approach .9.

Table 3.2
A Manifold of Abilities

	Grip	Push-Ups	Running Speed
G	1	.58	.54
P	.58	1	.65
R	.54	.65	1

The remark that extraction of a single factor from a positive manifold is not mathematically necessary, is true in two senses: one need not extract a single factor, and one need not extract any factor. Each point has been taken to show the arbitrariness of interpreting IQ as a measure of intelligence.

Table 3.3
Loadings on Power

	Grip	Push-Ups	Running Speed
Power	.69	.84	.78

3.7. PSYCHOMETRIC g VERSUS CLUSTERS

Consider first why no single factor need be extracted from a manifold. To continue with our example, the correlations in Table 3.2 can also be retrieved if G, P and R are taken to load on two uncorrelated factors, as in Table 3.4. The overlap of G and P on Factor 1 is .8 × .6 = .48, and .5 × .2 = .10 on Factor 2. The sum of the overlaps is .58, the original G/P correlation, and since the two factors have been assumed to be independent, there is no danger of counting twice. The rest of Table 3.2 can be retrieved in like manner. [10]

Tables 3.3 and 3.4 illustrate the general point that factor analysis, like any mathematical technique, is simply a method of representing data that cannot do our thinking for us. The two-factor representation may provide more insight in the example we have followed, for when Factor 1 is thought of as "strength" and Factor 2 as "endurance," it says that grip pressure depends mostly on strength, push-ups depend about equally on strength and endurance, and running with full pack depends somewhat more on endurance than strength. Factorization resembles the resolution of motion into vectors. The sliding of a penny northeast across the xy-plane can be represented by a single vector pointing in the direction of motion, or by two vectors, one pointing north and the other east; which representation conforms more closely to reality cannot be decided a priori. If someone has flicked the penny with his finger toward the upper right corner, the first representation is more realistic; if the penny is being tugged by two strings, one pulling northward and the other eastward, the second representation is. [11]

But this indeterminacy of factor analysis—it has been argued—removes any

Table 3.4
A Two-Factor Analysis

	Grip	Push-Ups	Running Speed	Factor 1	Factor 2
F1	.8	.6	.5	1	0
F2	.2	.5	.7	0	1

compelling reason to think that some one ability contributes to performance on all IQ subtests. Perhaps several abilities, possibly themselves correlated, are at work. The chief proponent of a multifactor approach, L. L. Thurstone, posited a cluster of "primary abilities," including memory, word fluency, numerical ability, and perceptual speed, and Spearman's followers certainly concede the existence of specific mental abilities and "group factors"—factors on which some but not all IQ subtests load—such as spatial reasoning.

The g/primary abilities debate is a technical one within psychology and applied statistics that I avoid henceforth, except to note here that g presently enjoys considerable support among psychologists (see Carroll 1991, 1993: 624) and on later occasions parenthetically noting data relevant to g. This agnosticism is possible because, contrary to what is often asserted, g is irrelevant to the interpretation and possible genetic causation of race differences in IQ.

Prima facie, the whole issue of group differences seems to dissolve should g be abandoned. How can there be group (or individual) differences in intelligence if there is no such trait? Gould's objection to all talk of variation in mental ability hinges on the indeterminacy of factor analysis (1981: 309), which he joins to the "reification" problem considered below, and his recent publications (see 1994, esp. 143–144) retain the assumption that talk of mental ability and speculation about its origin are meaningless without g. Fischer et al. (1996) also take the debate to turn on whether intelligence is a unified ability (see, e.g., 55). But this is a mistake. Allowing—which has yet to be discussed—that IQ tests are unbiased, the races must differ in whatever it is that IQ tests measure, whether that is a unitary g or something(s) else. For suppose IQ measures no one general ability, but an aggregate of Thurstone's primaries or some other, possibly quite numerous, set, perhaps with a different loading on each. All that would change is that, where we might once have inferred that the races differ in g, we would now infer that they differ in the primary abilities. What had previously been explained by variance in g would now be explained by (possibly heterogeneous) variances in the primaries. In particular, what might have been explained by low black g, such as poor academic performance, would now be explained by low black levels of some or all of these abilities, such as numerical reasoning. To the extent that genetic factors explain race differences in IQ test performance, IQ tests would now be understood to reveal race differences in the several (poly)genes controlling the various primary abilities rather than in the (poly)gene controlling g, and anyone disturbed by the prospect of genetic race differences in g should presumably be just as disturbed by genetic race differences in more specific abilities. Finally, from the moral point of view, if

enetic race differences in g relieve whites of responsibility for low black
tainment, so would genetic differences in the primaries. Grant that blacks
annot be genetically less intelligent than whites because intelligence is a
iction: nothing changes conceptually or morally if blacks are genetically less
uent and weaker in mathematics.

For this reason I will sidestep the Spearman/Thurstone controversy and
peak simply of "intelligence," whether it names a single across-the-board trait
r a composite.[12] Race differences in g when the positive manifold is analyzed
o yield a single factor must of mathematical necessity reappear when the
ositive manifold is analyzed multifactorially. In both cases, the analysands
re correlations between tests on which whites systematically outscore blacks,
nd this discrepancy must show up somewhere. To the extent that IQ test
erformance is heritable, some or all of the primary abilities must of
nathematical necessity turn out to be heritable when the manifold is analyzed
nultifactorially. It is hardly surprising that the abilities in the Thurstone
luster have in fact been found to be highly heritable (DeFries, Vandenberg,
nd McClearn 1976), a finding duplicated for other specific cognitive abilities
see DeFries, Vandenberg, and McClearn 1976; Nichols 1978; Pedersen et al.
992). Finally—again see chapter 4—the inference from high within-group to
igh between-group heritability for the primaries is no more but also no less
easonable than the corresponding inference from a high within-group to a high
etween-group heritability for g.

.8. VERBAL DISPUTES ABOUT "INTELLIGENCE"

I have already touched on some verbal issues; several more remain to be
iscussed before turning to the second factor-analytic objection to IQ tests. In
articular, the empirical hypothesis just mentioned that "intelligence" names a
luster of abilities must be distinguished from two similar-sounding ideas that
mount to little more than redefinitions of "intelligence." There is also the
dea, which it is convenient to treat first, that "intelligence" is not descriptive
t all, but evaluative. Perhaps the empirical data presented in forthcoming
ections obviate the need to dissect linguistic fallacies, but these fallacies are
o widespread that ignoring them would be neglectful.

"Intelligence is Inherently Normative"

[W]e can define and measure intelligence only if we can agree on which
sorts of thinking we value most. . . . [M]easuring intelligence . . . depends
on social convention. . . . [A] useful test is one that accurately mimics the
demands that some particular set of social conventions makes on us.
Whether we should call [the WISC] an intelligence test is a political
question. (Jencks 1992: 104–105)

Views of this sort confuse the subject of a judgment—here intelligence, or
earning ability—with what can be inferred from someone's making the
udgment. The fact that we measure intelligence shows that we are interested in
t, but this no more means that what we are measuring depends on our interests

than the fact that millions of people care enough about their weight to mount a scale each morning means that body weight depends on the sort of physique w value. Intelligence itself would exist whether anyone valued it or not, just a body weight would exist whether or not anyone cared about his figure. A "convention" is something people freely decide on, and could have decided differently, like using Roman numerals on clocks; no one's learning ability would change were it decided tomorrow that learning ability is unimportant just as no one's weight would change were it decided tomorrow that fitness is unimportant. Nor does the fact that IQ tests happen to be designed by (certain whites, so in that sense measure what (certain) whites want them to measure mean that the uses to which IQ tests are put reflect their designers' race Bathroom scales measure what their largely white creators want them to, the empirical quantity of weight, without distorting the weight of non-whites.

"Intelligent" does carry positive connotations, but the horse of reality must be kept before the cart of language. People don't admire intelligence because "intelligent" is a compliment; "intelligent" is a compliment because people admire intelligence—just as "obese" is derogatory because people dislike obesity. Apart from relatively pure commendations and derogations like "good" and "evil," most evaluative words acquire their force from the value placed on what they name. Let some tyrannical authority decree that flabbiness is henceforth to be labeled "beauty," and, instead of people coming to admire flab, "beautiful" will become an insult.

Consensus evaluations are often packed into descriptors, yielding what Nowell-Smith (1954) called Janus-words. "Rancid," for instance, does two communicative jobs. Calling butter "rancid" advises against eating it, and since everyone agrees and knows others agree on what sort of flavor to avoid "rancid" also denotes that flavor. "Intelligent" serves a parallel dual function calling someone "intelligent" conveys a definite idea of how his mind works, and, because that sort of mind is admired and known to be admired, praises him as well.

The descriptive element of a Janus-word is usually more central than the evaluative, for evaluative force, but not its descriptive meaning, can be canceled. You will raise eyebrows but not be accused of abusing language should you say you like rancid butter, whereas you *will* be accused of abusing language should you use "rancid" of butter that tastes sweet. Likewise, you can coherently criticize intelligence—Caesar utters no solecism in saying Cassius "thinks too much"—but you will not be understood if you insist that someone with an IQ of 40 is intelligent. Since it is thus possible to agree on what sort of thinking is intelligent without agreeing on its value, "intelligence" is definable independently of "the sorts of thinking we value most."

"There Are Different Kinds of Intelligence"

IQ tests are often said to measure only one sort of intelligence, the logico-mathematical. Gardner (1983) discerns "musical," "bodily-kinesthetic," and "personal" intelligences as well.

The trouble with such proposals is not that musical and social skills are

ctions; they plainly exist. Something is known of their location in the brain,
Gardner emphasizes (1983: 63, 212–213). The trouble is that calling them
intelligences" arbitrarily extends the word "intelligence" to abilities it did not
previously cover, in effect redefining it.

One litmus test for a disguised redefinition involves the contrast, mentioned
earlier, between meaning and reference. The *meaning* of an expression is its
dictionary paraphrase; its *reference* is the set of objects it is true of. The
distinction is patent on reflection. One can grasp what "world's busiest canal"
means without knowing which canal it refers to (the Kiel, according to
Guinness), and grasp what "Martian life-form" means without knowing
whether it refers to anything at all. Now, while the precise relation between the
two is disputed,[13] there is general agreement that meaning determines reference.
Expressions with different meanings may either differ or agree in their referents
"world's busiest canal" and "canal between the North and Baltic Seas" differ
in meaning, agree in referent), but expressions with the same meaning,
synonyms like "bachelor" and "unmarried man," or "horse" and "Pferd," must
agree in reference.[14] Consequently, words with different referents must differ in
meaning. Since "intelligence" as used by Gardner plainly differs in reference
from "intelligence" as ordinarily used, the two words differ in meaning. The
two "intelligences" are mere homonyms.

The basic point here is quite simple. As most people use "intelligent,"
Einstein is a paradigm of intelligence while Babe Ruth is not; as Gardner uses
Babe Ruth is as much a paradigm as Einstein. Normally, anyone who
announced, "On my definition, Babe Ruth is as smart as Einstein" would be
accused of playing with words. By this standard, the multiple intelligence
theory is wordplay, pouring new wine into old verbal bottles.

Of course, a dispute about words may express a substantive disagreement.
You and I might both use "intelligent" in its customary sense, yet you call
Smith "smart" while I do not because we disagree about his intelligence. But
Gardner's departure from received usage plainly does not join issues of fact. In
order to disagree in substance about Smith's intelligence, we must first agree
on the sort of data that bears on who is right. You would probably mention
Smith's physics degree, supposing he has one, on the assumption that we both
regard a grasp of physics as a sign of intelligence. But if I blithely deny that
mastery of physics counts—if we keep finding ourselves at odds about what
intelligent people are like—we probably are using "intelligence" in different
senses. We might conceivably both be talking about the same something (e.g.,
learning ability) yet hold such different theories of how this something
manifests itself that we seem to be working at cross purposes; but the longer
miscommunication persists the less likely this becomes. Gardner is talking
past, not disagreeing with, the rest of us because he evidently rejects the usual
criteria for intelligence.

Some extensions of familiar words, like computer "memory," are well
motivated. Generally speaking, a word may reasonably be stretched to cover
new cases that resemble its standard reference in some significant way(s). Talk
of "computer memory," for instance, is warranted by the similarity of literal
memory to data storage. Extended usages are metaphors, meant to point up
unnoticed or underappreciated analogies, and are frivolous absent such

analogies. How does Gardner's wordplay fare by that standard? Six of th
analogies he cites between learning and other abilities, as embodied in his fir
six criteria for a trait's being "an intelligence," are isolability by bra
damage, display to an exceptional degree by prodigies, presence of co
"operations," development in a recognizable sequence, possession of
plausible evolutionary function, and susceptibility of encoding in a "symb
system" (Gardner 1983: 63–67).[15] The last is highly misleading, since it seerr
plausibly, to associate "intelligence" with the capacity to manipula
abstractions, yet Gardner understands "symbol system" so widely that
boxing match becomes "a dialogue between bodies, a rapid debate betwee
two sets of intelligences" (207, citing Norman Mailer). Gardner has apparent
confused an activity's being *describable* by a symbol system with its requirir
the use of a symbol system for its performance. In the former sense, anythir
whatever, from the swaying of a wheat field to the motion of the planets, ca
be "encoded." Astronomical bodies need not know the laws of motion
conform to them. In the latter sense only activities mediated by manipulati
of symbols are encodable, but then athletic skill, and probably other talen
that Gardner specifies, are not "intelligences." (Does the forceful Alpha male
a wolf pack use symbols to attain leadership?)

   In any case, Gardner undercuts these first six criteria with his final on
"support from psychometric findings." "To the extent that psychometric resul
prove unfriendly to my proposed constellation of intelligences," he write
"there is cause for concern" (66). This seems to imply that an ability is a
Gardner-intelligence only when it correlates with intelligence as conventional
conceived, a requirement presupposing the validity of "intelligence" in i
ordinary sense. What is more, this new requirement commits Gardner
claiming that the talents he is concerned with are better predictors tha
previously thought of anything "psychometric results" predict. Yet he does n
(and cannot) make this claim, and he wrongly denies the positive manifold
IQ test correlations (see e.g. 18–19). Nor does he pretend to show th
leadership etc. and abstract thought are produced by the same underlyir
mechanism. So Gardner vacillates, sometimes relying on "intelligence" in tl
old sense—counting a talent as an "intelligence" only when it covaries wi
IQ—at other times changing the subject.

   There is no foolproof test for the significance of the similarities betwee
musical, athletic, and cognitive abilities, or, for that matter, between any tw
things. A good rule of thumb, though, is that a significant resemblance can l
described in ways that beg no questions, so that—in particular—a resemblan
between intelligence and something else is "significant" when it can l
described without use of the word "intelligence" itself. There is point to talk
computer memory because it is possible to say what computer storage an
conscious recall have in common *without* calling computer storage "memory
if Gardner's redefinition has a point, it should be possible to say what Bal
Ruth and Einstein have in common without calling it "a form of intelligence
The common element Gardner has in mind seems to be excellence at a comple
activity involving the nervous system—hardly sufficient to bring the two und
a common rubric.

   Homonymy sows confusion. It abets the unearned transfer of association from familiar words to new ones, and thereby the transfer of assent from familiar propositions to new ones that, undisguised, might be doubted. People persuaded to call athletic ability a "form of intelligence" will feel compelled to conclude, from their standing association of "intelligence" with "importance," that athletics are more important than previously thought. Those who believe athletics are undervalued should say so in just those words, arguing the case on its merits.

" 'Intelligence' Means Different Things in Different Cultures"

A position akin to Gardner's (1983: 60) is that " 'Intelligence' is not a fixed one-dimensional trait. Nor does it mean the same thing in all cultures" (Jencks 1992: 104). Block and Dworkin concur: " 'Intelligence' is a vague term, often used to refer to different things in different cultures" (1976a: 412).

Neither position is coherent. To begin with, it is blazingly obvious that monolingual speakers of languages other than English never mean or refer to anything by "intelligence," since "intelligence," a word of English not found in their languages, is a word they never use. The most that a language other than English can contain, and speakers of that language use, is a synonym for the English word "intelligence." Therefore, the most that can sensibly be asked about a tongue other than English is whether it contains a phrase which means what "intelligence" means, and, if it does, what that phrase is. That term, should it exist, is their word for "intelligence." But a question without point, because self-answering, is what their word for "intelligence" means. "Their word for 'intelligence' "—the word in their language, should there be one, that means "intelligence"—means "intelligence." Every culture's word for "intelligence" is by hypothesis synonymous with "intelligence," so the idea that "intelligence . . . does not mean the same thing in every culture" is an absurdity.[16] Consider "ehrlich," which is German for "honest." By that fact alone, "ehrlich" cannot be German for "intelligent." To insist that "ehrlich" does mean "intelligent," and then infer that Germans (unlike us) think intelligence includes honesty, would be akin to insisting that "pferd" is German for "dog" and concluding that Germans think dogs whinny.

Block and Dworkin might be understood as saying that a genuine synonym of "intelligence" in another language may differ in reference from (our) "intelligence." Perhaps all cultures agree (as they must) on what intelligence is, but differ in their beliefs about who has it, just as you and I disagreed about Smith. Once again, however, we must establish that a word has been translated correctly *before* taking what speakers apply it to as a guide to their beliefs, and, once again, broad agreement in reference is a criterion for correctness. The very fact that Germans use "pferd" to refer to horses means that "pferd" should not be translated as "dog." Likewise, as Germans use "ehrlich" to refer to promise-keeping, it would be both preposterous and question-begging to translate "ehrlich" as "intelligent" and announce that Germans use their word for intelligence to refer to promise-keeping. Who said "ehrlich" is German for "intelligent" in the first place?[17]

The mistake scouted here—construing a mistranslation as a deviant belief—is made in Wober's (1974) purported showing that Kigandan and

"Western" concepts of intelligence differ. Wober's evidence is that *obugezi*, which he describes as "the Kiganda equivalent to 'intelligence,' " is more closely associated with "social conformity" than "intelligence" is for Western speakers. Surely, however, the proper inference from this association is that *obugezi* is ipso facto not equivalent to "intelligence" after all, but is a near-equivalent better translated as "knowledge of cultural forms." Wober himself says as much when he remarks that "*obugezi* is more like wisdom than intelligence." If *obugezi* is more like wisdom than intelligence, "*obugezi*" does not mean "intelligence."

Mistranslation artifacts are confused with unusual beliefs because of the assumption that every language has a word for everything. Stated baldly, this assumption is plainly false; the ancient Greeks had no word for the Christian idea of sin, and few modern languages possess a word for neutrino. Consequently, the translator must often settle for that word in a target language which is closest (but not identical) in meaning to some word of his own. There is nothing wrong with settling for approximations, but trouble strikes when this closest counterpart is, absolutely speaking, pretty far from the word the translator is trying to match *and* the target language is tacitly assumed to have an exact equivalent. A rough counterpart having been taken for a synonym, the resulting loose translation looks like an odd belief. The closest classical Greek approximation to "sinfulness" was "kakotropia," which, unlike "sinfulness," connoted low birth. It is perfectly all right to identify "sinfulness" with "kakotropia," so long as this decision to equate non-synonyms is not taken to show that the Greeks had a non-Christian idea of sin. "Intelligence" can be paired with its closest counterpart in a foreign tongue—like *obugezi*, the Kagandan word for integration into society—so long as this decision is not taken to mean that intelligence is thought by some to be social integration.[18]

It is appropriate in this context to mention Flynn's doubts that IQ measures intelligence (Flynn 1984, 1987a). Flynn claims that IQ scores in the Western world have been rising .3 points per year for six decades, and that this rise is not an artifact of improved living conditions (which eliminate low IQ produced by extreme deprivation). What strikes him is the absence of any corresponding rise in what constitutes the intelligence of the population: there are not many more people who "find school easy and can succeed in virtually any occupation, [whose] achievements are so clear that they fill the pages of *American Men of Science*, [who] resemble the famous geniuses." Flynn concludes from this disparity that IQ measures "abstract problem-solving ability," only a correlate with a weak causal link to "the real-world problem-solving ability called intelligence" (1987a: 188). Flynn's interpretation of th his data is questionable because within generations IQ does correlate with these expected criteria; the Flynn effect is puzzling, and its full meaning, it seems t me, remains elusive.[19] What is to be emphasized here is that Flynn accord "intelligence" a clear fixed meaning, namely "real-world problem-solving ability," and recognizes fixed criteria for real improvements in intelligence. H complaint is that IQ tests do not test it, not that "intelligence" is in any wa ambiguous.

Calling athletic ability or leadership ability "intelligence" no more change the intelligence of athletes, or enhances the importance of leadership for life outcomes, than calling dogs "horses" will make them whinny. The world is what it is no matter how it is described. That is why, when dispute focuses on a word, the word is best dropped and the facts restated without it. Should someone insist that whether Albert Einstein and Babe Ruth were equally intelligent is culture-relative, it is best to say that Einstein was better than Ruth at abstract reasoning, and Ruth better than Einstein at hitting baseballs, whatever those traits are called. If IQ predicts academic success and brain function, say so in just those words. [20] Indulge interlocutors who won't call the ability measured by IQ "intelligence," or insist on calling that ability and athletic ability by the same name. As with "race," surrender the disputed word.

People who make a point in argument of not understanding "intelligence" invariably do understand it in all other contexts. They know an "intelligent" child is one who learns quickly, and that, of the two, Nobel laureates tend to be more "intelligent" that manual laborers. To ridicule brain size as a sign of intelligence, Gould observes (1981: 93) that the brain of "the great mathematician" Karl Gauss, a man of "genius," was not remarkably large—a fact he cites only because he agrees and expects his readers to agree that Gauss was highly intelligent. People pretend not to understand "intelligence," I suspect, to avoid embarrassment over race. Positing many kinds of intelligence or meanings for "intelligence" allows them to select an ability prominently displayed by blacks, dub it "intelligence," and announce that, in their different ways, the races are equally "intelligent"—another example of obscurantism about race differences.

3.9. REIFICATION VERSUS CORRELATION

Recurring to section 3.7, the second objection to the factor-analytic extraction of g is that clusters of correlations need not be factored at all. Factors are not new data; they are representations of data already known, and like any purely mathematical construct may be considered an artifact. Geographical position is representable by lines of longitude and latitude, but the Earth's surface is not literally crosshatched. Perhaps g is no more real than the equator.

Proponents of this objection accept (as they must) the mathematical adequacy of factor analysis. Gould for one concedes: "Since most correlations in the matrix [of correlations between mental tests] are positive, factor analysis must yield a reasonably strong first principle component" (1981: 251). The trouble, according to Gould, is that believing in either g or Thurstone's primaries commits the fallacy of "reification" (1981: 310). [21] Taking g—or any primary—as more than a mathematical fiction misconstrues the "wondrously complex" network of human capabilities as a thing (251) inside the head. He asks:

Can the plethora of causes and phenomena grouped under the rubric of mental deficiency possibly be ordered usefully on a single scale, with its implication that each person owes his rank to the relative amount of a single substance? . . . The principle error [is] reification—in this case, the notion that such a nebulous, socially defined concept as intelligence might be identified as a "thing" with a locus in the brain. (1981: 159, 239)

It should be apparent at once that dismissing g or anything else as an artifact simply because it is not an entity occupying space is to confuse objectivity or non-artifactuality with *being an object*. Intelligence is admittedly not stuff inside the brain, but it does not have to be for people, behavior, and thought-processes to be objectively characterized as intelligent, or objectively ordered by relative intelligence. Consider solubility.[22] The solubility of a sugar cube is not a thing or stuff residing in the cube, capable of being extracted and put on a shelf, yet it remains an objective fact that sugar dissolves in water. Talk of the solubility of sugar supported by this fact is perfectly clear to anyone not determined to misunderstand it. Solubility is not what makes sugar dissolve—the weakening of molecular bonds does that—but sugar is soluble nonetheless, and was known to be soluble before anybody knew about molecules. So, too, the objective fact that sugar dissolves more rapidly than quartz is what supports the statement that sugar is more soluble than quartz, or, a stylistic variant, that the solubility of sugar exceeds that of quartz. It would be obscurantism at best to deny that sugar is more soluble than quartz because solubility isn't a "thing."

Nouns like "intelligence" and "solubility" do not acquire meaning by naming substances, but through their occurrence in larger contexts well-defined as wholes. "Solubility" is meaningful because it occurs in contexts like "The solubility of sugar in water exceeds that of quartz," which means, as noted, "Sugar dissolves more rapidly than quartz." A particularly close parallel to "intelligence" is "kinetic energy," ascribed to bodies by physicists perfectly aware that kinetic energy is not stuff. "Kinetic energy" is what mathematicians call a functor, the name of a function,[23] its canonical context being "the kinetic-energy-in-ergs of body $b = n$." The value of this function for a body b, kinetic energy-in-ergs(b), is determined by the compression b induces in a test spring and can be estimated from other indicators, like the dent b makes in a resisting surface. For b to have more kinetic energy than c just means that kinetic energy-in-ergs(b) > kinetic-energy-in-ergs(c); for b's kinetic energy to change just means that kinetic-energy-in-ergs(b) at one time $\neq$ kinetic-energy-in-ergs(b) at a second. The only "things" there need be for bodies to have kinetic energy, or to differ in kinetic energy, or for kinetic energy to change, are bodies themselves. From this same analytic perspective, talk of intelligence is talk that uses the "intelligence-of" functor, which names that function from individuals to numbers whose value for an individual is determined by IQ tests and can be estimated from other indicators. Mr. A is more intelligent than Mr. B when intelligence(A) > intelligence(B); the only "things" that need exist for this relation to hold are A and B. Were "intelligence" ill-defined simply because "intelligence is not a thing," kinetic energy would be ill-defined also.

Notice, incidentally, that solubility can be well-defined without "ordering all substances on a single scale." The solvents of two substances might overlap only partially. One substance might dissolve more rapidly than another in solvent X, less rapidly in solvent Y. Is-more-soluble-than, that is, might be a partial ordering. Moreover, substances can be simultaneous ordered by other physical properties, such as hardness. It is thus not required, for intelligence to be a real trait, that it correlate with every human ability or that humans be "ranked" only according to it. A variety of weightings and partial orderings are possible. Gould's insistence that reifiers wish to assess all humanity along a single dimension is perfectly gratuitous.

Although he usually writes as if he has proven g a fiction, all Gould actually shows, claims to have shown, or can show is that—what no one would deny—the reality of g is not logically guaranteed by its factorial extraction.[25] To claim more would commit the textbook fallacy of arguing from ignorance, in this case inferring that a thing does not exist because it is not known to exist. Gould concedes that g could be well-defined, but insists—again indisputably—that acknowledging g requires "convincing, independent information beyond the fact of correlation itself" (1981: 251). But he goes far wrong in demanding that a direct physiological counterpart of g, a "concrete tie" between "neurological object and factor analysis" (1981: 310), would have to be found. A neurological tie would suffice to establish g, and below I mention some recent research that has begun to supply this sort of "independent information," but it is by no means necessary. As solubility reminds us, a trait can be identified absent knowledge of its structural basis. Sugar was known to be water soluble, and to differ from quartz in ways relevant to behavior in water, long before anyone tied dissolution to a "molecular object."

A less demanding but still sufficient condition for the reality of a posit, reminiscent of the physicist's criteria of invariance and symmetry, is that the posit correlate with a wide range of independent variables. For unless a multiplicity of reliable correlations are to be accepted as coincidence, something must be producing them. Consider why "AQ," one's height divided by street address, would be considered artifactual. The problem is not obscurity; AQ is clearly defined. The reason no one would think AQ corresponds to anything real, rather, is lack of predictive power. A person's AQ tells you nothing else about him. In contrast, the heat of a substance as measured by thermometer is considered physically real because it predicts physical state, radiance, demagnetization, and innumerable other phenomena. Were heat an artifact, how an object will feel to the touch could no more be predicted from its temperature than income can be predicted from AQ. Solubility was considered real before the advent of modern chemistry because objects that dissolve in water also tend to dissolve in other fluids, dissolve at rates that vary with the solvent's temperature, shatter in a characteristic way, and are stable in behavior over time. (AQ fluctuates.) A hyper-operationalist Gould of the preatomic era who protested that "solubility" merely names a manifold of correlations, not known to correspond to anything that actually underlies the plethora of causes and phenomena grouped under that rubric, would have been profoundly confused.

The substantial issue, then, is the number, variety and stability of external correlates of IQ. The more there are, the more likely it is that IQ along with these other variables manifest some underlying reality. Many critics of IQ (see, e.g., McClelland 1976: 46–48; Schönemann 1987: 324; Mercer 1984; Fischer et al. 1996: 35, 44) scant this criterion because they assume all the correlates of IQ relate to schooling and test-taking itself, thereby forming a tight circle dismissible in toto as socialization. But IQ correlates with many nonsocial variables, as we will see, and these critics (as we will also see) seriously misinterpret the role played by specific knowledge in test items that assume it. Multiple manifestation is of more than methodological importance, however. The hypothesis that IQ measures only white socialization predicts insignificant correlations between IQ and nonsocial variables. Significant correlations with such variables disconfirm this interpretation.

3.10. CORRELATES OF IQ

IQ associates with occupational attainment. The correlation between individual IQ and occupational status is .5–.7 (Jensen 1980: 341; 1986: 317), and the correlation between the prestige rating of an occupation and the average IQ of its incumbents exceeds .9 (Jensen 1980: 343). The correlation between IQ and on-the-job success increases with the IQ demands of the job (Hunter 1986: 342–346). Jensen cites correlations ranging from .0 to .19 between IQ and success in sales and packing jobs, up to .47 for professional positions (1980: 348); using the same database, Hunter reports correlations from .27 to .53 (1986: 343). He also reports correlations from .37 up to .87 between IQ and successful on-the-job training when the knowledge of trainees is controlled for. These correlations hold whether job success is judged by supervisor ratings (which might admit race bias) or anonymous work samples. Neisser et al. agree that IQ explains about 29% of the variance in job performance (1996: 83).

The .4 correlation between IQ and overall socioeconomic status (Jensen and Sinha 1993, Herrnstein and Murray 1994) is noteworthy for its intermediate size. It is large enough to show that IQ measures something that affects life outcomes, but too small to sustain the claim that IQ is merely a proxy for social status, as measured by sociologists. That the correlation between an individual's IQ and his parents' status is lower than that between his IQ and the status he achieves (Jensen and Sinha 1993; Neisser et al. 1996) further suggests that IQ is not a status artifact. IQ correlates with income in way independent of years of schooling (Currie and Thomas 1995; Hanushek 1986; R. Weiss 1970; Ashenfelter and Mooney 1968); a pure IQ × income correlation tends to be an underestimate because IQ correlates so highly with schooling but Jencks (1972: 337) estimates this figure at .35. In prestigious occupations those above the 90th percentile in IQ outearn those below by about 30%, finding replicated for nonprestigious occupations (Herrnstein and Murray 1994: 98, 685).

Herrnstein and Murray supply a wealth of data on the relation between IQ and a large number of variables in a 12,000-person mixed-race sample. Much of their analysis (127–266) is done on the white subsample only, which has the disadvantage for present purposes that within-race correlations bear only indirectly on between-race phenomena, but the advantage of forestalling the objection that race differences in IQ and its correlates are due to bias along both dimensions. Herrnstein and Murray do however present data in the

ppendix 4 for the entire sample. Two are of interest here: the correlation
etween IQ and being below the poverty line is about .32 in their sample, and
at (for women) between IQ and being on welfare one year after the birth of a
iild is .56.[26]
Herrnstein and Murray find an r of .58 connecting IQ with dropping out.
here is an overall correlation of about .5 between IQ and academic
:hievement (Jensen 1980: 316-317; Linn 1982), but this figure understates the
)nnection. At primary and secondary levels, where the widest range of
)ilities is present—and the race difference in performance most pronounced—r
es in the .6–.7 range (Jensen 1980: 319); it falls to .4–.5 for college students
id .3–.4 for graduate students. This shrinkage is due to "range restriction,"
hich is easy to explain with an example. The correlation between running
)eed and proportion of body fat among world-class sprinters is close to 0,
:cause all world-class sprinters are very lean. What makes one sprinter faster
an another—the sources of variance in speed in that population—must be
)mething other than their negligible differences in body fat. Similarly,
/eryone in graduate school is fairly smart, so primarily nonintellectual factors
ke industriousness differentiate success at that level. But just as the true r
etween running speed and body fat is the r for the most representative, i.e.
idest, population, the r most indicative for IQ and academic success is the
+ among elementary school children.
   IQ associates strongly with crime (Gordon 1975a, 1975b; Hirschi and
indelang 1977; Herrnstein and Wilson 1985; Jensen 1980: 359). The average
) of imprisoned offenders, white and black, is below 92 (Herrnstein and
/ilson 1985: 154), and the average IQ of rapists and murderers is lower than
at of "white collar" criminals—embezzlers and forgers—in the prison
)pulation. Low-IQ boys tend to be more aggressive (Wilson and Herrnstein
)85: 344). The correlation between IQ and being interviewed in a correctional
cility is about −.3 (Herrnstein and Murray 1994: 621).
   Cattell (1950) reports a significant correlation between "general ability" and
:ing "morally intelligent." Herrnstein and Wilson comment that "a person's
vel of moral reasoning is correlated with intelligence, particularly verbal
telligence" (1985: 169). Herrnstein and Murray (1994) report an r of .28
:tween IQ and a measure of prosocial behavior they call the "Middle Class
alues Index" (1994: 622). Lawrence Kohlberg, well known for his sequencing
' the stages of moral development, describes findings which "support what we
l know: you have to be cognitively mature to reason morally. . . . IQ tests
)rrelate with moral maturity" (1981: 138–9). Among (white) preadolescents,
[ussen et al. (1970) found correlations ranging from .32 to .62 between IQ,
:ruism and honesty. IQ also correlates slightly with sense of humor, stature
id myopia (Jensen 1980: 360–362).
   IQ is strongly concordant with the ordinary conception of intelligence; IQ
'edicts laymen's judgments of intelligence (Sternberg et al. 1981), and
fferences in intelligence are recognized cross-culturally (Reuning 1981). IQ
scriminates those conventionally viewed as "intellectually gifted" and
nentally retarded" (Jensen 1985c). This correlation is the source of a certain
ony, for many critics of IQ tests also oppose capital punishment, and
gularly cite the low IQs of convicted murderers when arguing against their

execution (columnist Anthony Lewis is a prominent example). IQ woul obviously be irrelevant to culpability if it measured only knowledge of whi culture; we see again how critics of testing themselves implicitly concede tl validity of IQ when testing per se is not the topic.

Egalitarians explain the linkage of IQ to academic and vocational succes law-abidingness, and morality as a cultural artifact: IQ measures socializatic to white norms that are reinforced in schools, the workplace, the legal systen and everyday moral assessment, so that, while there is an underlying factc tapped by IQ and its correlates, that factor is conformity to white culture, nc ability. (Intelligence, on this theory, is causally and correlationally isolatec and may not exist; see Fischer et al. 1996.) Some of the external correlate already mentioned make this explanation difficult to defend. For instance, stature correlates with IQ because poor nutrition reduces both IQ and height, a Gould suggests (1981: 109), one must conclude from the myopia-IQ correlatic that poor nutrition aids vision. More significantly, the socialization hypothes predicts an $r \approx 0$ between IQ and any variable unrelated to group differences i socialization, and is therefore disconfirmed by large positive or negative valu for r. As mentioned, large r's for nonsocial variables have in fact been found.

That increases in brain mass relative to body size parallel the evolution c intelligence has been recognized for almost a century (see Jerison 1973), but was difficult until recently to compare brain size to intelligence in man becaus brain measurements had to be post mortem and often indirect, leaving muc room for error. Van Valen (1974) used available data to suggest a correlatic of .3 between IQ and brain weight. However, the new techniques of magnet resonance imaging (MRI) and positron emission tomography (PET) hav allowed measurement of brain size in vivo and its comparison with IQ. Usin MRI, Willerman et al. (1991) found an r of .35 between brain size and IC among college students when body size is controlled for. Also using MR Andreasen et al. (1993) obtained r's ranging from .26 to .56 between IQ ar the size of specific brain structures, and an overall r of .38 between full-scal IQ and gray matter volume. Replications by Raz et al. (1993) and Wicke Vernon, and Lee (1994) found correlations between IQ and brain size of .4 and .47–.49. Egan et al. (1994) found an r of .32 between IQ and brain size i a sample whose SD for IQ was 9.3.[27]

Other studies have established a weak ($.15 \leq r \leq .25$) but statisticall significant correlation between IQ and head size (van Valen 1974; Weinberg c al. 1974; also see Rushton 1995b: 36–41). Jensen (1994) has established on data set independent of these earlier studies that the correlation between hea size (and therefore brain size) and IQ subtest performance increases with subte loading on the general factor g. The correlation between a subtest's g-loadin and its correlation with head size is about .64 (602). (This relation, nc implicit in the factorial definition of g, supports g's reality.) One should be in mind that head size is a crude measure of brain size, so IQ/head size r should be doubled to get associated IQ/brain size r's. The resulting estimate closely approximate those derived directly in the MRI and PET studies.

Far from all variance in IQ is explained by brain size, a relatively crud measure of neuroanatomical differences. "In all likelihood," suggest Andrease et al., the remainder of the variance is due to "aspects of brain structure tha reflect 'quality' rather than 'quantity' of brain tissue: complexity of circuitry, dendritic expansion, number of synapses [or neurotransmitter] efficiency" (1993: 133). A group at the University of California at Irvine is investigating some of these factors using PET scans. Haier et al. (1988) report that performance on the Ravens Advanced Progressive Matrices, nonverbal pattern completion tasks recognized as a very pure test of reasoning ability, correlates negatively with the rate at which the cortex metabolizes glucose, when controlled against cortical metabolic rates of nontest subjects passively attending to stimuli. Haier (1993) adds that the largest inverse correlation involves the left temporal lobe. In a follow-up study of learning, Haier, Siegel, MacLachlan, Soderling, Lottenberg, and Buchsbaum (1992), and Haier, Siegel, Tang, Abel, and Buchsbaum (1992) found that degree of improvement after practice in the computer game Tetris correlated with extent of decrease in glucose metabolic rate. (Improvement in Tetris with practice also correlated with IQ as measured by the Raven.) Parks, Loewenstein et al. (1988) found correlations exceeding −.5 between performance on a verbal fluency test and glucose metabolism in the frontal and temporal regions of the brain (also see the review article Parks, Crockett et al. 1989).

Along similar lines, Raz et al. (1993) found the most robust correlation to hold between performance on the Cattell Culture-Fair Intelligence Test and the asymmetry in size between the left and right hemispheres. Egan et al. (1994) categorized a variety of correlations between brain size and both cognitive tasks and electrophysiological responses into those involving white matter, gray matter and cerebrospinal fluid, suggesting that the first is particularly associated with verbal IQ, the second with performance IQ, and the third with full-scale IQ. MRI and PET technologies have become significantly more refined since these studies were done and could now be used to search for direct physiological evidence of race differences in brain function, although attempts to do so would surely generate fierce opposition.[28]

Head and brain size are unlikely to be determined by socialization. Head size is fixed by the closing of the cranial sutures in infancy; adult head size is predicted by head size at birth (Andreassen et al. 1993: 133), and the brain has virtually completed growth by the sixth year (Blinkov and Glezer 1968: tables 113 and 115, 335–336). One might conjecture that children with larger, rounder heads are socialized more diligently in the "white" cognitive style, and, given the correlations between head size, IQ, race and child-rearing practices (such as punitiveness of instructional style; see Moore and Erickson 1985), a weak correlation of this sort might well exist. But causation flowing from head size to socialization would not necessarily make the head-size/IQ correlation a socialization artifact. More intense stimulation of larger-headed children, should that occur, could be a genotype/environment correlation, with head size both contributing to intelligence and triggering an innate tendency in parents or adults generally to stimulate receptive, large-brained children.[29]

The socialization hypothesis seems unable to explain the connection between IQ and brain metabolism, since neither white nor black children are trained to retard glucose oxidation. It should be emphasized that if a "white" environment does somehow affect brain function, that would amount to the white environment increasing intelligence, and imply a phenotypic difference.

The IQ/attainment correlation would then not be spurious, but a measure of an environmentally induced race difference in intelligence. This supposition is likely to remain counterfactual, however, since (for reasons mentioned in the previous paragraph and chapter 4) brain development appears to be largely under genetic control.

Jensen has also explored the association of IQ with brain activity evoked by sensory stimuli. He describes the method:

> The subject, with an electrode attached to his scalp, merely sits in a reclining chair and simply hears a randomly spaced series of auditory "clicks," stimuli that cannot be regarded as "cognitive" or "intellectual" by any reasonable definition. The subject is not required to make any overt or voluntary responses during the session. But brain waves are recorded and averaged by computer over a given time-locked epoch marked by the occurrence of each auditory stimulus, yielding a highly distinctive waveform termed the *average evoked potential*, or AEP. The main features of interest are the average latency, intraindividual variability in latency, amplitude, and complexity of the AEP. . . . All of these measures have been found to be correlated with scores on various IQ tests. (1986: 323–324)

A similar experiment described by Jensen (1986) involves "EP habituation," the decrease of the evoked potential over time, found to correlate .59 with IQ as measured by the Wechsler Adult Intelligence Scale. (When the manifold is analyzed to yield g, the EP-habituation/subtest correlation increases with subtest g-loading. Since EP habituation is not a variable in the manifold from which a general factor is extracted, this finding supports the reality of g.)

Jensen and associates (see Jensen 1985c: 196, 209) have also studied "elementary cognitive tasks" involving a console with lights arranged in a semicircle, buttons just below them, and a "home" button on which the subject rests his finger. In the simplest experiment a light goes on and the subject must move his finger to the button below the light. In another, three lights go on and the subject must press the button below the one light not adjacent to another. This equipment is sensitive enough to separate "decision time" from "movement time." The Jensen group has found a correlation of −.5 between IQ and reaction time on these tasks, and correlations as large as −.7 between IQ and intraindividual variation. The higher an individual's IQ, in other words, the lower and less variable his reaction time. There have been challenges to the care with which Jensen has presented these data (Carroll 1993) and his interpretation of these data as showing a link between IQ and speed of neural processing (Sternberg 1988: 43), but the reaction-time/IQ correlation itself appears not to be in dispute (see Neisser et al. 1996: 83). This correlation bears on the socialization hypothesis because white and black children are not differentially rewarded for pushing buttons when lights go on, so the correlation should, on the socialization hypothesis, be near 0.

It is theoretically possible that IQ correlates with elementary task performance within but not between races. If elementary task performance does not predict IQ when race is allowed to vary, the within-race correlation between elementary tasks and IQ would not be evidence that the between-race difference in IQ expresses a real cognitive difference. However, Jensen (1987, 1993) has found that blacks and whites differ in reaction times and intraindividual variability on these tasks, and that the race differences are greatest on the most g-loaded tasks. Again, since white children are not reinforced more than black children for taking a uniform amount of time to push buttons when lights go on, the IQ difference between blacks and whites cannot indicate training differences. Jensen also found that black movement time was as short as white, indicating, as Herrnstein and Murray point out (1994: 284), that black levels of motivation were as high as whites, hence that motivation was not a biasing factor.

In other studies, Noble explored race differences in the psychomotor skills of "rotary pursuit" and "selective mathometer," the latter especially resembling Jensen's elementary tasks.[30] White children outperformed black children on both, with the white learning curve steeper than the black (Noble 1978: Figures 10.7, 10.8). (Noble notes two further points suggesting a genetic explanation of this difference. In both cases, the performance and learning curves of Negroid-Caucasoid hybrids were intermediate between Caucasoid and Negroid, and the interindividual heritability of performance on rotor pursuit early in life is high.)

Another variable intermediate between cognition and purely physiological response is attentiveness to new phenomena in infancy. There is an inverse correlation greater than .5 between IQ at four to six years and, at age six months, time spent watching an old picture when a new one is made available (Kolata 1987). It is difficult to construe this correlation as a result of (white) socialization, since the attention of 6-month-olds cannot be controlled by reinforcement.

The correlations reviewed in this section strongly suggest that IQ measures a basic property, or property cluster, of the mind. It might however be suspected (see Yee et al. 1995; Fischer et al. 1996: 34–37) that my larger argument eventually goes circular, since in chapter 8 I cite the race difference in intelligence to help account for race differences in many of these same correlates. Can a construct legitimately be used to explain data previously cited to support the reality of that very construct?

Yes. A phenomenon may so clearly be the effect of a hypothetical mechanism that observation of the phenomenon establishes the mechanism, allowing the mechanism to be assumed to produce the phenomenon. There is no circle in this reasoning so long as the mechanism predicts effects other than the ones that prompt its postulation. Such bootstrapping, common in science, is illustrated by the discovery of Neptune via the eccentricity of Uranus' orbit. This eccentricity could only have been caused by the gravitational tug of an unknown planet, so astronomers postulated such a planet, gave it a name ("Neptune"), and used it to explain the orbit of Uranus. This explanation would have been unfalsifiable had the Uranian eccentricity been taken to exhaust the evidence for Neptune. In fact, however, astronomers predicted Neptune's position and succeeded in observing it directly. In the same way, using the correlation between IQ and attainment to validate IQ and then turning around to use IQ to explain attainment is not self-sealing because IQ correlates with other phenomena, including physiological ones. In contrast, it is viciously circular to explain race difference in test performance as a bias effect, when bias is posited on no basis other than differences in test performance.

3.11. CORRELATIONS AND "POSITIVISM"

Critics such as Block and Dworkin who deride emphasis on correlation as "operationalist" show themselves blind to nonoperationalist accounts of correlation, and thereby find themselves embracing the weakest tenets of operationalism. This irony was noted earlier in Gould's attack on "reifying."

Operationalism (or "positivism"), whose heyday extended from the late nineteenth century to the middle of the twentieth, held that science is limited to describing relations between observables. Every measurement procedure was taken to define a distinct concept, with "several measurements of the same quantity" a misnomer for agreement between distinct procedures. On this view, distance-as-determined-by-parallax, for instance, becomes a mere correlate of distance-as-determined-by-ruler, the one yielding "fifty yards" when, as a matter of fact, the other does, distance itself as the source of the concordance being dismissed as superfluous. The view that IQ is "merely a predictor" exhibits this very refusal to infer multiple manifestation from correlation.

Students of scientific method eventually decided that operationalism multiplies concepts far beyond necessity, scants the role of unobservables in predicting new correlations, and arbitrarily forbids explanations of correlations themselves. If all there is to science, indeed reality, are relations between observables, it is pointless to ask why these relations hold—why water-solubility associates with alcohol-solubility, or IQ predicts income. Van Fraassen, whose views accord with positivism, replies (1980) that correlations can be explained by nonexistent theoretical entities, since the premises of an explanation need not be true. Such a view seems absurd; for an explanation to be correct, its premises must be true and the entities it posits real. It is surely self-defeating to say that sugar dissolves when the molecular bonds between sugar molecules weaken, and that molecules do not exist (see Levin 1990: 126–129; Cartwright 1983: Essay 5). If molecules do not exist, the most one can say is that sugar acts as if it were made of molecules, and being as if composed of molecules can't be what makes sugar dissolve.[31]

Turning to induction, consider the conjecture that IQ and its correlates tap the ability to process data. This conjecture predicts a smaller r between IQ and the recall of a string of digits in the order heard than between IQ and recall of the digits in reverse order, since reversing digits is a mental process. The prediction has been borne out (Jensen and Figuroa 1975), but the relevant point here is that it could not have been so much as conceived unless scientists had wondered what unobservable phenomenon IQ measures and how else this phenomenon might show itself. A mere list of known correlates, of IQ or anything else, suggests no new ones. The operationalist can add new correlates to his list once they are discovered by others, but he is limited to playing catch-up. As a heuristic, operationalism is pure impediment.

For these reasons, a realistic view of constructs has returned to favor. At the present time, an objection like "IQ tests predict, but so what?" would be dismissed as doctrinaire in any context other than psychometrics. Nobody belittles Maxwell's equations by saying "they predict radio waves, but so what?" Radio waves are taken to show the existence of the fields these equations describe. Likewise, the correlates of IQ are important not because correlation exhausts scientific knowledge, but for the opposite reason, that they indicate an underlying reality.

The connection between IQ tests and "intelligence" again involves the meaning/reference distinction. "Intelligence" does not and never has meant "what IQ tests tests"; for both layman and psychometrist it means learning ability. What remains to be discovered is its referent, whatever structure or architectural feature of the nervous system produces that ability. But let us suppose—what is in effect supposed by critics who deny any relation between psychometric "intelligence" and the everyday word—that "intelligence" did not exist before IQ tests, and was coined for what IQ tests measure. On the realist view, "intelligence" would still not have meant "what IQ tests test"; rather, IQ tests would have *fixed the reference* of "intelligence" (see Donnellan 1966; Kripke 1973; Putnam 1975). Many a term enters language via superficial traits which do not define it, but merely identify an unknown something for the term to name. "Star," for example, was introduced as a name for the points of light in the night sky, whatever they are. That is still how "star" is explained to children. Yet "star" has never meant "light in the night sky," for if it had, the shrewd surmise that the Sun is a star would have been a contradiction in terms, and, indeed there would have been no stars had the Earth not formed. "Star" denotes those things that happen to be visible as lights in the night sky, and anything else, including things not visible at night, scientifically similar to them. That is how the Sun, which is from a scientific standpoint similar to Arcturus, Betelgeuse, and the like, could turn to be a star, and how stars could exist without an Earth. Just so, psychometric "intelligence" even if a neologism is not defined by IQ tests; it names that trait, whatever it is, which displays itself in IQ test performance, which can display itself in other ways, and which would have existed had IQ tests never been invented.

Much misunderstanding of the importance of correlation is due to the circumspectness of psychometrists themselves. Not wishing to assert or imply that blacks are less intelligent than whites, they often describe their data as, simply, that "white IQ test performance exceeds black," or that "IQ test performance predicts school grades." This idiom, which treats IQ like a crystal ball, as *just* a predictor with no underlying mechanism, is consistent with and indeed invites interpreting IQ and academic performance as joint proxies for white socialization. If psychometrists refuse to commit themselves to anything beyond IQ for IQ tests to measure, they have only themselves to blame for being called positivists.

3.12. OUR CONCLUSIONS TO THIS POINT

In reviewing mental tests, the National Academy of Science was at pains to determine whether they penalize blacks. It cautioned that "Tests of general ability . . . can too easily be misunderstood to mean that intelligence is unitary ability, fixed in amount, unchanged over time" (Garner and Wigdor 1982: 28). Yet despite these and other qualifications, its conclusion was essentially the one defended here: "It *is* legitimate," they said, "to speak of general ability and to say that some people have more of it than others" (28).

Indeed, even the qualifications about unitariness and fixity are misleading. We have already seen that for many purposes unitariness is a side issue. Furthermore, IQ stabilizes in childhood (see Brody 1992: 232): r between IQ at 8 and 17 is about .8 (Brody 1992; also see Neisser et al. 1996: 81), with the correlation between factorial g at 8 and adulthood higher. In other words, the odds are 2 to 1 that an individual's adult IQ will fall within 3 points of his IQ at 8. IQ changes, but not nearly as erratically as critics, and the wording of Garner and Wigdor, suggest.[32]

Denying that IQ measures general mental ability involves a good deal of doublethink. I have already mentioned opponents of testing who also oppose the execution of low-IQ murderers, and Gould's (1981) acknowledgment of Gauss's genius. (At another point Gould describes his brother as "intelligent.") Isaac Asimov, a self-confessed aggressive "liberal," makes light of his Army intelligence tests, yet is careful that his readers know his score was 160 (Asimov 1979) (a score consistent with his remarkable literary achievements). Gould himself, although somewhat evasively, allows that a very low score on Binet's original scale did indicate retardedness and learning disability (1981: 155)[33]—and once low IQ is conceded to correspond to low levels of learning ability, why should it lose its meaning in its normal range?[34]

That IQ measures intelligence is also the reader's belief. I ask him to call to mind those ten acquaintances he considers the most intelligent and ask himself how likely he thinks it is that any of them have IQs below 90. If you protest that your judgment has been shaped by the very white norms that IQ measures, you can probe your attitude more deeply as follows. First, estimate the marginal monetary value to you of 50 IQ points. Now ask yourself whether you would take that amount plus $1,000,000 for letting your IQ be lowered 50 points. If the reader declines, as I suspect he will, he thinks IQ measures more than habits valuable in white society, for he is refusing more than the value of the habits he would lose. And I suspect the reason the reader will decline is that he regards those 50 points not as a useful asset, but as essential to his mind and personality.

3.13. TEST BIAS

A popular objection to gauging black intelligence by IQ is that IQ tests are biased. This charge overlaps the charge that IQ measures white socialization, but it also raises new issues.[35]

The bias objection loses most of its intuitive force on examination of actual IQ test questions. Some do assume background information, possibly more accessible to whites than blacks, but most do not. Items on the nonverbal Raven's Matrices, for instance, on which black performance is particularly poor, ask the subject to select the pieces missing from patterns. Given the ubiquity of patterns in experience, to say blacks are less familiar with them than whites virtually concedes a difference in cognitive function. Fischer et al. (1996) come quite close to this position, excusing poor black performance on backward digit span with the remark that it "requires familiarity and comfort with numbers" (189). Unfamiliarity with numbers on the part of anyone raised in the contemporary United States would certainly indicate a defect in mental functioning.

Yet perhaps intuition misleads and IQ tests are biased after all. The first order of business in deciding this is to define bias. Clearly, a test T for a given trait is biased against a group when group members score lower on T than nongroup members who possess the trait to the same degree. Measuring manual dexterity by having the subject trace a pattern with his right hand is biased against left-handers, because left-handers will do worse on this test than no-more-dexterous right-handers, or, equivalently, a left-hander must be more dexterous than a right-hander to score as well. IQ tests are racially biased if whites attain higher scores than equally intelligent blacks.

Figure 3.1
Bias

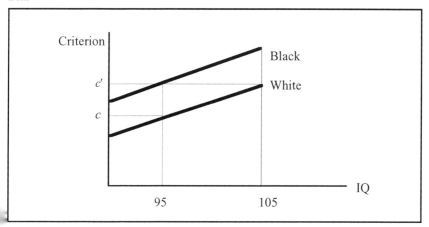

But talk of equally able individuals achieving different scores on T assumes some criterion for the ability independent of T: T can be shown to be biased against a group only via some further measure of the trait tested by T on which group members do as well as nongroup members with higher T-scores. Bias against left-handers in the tracing test is shown by the ability of left-handers to assemble bicycles, thread needles and perform other "criterion" tasks requiring dexterity as proficiently as right-handers who outscore them at right-handed tracing. The empirical meaning of bias, then, is underprediction of criterion variables. IQ tests can be shown to be biased against blacks just in case blacks outperform whites with the same IQ on other tasks requiring intelligence, or equivalently, blacks do as well on criterion tasks as whites with higher IQs. Graphically, in Figure 3.1, IQ tests are biased if the intercept of the regressor for blacks of the criterion variable on IQ exceeds that of the regressor for whites. The vertical line through IQ = 95 has blacks outperforming whites with the same IQ by $c'-c$, and the horizontal line through c' has blacks with IQs of 95 doing as well on the criterion as whites with IQs of 105. Thus, for instance, the well-known 200-point race difference in SAT scores (M.-W. Lee 1992; Belnick 1996: 128) does not, by itself and apart from the academic record of blacks and whites who achieve identical SAT scores, show bias. Without a criterion such as academic achievement, bias is undefined.[36]

Many people are surprised to learn that standardized tests predict performance on such criteria as academic achievement and vocational success as accurately for blacks as for whites. Blacks not only fail to earn higher grades than whites who have achieved the same SAT scores, they in fact earn somewhat lower grades than SAT-equivalent whites (Breland 1979; Herrnstein and Murray 1994), by definition bias against whites. To cite the National Academy of Science review once more:

Predictions based on a single equation (either the one for whites or for a combined group of blacks and whites) generally yield predictions that are quite similar to, or somewhat higher than, predictions from an equation based on data for blacks. In other words, the results do not support the notion that the traditional use of test scores in a prediction equation yields predictions for blacks that systematically underestimate their actual performance. If anything, there is some indication of the converse, with actual criterion performance being more often lower than would be indicated by test scores of blacks. Thus, in the technically precise meaning of the term, ability tests have not been proved to be biased against blacks: that is, they predict criterion performance as well for blacks as for whites. (Garner and Wigdor 1982, Vol. I: 77)

The NAS reported the same finding in a 1989 study of the U.S. Department of Labor's General Aptitude Test Battery (GATB), in which the criterion variables—job performance, training success, and supervisor ratings of job incumbents—measure vocational attainment: "Analysis of . . . 72 validity studies shows that use of a single prediction equation relating GATB scores to performance criteria for the total group of applicants would not give predictions that were biased against black applicants. That is, the test scores would not systematically underestimate their performance. A total-group equation is somewhat more likely to overpredict than to underpredict the performance of black applicants" (Hartigan and Wigdor 1989: 254). As a rule, ability tests overpredict black criterion performance by about .35 to .5 SD (Brody 1992: 288–289). In the most recent literature review known to me, Neisser et al. (1996) conclude that "Considered as predictors of future performance, the tests do not seem to be biased against African Americans" (93).

It might be objected that the criteria used to validate IQ tests are themselves biased, with test scores predicting grades and job performance as accurately for blacks as whites because discrimination depresses black academic and job performance along with test scores. Empirically, erstwhile criterion variable are found to be unbiased when measured against further criteria; supervisory ratings predict work sample evaluations equally validly for blacks and whites for instance (Schmidt 1988: 286–287). It will be replied in turn that these secondary criteria may also be racially biased, the whole "system" being so rigged that predictor and criteria march in lockstep. But this reply defeat itself, since, for the reasons explained, disqualifying every criterion deprives th bias hypothesis of empirical content. If the races are equally intelligent, should be possible to find a task intuitively requiring intelligence that black perform as well as whites. To insist time after time that black ability is preser but hidden by bias is like claiming that someone who apparently knows onl

English can also speak Pushtu, but never does so because circumstances are never propitious.

My impression is that egalitarians sense the danger of untestability, so do not assert categorically that all conceivable criteria are tainted, but at the same time they stand prepared to reject any particular criterion on which whites outperform blacks. Since the supply of potential criteria is limitless, the "bias" hypothesis hangs in limbo—not quite implicitly defining "biased test" as one on which whites outscore blacks, yet not quite purporting to state a fact. As a result of its virtual untestability,[37] the charge of bias is often indistinguishable from admission of real race differences. Hacker says blacks do poorly on the SAT because

> tests like the SAT now reflect not a racial or national corpus of knowledge, but a wider "modern" consciousness. . . . "Modern" now stands for the mental and structural modes that characterize the developed world. It calls for a commitment to science and technology, as well as skills needed for managing administrative systems. The modern world rests on a framework of communication and finance, increasingly linked by common discourse and rules of rationality. . . . [B]lack Americans spend more of their lives in segregated settings than even recent immigrants. One outcome of this isolation is that black Americans have less sustained exposure to the "modern" world than have many members of immigrant groups. . . . The fact that black modes of perception and expression, which are largely products of segregation, become impediments to performing well on tests like the SAT reveals that racial bias remains latent not only in the multiple-choice method, but in the broader expectations set by the modern world. (1992: 145–146)

"Mental mode needed for mastering rules of rationality" is just a roundabout description of intelligence, and talk of "impediments" imposed by "black modes of perception and expression" a windy way to say that blacks are less intelligent.

Or take Gould's (1994) distinction between "S-bias," statistical bias as defined in Figure 3.1, and "V-bias," or bias "in the vernacular sense." "I agree completely," says Gould (145), "that [IQ] tests are not biased—in the statistical sense" (an implicit retraction of chapter 5 of Gould 1981). The real question, he goes on,

> the source of public concern, embodies an entirely different issue, which, unfortunately, uses the same word. The public wants to know whether blacks average 85 and whites 100 because society treats blacks unfairly—that is, whether lower black scores record biases in this social sense.[38]

This distinction is muddled. When the public asks what to make of lower black IQ scores, it evidently has one of two questions in mind. It may want to know whether black scores really indicate a lower level of mental ability, or merely inadequate test preparation. What is then being asked, the source of public concern, is whether IQ tests are S-biased. On the other hand, the public may be conceding that black scores reflect lower intelligence—that is, that the tests are S-fair—and is asking whether blacks are less intelligent because of genes or unfair treatment (that is, V-bias). "Bias" as applied to tests continues to mean S-bias, the issue behind the second question being the processes that produce the race gap that IQ tests—faithfully—record. Bias in those processes, Gould's V-bias, has nothing to do with tests. To be sure, people who imagine unjustly caused effects are somehow unreal may well believe IQ tests are both S-fair (because they faithfully record a black deficit) and V-biased (because the deficit they record should not exist). But this confusion—abetted by Gould—is not an ambiguity in "bias."

Since IQ predicts black criterion performance as accurately as white, arguments purporting to prove test bias can be expected to fail. Three nonetheless repay discussion, two because they enjoy popular currency, the third because it has received attention in the technical literature.

The first argument is the absence of blacks from populations on which IQ tests are normed. Actually, reference groups have included blacks for some years,[39] but in any case the inference is a non sequitur. Including more blacks in a normative sample yields questions that blacks are better able to answer, but if these questions are easier for blacks because they tap black knowledge rather than reasoning ability, their inclusion makes tests less indicative of intelligence (and less predictive).[40] On the other hand, if whites are more intelligent than blacks, adding blacks to the reference group while continuing to reject nonpredictive questions simply lowers the mean performance of the reference group while raising the overall mean IQ of whites. Skinny men can be made more nearly "average" in weight by including more of them in the calculation of actuarial tables, but this reclassification of erstwhile average men as overweight makes skinny men no heftier.

The second popular allegation is that IQ tests measure white middle-class socialization. Some reasons to reject this charge were presented in connection with the reification issue. There are many others.

(*i*) Socioeconomic status as social scientists measure it—"SES"—explains relatively little of the race difference in test scores. A 9–to–12-point gap remains when status is controlled for (Jensen 1980: 44; also see Neisser et al. 1996: 94); white children have higher Peabody Vocabulary Test scores at all income levels (Currie and Thomas 1995). Lower-class whites consistently do as well or better than middle- and upper-class blacks (Jensen 1980: 44, Figure 3.1; Scarr 1981: 270–273). The mean SAT score of blacks from families whose annual income exceeds $70,000 is lower than those of whites and Asians from families whose annual income falls below $20,000 (Hacker 1992: 143). A study involving tens of thousands of students that maximized between-group similarity by controlling for race, sex, grade, and geographical locale found that "blacks and low SES whites have somewhat similar but far from identical [cognitive] profiles" (Humphries, Fleischman, and Lin 1977). Jensen (1980: 203–204) and Jensen and Reynolds (1982) compare comparisons: of black to white children, and high- to low-status children, holding full-scale IQ constant both times. Black and white children with identical IQs differ significantly in their performance on some subtests of the revised Wechsler for children, as do children with identical IQs from different status categories. However, the pattern of black-white differences with IQ held constant is unlike that of high-status/low-status differences with IQ held constant. In particular, blacks do as well as whites on tests of memory and vocabulary, markedly less well on tests of spatial ability and numerical reasoning. (Status differences associate with difference in verbal test scores.) This disparity would not occur if the black/white difference in test performance was a proxy for black/white status differences.

(*ii*) The cultural bias hypothesis predicts that whites will outscore Asians on IQ tests, whereas Asians outperform whites, particularly on mathematics subtests (Jensen 1981: 134; Lynn 1991a, 1991b). Japanese regularly score in the 103–107 range on IQ tests normed on American populations. As Lynn (1977) remarks, "it seems hardly plausible to advance test bias as an explanation for the high mean Japanese IQ." Some authorities dispute the overall Asian/Caucasoid difference, but all seem to grant a marked Asian superiority in the mathematical and spatial reasoning. Mathematical precocity is much more common in China than the United States (see Benbow and Stanley 1984). The non-verbal Raven's Matrices, a test described earlier on which blacks perform particularly poorly, was normed on a Scottish and English population, yet Eskimos do as well on it as whites (Jensen 1981: 134). It is hard to argue that the Eskimo way of life more resembles that of the white middle class than does the way of life of American blacks.

(*iii*) Most polemics against mental tests cite items, sometimes decades out of date, that assume information presumptively more available to whites. One such item concerns the proper appearance of a tennis court (Gould 1981: 211); another, repeated by Hacker (1992: 145) and regularly displayed by the anti-testing organization FairTest, queries the meaning of "regatta." (Also see Fischer et al. 1996: 189.) Greater familiarity with tennis and boating, it is argued, gives whites an advantage unrelated to mental ability.

To begin with, these critics often misunderstand the role played by information in IQ tests. Several critics cite the following question, similar to some found on the Air Force Qualifying Test: "If i and j are positive whole numbers and $j + k = 12$, what is the greatest possible value of k? (a) 6 (b) 36 (c) 32 (d) 11." Fischer et al. find this question to be "manifestly about *school* tasks . . . what students have been taught" (1996: 57–58), presumably because it involves arithmetical knowledge. In fact, the knowledge of elementary addition this question presupposes has been taught to (but not absorbed equally deeply by) every American over the age of 10. The question probes the ability to deduce a consequence of this knowledge, which is not a "school task," not itself knowledge, and, suitably generalized, is close to the intuitive notion of "thinking ability."

The performance of test subjects disadvantaged by ignorance should improve on tests not assuming culture-specific information. Yet—this objection to egalitarianism is the one most emphasized by Herrnstein and Murray (1994: 302–303)—the black/white gap is consistently wider on less culture-bound tests. Blacks do better on vocabulary items than on items probing numerical reasoning [41] (Jensen 1980: 552), and, when general factors are held constant, blacks outperform whites on what is distinctively measured by each subtest in the WISC-R battery (Jensen and Reynolds 1982: 435–436). These findings are contrary to and seemingly inexplicable under the bias hypothesis.

Whether black IQs reflect limited access to information can also be tested by comparing the items found most difficult by blacks to those found most difficult by whites. On the limited-information hypothesis, there should be questions of varying difficulty for whites, all of which assume knowledge unavailable to blacks and therefore uniformly difficult for blacks. On the other hand, if whites outperform blacks because of an ability to which knowledge is irrelevant, the questions found hard by whites should coincide with those found hard by blacks, and, in general, the ordering of items by difficulty for one race should parallel the ordering of items by difficulty for the other. Jensen reports (1980: 552–580) that the difficulty ordering of standard intelligence test questions is virtually the same for whites and blacks, a finding often replicated (see, e.g., Owen 1992).

The best test of the relevance of specific knowledge, comparing black to white performance when knowledge is controlled for, is not practically possible, but when similarly performing blacks, whites, and Asians are compared for age, black children match white children two years younger and Asian children three years younger (Jensen 1980: 554). To account for this lag, the discrepant information hypothesis must posit a mechanism that somehow keeps information available to Asian children from white children for one year and from black children for three, whereas the lag is unsurprising on the hypothesis that, after infancy, blacks trail whites in mental development. A parallel finding (Jensen 1980: 584) concerns choice of wrong answers on the Raven and the Peabody Progressive Vocabulary tests.[42] Whites and blacks of the same age differ on the frequency with which they choose plausible "distractors," suggesting that something besides reasoning ability is being tapped, but black children choose the same distractors as white children two or three years younger. Again a natural explanation is that the mental development of blacks lags behind that of whites. So far as I know, critics of test bias have not addressed the data about rank order of item difficulty or "pseudorace age groups."

Psychologists have labored to construct culture-fair tests, in many cases working with experts in bias detection.[43] One well-known effort was the Davis Eells test developed in the early 1950s, which involved no written work and was administered in what would today be called a nontraditional format—questioning children about cartoons depicting everyday situations. The race gap on this test was as large as that on conventional IQ tests (Ludlow 1956; see Jensen [1980: 635–714] for a fascinating discussion of the Eels and several other culture reduced tests). Deserving special mention are the pugnaciously named BITCH (Black Intelligence Test of Cultural Homogeneity) and SON OF A BITCH tests constructed by black psychologist Robert Williams, most of whose items concern sex, crime, black slang, and "black identity," and on which blacks outperform whites. Prof. Williams informed me, when I inquired about external validation of the BITCH, that it predicts sympathy for blacks. It correlates negatively with performance on cognitive variables like test performance, more so for blacks than for whites (Brod 1992: 292–293).

Anticipating the topic of chapter 4, I do not follow Herrnstein and Murray in taking the "profile difference"—the greater width of the race gap on abstract reasoning items—as evidence of genetic input. The profile difference shows that the overall race difference in IQ is not a socialization or bias artifact, but the difference might still in theory be due to environmental factors other than socialization. That the IQ gap is not an artifact means that it reflects a difference in some real property of the mind, but it leaves the origin of that difference open. Herrnstein and Murray rightly describe the profile difference as confirming "Spearman's hypothesis," that the races differ in g, but g is a phenotype about whose source the Spearman hypothesis is noncommittal.

(iv) Some evidence against the bias hypothesis resembles the data about physiological variables and elementary cognitive tasks, but is more theory-laden in assuming g.

Consider two tasks, neither reinforced by white socialization, which predict IQ scores with unequal accuracy (and on which g therefore loads unequally). If IQ measures socialization, the race gap should be the same on both; if IQ measures g, the race gap should be larger on the more g-loaded one. In another context I mentioned two such tasks, namely recalling digits in the order presented and recalling digits in reverse order, the latter the better predictor of IQ (although see Carroll 1993). Neither whites nor blacks are taught either skill, which differ only in that "backward digit span" requires some mental transformation of input. Assuming the race difference in g is spurious, any white advantage on backward digit span should be no greater than the white advantage in forward digit span. In fact, the race gap on backward digit span is twice as great (Jensen and Figueroa 1975; Jensen 1981: 163).

In a more systematic study, Jensen (1985b: 203–206) found a correlation of .59 between the factorially determined g-loadings of tests and black-white performance differences on these tests. Jensen further suggested that this r may be an underestimate because of range restriction, since the g-loadings of all the tests reviewed exceeded .3. A follow-up study (Jensen 1987) put the correlation at .78. It is possible to maintain that g is an artifact, and possible to maintain that IQ measures white socialization, but difficult to maintain that a mathematical artifact should so closely parallel white socialization.

(v) Members of the same family have the same social status. Hence, if IQ goes proxy for status, most of the variance in IQ explained by environmental variables should be explained by factors that distinguish families from each other, an individual family's so-called "shared environment," rather than factors which vary among members of the same family, the so-called "non-shared" environment of individuals.[44] Household income, for instance, should be more important than division of parental attention. Certainly, the environmental factors usually cited to distinguish whites from blacks, such as family income, level of education, marital stability, and exposure to racism affect all members of any single family equally.

Yet despite the naturalness of expecting variance in shared environment to explain more than nonshared environment, it is not so. Shared environment explains relatively little of the variation in IQ from childhood on (Tellegen, Lykken, et al. 1988: 1036, Table 5; Scarr 1992; Plomin and Thompson 1987; Plomin 1990a: 128; Bouchard, Lykken, Tellegen, and McGue 1992; Chipeur, Rovine, and Plomin 1990; Thompson, Detterman, and Plomin 1993): most of the environmental factors that affect a child's IQ are those that distinguish him from other members of his own family, not those that distinguish his family from others. The environmental traits usually cited to explain black-white differences explain very little of the individual differences in IQ within races—and the variance in nonshared environment appears to be as large among black families as among white (Jensen 1973: 188). Conceivably (see Plomin and Daniels 1987) the main environmental factors differentiating individuals of the same race are unshared while the only environmental factor differentiating the races is test-affecting bias, but this hypothesis seems ad hoc.

(*vi*) A race difference in IQ that vanished when status was fixed would still not necessarily be a status artifact, for the question would simply shift to that of why so many blacks occupy low status. In a recent study of low-birth-weight blacks and whites at age five, for instance, Brooks-Gunn, Klebanov, and Duncan (1996) found that controlling for family poverty, maternal education, maternal verbal ability, provision of learning experiences, and maternal warmth reduced an 18-point IQ difference to insignificance. They conclude from this that "ethnicity is not a predictor of IQ in this sample" (403). Fischer et al. (1996) offer a similar reanalysis of the Herrnstein-Murray sample, with a similar conclusion. The obvious question, though, is why black families are so much poorer than white and why black mothers provide their children with so many fewer learning experiences.

The conventional explanation, that the majority imposes low status, is doubtful. Whatever may have been true a century ago, in recent decades whites have not forced blacks to watch too much television or prevented blacks from buying books or visiting museums. Far from seeking to impede black attainment, whites have tried hard to promote it. As I argue in fuller detail in chapter 8, no one can responsibly explain current low black status by "racism." An obvious alternative hypothesis is that causality flows to status from intelligence as measured by IQ. An individual's IQ helps determine his attained status and correlates with his parents' status because his parents were likely to be near him in intelligence and to have created an environment commensurate with their common intelligence level. In other words, high-IQ parents have high-IQ children, for whom they create rich environments; blacks, being less intelligent, create poorer environments. Empirically, almost 25% of children will spend their adulthoods in a social class other than their parents', and a child's IQ in the elementary grades is a better predictor of his social class as an adult than the class of his parents (Jensen 1981: 194–196; Rubinstein 1993; Neisser et al. 1996: 87; see Herrnstein and Murray 1994: 130–137), suggesting that status is an effect of individual traits like intelligence rather than their cause. Even Brooks-Gunn, Klebanov, and Duncan, who seem unfriendly toward a genetic approach, admit that "maternal verbal ability test scores probably reflect genetics" (404).

This interpretation becomes most natural when we refrain from reifying status as a cause of life outcomes. "Status" is merely a collective label for income, occupation, number of books in the home, museum-going, and scholastic attainment, not a power producing those factors, so it is vacuous to attribute these factors to status. They must have some other explanation, an obvious one being individual traits like intelligence and personality.

Anticipating the discussion of genes in chapter 4, the persistence of the race difference in IQ within social classes is easily viewed, at least in part, as "regression to the mean." A child's status is defined by his parents' income and education. Assume that IQ influences status and that the race difference in

IQ is significantly genetic in origin; then the genes of high-IQ black parents of a high-status black child are rarer, relative to the black gene pool, than are the genes of high-IQ white parents of a high-status white child relative to the white pool. Each child's own genes are apt to be a more probable combination from his population, so the IQs of both children will tend to fall closer to their population's mean. As the black IQ mean is lower, the black child's IQ is apt to be lower. Since hypotheses are strengthened by explaining otherwise puzzling phenomena, this explanation of the within-class race gap in IQ confirms a genetic dimension of the overall gap. (Regression to the mean is unlikely to be the whole story, however, since mating is assortative—high-IQ men marry high-IQ women—and as a purely statistical matter blacks will tend to fall closer to the lower cutoff of any level of achievement; see chapter 8.)

The more technical argument to show test bias (see Seymour 1988; Hartigan and Wigdor 1989: 255–260) concerns the race difference in rates of selection error made by such purportedly unbiased predictors as the SAT and GATB. Let us suppose the criterion is job performance. If the minimum acceptable level of criterion performance is predicted by some cutoff score on a test, just those applicants will be hired whose test scores exceed the cutoff. Assuming the predictor nominally unbiased, blacks and whites who score above the predictor cut are equally likely to reach the criterion minimum. Equally likely but not guaranteed to: predictors are imperfect, so some testees who score below the cut would have succeeded on the job, while some testees who make the cut and are hired perform unsatisfactorily. Rejecting someone who would have succeeded is a *Type I selection error*, akin to rejecting a true hypothesis, and hiring someone who goes on to fail is a *Type II selection error*, akin to accepting a false hypothesis. Now, tests on criterion variables of blacks and whites who have failed predictors, and observation of the criterion performance of blacks and whites who have passed predictors, show empirically that standardized tests make proportionately more Type I errors about blacks and proportionately more Type II errors about whites. Since Type I errors disfavor their victims and Type II errors favor them, this appears to be double-barreled bias. Some statisticians (Hartigan and Wigdor 1989; Cole 1973) would replace identity of regression line with identity of selection error rates as the standard for test bias.

A little reflection shows that this anomaly—a valid test disqualifying disproportionately many competent blacks and passing disproportionately many incompetent whites—is due to the disproportionate number of low-scoring blacks. To illustrate the effect for Type I errors, suppose the criterion ability distributes normally among both blacks and whites, with the minimum acceptable level being the white mean. Also suppose that the criterion correlates .5 with IQ, so the regression is: criterion = .5IQ + 0 = .5IQ. An IQ of 100 thus predicts minimum job performance and serves as the cutoff. When both IQ and criterion are normalized to z, the minimum criterion ability and the IQ that predicts it lie at $z = 0$ as these variables distribute for whites. Now consider the biracial population whose IQ occupies the thin band from 84 to 86 centered on 85, or $-1z$.[45] Everyone in that subpopulation, black and white, is predicted to fail, since their expected value on the criterion is $.5 \times -1z = -.5z$. But actually—see Figure 2.1—criterion ability for this group distributes normally around a mean of $-.5z$, with an SD of .5. Hence a certain proportion, namely 16%, of those whose IQ is about 85 would have succeeded had they been hired. The situation is approximated in Figure 3.2, where the wide curve is the distribution of the criterion in the general (biracial) population, the narrow curve is the distribution of the criterion in the (biracial) subpopulation of IQ ≈ 85, and the shaded area the IQ ≈ 85 victims of Type I errors. There is no hint of bias so far, since 16% of blacks *and* whites with IQs of about 85 are victims. However, since the proportion of blacks with IQ 85 relative to all blacks exceeds the proportion of whites with IQ 85 relative to all whites, a greater proportion of blacks than whites will be IQ 85 Type I error victims. Specifically, about 4.7% of the black population and 2.9% of the white population have IQs in the band, so for this IQ level the Type I error rate of an unbiased test will be about 4.7/2.9 = 1.6 times greater for blacks.[46] Type II error rates present the mirror image. There are proportionately more whites than blacks at all IQs above 100, so proportionately more incompetent whites will pass the predictor.

Figure 3.2
Selection Errors

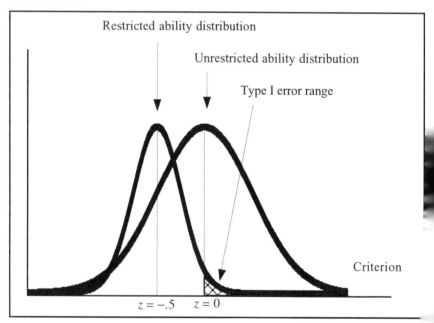

Consider a numerical example. Seventy percent of all dogs who bark a lc make good watchdogs, so using barking as a selector of watchdogs is fair t all breeds. However, every Yorkshire Terrier barks a lot while only 10% c Golden Retrievers do. This means that 3 out of 10 Yorkies will pass th barking test yet prove unsuitable watchdogs, while only 3 out of 100 Golde Retrievers will pass yet prove unsuitable. The test, even though unbiasec selects 10 times as many unsuitable Yorkies as Golden Retrievers. Bree differences in barking have caused an unbiased predictor to yield discrepant Type II selection error rates.

The discrepancy in Type I selection error rates is one of the more esoteric rationales for lower standards for blacks (Hartigan and Wigdor 1989). Ironically, the only way to equalize the error rates while maintaining predictive validity is to *raise* the IQ cutoff for blacks, in the case described to 104.5. 2.9% of whites lie in the 84-86 band, and 16% of those in the band will be victimized, so .464% of all white test-takers will be victims. Since 4.7% of blacks lie in this band, the test cutoff must be so set that only $.464/.4.7 = 9.8\%$ of blacks exceeds it. This corresponds to a z of 1.3 on the black IQ distribution, or 104.5.

3.14. TEMPERAMENT AND MOTIVATION

Herrnstein and Murray (1994) tend to describe blacks as if they were simply a less intelligent subpopulation of whites. These authors are aware that intelligence is not the whole story, noting that the races continue to differ in marital habits, illegitimacy, welfare dependency, and incarceration after IQ is controlled for (329–338), but they ignore possible noncognitive race differences as causes.[47] Such differences do appear to exist.

Many observers, including a number ready to attribute all black difficulties to racism, agree that there is a distinctive black behavioral style. Kochman (1983: 18) describes the white approach to argument as "low-keyed: dispassionate, impersonal and non-challenging . . . cool, quiet and without affect," while the black approach is "high-keyed: animated, interpersonal, and confrontational. . . heated [and] loud. . . . Blacks do not simply debate an idea [the white mode]: they debate the person debating the idea" (21–34).

Kochman observes that whites find the black style "dysfunctional" and often retreat into silence before it, although he considers the black approach more consonant with Mill's ideal of free speech.[48] Hacker, citing several black psychologists, remarks that "Black children are more attuned to their body and physical needs," so "need more leeway for moving around the classroom" 1992: 171). Kochman likewise mentions black "animation," "vitality," and "intense and spontaneous emotional behavior," and comments that the black "rhythmic style of walking" is "a response to impulses coming from within" 106–110). Boykin (1986: 61–65) describes the black personality as displaying verve," "affect," "expressive individualism," and a "social time perspective, n orientation in which time is treated as passing through a social space." He oes on to contrast the value placed by blacks on "a personal orientation oward objects" with the tendency of whites to "place reason above all else," white "possessive individualism and egalitarian conformity," and "an mpersonal (objective) orientation toward people."

Caucasoid classical music, emphasizing harmony and regularity, is atently unlike Negroid jazz, rhythm and blues, and rap, which feature mprovisation and syncopation—to judge by films made by anthropologists, aits common to African music and the musical games of black children in the nited States. Negroid dancing involves pelvic thrusting and other mimicry of intercourse absent from Caucasoid dancing. Nelson George (1992) generalizes these differences to a distinctive black "athletic aesthetic" common to activities as disparate as jazz, basketball and "sermonizing," involving improvisation, aggression, intimidation, arrogance, and the will to humiliate an opponent. (George concedes that this behavior can appear "untutored, undisciplined and immature;" see 1992: xviii.) Jones and Hochner (1973) agree that the black athletic approach involves "an individualistic dimension . . . black athletes place greater emphasis on the stylistic component of play [and] the expressive character of the performance." Majors and Billson, in their book *Cool Pose* (1992) write that the physical "posturing" or "styling" of black males is intended to "deliver a single, critical message: pride, strength and control."

The Minnesota Multiphasic Personality Inventory is a widely used personality test, race differences on which are described for a number of populations in Dahlstrom, Lachar, and Dahlstrom's *MMPI Patterns of American Minorities* (1986). Overall, blacks achieve higher average scores than whites on the Psychopathy, Schizophrenia, and Hyperactivity scales (Dahlstrom, Lachar, and Dahlstrom 1986: 29–31, 34, 36, 39, 41, 47, 95, 135), findings normally interpreted to show "estrangement and impulse-ridden fantasies . . . unusual thought patterns and aspiration-reality conflict" (Gynther 1968). Dreger and Miller (1960) also report higher black male scores on the Hypomania scale, taken to indicate "outgoing, sociable, and overly energetic patterns; tendencies to act impulsively and with poor judgment" (Dahlstrom, Lachar, and Dahlstrom 1986: 218). Black males tend to score higher on precisely the MMPI scales on which the prison population deviates from the population mean (Herrnstein and Wilson 1985: 189); elevated MMPI are equally predictive of criminal behavior for blacks and whites (Elion and Megargee 1975). It is striking, and surprised the investigators, that, while black males always exceed white males in Psychopathic Deviance and Hypomania, blacks who have attended integrated schools attain *more* deviant scores than those who have attended segregated schools (Dahlstrom, Lachar, and Dahlstrom 1986: 125–126, 245 Table D-11)—contrary to the hypothesis that segregation distorted the black personality. Shuey (1978) and others have also found that integration adversely affects the self-conception of black schoolchildren.

That black self-esteem is low is a virtual axiom of popular sociology (see e.g., Schoenfield 1988). In *Brown v. Board of Education* (1954) the Supreme Court cited damage to self-concept caused by segregation, supposedly shown by an experiment in which black children from segregated schools preferred white dolls, as proof that "separate is inherently unequal." The willingness of young black males to rob and kill for fashionable sneakers is attributed to desperate need for peer approval as a source of self-respect. Fischer et al. (1996) state it as fact that de facto segregation "lower[s black] youths' self-esteem and ultimately lower[s] their test scores" (197; also see 180). But, as in the MMPI study, black self-esteem is typically found to exceed white. According to Shuey,

[A]t the preschool level there seems to be some evidence of awareness of color differences and a feeling of inferiority associated with dark skin, but at the grade scho-

level and continuing through high school and college there is no consistent evidence of lower self esteem in Negroes; if there is a difference, it would appear to be more likely that Negroes have a *greater sense of personal worth*, rather than the reverse. (1966: 512; also see Shuey [1978] for a review of literature concerning blacks and whites aged 3 to 8)

A survey of adolescent girls commissioned by the American Association of University Women (American Association of University Women 1991) indicated—to its sponsor's apparent surprise, given the poor academic performance of black adolescents—that black females enjoy higher self-esteem than white. A study of body image found that while 90% of white females think they are too fat, only 12% of black females expressed dissatisfaction with their weight (Parker et al. 1995).[49] Black boys do "not [allow] their achievement to affect their view of themselves":

Bruce Hare, an educational researcher, has documented this process among fifth-grade boys in several schools in Champaign, Illinois [evidently Hare 1985]. He found that although the black boys had considerably lower achievement-test scores than their white classmates, their overall self-esteem was just as high. This stunning imperviousness to poor academic performance was accomplished, he found, by their de-emphasizing school achievement as a basis of self-esteem and giving preference to peer-group relations—a domain in which their esteem prospects were better. (Steele 1992, p. 74)

The doll experiment cited in *Brown* was duplicated with identical results for black children in integrated schools in the North (Seligman 1987), meaning either that integration harms black self-esteem as much as segregation, or that the doll test is invalid.

Literature reviews by Porter and Washington (1979), Crocker and Major (1989) and Baumeister, Smart, and Boden (1996) agreed in finding the self-esteem of black students equal to or higher than that of white students. Tashakkori (1993) and Tashakkori and Thompson (1991), in other review articles, also report that blacks consistently value themselves more highly than whites, the gap being smaller for blacks attending integrated schools. Like Hare and Steele, Tashakkori considers it likely that blacks do not use academic success as a standard of personal worth to the extent that whites do. At the same time, blacks in integrated schools are more aware of the shortfall in their achievements, given what is considered minimum performance in their surroundings. Many commentators explain low black self-esteem in terms of feelings of powerlessness; Tashakkori (1993: 597) reports that blacks do perceive themselves as more powerless than whites, but that this self-perception does not translate into lower self-esteem, apparently because blacks regard the self as less a locus of control.[50] Lefcourt (1965) found that, faced with a choice between two risky situations with equal expected payoffs, one of whose outcomes depends on chance while the outcome of the other depends on agent response, blacks are more apt than whites to choose the chance alternative. Coleman et al. (1966: 289, Table 3.13.14) found black high school seniors considerably more likely than white to agree that "good luck is more important for success than hard work."

Black students estimate their own academic competence more highly than whites despite their own objective and self-reported lower academic achievement (Hare 1985: Table 3; Tashakkori 1993: 597). Black high school seniors in the Coleman study were more apt than whites to classify themselves as "among the brightest," and less likely to agree that " 'Sometimes I feel I just can't learn,' " despite poorer academic performance (Coleman et al. 1966: 287–288, Tables 3.13.11, 3.13.12). Remarkably, southern rural blacks, whose academic performance fell below that of all other blacks as well as all whites, were more apt than other blacks to classify themselves as bright and able to learn. All told, "aspiration/reality conflict" is prominent in this literature. (Many teachers I have spoken to comment informally on their black students' unrealistic expectations of becoming doctors or lawyers.) Anderson (1994) reports that black gang members believe they deserve to be treated as superior, which Baumeister, Smart, and Boden (1996) interpret as manifesting high self-esteem.

Certain MMPI items reliably distinguish the races. Blacks agree more frequently with the following statements (Dahlstrom, Lachar, and Dahlstrom 1986: 229–231):

I am an important person.

It wouldn't make me nervous if any members of my family got into trouble with the law.

I am entirely self-confident.

If given the chance I could make a good leader of people.

I have often had to take orders from someone who did not know as much as I did.

Deficient self-esteem does not seem indicated. In addition, blacks more readily agree that:

Most people will use somewhat unfair means to gain profit or an advantage rather than lose it.

Most people make friends because friends are likely to be useful to them.

Most people are honest chiefly through fear of being caught.

It is not hard for me to ask help from my friends even though I cannot return the favor.

I am sure I am being talked about.

It would be better if almost all the laws were thrown away.

A person shouldn't be punished for breaking a law that he thinks is unreasonable.

These responses are interpreted to reveal greater cynicism, mistrust, conflict with authority, and "externalization of blame for one's problems" (Dahlstrom, Lachar, and Dahlstrom 1986). Race differences in MMPI scores diminish when social status, IQ and level of education are fixed, but once again this does not show MMPI differences to be status artifacts. Traits that affect scores may also affect status, as for instance friction with authority impedes advancement in hierarchies.

Employing Hans Eysenck's categories, Rushton finds Mongoloids to b more neurotic and less extraverted than Caucasoids, who are in turn more neurotic and less extraverted than Negroids (Rushton 1995b). An evolutionary theorist, Rushton is especially interested in sex drive strength and reproductive effort. He reports that both in the United States and internationally the mean age for first sexual intercourse for blacks is lower than that for whites (which is lower than that for Asians) (1995: xiv, 172; as reported in a survey of young adults in Los Angeles, blacks on average begin intercourse at 14.4 years of age, in contrast to 16.2 for whites; blacks also reported themselves as slightly more sexually active [Moore and Erikson 1985]). Black married couples have intercourse more frequently than white and have more permissive attitudes toward sex (Rushton 1995b: 171–178, a section dense with more data than can be summarized here). Rushton observes, as have others, that AIDS is spreading most quickly in Africa and the American heterosexual black population, while it has not broken into the white heterosexual population. Half of black sufferers in the United States acquire AIDS sexually, not through drug paraphernalia, which Rushton attributes to the greater promiscuity in African and American black populations (1995b: 178–183). Rushton notes that a black is almost 1.5 times more likely than a white to be confined to a mental institution (1995b: 157), and cites statistics confirming that alcohol and drug abuse are common in American black slums.

The theme connecting anecdotal and analytic studies of the black personality is focus on the near term. Blacks average 73 hours of television watching per week to whites' 50 (Kolbert 1993) and spend twice as much on movies as whites (Holly 1996). Since black income is only 2/3 that of whites, this means that blacks spend proportionally three times more of their income on movies. Black outlays for electronic games and gadgets is comparable to white, but blacks buy proportionally far fewer computers than whites or Asians (Marriott, Brant, and Boynton 1995). A pivotal incident in *The Promised Land,* Nicholas Lemann's study of the migration of a black woman from the rural South to the urban North, involves the subject's boyfriend blowing money needed to finance a house on a flashy new car (Lemann 1991: 105). The plot of Lorraine Hansberry's drama of black life, *A Raisin in the Sun*, turns on the husband impulsively squandering a windfall. In a classic series of studies (Mischel 1958, 1961a, 1961b), Walter Mischel found that black Trinidadian children given a choice between getting a smaller candy bar immediately or a larger one a week hence tended, much more than matched white children, to chose the smaller, immediate candy—the difference being "so great as to make tests for the significance of the difference superfluous" (Mischel 1961a: 6). Mischel reports undertaking the study because informants had suggested that 'Negroes are impulsive, indulge themselves, settle for next to nothing if they can get it right away, do not work or wait for bigger things in the future."

This attitude toward reward can be described in a variety of ways: more rapid decay of reinforcement, unwillingness or inability to defer gratification, "extreme present-orientation" (Banfield 1974: 54), impulsiveness, lower superego-dominance. As some of these descriptions contain value judgments or causal hypotheses, the economist's neutral notion of *time preference* may be best. (Banfield 1974 uses this term as well.) An individual's time preference is measured by the money he insists on getting tomorrow for foregoing $1 today.

The larger the return he requires, the greater his preference for the present over the future.[51] Low-time-preference individuals will endure deprivation today for benefits in the far future, the rate at which they discount future goods in comparison to present ones being small. Unlike intelligence, time preference does not cause behavior; it summarizes behavioral syndromes. To say that someone saves assiduously because his time preferences are low is to fit his tendency to save into a pattern of caution and foresight he follows in other situations. This understood, the central motivational difference between blacks and whites may be said to be higher black time preferences.

Talk of time preference avoids the metaphysical issue of whether blacks are subject to stronger impulses than whites, or are less able to resist impulses of equal strength, treating it as a distinction without a difference. Liebow (1967: 219–221) for instance argues that the "serial monogamy" characteristic of black sexual relations does not represent a distinctive cultural pattern, but rather failed intentions to conform to the white pattern of "durable, permanent union"; from a behavioral point of view this contrast is empty. Also, to note that lower time preferences are more conducive to accumulating wealth does not imply that they are better than higher ones. Whether investing $50 is wiser or more responsible than spending it on action movies is, from the economist's point of view, meaningless. Finally, high time preferences tend to characterize all lower-class individuals, as Banfield notes repeatedly (see 1974: 87), so the concept is race-neutral. However, disproportionately many blacks are in fact lower class, and Banfield often describes "class" attitudinal differences in overtly racial terms (e.g., 114).

It is often said that blacks ignore the future because oppression makes thinking about tomorrow pointless. That explanation is consistent with, indeed entails, the fact of higher black time preferences. Whether oppression or endogenous variables account for that fact is one of the topics of the next chapter.

NOTES

1. Let us call e *prima facie evidence* for h if P(e|h) > P(e|~h). If e is prima facie evidence for h, and for some h* such that P(h*) > 0 we have P(e|~h & h*) ≥ P(e|h), h* *explains e away* consistently with ~h. When there is no reason to believe h* except that it explains e away, h* is *ad hoc*. In these terms, there is some prima facie evidence of lower black intelligence but no prima facie evidence that black intelligence equals white (which is itself further evidence of lower black intelligence). The bias hypothesis explains this evidence away, but absent independent evidence for it, it is ad hoc.

2. The odds of correctly assigning n distinct traits to m distinct groups at random is $1/n^m$, and $1/n! - (n - m + 1)!$ if traits can repeat. When m = 8 and n = 10, the odds are 1 in 900 million and 1 in 3.6 million, respectively.

3. Narrowness of reaction range, in the language of genetics explained in chapter 4.

4. Sternberg (1994) also reports that the mean IQ of black females is several points higher than that of black males.

5. Peoples, Fagan, and Drotar (1995) also control for birth order; it has been argued that, since later siblings do less well on IQ tests than firstborns, the race gap might be due in part of the larger size of black families.

6. Flynn (1980: 211) is reasonably clear about this, but not entirely, since he doe

not quite commit himself, in 1980, as to what IQ measures (see 30–35, 197).

7. They later retreat. "Should people have refrained from using Galileo's air thermoscope as a poor but useful measure of temperature? Of course not. But neither should they have pretended that a finely honed supersophisticated air thermoscope is anything more than an air thermoscope. We do not object to IQ tests being *used*, but rather to their use under descriptions like 'intelligence tests' " (429).

8. A position in current academic philosophy called "eliminative materialism," championed by Churchland (1981), does deny the existence of beliefs and other mental states; Churchland explicitly compares belief in beliefs to belief in witches, and would presumably extend that comparison to belief in belief-acquisition. Churchland's main argument is that the belief-concept is itself hostage to false theories, primarily Cartesian dualism. Most materialists find it easier to deny a link between mentality and Cartesianism than to reject mental states.

9. Ryle asks: "Did Archimedes' discovery take more or fewer 'units of intelligence' than those taken by Shakespeare's composition of his last sonnet [is] a nonsense-question. By how many units is Sarah prettier than Tommy is well-behaved?" Why nonsense? Sample size aside, Shakespeare's IQ exceeding Archimedes' would be evidence that writing Shakespearean verse takes more intelligence than making Archimedean discoveries. Tommy's z score for rule-following can be compared to Sarah's z score for prettiness. Conversion to SD units, called "normalizing," is a standard method for comparing disparate phenomena (such as height and caloric intake in Figure 2.2). Goldberger and Manksi (1995: 769–770) chide Herrnstein and Murray for normalizing a number of dichotomous variables so as to regress them against IQ and SES. Instead of comparing the probability of (say) health against IQ and socioeconomic status, they propose comparing the improvement in health achieved by investing a fixed sum of money in raising IQ vs. the improvement achieved by investing that sum in raising SES. Since increasing IQ is now virtually impossible (see chapters 4 and 8), the Goldberger-Manski measure would minimize the importance of IQ no matter how much variance it explains.

10. Using unnecessarily involved computations, Schönemann (1987) makes the converse point that a null manifold can be made to yield a principal component, indeed one on which two ostensibly equal groups diverge. Schönemann blames Arthur Jensen's supposed failure to see this on "arrogance, ignorance and prejudice."

11. Competing criteria for adequacy yield different factorizations of the same data. One may seek to maximize explanation of the variance of the observed variables rather than retrieval of known correlations. One may or may not require the factors to be independent ("orthogonal"). Structural constraints, for instance that each observed variable load, and load heavily, on only some of the posited factors, yield different analyses when made mathematically precise. Thus, descriptions of the observed variables might be considered "simplest" when the sum of the fourth powers of the factor loadings is maximized (the "quatrimax" criterion), or when the sum of the squared factor loadings is maximized (the "varimax" criterion). See H. Harman (1976) for theory and computational techniques.

12. Flynn (1984) also adopts this usage.

13. On the classical Frege-Russell theory a word's meaning is a property possessed by exactly its referents; on the direct-reference theory the referents of a word are literally part of its meaning, conceived as a set.

14. That is, "is the denotation of " is a many-one function whose domain is word-meanings.

15. Early on (60–61) Gardner mentions solving and finding problems, but this trait plays little subsequent role.

16. This formulation is Steven Goldberg's. Jencks may have misled himself by omitting the quotation marks from "intelligence." Had he attached them, the resulting

statement—" 'intelligent' does not mean the same in every language"—would be self-evidently absurd. Without the quotation-marks the statement makes no literal sense, since intelligence is an ability, not a lexical item, so cannot mean anything.

17. The same-meaning/same-reference principle is preserved in both within-language and between-language disagreement. When speakers of English disagree about Smith, "intelligence" as they both use it has the same reference: either Smith is not intelligent, in which case one of the speakers falsely believes that Smith belongs to the reference of "intelligent," or Smith is intelligent and the other speaker falsely believe he does not. "Intelligence" and its synonyms must also have identical references across languages. At worst, speakers of different languages might hold different *beliefs* about the reference of (their terms for) "intelligence."

18. Attribution of strange beliefs creates the presumption of mistranslation on either of the main theories of translation. The Quine-Wittgenstein equation of meaning with use implies that an optimum translation maximizes agreement between speakers. Obviously, translating *obugezi* as "intelligent," which creates discrepant references, fails to maximize agreement. (Quine himself [1960, 1968] rejects objective translation, and with it the line between factual and verbal disputes, but the resulting "indeterminacy of translation" is quite narrow. "Rabbit" can be mapped to "undetached rabbit part" but not "buffalo," and mass terms and predicates stay put.) Lockean mentalism takes two words in distinct languages to be synonyms when they are attached to the same idea. Obviously, a word not attached to the idea to which "intelligence" has been annexed is not, on the Lockean theory, a suitable translation of "intelligence."

19. Its bearing on race differences is discussed in chapter 4.

20. Block and Dworkin dismiss this policy as "operationalism" (1976a: 413, 425), a charge considered below. Note here the captiousness of calling "intelligence" unclear, and then protesting its abandonment.

21. In his recent (1994), Gould professes to "support" multiple ability theories based on Thurstone's analysis (144). But by the end of this essay he is again standing in awe of "the wondrous variousness of human abilities, suitably nurtured."

22. The classic treatments of "dispositional predicates" are Carnap (1936) and Ryle (1949).

23. Functions are ultimately sets.

24. Eysenck's (1993) defense of the reality of intelligence, that, like gravity, it is a "concept," is inadequate. Concepts are ideas in people's minds, and would vanish if human minds did. Gravity itself, the force attracting the Moon to the Earth, does not depend for its existence on human minds. Eysenck's comparison of intelligence to phlogiston does the critics' work.

25. On 252 Gould (1981) asserts, correctly, that "principal components cannot be automatically reified," and on 269, again correctly, that "Automatic reification is invalid." But by 320 he is announcing that "The chimerical nature of g is the rotten core of Jensen's edifice, and of the entire hereditarian school."

26. Herrnstein and Murray restrict their sample in other ways besides confining it to whites.

27. Egan et al. report (1994: 363) that correcting for range restriction raises r to .66.

28. See Matarazzo (1992). Protest marches prevented inclusion of blacks in one 11,000 subject study of crime that included biological variables (Gibbs 1995: 106). For this reason, "all the biological and genetic studies [of crime] that have been conducted to date have been done on whites" (Adrian Raine, cited in Gibbs 1995: 106).

29. Jensen (1994) considers the "assortative mating" possibility that the gene for IQ associates with the gene for large-headedness not because head size is functionally related to intelligence, but because intelligent people marry large-headed ones. This hypothesis assumes that the heritability of IQ exceeds 0, hence would be suspect to environmentalists, but in any case Jensen shows the correlation to be statistically

significant within as well as between families. If the correlation were due to assortative mating, the association between genes for IQ and large-headedness should disappear when parental genes are redistributed randomly to offspring.

30. For descriptions and illustrations, see Noble (1978: 299–301).

31. One does talk of the Ptolemaic explanation of eclipses, meaning a theory *believed* to be explanatory, or that *would have been* explanatory had it been true.

32. Stability is also something of a red herring. A low correlation between childhood and adult IQ would not reduce the correlation between IQ and chievement, or the power of the race difference in IQ to explain the achievement gap.

33. This remarkable passage occurs in Gould (1994): "We must fight the doctrine of 'The Bell Curve' because it will . . . cut off all possibility of proper nurturance for everyone's intelligence. Of course, we cannot all be rocket scientists or brain surgeons, but those who can't might be rock musicians or professional athletes, while others will indeed serve by standing and waiting" (149). What is "everyone's intelligence" if intelligence is a reification? Why can't we all be scientists if no one is innately limited? (Are all nonscientists talented enough talent to earn a living as musicians?)

34. Environmentalists do not fear such ad hoc distinctions. "Genetic linkage analysis can ultimately lead to insight into the biology of disease processes such as schizophrenia," writes Baumrind, "but it is unlikely to contribute to an understanding of variations in attitudes or normal personality attributes" (1991: 387). In contrast, Plomin, Owen, and McGuffin remark that some common behavioral disorders are now thought to "represent the genetic extremes of continuous dimensions" (1994: 1735).

35. Coleman et al. (1966) and Garner and Wigdor (1982) contain numerous examples.

36. Fischer et al. (1996) badly misconstrue the function of criteria. They explain tests as simply a cheaper means of determining what criteria determine, implying that these racially biased devices are deployed simply to make a buck. It is not realized that, without criteria, it is logically impossible to say that a test does (or does not) measure what it is intended to.

37. A logician would say ω-inconsistency.

38. "V-bias" resembles the Cole and Moss (1989) "extra-validity issues."

39. The original omission is usually ascribed to "racism." In fact, Wechsler omitted blacks from his original standardization because "norms derived by mixing the populations could [not] be interpreted without special provisos and reservations," and finding enough blacks to guarantee reliability across groups was impractical (Wechsler 1939: 109).

40. See the discussion of the BITCH test below.

41. Low levels of numerical reasoning ability cannot be due to lack of a "knack" for numbers (Hacker 1992: 144), since mathematics is not essentially about numbers or "quantity." The defining properties of integers and the real continuum can be stated in non-numerical terms, and other domains are quantitative when they possess these properties. Mathematics is often characterized by its concern with abstract structure, but the role of numbers in expressing distinctions may be equally relevant to race differences. Only two possibilities are opened when the sun is said to be "near" or "far." Putting the sun's distance in miles opens the literal infinity of possibilities that it is n miles away, for any value of n. One definition of intelligence, it will be recalled, is the ability to discriminate.

42. I emphasize again that, while questions about English words assume a knowledge base, human beings differ in what they do with that base. What comes out, in terms of recognition of synonyms, antonyms and the like, far outstrips what goes in. Psychologists speak of the "poverty of the stimulus."

43. Because few black patrolmen were passing the test for sergeant, the New York Police Department created a new test in 1989, which black police officers and

consultants approved as unbiased. The pass rate for blacks on this test proved to be 2%, and it too was labeled biased (Pitt 1989). At this writing a similar sequence of events is unfolding in Chicago.

44. For reasons explained in chapter 4, the most direct estimator of the importance of nonshared environment for a trait is $(1-r_{MZT})$, r_{MZT} the concordance for the trait between identical twins reared together.

45. The band is $[-.06z, +.06z]$ for blacks, $[-.94z, -1.06z]$ for whites.

46. Proportionately more whites will be Type I victims in the IQ range from 100 to 92.5, where the white and black curves meet, covering 19% of the white population. But the range to the left of $-5z$ covers 71% of the black population, so Type I errors will be more frequent in the black population as a whole.

47. Herrnstein and Wilson (1985) do take up temperamental race differences relevant to crime.

48. Kochman consistently describes whites as the beneficiaries of racial asymmetries. Remarking that black women repulse the sexual advances of black men with sassy insults, and that black men perceive being ignored as a grave affront, Kochman suggests it is up to white women to meet black sexual advances with repartee, the fault lying with white women if embarrassed silence provokes black men to aggression.

49. 29% of significantly overweight black adolescent females are reported "satisfied" with their weight. Unfortunately, the authors do not cite white controls for the "overweight" variable.

50. The possibility of race differences in autonomy is explored in chapter 9.

51. Appendix A contains a more detailed treatment of time preference.

4

Genetic Factors

Like other biological processes, intelligence must have evolved
under the influence of natural selection.

—Jerison (1973)

4.1. WHAT IS A GENE?

As the cause of race differences proves to be morally pivotal, it is important to
estimate the extent to which genes are involved.

The idea of a trait being "due to genes" is said to be problematic, primarily
because genes by themselves are inert. Environment always mediates the
expression of a gene, however "near" its expression a gene may be. A height
gene produces one phenotypic height with one diet, another height with another
diet. Not that dependence on environment is unique to genetic causation; rubber
melts if heated to 400°C, shatters when struck if cooled to –200°C. Neither
genes nor chunks of rubber have causal powers in and of themselves. Still,
there is widespread mistrust of explanations using genetic factors, with some
writers cautioning so emphatically that all phenotypes are the joint product of
genes and environment as to imply the two are inseparable.[1]

Yet the causal role of genes can be isolated, as can the role of any single
reactant in a chemical process. Following a tradition that identifies things with
their causal powers,[2] we may identify a gene, or more generally a genotype,
with its capacity to produce different phenotypes in different environments. The
proposal, in other words, is that a gene or genotype is what is usually called
its range or norm of reaction. Suppose a gene expresses itself as height.[3] Figure
4.1 depicts its norm of reaction, with points on the x axis being environments
in calories consumed daily, and points on the y axis being heights, the
expressions of the gene in those different environments. The present proposal
takes this curve to be more than the rule by which the expression of the height
gene varies with calorie consumption; the curve, a particular function from
calories consumed to heights, is the gene itself.

Taking mechanisms as functions is not alien to behavioral science; Jerison suggests construing cognitive "functions" in the purely mathematical sense of mappings from stimuli to behavior (1973: 11). Should correlations seem too insubstantial to go proxy for bits of chromosome, however, a gene may be

Figure 4.1
Norm of Reaction for a Hypothetical Height Gene

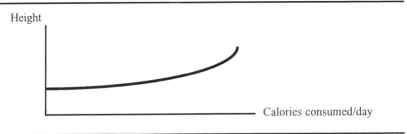

specified more concretely as the *physical basis* of an environment/phenotype correlation. Under this proposal, a "height gene" is whatever sustains an organism's tendency to attain different heights in different environments, the X, whatever it is, that explains Figure 4.1. (X has turned out to be deoxyribonucleic acid [DNA].)

A gene's environment, for its part, is the totality of factors impinging on it, commonly but not exclusively by way of its home organism. "Environment" is usually understood more broadly to include distal factors such as air temperature, but distal factors matter to a gene only insofar as they affect the home organism. Their inclusion in "environment" is justified pragmatically, by the correlations between distal and proximal factors in the evolutionary wild. The Herring gull hatchling's pecking response is cued by a suitably presented red dot, for instance, but because in the wild red dots are found only on the beaks of adult gulls, the hatchlings may be said to be programmed to peck at beaks. In this wider sense a gene's environment also includes neighboring genotypes, with which the gene interacts via the organisms these genotypes produce. The other genes in a gene's home genotype also belong to its environment, since they too affect the home organism, and thereby the gene's expression.

Environments may be momentary or prolonged, encompassing all that impinges on a gene during the life of its organism. When its critics say IQ reflects environment, they mean that IQ test performance reflects all that happens to testees from birth, and possibly conception. The claim that race differences in intelligence are environmentally caused means that these differences are due to differences in the treatment of blacks and whites from conception.

4.2. SOME NOTIONS EXPLICATED

Taking a gene to be a function from environments to phenotypes clarifie ideas often said to be empty or obscure. Most of these explications correspond closely to ordinary usage; they are formalized in Appendix B.

(1) That genes may express themselves differently in different environments becomes the truism that a function may map different environments to different phenotypes. This truism is closely related to the phenomenon of genotype/environment interaction (DeFries, Loehlin, and Plomin 1977; Falconer 1989: 135–136), the difference in response of two genes to the same environment. The corollary of interaction cited extensively in the nature-nurture debate is that the expressed phenotypic difference between two genotypes may vary with environments. I often use "interaction" informally, for the broad idea that a gene's expression depends on its environment.

(2) Two individuals may be said to have *the same gene* for a trait when they display the same values of the trait in all environments. Hence

(3) Two individuals may be said to have *different genes* for a trait when there are some environments in which the individuals (would) display different values of the trait. The physical basis for identity or diversity of genes is the DNA at the appropriate loci.

(4) A phenotypic difference between two individuals is usually termed (wholly) *genetic in origin* when both individuals have been raised in identical environments. A "genetic" difference need not exist in all environments: see (9) and (10).

(5) Genetically diverse individuals may be phenotypically identical. Genetic diversity, which by (3) means the existence of environments in which the individuals in question display different values of a trait, permits environments in which they display the same value for the trait. Smith and Jones may have different genes for intelligence and different phenotypic IQs in most environments, but there may exist environments in which their intelligence genes are or would be expressed as identical IQs.

(6) Genetically diverse individuals may differ phenotypically in all environments, yet be phenotypically identical. This happens when there are pairs of environments such that the value of the phenotype for one individual in one of the environments is the same as the value of the phenotype for the other individual in the other environment. Perhaps no single upbringing will make Smith and Jones equally smart, but there might be distinct upbringings such that Smith raised in accordance with one would become as bright as Jones raised in accordance with the other.

(7) Phenotypic identity does not imply genotypic identity. This is another truism, that display of the same phenotype by two individuals in one environment does not imply display of the same phenotype in all environments. Even more obviously, in light of possibility (6), the existence of a pair of environments such that Smith in one is phenotypically similar to Jones in the other does not imply that Smith and Jones would be phenotypically similar everywhere. Had Einstein grown up to exhibit average intelligence after a childhood spent in a sensory deprivation tank, his genotype still would have been unusual.

(8) Genotypic identity does not imply phenotypic identity, a point made in Chapter 2 and a corollary of (1). Genetically similar individuals may differ phenotypically because of exposure to different environments. While (8) allows that race differences in phenotypic intelligence need not be genetic in origin, it

also drives home the reality of phenotypic differences, in intelligence for instance, expressing the same genes in different environments.

(9) Despite the dependence of all gene expression on environment, some traits seem especially "in the genes," or "natural." There is no gene for speaking French, but there evidently is one for vocalizing. (a) What people seem to mean by calling a trait *genetic* or *in the genes* or *innate* is that it is displayed in all environments. Carnivory is innate in lions because lions everywhere eat meat. A related idea is the relative narrowness of reaction range of "genetic" traits; that is, that the gene for an "innate" trait expresses itself in roughly the same manner in all environments. (b) Calling a phenotype *natural* for an organism appears to mean, more specifically, that the organism will display it in environments like those in which its ancestors evolved. Hunting is natural for lions because lions hunt in the wild. Lions in zoos don't hunt, but lions did not evolve in zoos. So too, a contemporary environment is "natural" for an organism when it resembles the ones in which its ancestors evolved.

(10) Where P is a quantifiable trait, (a) individual A is *innately* more P than individual B when A is more P than B in all environments. (b) A may not be innately more P than B, but—a phrase that most makes sense for humans—A is innately more P than B *for all practical purposes* when A is more P than B in any environment that permits human society. A Tutsi baby would grow no taller than an Eskimo baby of the same size if both were left exposed on the Moon, but for all practical purposes Tutsis are taller than Eskimos. (c) In light of 9(b), A is *naturally* more P than B if A is more P than B in environments like those in which their ancestors evolved. Lions grow lethargic in zoos, but as a group lions are nonetheless naturally faster than zebras because they outrun zebras in the wild.

There are in fact two legitimate notions of "heritable influence" (see Whitney 1990a). One, captured in (1)–(8), is the role of genes in individual and group variability. The other, captured in (9) and (10), is that of ontogenetic stability and interindividual *in*variance. The two tend to coincide, since the more "canalized" a phenotype is, the less its expression varies with environment. However, some traits are highly heritable in the second sense but not the first. Ear number is developmentally fixed, yet, since individual differences in this trait are always due to environmental accidents (cf. van Gogh), it is not heritable in the first sense. As our subject here is racial variation, we will deal mainly with "heritable influence" in this first sense.

Points (8) and (3) together deserve special notice. (8), I said, allows race differences in the intelligence polygene to be denied despite the race difference in phenotypic intelligence. At the same time, the dependence of phenotypes on both genes and environments undercuts a common argument against genetic accounts of race differences (see, e.g., D'Sousza 1995). It is observed that some aspect of black behavior, criminality for instance, has changed appreciably over a period—from 1930 to 1990, say—too short to permit genetic change, from which it is inferred that the cause of the race difference in that behavior is environmental. This is a non sequitur. The question of individual or group *differences* is not whether environmental variation can alter a phenotype, which it plainly can, but whether two individuals or groups exposed to the same environmental variation react to it similarly. If their reactions diverge, then, while a change in environment may explain the change in the phenotype of each, the phenotypic difference between them remains genetic in origin.

It is pertinent in this connection that in recent times the most marked changes in black behavior have been away from white means, while accompanying environmental changes have brought black environments closer to the white. Black criminality has increased, yet blacks and whites are treated more similarly now by the criminal justice system than sixty years ago. By (3), this divergence of black and white behavior as environments have converged implicates genetic diversity. A shift in environment, not in genes, is the most probable trigger of changes in black criminality, but the change is such as to suggest an underlying innate race difference. Diversity of environments can not only create the illusion of genetic differences where none exist, it can mask genuine genetic differences. It is irrelevant that (by [5]) there might be common environments in which black and white criminality also converge; so long as blacks become less like whites as there is convergence to some single environment, the races must differ genetically.

4.3. GENE/ENVIRONMENT CORRELATION

Genes do not produce phenotypes merely by producing enzymes, which produce proteins, which then form tissues, ultimately forming organisms. Genes influence phenotypes indirectly, through expression as the ability to affect the environments in which organisms develop. The ability of men to fly faster than birds seems at first environmental in origin, since humans outfly birds by taking airplanes, not by flapping stronger wings. But the airplane is a product of human inventiveness, which exceeds that of birds. The ability to fly is genetic after all, in the sense that human genes make the human brains which make airplanes.

Speaking more generally, it is no accident that certain genes, such as those of humans, regularly occur in certain environments, such as airline passenger seats. Genetic shaping of the environment is called *genotype/environment correlation* (DeFries, Loehlin, and Plomin, 1977; Falconer 1989: 134–136), usually divided into three types. A correlation is "active" when a gene affects its environment via its own phenotypic expression. Someone genetically inclined to exercise may seek out environments already conducive to it, such as mountainous country, or construct a suitable environment by building a tennis court (see Scarr and McCartney 1983); the human genotype/airplane correlation is active. A correlation is "passive" when the gene that causes it is not a correlate. For instance, the genes of baby robins are usually found in nests because the genes of its parents program both reproduction and nest-building. The hypothesis that genetically high-IQ parents produce high-IQ children and surround them with books says that the IQ/status correlation is passive genotype/environmental. Finally, a genotype/environment correlation is termed "reactive" or (Scarr and McCartney 1983) "evocative" when the phenotypic expression of a gene causes others to modify their behavior toward the gene's bearer: the standard example is that of a child's innate curiosity prompting his parents to buy him a chemistry set. The particular kind of correlation in this

example might be called genotype/environment feedback: the reaction caus
by the phenotype so changes the phenotype that it elicits a further reactio
Feedback may be positive, amplifying the phenotype provoking the reactic
as when encouraging a curious child stimulates his curiosity, which evok
more encouragement. Or it may be negative, dampening the phenotype. Ma
environmentalists maintain that race and academic performance form
negative loop wherein race predicts poor grades because it lowers teach
expectation, which depresses performance, which lowers expectations furth(
and so on.

4.4. THE THREE-TIERED CHARACTER OF INTELLIGENCE

The (poly)gene for intelligence underlies a disposition to acquire vario
degrees of intelligence in various environments. However, the dispositior
character of intelligence itself requires that three levels of dispositions I
distinguished.

On the first level are specific cognitive abilities, such as those involved
sorting shapes, speaking French, or recognizing odd numbers. "Knowing hov
is often distinguished in this context from "knowing that," and there is a cle
difference between knowing the proposition that Chile exports nitrates ar
being able to type without awareness of how one does so. Memory itself is a
ability, however, and most knowledge beyond memorized facts has an abili
component. Grasping the laws of physics, a paradigm of declarati·
knowledge, involves being able to solve problems with those laws. Thu
virtually all of what we know, hence of what we learn, is found at this fir
level.

At the second level is the ability to acquire first-level cognitive abilitie
This—the ability to learn French or physics—is phenotypic intelligence, wh
IQ measures. People differ in phenotypic intelligence when they differ in tl
ability to acquire these first-order abilities. Capacities to acquire capacities a
met elsewhere, if not by that name. Rubber at $30^{\circ}C$ is not disposed, n·
"able," to shatter, but it becomes disposed to shatter if struck when cooled ·
$-200^{\circ}C$. Since rubber at room temperature can be cooled to $-200^{\circ}C$, it has tl
capacity to become brittle.

At the third level is the ability to acquire the ability to acquire specif·
cognitive abilities, or, more simply, the ability to develop learning ability, c
most simply, the ability to become phenotypically intelligent. This capacity
genetic intelligence. There can be no question of its existence, and distinctne
from phenotypic intelligence, for human beings at birth are not as clever a
they will be as adults, so must be born capable of becoming cleverer. Wheth·
two newborns are equally genetically intelligent—whether they would develc
equal phenotypic intelligence if raised alike—is an empirical question.

Once again, while talk of third-order abilities is cumbersome, its referent
straightforward. Rubber at room temperature, which becomes brittle if froze
has the second-order disposition to become brittle. Other substances lack th
second-order disposition: ordinary helium does not become brittle at ar
temperature. However, if helium passed through a magnetic field so change
that it did become brittle when cooled to $-200^{\circ}C$, unmagnetized helium a

room temperature would have the third-order capacity to become, through magnetization, capable of becoming brittle. And, just as things differ in their third-order dispositions—some but not all substances can become capable of becoming brittle—some but not all newborns may be capable of becoming capable of learning calculus. A difference here would be one in genetic intelligence. Strictly speaking, genetic capacities belong not to newborns, who have experienced uterine environments, but to zygotes at the genotype's moment of creation. Differences at birth in the ability to acquire learning ability may reflect environmental as well as genetic differences.

4.5. THE INNATENESS OF INTELLIGENCE; ENVIRONMENTALISM

Implicit in ordinary usage, according to Derr (1989), is the idea that "Intelligence is innate . . . a characteristic which is present at birth." Intelligence may change, according to the ordinary conception, but only as "the unfolding or realization of what was present at birth" (115). Assuming this description accurate, ordinary ideas about intelligence seem to ignore the dependence of all phenotypes on environment, and Derr himself contrasts the ordinary conception to the "stipulative" one of scientists (117).

In fact, far from being confused, the everyday notion of innateness—of intelligence and other traits—resembles its scientific counterparts in the most salient respects, including the element of presence at birth. What the man in the street has in mind in calling intelligence "innate" is more or less the cluster of ideas (3), (9), and (10) just defined. He means that people raised in identical environments need not end up with the same intelligence; and, when he calls Einstein's genius "genetic," he means that the average person raised as Einstein was would not turn out a scientific genius. The claim made in popular debate that the race difference in intelligence is genetic should be taken to mean that black and white children would not develop the same mean intelligence if raised in the same environment, and, perhaps, the stronger thesis that blacks would develop lower intelligence than identically raised whites in all practically possible environments.

"Hereditarianism" is a convenient label for the latter view, and its denial, he claim that identically raised blacks and whites would develop identical mean intelligence, may be called "environmentalism." This is certainly how environmentalists understand the distinction:

T]here is an easy rebuttal to research and reasoning based on racist assumptions. In implest terms, it is the environmental answer. . . . Of course, races differ in some utward respects [but] despite more than a century of searching, we have no evidence hat any one of those pools of race-based genes has a larger quotient of what we choose o call intelligence or organizational ability or creative capacities. If more members of ome races end up doing better in some spheres, it is because more of them grew up in nvironments that prepared them for those endeavors. If members of other races had milar upbringings, they would display a similar distribution of success. (Hacker 992: 27)

The balance of this chapter tacks closely to empirical data and accepted statistical methods, but it is also important to keep in mind a broader perspective from which the influence of genes on intelligence and motivation seems undeniable. Everyone knows that, no matter what efforts are made, even if raised from birth in the most stimulating environment imaginable, a chicken cannot learn calculus. Therefore, the ability of humans to learn calculus cannot be due solely to differences in the environments in which humans and chickens develop, but must involve human and chicken genes. And once one grants that genes determine interspecific intelligence differences, there is no clear point at which groups become so closely related that genes cannot be relevant to variation. People are not only smarter than chickens because of genes, they are smarter than chimpanzees for the same reason. Different breeds of dogs differ in intelligence as judged by speed of learning (Coren 1994). It seems arbitrary to declare that human groups cannot differ in intelligence for genetic reasons.

A more systematic reason to expect genetic influences on intelligence is the linkage between mind and body. Materialists think mental processes literally are processes in the brain; Cartesians who consider mind distinct from matter nonetheless admit the (to them mysterious) dependence of events in consciousness on neural events. "Functionalists" compare the mind to computer software, the brain's program, but they too admit that the software must be embodied in the brain's hardware. All accept as axiomatic that individuals whose brains were in atom-for-atom identical physical states would think the same thoughts and undergo the same emotions.[4] But the brain is a physical structure, and physical structures are controlled by genes. Whether an organism grows arms rather than wings, or short arms rather than long ones, plainly depends on its genes, so it would be ad hoc to deny that the kind of brain an organism develops depends on its genes. And everyone accepts the dependence of the functional capacities of an arm or brain on its physical properties. So once it is agreed that the difference between what your arm and mine can do (like throw curve balls) is under genetic control, it becomes arbitrary to deny that differences in mental powers might also be genetically controlled.

The premise of the popular objection (see Horgan 1993) that there is no specific gene for intelligence is likely true, but irrelevant to the issue of genetic influence. Most traits are polygenic, and a trait influenced by numerous genes will vary as each one does: that no single gene determines intelligence hardly makes genetic variation irrelevant. It is also objected that the precise manner in which genes produce intelligence is unknown, but a mechanism need not be identified to be known to exist. It does not take a chemist to realize that cars are fueled by gasoline, because cars stop when their gasoline runs out. In any case this appeal to ignorance is becoming passé, as race differences in specific genes are being discovered. For instance, the A1 allele of the D_2 dopamine receptor gene, implicated in alcoholism, occurs more frequently in blacks than whites (Blum et al. 1991). This gene is not known to bear on intelligence directly, but it marks a race difference at the molecular level with respect to an overt phenotype (and one in the direction of everyday perception). Ebstein et al. (1996) and Benjamin et al. (1996) have traced some variance in "novelty seeking" to a single gene site; the sample in Benjamin et al. (1996) include both blacks and whites, although whether the allele associated with greater novelty seeking was found more frequently in one race was not reported, and its authors have not responded to requests for further information.

These heuristic arguments do not establish, and should not be expected to establish, that everything is genetic. But they do make it plausible that genes matter to intelligence and personality.

4.6. HERITABILITY

Genetic causation is not all or nothing. It is important theoretically and, it turns out, morally, to say how much of the difference between two phenotypes is "due to genes." Here again the ordinary conception of intelligence as described by Derr might appear to ignore gene-environment interaction, but, again, ordinary ideas prove surprisingly sound.

When each of two causal factors are inert without the other, we cannot speak of what either does by itself. We can however apply Mill's method of concomitant variation to ask what happens when one varies while the other remains fixed. A phenotype like body weight is not the sum of a part due to genes and a part due to environment, but one can measure variations in weight when genes are held constant. Should genetically similar organisms weigh about the same in dissimilar environments, genes are important for weight; should genetically similar organisms in dissimilar environments differ in weight by about as much as unrelated organisms do, genes are unimportant.

Phenotypic similarity cannot be measured absolutely, but the separation of two values of a phenotype in a population can be compared to the phenotype's standard deviation. Letting the standard deviation for weight among humans be 21 lb, individuals differing in weight by 63 lb are $3z$ apart. Similarity of environments can likewise be expressed in terms of the variation in the environments to which the population is exposed. Given these scalings, it becomes possible to define a phenotype's *heritability*, the geneticist's explication of "degree of innateness." The heritability of a phenotype P in a population, usually written h^2, or $h(P)^2$ when P is made explicit, is the ratio of the genetic variance in the population to the variance of P. Algebraically, if V_G is the genotypic variance and V_W is the variance of weight, h^2 for weight is V_G/V_W. To find this number, square the proportion of an individual's deviation from the weight mean that would vanish if he did not deviate from the genetic mean. Thus, h^2 for intelligence is the extent to which genetic differences between individuals explain individual differences in intelligence.

What still needs defining is "genetic variance." Ideally, the value of a genotype, and with it the mean value of an ensemble of genotypes and thence genetic variance, would be measured by some sort of direct inspection of the DNA. Bouchard, Lykken, and their group speak of "emergenesis," the capacity of unique genotypes to produce unique phenotypes. But for now the value of a genotype must be construed indirectly, in terms of its phenotypic expressions. Specifically, the value of a genotype for a phenotype P is defined as the mean value of P for all organisms with that genotype. The genotypic mean of a population is then the average value of each genotype weighted by its frequency.[5] With all genotypic values and the genotypic mean so defined, genetic variance is derived as in chapter 2, and therewith heritability.

To work through an example, suppose body weight is controlled by two alleles A and a, each occurring with a frequency of .5, so that 25% of mankind is AA, 25% is aa, and 50% is Aa. Now consider a "Moderate" environment where everyone gets the same sensible diet, in which all AA's weigh 160 lb, all Aa's weigh 150 lb, and all aa's weigh 140 lb The genetic mean—like the population mean—is 150. The genetic variance is $.25(150 - 140)^2 + .25(150 - 160)^2 + .5(150 - 150)^2 = 50$. The variance of phenotypic weight is also 50, so the heritability of weight in the Moderate environment is 50/50 = 1. Among Moderates, in other words, weight differences are determined solely by genes.

Quantitative statements about innateness now become feasible. Intelligence is "mostly innate" if differences in intelligence between individuals are due primarily to differences in their genes, not their upbringings—if, that is, h^2 for intelligence is large. (Unless otherwise stated, h^2 is reserved henceforth for the heritability of intelligence.) Setting h^2 at .7, for instance, two genetically unrelated individuals from randomly selected backgrounds who differ in IQ by 15 points would on average have differed by $15 \times \sqrt{.7} = 12.5$ points had they been raised identically. A large h^2 means that individuals with dissimilar IQs would still on average have been dissimilar had they been raised identically. The mean race difference is "mostly innate" if it is due more to genetic differences between the races than to differences in their environments.

"Hereditarianism," then, predicts that some race difference in IQ would remain if blacks and whites were raised in an identical range of environments (the size of the difference indicating how much genes explain), while environmentalism predicts that the difference would vanish. Environmentalism, all of whose eggs are in a smaller basket, is more adventurous. Hereditarianism does not attribute all differences to genetic factors, although it is compatible with that outcome, whereas environmentalism does attribute all differences to environmental factors not themselves correlated with genes. Environmentalism rules out any genetic contribution to the race difference, while hereditarianism allows environmental contributions.

It aids clarity to ignore the environmentalist's own uncompromising words and order the various versions of environmentalism and hereditarianism by strength. Following Cohen's (1983) classification of effect sizes, "extreme environmentalism" may be defined as the claim that all race differences in (phenotypic) intelligence are due to environmental variation, and "extreme hereditarianism" as the claim that all race differences in intelligence are due to genetic variation. "Strong environmentalism" is the hypothesis that genes explain less than 20% of the difference, and "weak environmentalism," the denial of extreme hereditarianism, the hypothesis that environment explains some unspecified portion of the difference. I will understand "strong hereditarianism" as the hypothesis, not quite the counterpart of strong environmentalism, that genetic variation explains more than half of the race difference, and "weak hereditarianism" as the hypothesis that genes explain some but less than half of the race difference. I note again that authors like Hacker and Lewontin, Rose, and Kamin (1984: 88) consider the assignment of any weight at all to genes to be extreme, morally as well as scientifically, and would equate "environmentalism" with what I am calling "extreme environmentalism."

As (extreme) environmentalists stress, genetic variance and heritability are relative to populations and environments. In fact, this dependence is an immediate consequence of interaction: since the value of a genotype depends on its environment, the mean value and variance of an ensemble of genotypes, as well as the variance of the trait that expresses them, depend on the particular ensemble of environments the genotypes find themselves in. And, as the variances of a genotype and its phenotype change, so does the phenotype's heritability. Recurring to our numerical example, h^2 was 1 in the Moderate environment. But suppose there is also a bifurcated Extreme environment: the population is randomly divided into those who are starving and those with unlimited access to food. Further, both extreme want and extreme plenty swamp genetic effects. All starvers, whatever their genotypes, weigh 100 lb, and all gorgers weigh 200 lb. The variance for weight among the Extremes is 2500^6 but the genetic variance is 0, since all three genotypes have a mean of 150. (Since half the AA genotypes are starving and half are gorging, half the AAs weigh 100 and half weigh 200; the same is true of the other genotypes.) Hence, the heritability of weight among the Extremes is $0/2500 = 0$.

Indeed, a trait may be highly heritable in each of two populations, yet their mean difference be due entirely to environment. Imagine a third nutritional environment, the Athletic, differing slightly from the Moderate, in which all AA's weigh 140, all aa's weigh 130, and all Aa's weigh 135. The heritability of weight for the Athletics is also 1, but the Athletic mean is 135, and the 15-lb Moderate/Athletic difference in mean weight is due entirely to environment.

It is therefore an error to deduce a genetic origin for between-group differences in intelligence or anything else from high within-group heritability. It is also an error hereditarians are constantly accused of making (by Asmin 1995; Block and Dworkin 1976a: 476, and, more cautiously, 531; Gould 1981: 156–7; Lewontin 1976a; Thoday 1976: 133; Goldberger and Manski 1995: 770-771), although I know of no hereditarian who makes it. Neither Lewontin, Thoday nor Gould cite any offender; Asmin names Jensen without textual reference. Block and Dworkin cite Jensen's use of a large h^2 as "one support" for an explanation of part of the between-race variance in IQ by genetic variance, but what they call "fallacious" and "in no respectable sense . . valid" is a *deduction* of between-group heritability from large h^2, which Jensen never proposes.[7] In fact, as we will see in section 12, within- and between-group heritabilities are related, but the evidence that bears on the one is easily distinguished from that bearing on the other, and hereditarians routinely observe the distinction.[8]

To avoid misunderstanding on a point which occasions so much heavy weather, I will use H^2 for the proportion of the variance in IQ *between* races explained by genetic variation, i.e. how far genetic differences between groups explain group differences. Whether $H^2 = h^2$, and whether $H^2 > 0$ when $h^2 > 0$, are questions to be decided empirically for each phenotype. The various versions of hereditarianism and environmentalism are distinguished by the values they assign to H^2. According to extreme environmentalism, $H^2 = 0$; according to strong environmentalism, $H^2 < .2$, while according to weak environmentalism, $H^2 < 1$; according to extreme hereditarianism, $H^2 = 1$;

strong hereditarianism sets $.5 \leq H^2 < 1$, and weak hereditarianism sets $0 < H^2 < .5$.[9]

As populations do not normally experience all possible environments, the heritability of a phenotype for a population is determined by the variation in the phenotype over the environments to which the population has actually been exposed. The same genes might behave differently in a different range of environments, yielding a different heritability, as happened with the genes for body weight in the Moderate and Extreme worlds. That gene/environment interaction limits heritability estimation is assigned great prescriptive force by environmentalists, since, they argue (see chapter 8), the fact that intelligence *is* highly heritable does not mean it *must* be. Feldman and Lewontin (1975) go so far as to say that interaction makes heritability a useless number: "no statistical methodology exists that will enable us to predict the range of phenotypic possibilities that are inherent in any genotype" (1168; Block [1995] calls the heritability notion "lousy"). Lewontin explains further:

[T]he linear model [that phenotypic variance = genetic variance + environmental variance] is a *local analysis*. It gives a result that depends upon the actual distribution of genotypes and environments in the particular population sampled. Therefore, the result of the analysis has a historical (i.e., spatiotemporal) limitation and is not in general a statement about *functional* relations. . . . [T]he particular distribution of genotypes and environments in a given population at a given time picks out relations from the array of reaction norms that are necessarily atypical of the entire spectrum of causative relations. . . . The analysis of causes in human genetics is meant to provide us with the basic knowledge we require for correct schemes of environmental modification and intervention. . . Analysis of variance can do neither of these because its results are a unique function of the present distribution of environment and genotypes. (Lewontin 1976c: 183–192)

This argument moves much too far much too quickly. The ratio of genotypic to phenotypic variance in a limited ensemble of environments is perfectly well defined.[10] It does not tell us everything we want to know about a genotype, but it tells us something. Without infallibly predicting the behavior of a genotype elsewhere, a heritability estimate over widely varied environments permits informed guesses about phenotypic variation in unexamined ones. When a trait—such as intelligence—has appeared in the same form in every ecological niche so far colonized by humans, it may reasonably be expected to emerge in like form in new niches, and it is reasonable to expect genetic variation to continue to be roughly as important as it has been. There is no reason whatever to consider all selections of environments "necessarily atypical."[11]

Just as within- and between-group heritability must be distinguished, heritability must be distinguished from biological causation. Pursuing our earlier heuristic argument, it seems clear that the causes of all behavior are biophysical events in the gene-constructed nervous system. However, if all individuals and groups have the same nerve-building genes, all individual and group differences in these immediate causes of behavior will be due to environmental factors. One can therefore be both a physicalist and an extreme environmentalist. Such a position is taken by R. Wright (1995; also see Da-

1996), who argues that self-control is facilitated by the neurotransmitter serotonin, and speculates that white serotonin levels exceed those of blacks, but is confident that this difference is due entirely to oppression. Thus, while Wright takes serotonin to inhibit violence, his attribution of all race differences in serotonin level and hence violence to the social environment is strongly environmentalist.

So too, one can view all phenotypes as expressing a genetically programmed universal human nature yet be an environmentalist about group differences, as are Tooby and Cosmides (1990: 35–36, 43). They speculate that all humans may have innate capacities for several organized life strategies, but which one is triggered depends on environmental contingencies—which, they say, creates the illusion of several personality "types." Similarly (although Tooby and Cosmides do not argue this explicitly), one could hold that the systematic differences between blacks and whites are due to the innate strategies triggered by their different experiences.[12]

Historically, the empiricist John Locke put discussion of innateness on the wrong track by insisting that any innate feature must be present at birth.[13] By that stringent standard, only transient features like birth weight and trivial ones like having toes are innate—the very conclusion Locke deployed against Cartesian rationalism. Furthermore, as Locke triumphantly observed, it is pointless to define a trait as innate if the capacity to develop it is inborn, since—where "being born with the capacity to develop phenotype P" means being born such that there are environments in which P will emerge—every phenotype is innate. Anyone who knows French must have been born able to learn it on exposure to the education he was in fact exposed to, which, by the developmental capacity test, makes knowledge of French innate. A medical man, Locke must have been aware of age-dependent phenomena such as puberty, but he would no doubt have responded that castrati never mature, proving that puberty, like learning French, requires a suitable environment, and is thus no more or less innate. Locke may be said without anachronism to have appreciated the dependence of phenotypes on environment.

What Locke did not anticipate was evolutionary biology, which recognizes nontrivial sense in which a trait P may be absent at birth, and sometimes never appear, yet be inborn. This is the sense explicated as (2) in section 4.2: differences in P's developmental trajectory are due to differences in genes. Male sexual maturation is innate, even if baby boys lack beards and hormonal imbalances sometimes obstruct it, because little girls born without the Y chromosome develop differently. Knowledge of French rather than English, dependent as it is on the rearing environment, is not an inborn difference. Sexual maturation is also "natural" because it continues to appear in environments like those in which it evolved, whereas French is not a holdover from the evolutionary wild.) Even Locke's demand for presence at birth is met, since something relevant to later phenotypes, namely the genotype, is congenital. Every organism is born with chemicals which subsequently produce various phenotypes in various environments.

4.7. DETERMINING THE HERITABILITY OF INTELLIGENCE AND TEMPERAMENT

An index of the heritability of a trait is the similarity with respect to that trait of genetically identical individuals raised apart, in randomly varying environments. The more similar these pairs are, as compared to pairs of unrelated individuals raised separately, in environments that presumably vary at random, the higher the heritability. Monozygotic ("identical") twinning is frequent enough in humans for there to have been several hundred recorded cases of identical twins put out for adoption early in life and reared apart. The possibility always exists that placement agencies choose adoptive families similar to the adoptees' biological ones, disguising environmentally caused similarities between reared-apart twins as effects of genes. Kamin (1974: 50), Lewontin (1976a: 87n.) and Horgan (1993, esp. 125) argue that this problem leaves 0 as a reasonable estimate for h^2. Yet identical twins reared apart are so much more alike than individuals paired by any nongenetic criterion—indeed, they are almost as alike as identical twins reared together—that most experts agree that genes must explain some of the variation in many, if not most human traits.[14]

To turn twin comparisons into heritability estimates, of body weight for example, let D_T be the average difference in weight between identical twins reared apart and D be the average difference in weight between randomly chosen pairs of individual. It can be shown that, for a normally distributed variable, D is 1.128 SD.[15] Then D_T/D is that proportion of the average difference between people's weights that remains when environment alone varies. That is to say, $(D_T/D)^2$ is the proportion of the variance of weight explained by environmental factors. It follows that $1 - (D_T/D)^2$ is the proportion of the variance explained by non-environmental factors, that is genes. So heritability is $1 - (D_T/D)^2$.

To run an example, let SD for weight be 21. Thus, pairs of people chosen at random will typically differ in weight by $21 \times 1.128 = 24$ pounds. Now suppose identical twins reared apart typically differ by 8 pounds. Environment working alone has made an 8-pound difference, or $8/24 = 1/3$ the difference made when genes and environment work together. This ratio is the empirical core of heritability. To transform it into standard form, square 1/3 to get 1/9 then subtract 1/9 from 1 to get an h^2 for weight of 8/9 = .88.

To be sure, $1 - (D_T/D)^2$ may overstate heritability. As noted, it ignores the time identical twins reared apart spend in a roughly (but not exactly) similar uterine environment, which may induce some phenotypic similarity. Also adoptees are not given to the least favored families, so adopted-apart twins are not exposed to the full range of environments that contribute to variance in the general population. In addition, IQ among identical (and dizygotic) twins tends to vary less than in the general population, so the closeness of the IQs of reared-apart co-twins may be due in part to the restricted range of IQs among twins generally. This latter effect comes out more clearly when estimates for the heritability of a trait are expressed, as they often are, as the correlation for the trait between identical twins reared apart.[16] Ideally, heritability would be estimated from clones separated at the moment of creation. Yet, despite these and other limitations, $1 - (D_T/D)^2$ is a reasonable estimator.

Turning to intelligence specifically, since SD for IQ is 15, D is 15×1.128, or roughly 17. In the six studies that have been done of the IQs of identical twins reared apart, namely Newman et al. (1937), Shields (1962), Juel-Nielsen (1965), Burt (1966), Bouchard et al. (1990), and Pedersen et al. (1992), twin pairs have been found to differ by about 7 IQ points. As a first approximation, then, environment explains $(7/17)^2 = 17\%$ of the variance in IQ, and genetic variation explains the remaining 83%, that is, $h^2 = .83$. When this figure is refined in accordance with the complications mentioned earlier and factors unique to each study, the estimates of h^2 that emerge are given in Table 4.1. Pedersen et al. studied reared-apart monozygotic twins late in life; that their estimate is the highest suggests that the heritability of IQ increases with age, perhaps as individuals become freer to chose their own intelligence-influencing environments (see McGue, Bouchard, Iacono, and Lykken 1992). Reviewing the first three studies and their own, Bouchard et al. (1990) find the data to suggest "under the assumption of no environmental similarity, that genetic factors account for approximately 70% of the variance in IQ." They add the intriguing observation that the average difference in IQ between identical twins reared apart is almost as small as the average difference between IQ test performances of the same individual at different times, making identical twins act like single individuals with respect to intelligence.

Table 4.1
Estimates of the Heritability of Intelligence

Study	Monozygotic Twin Pairs	Estimates of h^2
Newman et al. (1937)	19	.71
Shields (1962)	38	.75
Juel-Nielsen (1965)	12	.69
Burt (1966)	53	.77
Bouchard et al. (1990)	48	.75
Pedersen et al. (1992)	303	.78–.80

Burt's work requires separate mention because Kamin (1974) and others have charged him with fabricating data, a charge now treated as fact in some textbooks (e.g., Fogelin and Sinnott-Armstrong 1997: 342–343). The main evidence of fraud has been the improbably high concordance of Burt's figures over time and the elusiveness of some of his co-authors. However, more recent research (Joynson 1989, Fletcher 1991) indicates that Burt's figures were stable because he recycled the same data—not the best science, but not fraud—and that witnesses do exist who remember Burt's collaborators. (It also appears that some of Burt's papers were destroyed immediately after his death.) Burt's case is of more than historical interest because of the widespread impression that non-zero estimates of the heritability of intelligence depend on his work. As is clear from Table 4.1 (and see Plomin 1990a), however, omission of Burt's work changes nothing even in the narrow context of twin studies. The unweighted mean of the other five estimates of h^2 is .74; their weighted mean, .775, is virtually Burt's .77.

Comparing identical twins is not the only way to estimate heritability. Siblings, including fraternal (dizygotic) twins, and parents and offspring, share half their genes. If members of such pairs are reared apart, the average difference between them for a trait should be twice the mean difference between identicals reared apart, which comes to twice the difference in the correlations between separated fraternal twins (with age held constant) and separated identical twins: 50% of the difference in IQ between separated fraternal twins is due to environmental effects and 50% = 100%/2 to genetic effects. So an independent estimator of the impact of environment on a trait is half the ratio of the average difference between reared-apart siblings to the expected difference.

Heritability estimates based on correlations between fraternals do not dovetail perfectly with estimates based on correlations between identicals. In particular, correlations between siblings and between parents and children tend to run less than half those between identicals because of dominance, the interaction of genes at different loci (epistasis), and other "nonadditive" effects. Since deviation from a phenotypic mean is determined by the precise character of the underlying genotype, removing half the genotype generally removes more than half the deviation, so non–identical-twin kinship studies tend to underestimate heritability. For this reason, what I have been calling heritability is often called "broad" heritability; "narrow" heritability is the ratio of "additive" genetic variance to phenotypic variance, where "additive" variance excludes dominance and epistatic effects (see Falconer 1989: 126–134). The monozygotics-reared-apart design measures broad heritability while other kinship designs pick up narrow heritability.

A further check on kin-reared-apart studies, in fact an independent method for estimating heritability, is the effect of varying genes when environments are fixed. This effect is gauged by comparing correlations between adopted children and their adoptive siblings and parents to correlations between biological siblings and between parents and birth children. Correlations between adoptive children and their biological parents are further points for triangulation, for the more heritable a trait, the greater should be the correlation between adoptive children and the natural parents they never met.[17] Extreme environmentalism about interindividual IQ variation, on the other hand, predicts correlation between adoptive family members equal to those between members by birth.

Table 4.2, adapted from Plomin's (1990b) Table 4.2, indicates the heritability of IQ as estimated by six comparison designs. One robust finding predicted by $h^2 > 0$ is that r between the IQs of adopted offspring and their natural parents, typically .2, exceeds r between adopted offspring and adoptive parents, typically .15 (see Scarr and McCartney 1983: 430). Environmentalists call this a placement effect, postulating that adoptees are given to parents that resemble their natural parents, but it is hard to see how adoptive parents, no matter how similar to the natural parents, can socialize children to be more like the natural parents than themselves.

Table 4.2[18]
Estimates of h^2

Comparing	Estimate of h^2
Identical twins reared apart	.72
Identical twins reared apart to fraternal twins reared apart	.52
Biological parents to adopted-away offspring	.44
Biological siblings reared apart	.48
Adoptive parent/adoptive offspring correlation to biological parent/biological offspring correlation	.46
Adoptive to biological siblings	.30

Source: After Plomin 1990b

Plomin's .5 estimate for h^2, taking all these studies into account, is consistent with Bouchard's .7 (also see Plomin and DeFries 1980), since Bouchard is estimating "broad" heritability while Plomin is estimating "narrow." If "IQ" turns out to name a set of abilities, h^2 for the abilities in his set, or the dimensions around which they cluster, typically reach .7 DeFries, Vandenberg, and McClearn 1976).

In recent years there have also been studies of the heritability of personality traits. The impression conveyed by this literature is that heritability estimates or personality fall in the .4–.5 range, also the all-in estimates of Bouchard 1994) and Plomin, Owen, and McGuffin (1994), with a larger variance than or h^2—perhaps because personality is more difficult to operationalize (but see Nichols 1978). Table 4.3 lists heritability estimates for various traits. Genes apparently play a major role in differentiating individuals with respect to impulsivity, self-control, excitability and other traits discussed in chapter 3 as differentiating the races.

The reader exposed to this material for the first time may feel at sea, not knowing know what to believe. He is urged only to realize that a large body of evidence supports the view that genes are important determinants of the human psyche. No one can credibly say otherwise.

8. GENE/ENVIRONMENT CORRELATION AGAIN

One can now see in numerical terms how gene/environment correlations can lead to underestimation of heritabilities. Suppose identical twins reared apart show a correlation of .4 for running speed. But suppose 10% of this similarity can be explained by the environmental factor of nutrition: the eating habits of identical twins tend to be similar, and the correlation for running speed

Table 4.3
Heritability of Personality

Trait	Source	Heritability
Neuroticism[a]	Plomin 1990b	.3–.5
	Plomin 1990b	.4
	Floderus-Myrhed 1980[b]	.5
	Henderson 1982[b]	.3
	Bouchard 1984[c]	.58
Impulsiveness	Bouchard 1984[c]	.38
Emotionality[a]	Plomin 1990b	.4
Positive Emotionality	Tellegen, Lykken, et al. 1988	.4
Negative Emotionality	Tellegen, Lykken, et al. 1988	.55
Activity Level[a]	Plomin 1995	.25
	Rushton 1992a	.25–.5+
Extraversion[a]	Plomin 1990b	.3–.5
	Bouchard 1984[c]	.54
Sociability[b]	Plomin 1990b	.25
	Bouchard 1984[c]	.52
Sense of well-being	Plomin 1990b	.48
	Tellegen, Lykken, et al. 1988	.48
Social Potency	Plomin 1990b	.56
	Tellegen, Lykken, et al 1988	.54
Achievement orientation	Plomin 1990b	.36
	Tellegen, Lykken, et al 1988	.39
	Keller et al. 1992	.68[c]
Alienation	Plomin 1990b	.48
	Tellegen, Lykken, et al. 1988	.45
Aggression	Plomin 1990b	.46
	Rushton, Fulker, et al. 1986	.5
	Tellegen, Lykken, et al.1988	.44
Stress reaction	Tellegen, Lykken, et al. 1988	.53
	Rushton 1992, after Bouchard, et al. 1990	.61
Altruism	Rushton, Fulker, et al. 1986	.5
	Keller et al. 1992	.37[d]
Cautiousness	Plomin 1990b	.5

Table 4.3 continued

Trait	Source	Heritability
Constraint	Tellegen, Lykken, et al .1988	.58
Control	Tellegen, Lykken, et al. 1988	.44
Following rules and authority	Tellegen, Lykken et al. 1988	.53
Traditionalism	Tellegen, Lykken, et al. 1988	.45
	Bouchard 1984	.42
Dominance	Carey et al 1978[c]	.56
Emotional Reactivity	Floderus-Myrhed 1980[b]	.56
Job Satisfaction	Arvey, Abraham, Bouchard, and Segal 1989	.31[d]
Work Values	Keller et al. 1992	.4[d]
Comfort	Keller et al. 1992	.42
Autonomy	Keller et al. 1992	.38
Locus of Control	Miller and Rose 1982	≈.5

a. Listed as "superfactor."
b. Cited in Rushton 1992.
c. Citing data from Nichols 1978, using $h2 = 2(r_{mz} - r_{dz})$.
d. As a Work Value.

between identical twins with dissimilar diets is .3. We might then be tempted to set the heritability running speed at .3. But also suppose that variance in the genes for running speed explains 30% of variance in nutrition, either because those genes are also expressed in good eating habits, or because healthy eating is an unconditioned reinforcer, or because athletic parents make their children eat well. In any of those cases genes explain an extra $.1 \times .3 = 3\%$ of the variance in running speed, whose true heritability is .33. Plomin and Niederhiser (1992) estimate within-race heritabilities of about .3 for nominal measures of environment. In a more systematic article, Plomin and Bergeman 1991) cite heritabilities ranging from .25 to .4 for "family environment."[19]

That environment is itself partly phenotypic bears significantly on race differences, for many disparities between black and white environments may be reactive correlates of genetic factors. If so, standard kinship and adoption studies may understate the between-group heritability of intelligence and other traits. Imagine a cohort of identical black twins separated at birth, with one twin of each pair placed in a randomly selected white family and the other in a randomly selected black family. Assume the mean IQ of the white-adopted co-twins at age 18 is 90 while that of the black-adopted co-twins 85, suggesting a H^2 of .44.[20] Should IQ-relevant race differences in environmental factors such as culture in the home themselves be influenced by genetic differences that correlate with race, however, the true value of H would be higher. Thus suppose, for example, that in our hypothetical study, controlling for culture explained 25% of the race gap remaining. Extending the Plomin-Bergeman within-race heritability estimate of .4 for this variable (Plomin and Bergeman 1991: 376, Table 1) to between-race differences, so that genes explain 40% of the between-race variance in culture in the home, raises H^2 by $.4 \times .25 = .10$ to .54. Part of the observed reduction in the race difference would be an effect of factors in the environment of the white-reared co-twin that correlate with white genotypes, and therefore count as a genetic effect. Exclusive attention to sociological variables can miss this point. When Mischel found preference for immediate reinforcement among Trinidadian children to be associated with father-absence, he concluded that father-absence was the cause (1961b). One might alternatively suppose genetic factors affect father-presence and, as a taste for immediate gratification passed on to children, impulse control. Likewise, Fischer et al. (1996) repeatedly cite social-environmental variables to explain social outcome differences, while mentioning only in passing (195–196) that social milieu might itself be a partially genetic effect, and offering no data to contest this interpretation.

Block (1995) wishes to deny environmental correlates of genes the title of genetic effect because they might be negative reactive feedback triggered by race. His analogy, apparently seriously meant, is a spurious correlation between low IQ and a gene for red hair in a society where redheads are regularly beaten on the head. His overall point is that h^2 and H^2, being *correlations* (between genetic and specified phenotypic differences), do not imply causality. Race-specific genotypes may cause race-specific phenotypes, or genotypes and phenotypes may covary as effects of an underlying cause.

It is possible that, in the case of race, the environmental effects of genes are reactive. It is also possible (to say the least) that genetic influences are active or passive. The proper interpretation of a large H^2 is an empirical question, to which, as we will see in 4.19, extant data suggests an answer. With that observation, in fact, it is time to turn to the data bearing on hereditarianism.

4.9. GENETIC FACTORS IN RACE DIFFERENCES

I doubt that many geneticists would dispute much that I have said so far. There may be less consensus on the heritability of personality, but even determined skeptics agree that intelligence is highly heritable. The critics cited in Horgan's attack on behavioral genetics (Horgan 1993) accept an h^2 of 50% for intelligence, and agree that "there may well be a significant genetic component" (Horgan 1993, 126, 127). Gould himself expresses "no doubt that IQ is to some degree 'heritable.' . . . It is hard to find any broad aspect of human performance or anatomy that has no heritable component at all" (1981, 155).

The situation is otherwise with regard to between-group differences, with most people who grasp the issue professing environmentalism. Yet their sincerity may be doubted. The Rothman-Snyderman (1987) poll (cited chapter 1) of several hundred experts on intelligence—"expertise" defined by membership in such organizations as the National Council on Measurement in Education and the Developmental Psychology section of the American Psychological Association—and found 53% agreement that both genetic and environmental factors are involved in the black/white intelligence difference (285). Only 17% of respondents attributed the difference entirely to environment, with 1% attributing it entirely to genes. Of 159 editors and journalists surveyed, 27% said they believed both genetic and environmental factors are involved in the race difference, and only 38% committed themselves to extreme environmentalism. One would not guess from the public debate that most experts and a significant minority of opinion-makers are hereditarians. Perhaps more people profess environmentalism than believe it because the profession is thought virtuous, or prudent. Less is said of the cause of race difference in time preference, probably because even granting that phenotypic difference is taboo. Usually a black personality style is simultaneously denied and attributed to oppression.

According to Neisser et al., "what little [direct evidence] there is fails to support the genetic hypothesis. . . . There is certainly no [direct] support for a genetic interpretation" (95–97). Yet the step from phenotypic race differences to genetic variation is a short one, and a number of lines of evidence converge on it: time of onset of race differences, relevant physiological race differences, adoption studies, intervention studies, comparison of of African and Eurasian attainment, and the positive correlation between the heritability of an IQ subtest with the race difference on that test.

4.10. ONSET OF RACE DIFFERENCES

While the IQs of young children are hard to measure, and measured IQ does not begin to correlate firmly with adult IQ until about age five, several studies (Broman et al. 1987, Montie and Fagan 1988, Peoples, Fagan, and Drotar 1995, Brooks-Gunn, Klebanov, and Duncan 1996) report a 1 SD race gap present by age three. Brody concludes that "the black-white difference is present somewhere before the third year of life and remains more or less constant through the adult life span" (1992: 283). Writers like Neisser et al. (1996) would doubtless contend that this is not "direct" evidence, since it is consistent with differentiating environmental factors entering earlier, but Broman et al. 1987) found that the usually suspect variables, such as care during the first year of life, accounted for relatively little (within-race) variation.

A second possible (and quite common) reply to the time-of-onset data is that the relevant early environmental factors have simply not been identified, but there are two countervailing considerations. First, it is an ethological rule of thumb that the earlier and more regularly a phenomenon appears in a population, the likelier it is to be genetically controlled. Earliness and uniformity rule out any factors whose time of impact varies, which excludes many environmental factors and therewith reduces the prospects for an environmental explanation.[21] Environmentalists must posit a factor that affects all black and white children between birth and the third year.

Second, anticipating normative issues, the earlier in life differentiating environmental factors enter, the less likely they are to be the fault of whites. Low teacher expectation cannot quash the intelligence of blacks before they reach school. Hence, whatever early environmental factors lower black IQ are less likely to create white liability for the intelligence gap or its consequences.

4.11. PHYSIOLOGICAL RACE DIFFERENCES

Races differ in a number of physiological correlates of intelligence and temperament that offer plausible proximate causes of phenotypes and are almost certainly under genetic control. Many of these differences are present at birth, so are unlikely to be due to social-environmental causes.

Table 4.4
Cranial Capacities by Race (in cm^3) [22]

Study	Black	White	Asian	Sample N	Black/White
Ho et al. 1980	1267 [a]	1370 [a]	—	1261	.92
Gould 1981, after Morton	1356 [b,c]	1426 [b]	1426	Ca. 600	.95
Beals et al. 1984	1276	1362	1380	20,000	.93
Rushton 1993, after Herskovits	1295	1421	1451	50,000+ [d]	.91
Rushton 1992, Army Personnel	1346	1361	1403	6,325	.98
Rushton 1994, I. L. O. data	1228	1284	1312	"Tens of thousands"	.95

a. Converted from gms by equation cm^3 = 1.036 g.
b. Converted from $in.^3$
c. Black sample said to yield "an average value between 82 and 83, but closer to 83," taken here to be $82.75 in.^3$
d. Males only.

The reality of IQ (g need not be assumed) and the correlation of .4+ between IQ and brain size discussed in chapter 3 predicts race differences in brain size when body size is fixed. These have been found, with Asian brains larger on average than white brains, and white brains larger than black brains. Table 4.4 displays the differences in a range of studies, all but one of which control for sex. The rightmost column indicates the black/white ratio. The variation in the estimates is due primarily to the variety of techniques used: Ho et al. (1980) weighed brains at autopsy, whereas the other studies inferred cranial size from amount of birdseed or shot held, or from measured circumference.

Not displayed in Table 4.4 is an underreported study by Persaud et al. (1994) which found by MRI that, among 108 white, African, and West Indian schizophrenics, victims of affective disorders and normal controls, intracranial volume was related to ethnicity at a significance level of .007. Further data are not given (and Persaud has not responded to my request for more), but significance at the .007 level for a sample as small as 108 indicates a sizable ethnic difference. A complete review of the brain-size/IQ literature appeared (Rushton and Ankney 1996) as this book was going to press.

The weighted average of the black/white ratio, excluding the Persaud study, is a bit below .95. According to Jerison (1973: 66, eq. 3.10), the number of neurons in the brain is proportional to the 2/3 power of its weight.[24] There are therefore about 96% as many neurons in black brains as in white. As the SD of interindividual brain measurements runs at about 6% of the mean, a 4% difference is considerable. Although the functional importance of the head width/head length ratio, or cephalic index—indicative of roundness and volume—is disputed, it is slightly higher in whites than blacks (Harrison et al. 1964: 209), suggesting more mass in the regions controlling abstract thought.[25]

One might conjecture (Schwartz 1991; also see Halpern 1995) that racism reduces head and brain size. The fixing of brain size in infancy makes this unlikely, and mean black head perimeter at birth is less than white (Broman et al. 1987). Black neonates are also smaller and lighter, and gestate for a shorter time, but the size difference vanishes when gestation period is fixed, and, while black infants overtake white infants in stature, they do not catch up in head size.

The IQ-head size correlation has been empirically established for both whites and blacks (Jensen and Johnson 1994, Jensen 1994). Since head size correlates with brain size, an IQ-brain size correlation exists. It is theoretically possible that IQ regresses differently on brain size for blacks and whites, and therefore that brain size plays a different role in explaining between-race and within-race intelligence differences. Note that a race difference in the significance of brain size would actually favor hereditarianism, as indicating a biologically based difference in neural organization. In any event, roughly similar correlations between IQ and cranial circumference have been found for both blacks and whites in a study involving 14,000 children at ages four and even (Jensen and Johnson 1994). Virtually no difference in cranial circumference was found when IQ was controlled for, so the intersects of the regressors must be quite close. A slight difference in IQ remains when circumference is controlled for, perhaps an effect of the race difference in cephalic index. Paralleling the increase of an IQ subtest's correlation with head size with its g-loading, Jensen notes a correlation of .7+ between the size of the black-white difference on a subtest and its correlation with head size.

An important noncognitive physiological race difference concerns the hormone testosterone, a facilitator of aggression and libido. Ross et al. (1986) report testosterone levels in black males 19% higher than those of white. Ellis and Nyborg (1992) report a difference of 3.3%, attributing the discrepancy to the Ross et al. sample's having been younger. Testosterone levels fall with age within-race, and "There appears to be more of a decrease in testosterone values with age in black men" (1992: 74), a decrease perhaps connected to the convergence of crime rates for black and white males after age 40. Lynn (1990b) argues that the greater incidence of prostate cancer in blacks is further evidence of higher black levels of serum testosterone, as prostate cancer victims exhibit higher levels of serum testosterone than healthy controls.

Elevated testosterone in black males is unlikely to be an effect of racism, since stress and loss of status typically reduce testosterone levels in primate and human males (Kemper 1990: 23–24). For the same reason the black testosterone advantage works against explanations of black behavior in terms of low self-esteem. In any event, "recent evidence has shown that black men exhibit biochemical responses to stress that are, on average, distinct from white men" (Ellis and Nyborg 1992: 74). These differences in response-readiness are presumably nonsocial in origin.

Neisser et al. (1996) omit the evidence relating to race differences brain size or serum testosterone, presumably because it is indirect.

4.12. ADOPTION STUDIES

The closest approximation to a crucial experiment pitting hereditarianism against environmentalism is cross-racial adoption. I know of no studies of white children adopted by blacks, but two studies have been done of black children adopted by whites. One involved black/white hybrids fathered by black American soldiers stationed in Germany and reared by their white biological mothers. I have relied on secondary sources (Brody 1992, Brody and Zuckerman 1988: 1028), which report that the IQs of the mixed-race preadolescent adoptees were about the same as those of illegitimate children of white American soldiers raised by their biological mothers. This study did not control for differences between the IQs of the fathers and the general black population or follow the mixed-race children through adolescence. Also hereditarianism predicts higher IQs for hybrids than for blacks with more African ancestry. This German study nonetheless supports environmentalism.

A more careful study (Scarr and Weinberg 1981; Weinberg, Scarr, and Waldman 1992) followed 101 white families who adopted 130 black or mixed race children, 25 white children and 21 "Indian/Asian" children, and had 14? biological offspring. Table 4.5, adapted from Weinberg, Scarr, and Waldman (1992) and reproduced from Levin (1994), summarizes the mean performance of family members and adoptees on various measures of mental ability and achievement in both 1975, when the average age of the adoptees was seven and that of the birth children ten, and ten years later in 1986. B/W denotes the mixed-race cohort, B/B the adoptees with two black parents, or "black," and W the white adoptees; the Asian/Indian cohort is omitted. The fusion of the black and mixed-race cohort is slightly smaller than what the authors call the "Black/interracial" cohort, since the parentage of some "socially classified black children" was not known (Scarr and Weinberg 1981: 122). I have disaggregated where possible, using figures for the "Black/interracial" coho only when separate figures for black and mixed-race are not given.[26]

As the authors assert, these data do "demonstrate the persistent benefici: effects of being reared in the culture of the schools and the IQ tests" (Weinber;

Table 4.5
Performances of Cohorts on Measures of Ability and Attainment

Measure	Parental	Biological Offspring	B/W	B/B	W
N [a]	171	118	(105)[b]	(105)[b]	18
IQ 1975	119.5 $N = 198$	116.7 $N = 143$	109 $N = 6$	96.8 $N = 29$	111.5 $N = 25$
IQ 1975[a,c]	120.3 $N = 158$	116.4 $N = 104$	109.5 $N = 55$	95.4 $N = 21$	117.6 $N = 16$
IQ 1986[d]	115.35	109.4	98.5	89.4	105.6
Vocabulary 1975[e]	—	73	(57)[b]	(57)[b]	NG
Vocabulary 1986[e]	—	70	60	54	62
Reading 1975[e]	—	74	(55)[b]	(55)[b]	NG
Reading 1986[e]	—	73	59	48	56
Math 1975[e]	—	71	(55)[b]	(55)[b]	NG[f]
Math 1986[e]	—	69	50	36	56
Aptitude 1986[e]	—	66	42	17	61
Class Rank 1986[e]	—	64	40	36	54
G.P.A. 1986	—	3	2.2	2.1	2.8

a. Of participants in the follow-up study.
b. B/B and B/W not disaggregated.
c. Stanford-Binet, WISC, or WAIS.
d. WISC-R or WAIS-R.
e. In percentile rank.
f. Not given.

Scarr, and Waldman, 1992: 131), for in both 1975 and 1986 the IQs of the black children exceeded national black averages. The mean IQ of the seven-year-old black group in 1975 was 96.8, .78 SD above the national black mean, although still a significant .22 SD below the national white mean. It should also be noted that the mean IQ of Minnesotan blacks is considerably above the national mean (Shuey 1966: 155, 183). (I know of no national data for mixed race children.) Weinberg, Scarr, and Waldman also take the data to show that "the social environment maintains a dominant role in determining the average level of black and interracial children"; in 1981 they had concluded that "putative racial genetic differences do not account for a major portion of the IQ performance difference between racial groups," and in 1992 they reiterate that "genetic background" is not the sole determinant of the race difference in IQ. However, they stop short of endorsing extreme environmentalism, and in fact, contradicting their words, their data support hereditarianism, indeed strong hereditarianism.

The most pertinent comparisons are those between the black cohort and the birth children, and between the black and white adoptee cohorts. The natural statistic by which to compare these groups is d, a pooling of the SDs of the groups being compared as weighted by group size.[27] Weinberg, Scarr, and Waldman compute a pairwise d of .99 for the birth children and the combined black/interracial group. However, we also see that d for the birth cohort-black cohort in 1975 was 1.63 and 1.58 in 1986. (These numbers are inflated by the small sample SDs; assuming a "true" pooled SD of 15 reduces both ds to about 1.3.) Such large ds are not predicted by environmentalism. If environment accounts for all between-race differences in IQ, one would not have expected the difference between the black adoptees and the birth children to exceed the deviation of the birth children from the white genetic mean (see below). In addition, environmentalists would presumably expect preadoption experience to be a significant source of variance, but Weinberg, Scarr, and Waldman (1992) report that, so far as can be determined given its limited range of variation, time of adoption explained 17% or less of the variance in adoptee test scores.

What complicates matters is that the birth cohort was evidently atypical of whites. "Clergyman, engineer and teacher" are given as typical occupations of the adoptive fathers, who average 16.9 years of schooling (Scarr and Weinberg 1981: 116). The adoptive families are characterized overall as "highly educated and above average in occupational status and income." It is reasonable to assume that the birth cohort is above the white mean genetically, so that a large d separating this birth cohort from the black adoptees does not show that $H^2 > 0$ (an objection that is unavailable, incidentally, if $h^2 \approx 0$.) The white adoptees, presumably not above the white average in genetic makeup, are appropriate controls. In 1975, d for the white and black adoptee cohorts was 1.72, and in 1986 it was 1.28. Again assuming a true SD of 15 for both populations in both years reduces d to 1.48 and 1.08 respectively. Presumably genetically typical whites and blacks raised from birth in the same (very favorable) environments thus differ at the end of adolescence by a bit more than 1 SD, the usual gap.

A sharper focus is provided by the common environment of the adoptees. Environments within families whose parental IQ is 115 and whose children's mean IQ is 109 (to use the more recent norms) are presumably quite rich whether because rich environments produce high IQs or (the gene/environment correlation hypothesis) high IQs produce rich environments. That the adoptive environments were indeed favorable to cognitive development is further indicated by the above-average performance on all measures by the white adoptees, as noted a sample probably not biased toward the high end of the white scale. It may be concluded that rearing in a rich environment up to age seven will bring the IQs of typical black children to within .2 SD of the white mean, and that rearing in such an environment up to age 17 will bring the IQ of blacks to within −.8 SD of the white mean. This .6 SD increase in the gap over time fits the known increase of h^2 with age.

The 1986 black mean was 89.4 on tests on which the national black mean is 85 (Jensen and Reynolds 1982). Assuming the black adoptees represent the overall black population, blacks raised to adolescence in a white environment did about .3 SD better than they would in a black environment, less improvement than is predicted by black and white agreement in the intelligence polygene. An hypothesis that better fits the data is that about 70% of the deviation of the races from the between-race mean is due to genetic variation, i.e. that H^2 is $.7^2 = .49$. However, because of the superior richness of the adoptive environments, .49 may understate H^2. The authors do not given an SES index for the adoptive families, so attempting to measure the extent of this superiority may appear an exercise in arbitrariness, but three assumptions about within-family environment suggest themselves.

First, a family's environment might be reflected by the IQ of its birth children, in this case 109 at age 17. By this measure, the black adolescents were raised in environments whose mean is $(109 - 100)/15 = .6$ SD richer than the white mean, and on the same assumption, 1.6 SD richer than the black mean. It follows that a 1 SD improvement in environment makes a $.3/1.6 = .1875$ SD difference in IQ for the black adoptees, yielding an H^2 of $(1 - .1875)^2 = .66$.

Second, richness of family environment might be reflected by parental IQ, in this case about +1 SD. Estimated H^2 is then $(1 - .15)^2 = .7$.

Third, it might be most natural to measure family environment by the IQ of white adoptees reared within it, although so doing ignores genotype/environmental correlations. In that case, computations like the foregoing yield an estimate for H^2 of .59. All three estimates resemble those for h^2 derived from twin studies.

The trend found for IQ holds for measures of academic achievement, including mathematics, general aptitude, and class rank. The overall academic performance of the blacks by late adolescence as indicated by a mean class rank in the 36th percentile is well below the mean of 50. I have been unable to find data on the mean class rank for blacks nationally in elementary and secondary schools, but reasonable inferences are possible that yield estimates of the role of genes in race differences in academic performance. As class rank generally parallels IQ, the mean class rank for blacks nationally should be near the 16th percentile. (Expressing the white mean as $z = 0$, the black mean lies at $z = -1$, above 16% of the total population.) This assumption is consistent with the 2- to 3-grade race difference on typical tests of achievement. A mean class rank at the 36th percentile corresponds, by coincidence, to a z of $-.36$. Thus, blacks reared in an environment 1.6 SD above the black environmental mean show a .64 SD improvement in class rank. One might infer that race explains 36% of the between-race variance in class rank.

Again using data from Weinberg, Scarr, and Waldman (1992), Table 4.6 displays the differences between the birth cohort and adoptive cohorts on the various indicators of ability and achievement. The bottom row gives the average of the ds in each column. Except for reading in 1986, the white group consistently outperforms the mixed-race group and the mixed-race group consistently outperforms the black group. In light of the means of the ds, the

Table 4.6
Between-Cohort Performance Differences

Measure	d(*Birth-B/B*)	d(*Birth-B/W*)	d(*Birth-W*)	d(*W-B/B*)	d(*W-W/B*)
IQ 1975	1.63	0.53	−0.09	1.72	0.63
IQ 1986	1.58	0.86	0.3	1.28	0.56
Vocabulary 1975	.94a	.94 a	b	b	b
Vocabulary 1986	.62	.39	.31	.31	.08
Reading 1975	.71a	.71a	b	b	b
Reading 1986	.93	.52	.56	.37	−.03
Math 1975	.65a	.65a	b	b	b
Math 1986	1.3	.76	.52	.8	.24
Aptitude 1986	2	.98	.2	1.8	.78
Class rank 1986	.96	.82	.34	.62	.48
G.P.A. 1986	1.28	1.14	.28	1	.85
$\bar{d}$	1.14	.75	.26	.99	.45

a. Black and mixed-race not disaggregated
b. Data not available

mixed-race outcomes lie just above halfway between those of the black and th
birth cohorts, and white adoptee outcomes lie about three-fourths of the wa
toward the birth children. Calling white performance 1 and black performanc
0, the mixed-race cohort lies at .545. Although interaction precludes precis
numerical prediction, this is the order and the magnitude of the ga
hereditarians expect. If the mean white genotype produces higher intelligenc
than the mean black genotype when environment is fixed, the intelligence c
black/white hybrids should be intermediate and that of genetically typic
whites should lie between that of the hybrids and the genetically favor
whites. If, on the other, hand, mistreatment because of race complete

explains these discrepancies, lighter but still plainly Negroid blacks should not outperform darker blacks. The modest .21 SD for the distribution of birth-mixed-race ds indicates uniform performance across variables for the birth and mixed-race cohorts. Greater white dislike of darker-skinned blacks may explain part of the black/mixed-race performance gap, but why should the effects of this animosity so closely mimic the hereditarian prediction?

In reply to Levin (1994) and Lynn (1994a), Waldman, Weinberg, and Scarr contend that adoption experiences for the white and black cohorts did differ, but seemingly concede that controlling for that variable does not reduce the IQ gap by more than about 15% (1994: 35, 36). They also observe, what is true enough, that previous studies using blood group frequencies have not found significant correlations between degree of African ancestry and IQ, although one might reply that theirs is then the first study that does. The hypothesis Waldman, Weinberg, and Scarr favor is that the black adoptees' race triggered "racially based environmental effects" (40), so that what their study reveals is "the pervasive effects of racism in American life" (41; cf. Block 1995).

Now, subtle forms of racism *may* explain some or all of the deterioration in black adoptee performance. However, Waldman, Weinberg, and Scarr cite no evidence that this is so, and, as I cautioned in chapter 3, the sheer possibility of explaining away evidence against a hypothesis, in this case $H^2 \approx 0$, does not support it. There also remains the unexplained sensitivity of racism to degree of African ancestry. In the terminology of n.1 of Chapter 3, the Minnesota adoption study is prima facie evidence that $H^2 > 0$, the Waldman-Scarr-Weinberg racism hypothesis to save $H^2 \approx 0$ is ad hoc, and the adoption study is not even prima facie evidence for $H^2 \approx 0$.

Furthermore, while adoption studies may ignore racism, they may also understate H^2 by ignoring gene/environment correlations, since the "culture of the schools and the IQ tests" that benefits black IQ may itself be a correlate of white genotypes. To the extent that family background reflects genetic factors, exogenous environment explains less of the between-race variance that it immediately appears to.

Finally, two-thirds of the transracial adoptees also scored in the clinically deviant range on the MMPI (DeBerry 1991), as did about the same proportion of the biological offspring. This suggests a significant genetic component in he race difference in temperament, and, perhaps, that rearing in close proximity to black children adversely affects white children.

That Neisser et al. (1996) omit the Minnesota study, despite citing items hat report it, warrants caution about their negativity toward hereditarianism.

4.13. INTERVENTION STUDIES

Interventions to improve the academic performance of poor children, redominantly and sometimes exclusively black, also test the theory that the ace difference in IQ is due entirely to poor environments. This theory predicts hat exposure to rich environments of the sort such programs provide will markedly improve the IQs of "at-risk" black children. The failure of these rograms is described in Spitz (1986) and Currie and Thomas (1995).

Three characteristic efforts are the Perry Preschool Program in Ypsilanti, Michigan, the Milwaukee Project, and Head Start. In the Perry Program, black children with IQs from 50 to 85 were given special classes for five years and tracked until age 19 against controls. In the Milwaukee Project, children six months of age or younger of low-IQ black women spent large parts of each day for five years in an Infant Stimulation Center. There were also extensive home visits and remedial education for the mothers (also see Garber 1988; and Jensen 1985a, 1985b, 1989). Incidentally, the director of the Milwaukee Project was convicted of embezzlement; so far as I know, his demonstrated dishonesty has never been used to discredit environmentalism, although the evidence of fraud against Burt, widely taken to discredit hereditarianism, was much flimsier.

In each case, the experimental children made large gains in IQ and academic performance while the program lasted, but these gains disappeared after the program came to an end (Spitz 1986: 90, 103–108). For instance, by the fourth grade, the experimental children in the Milwaukee group were at the 10-11th percentile in mathematics, only 2 percentiles above the controls. In each case, the IQs of the experimental children had fallen to the level of the control children by the age of 15. Neisser et al. (1996) agree that in such studies "long run gains have proved more elusive" (88). Ironically, the early gains of the experimental groups in these studies appear to result from "teaching to the test," the common criticism of white IQ test performance.

Head Start, the best-known enrichment program, has at this writing become something of a scandal. Project Follow Through, introduced when Head Start children were not maintaining their early gains, also failed to produce lasting differences either in academic achievement or IQ as measured by the Raven's Matrices (Spitz 1986: 91–93). In a study comparing the performance of Head Start children with siblings who did not participate in the program, Currie and Thomas (1995) found that, while at age ten white children retain a gain of 5 percentile points on the Picture Peabody Vocabulary Test, "by age 10 African-american children have lost any benefits they gained" (359), a dissipation of benefits uncorrelated with black home environments. Likewise, a small but statistically significant inverse correlation for whites between Head Start participation and repeating a school grade was not matched by blacks.

Its advocates (e.g. Zigler and Berman 1983) now characteristically endorse Head Start for its positive effects on social behavior rather than intellectual ability. An article in *Scientific American* purporting to show it "abuses science" (Beardsley 1995) to argue that intervention does not boost IQ ends by walking away from IQ, arguing instead that "providing education can help [low-IQ] individuals in other ways." Health benefits are often cited. However Currie and Thomas found that, as measured by height-for-age, participation in Head Start had no effect on health (356). In any case, such non-intellectual benefits of Head Start as there may be are irrelevant to the main point. Contrary to environmentalist predictions, intervention beginning at age three makes no difference to the intellectual development of blacks. Perhaps surprisingly, intervention for whites does, indicating a possible nonsocial racial difference in receptiveness to stimulation.

Currie and Thomas neglect to consider this possibility, limiting their hypotheses to "heterogeneity in program delivery or the types of schools that whites and African-Americans attend once they leave the program" (359). However, on their own showing Head Start programs tend to be uniform, and the conventional environmentalist explanation of "fade-out" is that enrichment programs begin too late, after environmental factors have affected black children.

But let us look at one specific factor very often proposed, namely malnutrition, in turn attributed to racism[28]—the denial by whites of child-care information for blacks, or black prospects so dim that black parents ignore their children's well being. Now, a nutritional deficit might equally well be a passive environmental correlate of genes. After all, no external source provides whites with child-care information. On the correlate hypothesis, low black parental intelligence is transmitted to children, and causes black parents to select poor diets for their children. But in fact black children eat as well as white. According to a 1985–1986 survey by the U.S. Department of Agriculture described in Rector (1991) and Rector and McLaughlin (1992), black preschool children consume 56.9 grams of protein daily to 52.4 grams for white preschool children, both quantities more than twice the U.S. Daily Recommended Allowance (USRDA). Children in families 75% below the poverty line, a disproportionately black cohort, consume about as much vitamins A, B-6, B-12, C, and E, thiamin, riboflavin, niacin, phosphorus, and magnesium as children in families 300% above the poverty line, and more than the USRDA. Both poor and nonpoor consume less than the USRDA of calcium, zinc, and iron by about the same amount. Banfield (1974: 131) cites a 1955 Department of Agriculture survey showing that the diets of the lowest income third in cities receive more than the USRDA in calories, protein, minerals, calcium, iron, and vitamins. (Duncan and Currie 1995, on the other hand, cite evidence of greater iron anemia among the poor.)

Since malnutrition retards bone ossification, the relatively greater speed of ossification of cartilage in black children (Jensen 1973: 338–339, citing Taylor and Myrianthopolous 1967) indicates adequate nourishment. Such black precocity is in fact characteristic. Up to 18–24 months, black infants outpace white infants in muscular and neurological development (Bayley 1965), playing pat-a-cake, walking with help, standing alone, and walking alone, and exceed white infants in the Gesell "Development Quotient" when age is controlled for. (Within-race, the Gesell Quotient correlates slightly negatively with later IQ.) This offset favoring blacks is inconsistent with greater black malnutrition. The simplest explanation of the failure of intervention, then, is that nonenvironmental factors account for black IQ.

Hopes for intervention programs obviously assume that intelligence is teachable, a proposition in whose defense Fischer et al. commit a fallacy of remarkable grossness (1996: 47–51). "If intelligence is defined as mental self-management," they argue, "intelligence can be taught. The very existence of business schools attests to our confidence that management skills can be taught" (47). Similarly, if intelligence is a matter of identifying relations, "then students can certainly be taught them. We regularly teach small children to recognize thousands of relationships among a small set of signs" (49). Finally, they reason, "because every expert was once a novice, somehow novices must *learn* to think like experts. Differences in how novices and experts process information could therefore be taught, as part of mental self-management" (50). Yet the very question at issue is whether every individual or every group can learn mental self-management, relations, or anything else *at the same rate*—whether, that is, all individuals and groups are equally phenotypically intelligent—and whether any individual or group differences in the capacity to learn mental self-management and relations are due to genetic variations. Why are white children better than black children at this skill by the fifth year of life? The "alternative paradigm" Fischer et al. present does not even address this issue. Forget that in this context appeal to business school "management" techniques is an equivocation; defining "intelligence" as something learnable invites us to ask about the comparative abilities of blacks and whites to learn to be intelligent. A difference in *this* ability, whatever it is called, leaves "intelligence" beside the point.

4.14. BLACK ATHLETIC ABILITY

The hypothesized malnutrition of blacks also conflicts with blacks' greater mean mesomorphy (Herrnstein and Wilson 1985: 469), and success in sports: undernourished children should not grow up to dominate sports at all levels. In fact, the remarkable athletic attainment of blacks merits fuller discussion.

American blacks are widely perceived to be most prominent in sports requiring fast reflexes and sudden exertion such as boxing, sprinting, jumping, hurdling and basketball—a stereotype acknowledged in the title of the popular movie *White Men Can't Jump*. In a literature review, Malina (1988) confirms that blacks do excel in the dash, the long jump, and vertical jumping. Blacks are overrepresented in baseball, but particularly so in the outfield, where a premium is placed on fast starts, speed, and power hitting. Blacks in football are typically running backs, seldom quarterbacks. Worthy and Markle (1970) argue that blacks dominate positions that require quick reaction, while whites dominate positions from which actions are initiated, such as quarterback and pitcher. By contrast, blacks are underrepresented in sports requiring concentration and control, such as tennis, golf, bowling, and, with the exception of East Africans, distance running, despite the wide availability of facilities for most of these activities, and are virtually absent from competitive chess, bridge, and Scrabble, games heavily dependent on mental ability.

The scarcity of black swimmers can be explained by blacks' greater mesomorphy, which implies a lower ratio of fat to denser muscle tissue and consequent lower buoyancy. The advantage in this case of explaining a sports difference physiologically rather than psychosocially, as by low self-esteem and racism (see, e.g., C. Walker 1995), is characteristic. Unlike psychosocial variables, physiological variables can account for the *pattern* of black achievement. Ama et al. (1986) have verified that African blacks have relatively more "fast-twitch" muscle tissue than whites, a trait American blacks presumably share. In addition to having less body fat, American blacks are on average stronger and have heavier bones, longer extremities, and narrower hips (Noble 1978; Jordan 1969; Malina 1969, 1973, 1988), differences sufficient to explain attainment where jumping and speed are needed. Moreover, race differences in jumping and running are present by age five or six, and many observers unaware of the Gesell Scale have commented on the "motor precocity" of black children (again see Malina 1988). These differences persist when social class is controlled for (Malina 1988), further weakening socioeconomic explanations of black sports attainment. The popular theory that blacks choose sports as a way out of poverty leaves it mysterious that blacks should chose a path through basketball rather than tennis—or mathematics, said by the same theory to be the Chinese route into the American mainstream.[29] Furthermore, the flight-from-poverty theory provides no mechanism by which a desire to excel creates ability. Ambition inspires practice, but it also inspires whites to practice, with less impressive results. I would add my personal impression, confirmed by several athletic coaches I have consulted, that practice is relatively unimportant for the activities at which blacks excel, such as sprinting, jumping, and shooting a basketball. Training does not increase sprinting speed as it improves endurance and time in distance events.

A rather fanciful sociological theory has it that whites let blacks excel in sports to confirm the stereotype of black physicality. The obvious difficulty with this theory is the absence of evidence for and the abundance of evidence against the motive it ascribes: colleges and professional teams certainly appear to recruit black (and white) athletes for the sole purpose of winning. A less obvious but equally serious objection is that white desires for black excellence could not enable blacks to excel, nor direct black efforts into the particular channels they take. Why should whites wishing to confirm black physicality make blacks outstanding boxers but mediocre swimmers? Again, a sociological hypothesis fails to explain the *pattern* of black performance.

Far from supporting racism as an explanation of overall low black attainment, the exclusion of blacks from professional sports until the 1940s undermines it. What the late integration of sports shows is that as soon as blacks were allowed into professional athletics, where their talent was equal to the opportunity, they rose rapidly. If American whites are racists, why do they avidly follow the 90%-black National Basketball Association and the 66%-black National Football league? At present, affirmative action makes it easier for blacks to enter science, business, and higher education than it ever was to enter sports. No money is spent to increase black representation in basketball or football: the great number of athletic scholarships awarded blacks are merit-based.[30] At the same time, the National Science Foundation and the National Institute of Health alone have spent over $2 billion since 1972 on programs to increase black representation in science (Gibbons 1992: 1181); despite Medical College Admission Test scores much lower than those of whites, over 80% of blacks in medical school receive scholarships (W.-M. Lee 1992), and the acceptance rate of blacks to elite colleges is much greater than that of whites with superior records (Zelnick 1996: 132–137). That blacks fare better in sports than other fields suggests that black absence is not due to lack of opportunity.

It is a matter of observation that, as George (1992), C. Walker (1995), and Jones and Hochner (1973) assert, blacks play with a distinctive flamboyant style unlike that of whites. The black approach emphasizes individualism, flair, and expression, as opposed to the white approach of subordinating the individual to the team and the goal of winning (Jones and Hochner 1973). Larry Bird did not shoot like Julius Erving or any of dozens of other black basketball stars. In the 1930s, the heyday of Joe Louis, sportswriters speculated that blacks were successful in sports like boxing because they were emotionally geared to short bursts of tremendous effort that whites could not match. These differences appear to be extensions of the differences in aggression and self-assertion discussed chapter 3.

4.15. AFRICA AND ELSEWHERE

Waldman, Weinberg, and Scarr (1994), Bodmer and Cavalli-Sforza (1970: 27), Jencks (1992: 99–100), Pettigrew (1964), and Crow (1969) all observe–a version of the environmental-reaction theory–that adoption and intervention studies do not control for perception of race. Hacker puts the point generally:

[T]here is no way to factor out whether any part of the results reflect "racial" elements in some genetic sense, since we would have to adjust for every specific environmental influence as it has affected each individual. (Indeed, even when a white family adopts a black infant, the child knows that she is "black" and that image of herself will affect how she adapts to a "white" environment.) (1992: 27; again see Block 1995: 107)

In fact, adoption of white-looking Negroid infants by whites unaware of their ancestry would factor out white attitudes toward color, although this design is unlikely to be realized. Another (impracticable) design is four-way comparison of the performance of light- and dark-skinned blacks whose proportion of sub-Saharan ancestry, as determined by family trees or DNA testing (see Cavalli-Sforza 1994), is genetically low and genetically high. Environmentalism predicts that light-skinned blacks of any degree of African ancestry will outperform dark-skinned blacks of any degree of African ancestry, while hereditarianism predicts that individuals with relatively little African ancestry will outperform individuals with more irrespective of skin color.

White perceptions are however factored out in Africa, where blacks have been dominant for millennia. Chapter 3 mentioned some medieval Arab characterizations of Africans. The first European explorers to encounter African Negroids also found their most salient traits to be present-orientation and absence of intellectual curiosity. Baker cites some characteristic descriptions:

Livingston writes of the Bechuana, for instance, "No science has been developed, and few questions are ever discussed except those which have an intimate connection with the wants of the stomach." Generalizing more widely he remarks, "All that the Africans have thought of has been present gratification." Fynn, too, tells us that he seldom found any Zulu gifted with the inquisitiveness that makes people interested in knowledge apart from the possibility of its immediate application to the practical affairs of their lives. . . . Livingston found that the Barotse had retentive memories on which he could rely for information about events that they themselves had actually witnessed long ago but he found little evidence anywhere of *foresight* or thought about the distant future. . . . Du Chaillu writes of the "utter improvidence" of the Gabon tribes. Speke expresses the same idea when he says that the Negro "thinks only for the moment." (1974: 396–397)

It would be a remarkable coincidence if oppression in the United States produced the same traits found among indigenous Africans unexposed to whites. The principle of parsimony counsels rejecting such an interpretation: the cause of the present-orientedness and lack of intellectual curiosity of blacks in Africa may be assumed to be the same as the cause of comparable traits found in American blacks.

Lynn (1991a) surveys 11 studies conducted between 1929 and 1991 on approximately 6000 African children and adults,[31] most using some form of

Table 4.7
Zimbabwe-English Performance on WISC-R and Raven's Progressive Matrices

	Zimbabwe N = 204		English N = 202		d a
	Mean	SD	Mean	SD	
WISC-R Verbal IQ	69.71	10.82	95.3	12.65	2.18
WISC-R Performance IQ	69.76	13.45	95.22	13.14	1.91
WISC-R Full Scale	67.09	11.65	94.7	11.77	2.35
Raven's	72.36	12.18	96.71	12.48	1.97

Source: Zindi (1994)
a. In weighted mean SD units

the nonverbal, culture-fair Raven's Matrices. The weighted mean IQ for blacks in these studies is 73.3, with the mixed-race group performing intermediately. Kamin (1995) criticizes Lynn for converting Raven scores into IQs, but conversion is unobjectionable for tests like the Raven that load as heavily on g as other IQ tests do. A z score on the Raven corresponds to the identical z score on an IQ scale; the size of a gap in SD units is what matters, not the essentially arbitrary numbers expressing it. Kamin is also right that, in some cases, the raw scores are not normally distributed. However, as explained in chapter 3, normality per se is irrelevant to group differences.[32]

In a separate study of 1093 South African black secondary school children, 1056 South African white children, and 778 mixed-race hybrids on a form of the Raven's for which the white mean and SD are 50 and 10, the black/white d is 2.78, and the white/hybrid d is 1.35 (Owen 1992), corresponding to a black IQ in the 60s. Two more recent studies have yielded similar results. Zindi (1994)[33] administered the WISC-R and Raven to 202 Zimbabwean secondary school children 12–14 years of age, and 204 white London children matched for age and, as far as possible, background. Their scores are displayed in Table 4.7. The two groups differ by about 2 SD on the Full Scale WISC-R, Verbal and Performance IQ, and the Raven, and the black means and SDs for the Raven, as reported by Zindi, correspond to IQs in the 69–75 range. Kaniel and Fisherman (1991) found that 15-year-old Ethiopian Falashas in Israel score 2 SD below Israelis of the same age on the Raven. They place the Ethiopian scores between the fifth and tenth percentiles of the Raven (corresponding to

IQs between 73 and 80); Lynn (1994b) contends the Ethiopian scores lie at the second percentile for an IQ of 69. Either way a large gap is indicated.[34]

An obvious issue is cultural bias and familiarity with tests. With regard to the relative difficulty of test items, Owen reports a similar ordering for all groups, and that "the items that best measure differences within each ethnic group are the same items that discriminate the most between the ethnic groups" (1992: 154). Kamin (1995) objects that "blacks reared in colonial Africa have . . . been subjected to discrimination," and a photograph accompanying his article of African children in a decrepit schoolroom suggests disadvantages relative to Western children. However, as Zindi points out, "one begins to wonder why group differences were still observed even on the Performance subtests of the WISC-R as well as on Raven's Progressive Matrices, which is said to have less cultural bias since the use of language is not necessary for its administration" (1994: 551). It must also be considered that the schoolroom pictured is of Western construction, and, while decrepit by Western standards, superior to any corresponding facility built by black Africans on their own, hence an unnaturally *stimulating* exogenous environmental influence. Neutral observers appear to agree that highly reliable tests probing the ability to "abstract a rule" can be constructed even for Kalahari Bushmen (Reuning 1988: 466, 477), and that the factorial structure of tests administered to literate blacks in South Africa in particular resembles that found for Western populations (Kendall, Verster, and Von Mollendorf 1988: 323).

Lynn (1991a) and Lynn and Holmshaw (1990) administered Jensen's reaction time tests (see section 3.10) to British, Irish, Japanese, Hong Kong and South African nine-year-olds. As in the American samples, decision time correlated negatively with IQ as determined by Raven's Matrices. Negroid movement time approximated Caucasoid and Mongoloid, but Negroid decision time exceeded that of Caucasoids, which exceeded that of Mongoloids, and intraindividual variability for Negroids significantly exceeded that for Caucasoids and Mongoloids. One anomaly was that black reaction time was faster on those tasks more highly correlated with Raven scores. (The mean of the black sample on the Raven's was 12.7, corresponding to a white IQ of 65.)

Since the IQs of contemporary African blacks are about 2 SD below those of contemporary American whites (with American blacks intermediate), the phenomenon of low black IQ, in Africa or the United States, is unlikely to be due to an oppressive environment. To attribute the white/African difference to colonialism implies that whites have had a large effect on blacks in a short time, implying in turn a marked black/Asian difference. Contact with Europeans, including extensive colonization, did not lower Asian IQ 2 SD below white norms. Further, appeal to colonialism overlooks the need to explain Europe's total domination of Africa. To say that Europeans had better weapons begs the key question, namely, why European weapons were better. Why hadn't Africans invented explosives or the science of ballistics? Medieval European weapons, tactics, and discipline were inferior to those of the armies of Ghengis Khan, and overtook Asian war-making capabilities only after the rise of modern science—itself a development that may reflect genetic differences.[35] If biological explanations of cultural differences are "reductive"—

an issue taken up in the next chapter—we begin to see why cultural explanations of cultural differences are circular.

There are also striking parallels on noncognitive factors between blacks in various parts of the world. In Chapter 3 I mentioned that, when allowed to choose between getting a small candy bar immediately or a large one later, black Trinidadian children tended more than nonblack children to choose the near-term reward (Mischel 1958; also 1961a, 1961b). Africans display the pattern of loose pair bonding and limited parental involvement found in American slums:

[T]here persists high fertility and a pattern of parental investment in which both mothers and fathers invest, by Western standards, relatively little in each offspring and pursue a pattern of delegated parental responsibility . . . a mating pattern that permits early sexual activity [and] loose economic and emotional ties between spouses. . . . The parents themselves do not expect to be the major providers for each offspring throughout the children's lives. . . . The norms regarding male domestic roles do not emphasize conjugal interdependence nor intimate involvement by fathers in the rearing of children. . . . This pattern [continues] in cities and shows few signs of changing under the influence of modernization. . . . The psychological, social, and spatial distance of husbands/fathers, together with their freedom from direct economic responsibility relieves them of most aspects of the parental role as Westerners understand the term. . . . The literature on the Nilotic and Niloticized Bantu-speaking tribes of East Africa and the Fulani peoples is fairly consistent. Domestic arrangements preserve [this] pattern. . . . [T]he typical African pattern is to terminate intense care of the child early (by Western standards) in the child's life. . . . As a woman enters her post-reproductive years, she is less and less likely to have a man who is making financial contributions to her support. (Draper 1989: 147–157)

These patterns are found "among many different ethnic groups with differing levels of modernization and urbanization. . . . [E]conomic and social forces that in a Western, middle-class context lead to reduced numbers of more intensively nurtured offspring do not have the same outcome in many African societies" (Draper 1989: 157–159).[36] Common-law marriage and prolonged father-absence is also common in Trinidad and Grenada (Mischel 1961b). Parents were unmarried in 56% of a sample of West Indian households in an British Midlands town (Scarr, Caparulo, et al. 1983).

Although difficult to quantify, the character of an isolated society, determined as it is by the genotypes of its members expressed in a given physical environment, indicates its member' innate qualities. There are obvious differences between European, Asian, and African indigenous cultures. Lynn (1991a) remarks that not one of the 1500 discoveries listed in Asimov's *Chronology of Science and Discovery* (1989) was made by a Negroid people. (As noted, Asimov was a self-described "liberal," so this omission is unlikely to be due to "racism.") None of the 200 most important persons in history in Michael Hart's list (Hart 1992) is Negroid.[37] Lynn and Rushton both cite Baker's criteria for civilization, which include the wheel, metallurgy, building with stone, cultivation of food plants, roads, domestication of animals, money, laws ensuring personal security, recognition of a right of the accused to defend himself, written language, abstract knowledge of numbers, a calendar, schools, appreciation of art and knowledge as ends in themselves, and the absence of gross superstitions, cannibalism, torture, and self-mutilation. Laying aside the honorific term "civilization," it is an objective fact that only "Europid and Mongolid peoples" (Baker 1974: 520) have displayed these features.[38] The African physical environment contains materials from which to fabricate wheels, axles and yokes; that no African group developed these devices, mathematics, or a calender suggests inability to do so.

Baker (1974: 525) emphasizes the dependence of a civilization's level of development on the innovations of a talented elite, but average capacities matter as well. A lone genius may distinguish twoness from pairs of hands and pairs of feet, but unless he is surrounded by a sufficient number of individuals who can learn this concept from him and transmit it to the next generation, his insight will not lead to the development of arithmetic. The absence of science and mathematics from Negroid cultures indicates not only the presence of fewer geniuses, but also of fewer non-geniuses able to preserve and amplify innovations. Afrocentrists describe the industrial revolution as "happening to" Europe, but technological innovations don't happen; they are produced by human ingenuity.

That the absence from Africa of advanced material culture is more than an accident is confirmed by the failure of postcolonial Africa to sustain the technology left by whites.[39] A journalist writes:

While the Belgians were often consummately patronizing to their African subjects, they installed an efficient colonial administration. In time, they introduced health care, water projects, education, telephones and power lines, helping to turn this once isolated village [Kikwit, Zaire] into one of the most affluent and best-tended cities in the core of equatorial Africa. Today, the legacy of Kikwit's colonial past is swiftly disappearing. "Civilization is coming to an end here," said Rene Kinsweke, manager of Siefac, a chain of food stores, as he spoke of how Kikwit has become a dispiriting tableau of chaos and catastrophe. "We're back where we started. We're going back into the bush." (K. Noble 1991; also see K. Noble 1994; French 1995)

According to another,

[W]herever there is an African society that works on a Latin American or a south Asian level, it is because colonial influence has been prodigious. . . . By the 1980s there were five times as many Frenchmen in key jobs in Cote d'Ivoire as there were at independence in 1960, with French nationals holding 80 percent of all posts requiring a university degree. Now, however, the French are trickling out. . . . Perhaps a democracy can take hold in a country like this, but few experts are betting on it. . . . Different parts of [Sierra Leone] are run by different tribal commanders. The official government is composed of young soldiers who control the capital of Freetown and some territory beyond. The smell of dope is said to be common in the palace. The streets of Freetown aren't paved. Water and electricity services [installed by Europeans] are sporadic. The capital's few photocopy machines and faxes are therefore unreliable. Crime is rising. A Lebanese expatriate community, like the Asian one in Kenya, props up the economy, and is therefore begrudged for it. . . . The empirical evidence, whether it be in literacy rates, infant mortality rates, economic statistics or whatever, proclaims a grim truth that the whole issue of race in America makes us afraid to utter: African culture, in a modern political sense, is simply dysfunctional. (Kaplan 1992)

A third writes of Liberia:

Liberia has come to bear little resemblance to a modern state, becoming instead a tribal cauldron governed less by commonly understood rules than at any time since it became Africa's first republic in 1847 under freed American slaves. [A legal authority is cited:] "Things have deteriorated to the point where what we are seeing emerge nowadays are sub-warlords, each of whom is a law unto himself." (French 1994: A12)

Food production in the poor African countries in 1993 was 20% lower than in 1970; malnutrition in Zambia increased from 5% to 25% (Darnton 1994a). Gross National Product declined by 2% annually through the 1980s, and Africa's share of world trade fell from 4% to 2%. "[I]t is almost as if the continent has curled up and disappeared from the map of international shipping lanes and airline routes that rope together Europe, North America and the booming Far East. Direct foreign investment in Africa is so paltry it is not even measured in the latest World Bank study" (Darnton 1994b). Several other countries have also sunk into chaos since these words were written.

It again begs the question to explain the disintegration of postcolonial Africa on the failure of Europeans to prepare it for independence. Who prepared Europe for independence? Why aren't England and France still divided into feuding tribes? Why were Africans "unprepared" during the millennia before contact with whites? [40]

The situation in the black Caribbean resembles that of Africa. Before a mulatto-led rebellion in 1804 made Haiti independent of France, the French had installed roads, bridges, and an irrigation system. These had fallen into disrepair by 1915, when the United States intervened to stabilize Haiti economically and politically: in the 72 years preceding, there had been 102 revolutions and coups (Maclean 1993). The Marines left in 1934 after rebuilding the infrastructure, setting the telephones working again, and training young Haitian men in farming skills (which the Haitians resented). American banks took over the economy. But by 1958 these improvements had been undone; a visiting historian found "telephones gone, roads approaching non-existence . . . ports obstructed by silt, docks crumbling . . . sanitation and electrification in precarious decline" (cited by Maclean 1993: 55). The situation has deteriorated since, and another American intervention has recently ended. Like Africa, Haiti has been unable to sustain Caucasoid technology.

It may be added that Scarr, Caparulo, et al. (1983) found that West Indians in Britain perform far below white norms on school examinations. In one study, although blacks comprised 10% of the school population, none had passed any A-level examination.

One can maintain that Negroids are as genotypically intelligent as whites or Asians only by claiming that African achievements did equal Eurasian, and that no one realizes this fact because it is suppressed. Environmentalism thus leads unavoidably to Afrocentrism. Hacker nearly endorses Afrocentric claims about African technology, although he obliquely concedes that "compared with other continents, Africa remains most like its primeval self" (1992: 28). More circumspect environmentalists confine themselves to the American situation, avoiding cross-cultural comparisons (but see the discussion of Fischer et al. in section 4.20).

4.16. AFRICA, CONTINUED

The IQ of Africans bears on three further arguments against hereditarianism. The last of these is somewhat technical, and I would bypass it but for the currency it has gained.

The first: by choosing a heritability of .5 for intelligence, Bodmer and Cavalli-Sforza (1970) calculate that, for genetic factors to explain the present 1 SD race gap in the United States, the blacks above the 85th percentile in intelligence in each generation since the start of slavery would have had to fail to reproduce. They rightly doubt so high a level of selection against IQ.

What Bodmer and Cavalli-Sforza must assume, however, is "that there was no initial difference in IQ between Africans and Caucasians . . . and that the divergence of black Americans from Africans started from slavery about 200 years ago" (28), which conflicts with the the low IQ of present-day Africans. If the first slaves were as intelligent as whites, so too was the African stock from which they came. Assuming a mean African IQ of 100 at the start of the slave trade, the present African mean of 73 requires more intense selection against IQ in Africa than Bodmer and Cavalli-Sforza find plausible for the United States Not only does their assumption (and objection) collapse, this assumption is not needed for the IQ difference in the United States to be genetic. The simplest hereditarian position is that the European/African difference already existed when slavery began.

Second: Cavalli-Sforza, Menozzi, and Piazza (1994) argue that the typical black and white share 99.84% of their genes.[41] How then can whites and blacks differ with respect to genetic intelligence when their genetic separation is so small? Mentioning similar findings, Tooby and Cosmides dismiss all "folk beliefs" in ethnic group differences (1990: 34–35).

I have already addressed this question in a general way; .0016 is small in absolute terms, but even different species are surprisingly similar by this standard. Humans and chimpanzees share 98.4% of their genes, for instance (Caccone and Powell 1989: 932, Table 4, and 933; Gibbons 1990), because most human and chimp genes go to building hearts, lungs, and other shared gross structures. Fine differentiation, such as that between the coordination of a championship golfer and a duffer, is the work of a minute portion of the genome. There is no a priori reason a .16% difference in genes could not manage the nuance needed to differentiate 70-IQ from 100-IQ human brains. More important as an empirical matter, one cannot assume, as Cavalli-Sforza, Menozzi, and Piazzi implicitly do, that genetic effects are linear. That the black-white gene overlap is 1000 times the chimp-human does not make blacks 1000 times more like whites than chimpanzees are like humans. The human/gorilla difference, at 2.5%, is twice the human/chimp difference (Caccone and Powell 1989), but chimps are not midway between gorillas and humans in phenotypic intelligence. The phenotypic manifestation of a genetic difference depends upon the precise constellation of all the alleles in a genome The significance of the .0016 black/white difference must be established empirically. Moreover, Cavalli-Sforza, Menozzi, and Piazza calculate the mean genomic difference between human groups as .0012, smaller than the black-white difference, indicating (by this measure) a relatively large distance between blacks and whites.

The third objection, urged by Lewontin (1982), Gould (1984), Wicker (1987), and Holt (1994), is that the races parted too recently to diverge in genes.

This is correct in the strong sense Gould assigns "genetic divergence," namely the existence of a " 'race gene' . . . present in all members of one group and none of another" (1984: 32), the same constraint imposed by Tooby and Cosmides on "discrete" human groups (1990: 35, 42). However, there need not be such a gene for groups to differ genetically. Divergent combinations of the same genes may diverge in their expression, so the same genes occurring with different frequencies in different populations can produce mean phenotypic differences. As Gould is constrained to admit, "Frequencies vary, often considerably, among groups." (Tooby and Cosmides 1990: 35 make the same concession; also see 48.) The question is whether enough time has passed for changing gene frequencies to have produced the phenotypic race difference.

The short answer, for readers willing to take it on faith, is "yes." A more detailed proof follows.

Computational experiments in Loehlin, Lindzey, and Spuhler (1975: 45–47, 270) yield a partial result. They show (a) that selection at the rate of .01 can increase a gene's frequency from 1% to 99% in 1000 generations,[42] and (b) (assuming a value of .64 for h^2) that initially identical populations evolving in, respectively, environments that deselect IQs of 65 and 60 at 2.5%, will differ in mean IQ by 40 points after 1000 generations.[43] Result (b) calls the the black-white gap to mind, and the parameters chosen are consistent with natural processes, but two of them, the number of generations involved and the required divergence in IQ, are arbitrary. Current estimates of the time of the African-European split and the African-Caucasoid IQ difference allow a more realistic derivation.

For the sake of simplicity, let us assume an additive model in which IQ is controlled by 100 gene loci. The two alleles for each loci may simply be thought of as A and a; the average phenotypic value of the aa homozygote genotype is 25, and every A adds 1 IQ point. (I'll call the A "dominant" and the a "recessive" for convenience, although the value of heterozygote is the homozygotic mean.) Next, suppose all 200 alleles are present in both of two populations B and W, but with the frequency of each a being .76 in B and .625 in W. The mean IQ of B will then be 73 and that of W will be 100. (If the frequency of each a is q, the probability of two As at a diallelic locus is $(1 - q)^2$, and the probability of just one dominant is $2q(1 - q)$). Double-AA loci contribute 2 points, dominant-recessive contribute 1. Should a single population in which q is .76 split into two subpopulations, a decrease of .76 − .625 = .135 in the frequency of the a in one while the frequency in the other remains constant yields a 27-point (= 1.8 SD) increase in mean IQ. The question is whether the 4400 generations since the Negroid/Caucasoid split permit that change in frequency.

We can sharpen the question by introducing s, the coefficient of selection against recessives, defined as the difference in fitness between the double-"dominant" and double-"recessive" genotypes; an s of .1 means that, holding all but one gene locus constant, 9 aas reproduce for every 10 AAs that do. The larger s is, the faster gene frequencies change and the lower the frequency of the recessive alleles falls in each generation. In general, if q_k is the frequency of the recessives in the kth generation, $q_{k+1} = [q_k^2 (1 - s) + q(1 - q)]/[1 - sq_k^2]$. The one-generation frequency change $\Delta q = q_k - q_{k-1}$ is $[sq^2(1 - q)]/(1 - sq^2)$. (I assume for simplicity that selection works only against double recessives.) Gene frequencies can change fast enough to produce the B/W difference if the s associated with the required rate of change is small enough to occur naturally.

To determine this, note that there is selection against recessives, so that $s >$ 0, when the frequency of the aa genotype exceeds its reproductive rate. The mean phenotypic value of the reproducing population then exceeds that of the population as a whole. Expressed in units of the phenotypic SD, the difference between the population mean and the mean of parents is the *intensity of selection*, or i. Thus, if the mean IQ of the parents in a generation is 100.75, i = (100.75 − 100)/15 = .05. The change in the phenotypic mean of the offspring of reproducers, that is, the next generation, depends on both i and h^2; it is in fact ih^2, and the cumulative change after k generations is kih^2. Conservatively assigning h^2 the value .5, we can now estimate the i, hence the s, needed to produce the observed race difference. Given that Caucasoids have evolved independently of Africans for 4400 generations, k = 4400. The white-African IQ difference of 100 − 73 = 27 is, in SD units, 1.8. Hence 1.8 = 4400 × i × .5, yielding an i = 1.8/2200, or .00082. To retrieve s, we use the intuitively obvious proportionality of s to i, and in fact s = .13i.[44] The s necessary to produce the observed race difference is thus .13 × .00082 = .000106. Assuming a generation to be 30 years raises s to .00013; a possibly more realistic figure of 20 years reduces s to .000085. Setting the African-white gap at 2.4 SD (a 30 point gap normalized when the black SD = 12.5) and a generation   at 30 years, s is .000173.

An alternative method for calculating s, adapting the one-parameter case discussed in Crow and Kimura (1970), is to approximate q's rate of change as $dq/dt = -s(1 - q)q^2$. (The rate is negative because $q_i > q_{i+1}$.) Then $\int_0^{q_{4400}} s \, dt =$ 4400s = $-\int_{q_0}^{4400} dq/(1 - q)q^2$. The function $f(q)$ such that $df/dq = 1/(1 - q)q^2$ is $\ln(1 - q) - \ln(q) + 1/q$. The mean value of s is consequently $4400^{-1}[f(q_0) - f(q_{4400})] \frac{.625}{.76}$, or .00035.[45]

In nature, an s of .000106 (or .00035) is *extremely small*. For the well-known peppered moth case s exceeds .3 (Clarke and Sheppard 1966), and Falconer (1989) reports values in the .2 range as typical for laboratory experiments. Operationally, s = .000106 means that, in two populations genetically alike but for a single locus at which members of the first are AA and members of the second aa, 99,989 members of the second population reproduce for every 100,000 reproducing members of the first. Hence, if after reaching Europe just 11 more individuals per 100,000 with an a allele failed to reproduce, gene frequencies would have changed fast enough to raise the mean IQ of the pioneers 27 points in 4400 generations. Equivalently, 4400 generations accommodate observed IQ difference if IQ of parents in each generation exceeds the population mean by just .00082 SD. The races have been apart long enough to permit changes in gene frequencies large enough to produce the observed race difference in IQ.

An independent craniometric argument yields the same conclusion. Beals, Smith, and Dodd (1984: 306, Table 2) report a black/white brain size difference of $1362 - 1276 = 86 cm^3$. From New World fossil evidence they estimate that climatic pressure changes cranial size at 3000 cm^3 per million years. At that rate 110,000 years suffices by a factor of 4 to produce the known race difference in brain size.

4.17. WITHIN- VERSUS BETWEEN-GROUP HERITABILITY

It is agreed that high within-group heritability does not imply a genetic cause for any between-group differences. Genetically identical groups can diverge phenotypically because of environmental differences, as did the Moderates and Athletics in 4.2 with respect to weight. Indeed, positive *within*-group heritability does not imply that phenotypically different *individuals* differ genetically. On average such individuals do, but genetically identical individuals might differ because of exposure to different environments.

However, that high between-group H^2 cannot be deduced from high within-group h^2 does not make the second irrelevant to the first. It is an error to say, as Gould does, "Within- and between-group heredity are not tied by rising degrees of probability as heritability increases within groups and differences enlarge between them. The two phenomena are simply separate" (1981: 157). In fact, the connection between the two is sometimes obvious. Not only is height heritable among the tall Tutsis and the short Eskimos, the height difference between them is also almost certainly genetic. No environmentalist however adamant expects Tutsi babies raised in Alaska to grow up squat.

An inference from nonzero h^2 to nonzero H^2 is natural in the Tutsi/Eskimo case because Alaska and Africa, despite their contrasts, do not seem dissimilar enough to produce a height disparity as great as that separating Eskimos from Tutsi. This is the central lemma: The larger h^2 is for a trait, the greater must be the difference in the environments of two genetically identical groups to produce a given mean group difference. Hence, as h^2 grows, it becomes less likely that any given between-group phenotypic difference can be explained wholly environmentally. Differences in environments accounted for the 15-lb Moderate/Athletic weight gap, but those differences would not have explained 75-lb gap. This is the tie of rising probability between h^2 and H^2.

Suppose the mean Eskimo-Tutsi height difference is .4 SD. To determine how much their environments would have to differ to explain this discrepancy fully, imagine first that the *within*-group heritability of height (for both groups, for simplicity) is .2. This is to suppose that identical twins—or two representatives of genetically identical groups—whose environments differ by 1 environmental SD typically differ in height by .89 SD. Environment explains $1 - .2 = .8$ of the variance in height, hence $\sqrt{.8} = .89$ of any difference in height.[46] Thus, a Tutsi environmental advantage of $.4/.89 = .45$ environmental SD would yield the observed .4 SD mean population difference in height, assuming Tutsis and Eskimos are genetically alike with respect to height. If however the within-group heritability of height is .95, the mean Tutsi environment must be 1.79 SD ahead of the Eskimo in the direction favorable to growth (again on the assumption that for both groups H^2 with respect of height is 0). If Tutsis and Eskimos are randomly distributed across environments, the probability that the Tutsi environment is that much "better" is about .1.[47] Hence, while the hypothesis that the two environments are 1.79 SD apart cannot be rejected, the odds are against it.

Figure 4.2
Explaining ΔP by ΔE

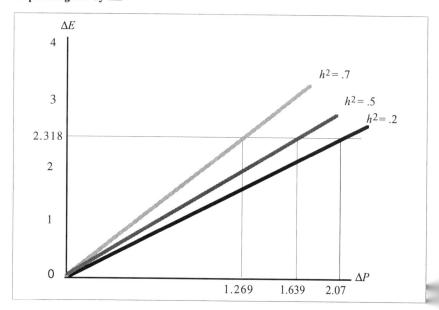

Generalizing this example, let the normalized mean difference in trait P for two populations be ΔP and the difference in mean environments be ΔE. Then if $H^2 = 0$, $\Delta E = \Delta P/\sqrt{(1 - h)}.^2$ Clearly, ΔE increases with h^2. The less likely the value assigned by this equation to ΔE, the less likely H^2 is to be 0. Figure 4.2 shows the connection between h^2 and E for the heritabilities .2, .5, and .7, as the phenotypic difference is plotted against the environmental difference needed to explain it, given that environment explains all phenotypic variance. As 95% of all environments lie within 2.318 SD of each other,[48] the probability that two randomly chosen points are more that $2.318z$ apart is less than .05. Note in Figure 4.2 the horizontal line $\Delta E = 2.318$ and the Δ intercepts. At the confidence level .05, the hypothesis that $H^2 = 0$ is rejected when ΔP reaches 2.07 for $h^2 = .2$, 1.639 when $h^2 = .5$, and 1.269 when $h^2 = .7$. In sum, an intragroup heritability of .7 for IQ makes it almost but not quite irrational to believe that the interracial IQ difference of +1 SD can be completely explained by differences in the black and white environments. The

does rising within-group heritability disconfirm environmental causes of group differences.[49]

Given $h^2 = .7$, an average 1 SD race difference in IQ for genetically identical races—in effect, identical twins reared apart—requires the mean black environment to lie $1/\sqrt{.3} = 1.85$ SD below the white. For $h^2 = .8$, the best estimate for h^2 late in the life cycle, black and white environments must lie 2.4 SD apart. Jensen (1973: 161–165), correcting for the unreliability of IQ tests, estimates the necessary environmental difference to be 3.2 SD. This inverse variation of h^2 with the difference between reared-apart "twins," which in turn varies inversely with IQ changes produced by shifts in environment, is the rising tie of probability between h^2 and H^2. The probability that the races lie 1.85 environmental SD apart is about .1; that they lie 2.4 SD apart is .045, below the rejection threshold; that they lie 3.2 SD apart is .011.[50]

This mathematical argument must of course be tested against data about the actual separation of black and white environments. Citing data from the the 1960s, Jensen (1973: 168–173) found blacks and whites to differ in employment rate by .33 SD, in completing high school by .52 SD, in family income by .8 SD, in children living with both parents by .87 SD, and in living below the poverty line by 1 SD. In a 1982 study (Jensen and Reynolds 1982), the SES difference was .67 of the average of the black and whites SDs. There is no reason to think they differ by significantly more now. Black and white children are about equally well nourished, as I noted, speak the same language, see the same movies and TV shows, and study the same subjects in the same schools. Herculean affirmative action efforts have been made to increase black representation in colleges. Jaynes and Williams (1989: 43–45) argue that the data present a mixed picture, with black educational status rising, but economic status, particularly among males, falling slightly since the 1960s. It is often asserted (by, e.g., Kozol 1992) that black schools are poorer than white schools, but, primarily because of special and remedial programs and the need for psychologists and social workers, public schools now spend more per capita on black children than on white. In Connecticut as a whole, for instance, where 74.3% of the students are white and 12.8% black, the average annual spending per student is $7330, but in Hartford, where 42% of the students are black and 8% white, spending per student is $7937. In Farmington, where 90% of the students are white and less than 6% black, average annual spending is $7194. (Judson 1993; 26. 39% of Hartford students read above remedial levels; the figure for Farmington is 93%, and presumably higher among the whites.) In New York City, "School districts serving poor students . . . get about the same government resources as middle-class districts in the city, even before special state aid for the poor is counted" (Barbanel 1993: B3).[51]

It is important to remember, when comparing black and white environments, that ΔE may be inflated by correlations between genes and ostensibly environmental differences like the .8 SD spread in family income. For instance, if 2.4 SD is the spread required for H^2 to be 0, and r for genes × family income is 0, fixing income reduces the IQ gap by $.8/2.4 = .33$ SD. But genes explain 30% of between-race variance in income, fixing the nongenetic part of income reduces the gap by only .18 SD.

To end this section on the note with which it began, the significance of the within/between distinction has been exaggerated. The higher the within-race heritability of intelligence, the less likely a between-group difference is to be wholly environmental in origin, precisely the situation in the within-group case. If the within-race heritability of intelligence is .7, it is possible but unlikely that two whites who differ in IQ by 15 points do so entirely because of environment, and this possibility lessens as h^2 rises.

4.18. THE FLYNN EFFECT

Flynn challenges the inference from high h^2 to a nonzero probability for nonzero H^2 on the grounds, mentioned in chapter 3, of an apparent rise in the mean IQ of Western populations over the last 60 years (1984, 1987a, 1987b). Because of this increase, he argues, "the mathematics of h^2 estimates can *not* render unlikely an environmental explanation of large IQ differences between groups" (1987b: 229). The argument as Flynn states it is a non sequitur. Suppose that mean IQ in the West has been rising .2 SD per decade, a change unlikely to be due more than marginally to genetic factors. All that follows, when .2 is put for ΔP and .7 for h^2 in the equation $\Delta E = \Delta P \sqrt{(1 - h^2)}$, is that environment as a whole has been changing at .36 SD per decade in a direction favorable to IQ. This figure may be surprisingly, even anomalously large, but it does not disqualify the equation from calculating environmental variation when H^2 is set at 0.

Sowell improves the intended argument. What is said to have "devastating implications . . . for the genetic theory of intergroup differences" (1995) is the corollary of the Flynn effect that contemporary blacks are as intelligent as the whites of two generations ago. Since whites have not changed genetically in that time, it is concluded that the present race gap is probably not genetic in origin:

If race A differs from race B in IQ, and two generations of race A differ from each other by the same amount, where is the logic in suggesting that the IQ differences are even partly genetic? . . . When any factor differs as much from A1 to A2 as it does from A2 to B2, why should one conclude that this factor is due to the differences between A in general and B in general? (35)

That is, if a difference across two white generations is (as it must be) due to changes in environment, a synchronous black/white difference of the same magnitude is probably due to environmental changes as well. Neisser et al (1996) say that "we cannot exclude the possibility" (94) that the black/white difference is to be explained this way; Block (1995) deploys a similar argument, although he vacillates between concluding that the black/white difference is probably environmental, and simply that environment can have significant interactive effects on even highly heritable traits.

But Sowell's conclusion does not follow. The environment shared by black and whites may indeed have so changed as to yield higher IQs for both groups but so far as the Flynn effect goes, the white IQ always exceeds the black i any one environment. Assuming this so, the data, as shown graphically in Figure 4.3, means that the race *gap* is due to genes. As the environment shifts from E_1 at time 1 to the superior E_2 at time 2, the IQs of blacks and whites both rise. But in both environments black IQ is lower than white, so, although the mean IQ of blacks in E_2 equals that of whites in E_1, the race difference in IQ at E_2, Δ_2, is no smaller than Δ_1, the difference at E_1.

One could of course argue that the mean black environment is always two generations "behind" the white, but this interpretation seems forced. Since the three points/decade increase is said to be constant across the Western world (Flynn 1987a), the cause must be very widespread, affecting everyone

Figure 4.3
The Flynn Effect with Constant IQ Differences

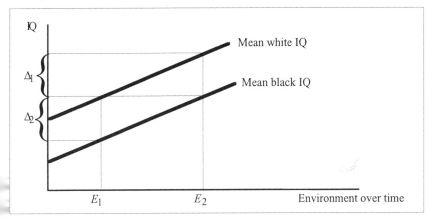

uniformly. It is therefore likely to be touching American whites and blacks, simultaneously shifting the environments of both in a favorable direction. Hence, if the racial difference is environmental, the Flynn factor is somehow being added in roughly equal amounts to an underlying, and stubbornly unchanging, environmental difference. Since, apart from the hypothetical Flynn factor, black and white environments have been converging in other ways, the persistence of the IQ gap makes it more likely that, while neither the white nor black IQ is fixed by genes, the gap may well be.

A related error about gene/environment interaction may appropriately be clarified here. It is often argued that, since genes change slowly, rapid changes in the distribution of a trait cannot be due to genetic factors, and for this reason sudden divergences between blacks and whites, as in criminal behavior or illegitimacy, cannot be genetic in origin. Having just seen that differences preserved across an environmental change common to both races are not only consistent with but suggest a genetic explanation, we can understand how rapid change in a phenotypic difference may sometimes reveal genetic influence. Suppose two groups become more diverse in phenotype P as their environments become more similar: then the differentiation must be due to genetic differentiation, and their prior similarity due to environmental differences. Such a situation is represented in Figure 4.4. At time t_1, genotype G_1 is exposed to environment E_a while genotype G_2 is exposed to environment E_b. Because E_a is less favorable to P than E_b, the phenotypic difference at t_1, namely Δt_1, is small. When the environment is made the same for both, as at E_c, the gap Δt_2 is wider.

Figure 4.4
Cloaked Genotypic Differentiation

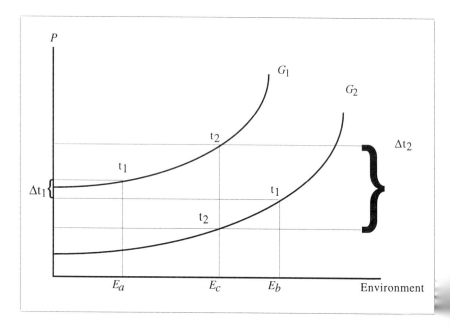

G_1 may be thought of as representing blacks, G_2 as representing whites, and P as representing such traits as criminal behavior, welfare dependency, and illegitimacy. Over the past half-century the behaviors of blacks and whites in these traits have diverged when, as a result of civil rights legislation and a shift in social ethos, the environments of blacks and whites have become more similar. The sudden phenotypic divergencies, far from being environmental effects, evidence a genetic difference previously hidden by dissimilarities in black and white environments. Black crime rates are higher now than in 1920 when blacks faced not only a more daunting social environment, but rules other than those applied to whites. (A black who killed a white, particularly in the South, faced almost certain death.) At present, the same, more lenient rules apply to both. Possibly this equalization of environments has released an innately greater black aggressiveness.

4.19. THE RACE GAP ON TESTS IS PREDICTED BY THEIR h^2.

If mistreatment triggered by skin color is what lowers black IQ, a suggested by Block, Hacker, Jencks, Weinberg, Scarr, and other commentators, black performance on particular IQ subtests should be unrelated to subtest heritability. Racism should depress performance on all subtests uniformly. Yet the race gap on IQ subtests correlates strongly with (within-group) subtest heritability, as determined by correlations between sibling (see Jensen 1973: 107–119; also section 4.7). The more heritable a test is, for both the black white populations, the larger the race gap, a datum which very strongly suggests genetic involvement in the race difference.

This datum also intersects a finding about inbreeding depression, the production of feeble offspring when relatives mate (Rushton 1989a, 1995b; Jensen 1991b). This finding does not decisively refute environmentalism, as Rushton and Jensen appear to believe, but it does underline how far environmentalism must be stretched to accommodate the facts.

Inbreeding depression is an indirect measure of genetic influence. To see why, let the two alleles for trait P be A and a, each with a frequency .5. Let the value of AA be 125, that of aa 75, and that of Aa 110 (so A dominates). The population mean $\bar{P}$ is then 105. Gene frequencies and $\bar{P}$ are constant when mating is random, but the mating of relatives reduces heterozygosity and, by excluding certain combinations, $\bar{P}$ as well. If sibs mate only with sibs, for instance, the 1/16 of the population that is aa from $aa \times aa$ matings in the previous generation will not breed with the 1/16 of the population that is AA from $AA \times AA$ matings in the previous generation, and the 1/256 of the population that would have been Aa from those $aa \times AA$ matings will be lost. In one generation, the frequency of heterozygotes will fall from .5 to .37 and the frequency of each homozygote will rise from .25 to .315, reducing $\bar{P}$ to 104. The invigorating presence of heterozygotes has been partially lost.

Clearly, inbreeding affects heritable traits only, so the degree to which it depresses a trait indicates the extent to which the trait is controlled by genes. Genetic variation explains more of the variance in vocabulary than picture arrangement in a population whose mean for vocabulary and picture arrangement are both 100, but in which offspring of first cousins average 95 on picture arrangement and 92 on vocabulary. Now, there is a robust correlation of .22 between extent of inbreeding depression on IQ subtests and black/white differences on these subtests (Rushton 1989a). The more that performance on a test is under genetic control as gauged by response to inbreeding, the larger the race difference. (Inbreeding depression of test performance was measured on a Japanese population, so the correlation with black/white differences is not built into the measure of inbreeding itself.) Here is strong evidence for positive H^2.

To reconcile environmentalism with the correlation between the size of the race gap and heritability (whether measured by inbreeding depression or a conventional sibling design), it is not enough to assume the races differ on environmental variables that affect traits also influenced by genes. The races must be assumed to differ by varying amounts on these variables, and, in particular, the black/white environmental gap must widen as heritability increases. Environmentalism must maintain that, although blacks and whites are alike in all genetically relevant ways, blacks perform increasingly poorly tests under increasing genetic control because the environmental factors affecting performance on these tests grow increasingly unfavorable to blacks.

Thus, the races would have to be so much further apart on the environmental factors affecting vocabulary than on those affecting picture arrangement that, even though environmental factors affecting vocabulary are relatively less important, the vocabulary gap is greater than the picture arrangement gap. Environmental gaps, in other words, must vary inversely with their importance. Such preestablished disharmony, while possible, is unlikely.

4.20. THE PERFORMANCE OF OTHER MINORITIES

The claim that racism harms black mental development rests on two premises: that subtle forms of racism are omnipresent, penetrating even intervention and transracial adoption, and that racism of this sort retards intelligence. Environmentalists have generally assumed the second without argument, but the superior performance of minority groups that have been less advantaged socially than blacks suggests that it is untrue.

Jensen summarizes the case of American Indians:

> [O]n a composite of twelve SES and other environmental indices, the American Indian population ranks about as far below black standards as blacks rank below those of whites. Within each ethnic group these indices are correlated with IQ and scholastic achievement. But it turns out that Indians score *higher* than blacks on tests of intelligence and scholastic achievement, from the first to the twelfth grade. On a nonverbal reasoning test given in the first grade, before schooling could have had much impact, Indian children exceeded the mean score of blacks by the equivalent of 14 IQ points. . . . Thus the IQ difference between Indians and blacks . . . turns out opposite to what one would predict from the theory that ethnic group differences in IQ merely reflect SES differences. (Jensen 1981: 217; see Jensen 1980: 479 for details)

Jensen notes a similar finding about Mexican-Americans. As is well known Asians and Jews in all cultures have succeeded in the face of sometimes murderous hostility. Jews in medieval Europe were legally confined to ghettos and forbidden to engage in certain trades. Pogroms occurred regularly in Czarist Russia, and the Nazis tried to exterminate Jews, yet Jews dominate the list of Nobel laureates in science and Field medalists in mathematics, are overrepresented by a factor of 10 in American college teaching, medicine and law, and make up 25% of the former Soviet Academy of Science. They have formed resurgent, prosperous communities in Poland and Germany. Before 1930 Jews were a marginal presence in my own field, philosophy, yet by the mid-1990s the majority of the most influential academic philosophers were Jewish.[52] Fischer et al. (1996) make much of (and considerably distort) the fact that some Jewish immigrants to the United States were once thought "dull." However, any period of supposed Jewish "dullness" was quite short, and every society in which they have participated, Jews have eventually been recognized (and disliked for) their exceptional talent. Blacks have not made the transition to brightness in the American context despite great efforts to help them do so, and have never been thought of as elite in any mixed-race culture.

Many environmentalists appeal to culture to annul intergroup comparisons, explaining Jewish achievement in law and intellectual pursuits as a residue

Talmudic scholarship, and Asian academic success similarly, in terms of academic values in Asian families. Yet such explanations sidestep the basic question, namely, why Jews developed a tradition of complex legal reasoning in the first place, and why Asians value academic success. (To say oppression selected for high Jewish intelligence treats group intelligence as a biological adaptation.) Such question-begging is endemic to cultural explanations of group variation: they do not tell us what we really want to know, namely why different groups have developed the cultures they have.[53]

Bolder environmentalists deploy "neutralizing by clear statement"—the rhetorical device of presenting a trenchant objection to one's position with no attempt to meet it, suggesting by this sangfroid that there is a reply too obvious to mention.[54] Thus Hacker's apostrophe to a black:

[Y]ou find yourself continually subjected to comparisons with other minorities. . . . Most stinging of all are contrasts with recent immigrants. You hear people just off the boat (or, nowadays, a plane) extolled for building businesses and becoming productive citizens. Which is another way of asking why you haven't matched their achievements, considering how long your people have been here. Moreover, immigrants are praised for being willing to start at the bottom. The fact that so many of them manage to find jobs is taken as evidence that the economy still has amply opportunities for employment. You want to reply that you are not an immigrant, but as much a citizen as any white person born here. Perhaps you can't match the mathematical skills of a teenager from Korea, but then neither can most white kids at suburban high schools. You feel much like a child being chided because she has not done as well as a precocious sister. However, you are an adult, and do not find scolding helpful or welcome. (1992: 44–45)

Here Hacker drops the subject. Yet, while "scolding" may not be appropriate, the question he seeks to deflect is. If blacks are as innately able and diligent as Asians, why are they less successful? Hacker is pleased to note Asian mathematical superiority to whites as well as blacks (but not that the Asian/black gap is much larger), but the jibe has a point only because mathematical ability is important. That granted, the failure of blacks to match Asians in mathematics suggests an important race difference. Calling the question unwelcome does not answer it.

When preparing this chapter I came across a paper by John Ogbu (Ogbu 1987) purporting to explain in environmental terms why other minorities outperform blacks. This paper was so flimsy that I ignored it, fearing that criticizing it would weaken my own case. Yet I have since encountered several respectful references to it (by, e.g., Fischer et al. 1996 and Neisser et al. 1996), so some discussion of it is necessary.

Ogbu (1987) claims that because blacks unlike other immigrant groups were brought to the United States forcibly, they have no incentive to adopt such American values as hard work in school (325). Blacks distrust the majority culture and see "cultural differences as . . . *markers of identity* [instead of] *barriers to be overcome*" (327, also 340). The most obvious flaw in this theory is that, while slaves were an "involuntary minority," American blacks have been free to emigrate for five generations. During the last century and the earlier parts of this one, the "back to Africa" movement urged by both white activists and black nationalists had few takers. Judging by their behavior, contemporary American blacks are where they are voluntarily. Moreover, like many environmentalist hypotheses, Ogbu's suggests no mechanism by which involuntary minorityhood reduces IQ, nor does Ogbu present any evidence linking inability to "return to the homeland" to academic failure and poor test performance. Finally, the involuntary-minority hypothesis fails to explain the most striking features of the black/white gap: the increase in the gap on ability tests as their g-loadings and heritability increase, and the transracial adoption results.

Fischer et al. (1996) seek to flesh out Ogbu's account by citing other involuntary minorities who commit more crime, do less well academically than the majority population, and otherwise behave in their societies as do American blacks. They particularly emphasize the Burakumin and Korean minorities in Japan, contrasting the record of the latter group with that of Korean-Americans, and also cite Mexican-Americans and Latinos in the United States, Australian Aborigines, Maoris in New Zealand, Arabs in Israel, and low-caste groups in India. Those of their sources I have consulted (particularly Lee 1991, De Vos, Wetherall, and Stearman 1983, and Shimahara 1991) do confirm that the Burakum resemble American blacks, although absence of quantitative information makes it difficult to compare the roles of blacks and Japanese Koreans. (Their more theoretical sources, such as Clark and Halford 1983 and Klich 1988, inadvertently corroborate profound cognitive differences between [in this case] whites and Australian aborigines.)

Now, such comparative data cannot confirm that black difficulties in the United States are due to subordination unless, first, genetic relatives of groups supposedly oppressed in some societies function autonomously elsewhere. Otherwise the direction of causation is unclear; if all members of a group are a minority that performs poorly and enjoys low status, their status might be due to their performance. Maoris, untouchables, Burakum, and Aborigines, who exist only as native minorities, are thus irrelevant to the argument. In addition members of the sometime minority group when on their own must do as well as the allegedly oppressive majority culture. This rules out Arabs in Israel and Mexicans in the United States, since by most measures Arab countries and Mexico are less successful than Israel or the United States The only subgroup that seems to meet both conditions are Koreans, as South Korea is prosperous technological country. However, as Japanese and Koreans generally outscore Caucasians on IQ tests, one would expect Koreans to do well on their own and well in Caucasoid societies, but, as they do not outscore whites by a much as Japanese do, not quite so well in Japan. So the Korean experience does not show low status and low tested intelligence to be joint artifacts.

Most important, the involuntary-minority hypothesis requires blacks on their own to do as well as American whites, and they do not. As we have seen American blacks outperform African blacks on standardized intelligence test and, with the exception of South Africa, until recently run by whites, majorit black countries are the poorest and most crime-ridden in the world.

A measure of the lengths to which environmentalists will go is Fischer al.'s treatment of Puerto Ricans and Mexican Americans, who certainly see to come to the U.S. of their own choice. This is alleged to be a misperceptio the Mexican-American War of 1848 left many Mexicans in the conquere territories. "Later immigrants to the United States, although voluntary, were absorbed into a conquered group. . . . The case of Puerto Ricans . . . bears many similarities to that of Mexican Americans" (176). The authors do not provide the date of America's war with Puerto Rico.

4.21. THE KLINEBERG EFFECT

A major objection to hereditarianism is the relatively higher IQs of Northern urban blacks. The best-known proponent of this argument was Otto Klineberg (1935a, 1935b; also see Kimble 1956: 93), who took the superior performance of blacks in a more hospitable social climate to show that they would achieve parity with whites if treated equally.[55] Klineberg argued that his data were not an effect of selective migration, since the IQ of black youths correlated with the length of time they had lived in the North, as shown in Table 4.8.

Table 4.8
Black IQ in the North

Residence in North	IQ of Black Children in Study I	IQ of Black Children in Study II
1 – 2 years	72	81.4
3 – 4 years	76	84.2
5 – 6 years	84	84.5
7 – 8 years	90	88.5
9 –11 years	94	81.5
Northern born	92	87.3

ource: Klineberg 1935b: 186–187.

Klineberg also took these data to show that the correlation is not due to ybridization, although his controls for this variable may have been ïadequate, since northern-born blacks, who have higher IQs than southern, so have the highest proportion of white ancestry (Reed 1969).

In fact, Klineberg's data do not significantly support environmentalism, for ve reasons. First, they are not longitudinal. No single black child's IQ was)served to increase as he moved north, or with continued residence in the orth. Second, white as well as black IQ increases with latitude, and igration leaves the usual 1 SD race gap intact when state of residence is :ed. Whatever causes the geographical trend does not differentially affect acks. Third, selective migration may have been involved after all. The data, >w 60 years old and gathered from the vanguard of the northward migration, ght reflect the greater ability and initiative of pioneers. Klineberg's control r this possibility, two migrant cohorts in successive years (1935b: 188–189), would have missed longer-term trends.

These methodological points do not disprove substantial equalizing effects for a "Northern" environment. However—fourth—the very low IQs given for blacks residing in the North for five or fewer years in Study I (which used the now-abandoned National Intelligence Test), approximately those of test-wise Africans, may have reflected the new arrivals' unfamiliarity with tests. Setting the true IQ of newly arrived blacks at 81.4 (see Kennedy et al. 1963), as in Study II, the Klineberg effect shrinks markedly; the increase from South to North is then $(87.3 - 81.4)/15 = .4$ SD, about the phenotypic change a 1 SD environmental change would induce if H^2 is .8. Fifth and most important, improvement flattens out as black IQ approaches the familiar −1 SD. Parity of treatment never brings black IQ closer to white.

I believe the Klineberg data does reveal some genuine rise in black IQ as an interaction phenomenon. Black genes express themselves as higher IQs in the North than in the rural South. Environment explains a relatively large portion of the race gap when environments vary from "Southern-black" to "Northern," and H^2 over that range is relatively small. But once the environment gap is closed in the more equitable North, the race difference in IQ expresses itself as a stable 1 SD, and the contribution of genes to the difference is much greater.

The Klineberg so understood data show how equalizing environments tends to raise heritability. In unequal environments, as the environments for blacks and whites in the South were many decades ago, some variance may be due to environmental differences. When environments are the same, more variance is due to genetic differences. In fact, equalizing environments will increase phenotypic race differences if black and white genotypes diverge most markedly at the equalization point. I have already noted that other hypotheses along these lines suggest themselves for other short-term changes in black behavior.

4.22. WHY IS $H^2 > 0$?

Environmentalists who credit race differences to "culture" must give noncircular account of why black and white cultures differ. Hereditarians are equally obliged to give a noncircular account of genetic race differences.

Speculation has long focused on the different pressures exerted by the African and Eurasian climates. Survival in the colder climates of Europe and Northern Asia requires technologies unnecessary in Africa: clothing has to be fabricated, fires sustained, food hunted and stored. These constraints favored the ability to plan, in turn entailing ingenuity and low time preference. Planning is less adaptive in warmer climates where food is easier to get and spoils when stored. (Lynn 1987 speculates that tracking game in snow selected for spatial ability, at which Mongoloids excel.) Hunting also selects more strongly for cooperativeness and reciprocity than does individual gathering and harvesting.

Rushton (1988a, 1991d, 1995b) conjecturally organizes this differentiation around two reproductive strategies. Reproductive-rate, or "r," strategists such as fish produce numerous offspring, few of whom survive. Across species, the strategy associates with lower intelligence, greater investment in reproduction than in postnatal care, short gestation periods, an accelerated life history opportunistic feeding, little interindividual cooperation, lax social structure, and boom or bust population cycles. Carrying-capacity, or "K," strategists, typically large mammals, produce a few offspring in widely spaced litters and care for them long after birth. The K strategy associates with higher intelligence, regular feeding habits, pair-bonding, cooperation, complex social structure, and longevity. Rushton argues that the greater adaptiveness of the r-strategy in Africa made Negroids more r, which would explain the lower mean black levels of intelligence, self-restraint and social organization (as evidenced by failure to form stable political units beyond the tribe, or, in the United States, the gang), and a stronger black reproductive drive as measured by illegitimacy, age of menarche, age of first intercourse, age of first pregnancy, frequency of intercourse, and marital instability. An intriguing phenomenon resistant to environmental explanation emphasized by Rushton is the race difference in litter size: there are 4 pairs of dizygotic twins per 1000 births for Mongoloids, 8 per 1000 for Caucasoids, 16 per 1000 for Negroids. Black infant mortality remains twice that of whites even when social factors are controlled for (Schoendorf et al. 1992).[56] An accelerated life cycle is suggested by the greater maturity of black babies when gestation period is controlled for, their greater developmental precocity, and the constancy of the race difference in life expectancy during the twentieth century.

One need not accept the r/K analysis to expect evolution to diverge in environments as dissimilar as Africa and Eurasia.[57] Knowing only that blacks, whites and Asians differ genetically, conversely, one would expect that they had developed in different environments. The two inferences support each other.

Classical quantitative genetics deduces properties of genes from phenotypes and environments. But just as progress in *in vivo* study of the brain has allowed more direct observation of cognitive activity, progress in mapping the human genome will provide more direct observation of genetic differences. There will someday be detailed knowledge of the molecular biology of human groups, and the ways in which gene-environment combinations produce phenotypes. Assuming investigation is permitted, it will then be possible to determine directly whether, under the same conditions, black and white genotypes produce identical nerve tissue.

NOTES

1. "If from the moment of conception each child's nature helps to determine his nurture, and each child's nurture helps to determine his nature . . . it is not meaningful to calculate what proportion of a child's aggressive behavior . . . is due to his genes or his environment" (Baumrind 1991: 386).

2. See Locke's *Essay Concerning Human Understanding*, II, xxiii, 10. Some things must also have occurrent properties; since a power is the power to bring something about, nothing would exist in a universe consisting only of powers to affect other powers. Consequently, not all substances can be identified with their powers; some must be permitted nondispositional properties. In the present context, genes can be taken powers to create occurrent phenotypes.

3. I ignore pleiotropy, the expression of a gene in several phenotypes

4. Compare Quine (1981: 97–98). Davidson (1970) and others deny brain correlates contextually individuated mental states. For suppose Fred loves Sue. How can

Fred$_2$, an atom-for-atom laboratory copy of Fred, also love Sue, never having met her? The thing to say (Levin 1979: 221–223) is that Fred$_2$'s feeling *is* just like Fred's, but can't be *called* "love" because of the linguistic rule that you can't love someone you have never met. Fred$_2$'s state (for which there may be no English word) *would* have been love had Sue caused it. Whether twin-Earth copies of humans can thirst for water, which has also generated a gigantic philosophical literature, is to be handled similarly.

5. The reader might test his understanding by verifying that genotypic mean = phenotypic mean.

6. The mean is 150; everyone deviates from the mean by 50, so the square of the mean deviation is 2500.

7. The cited passage (Jensen 1973: 162) runs: "While it is true that heritability *within* groups cannot *prove* heritability *between* groups, high *within* group heritability does increase the a priori likelihood that the *between* groups heritability is greater than zero."

8. "Of course," says Noble (1978), "the group differences may not arise from the same sources as the individual differences do."

9. Waldman, Weinberg, and Scarr (1994: 31) claim the present terminology implies that race differences "are entirely genetically based or entirely environmentally based." I hope it is clear that this is not so.

10. If a genotype is conceived as a function, restrict the function to the ensemble and define heritability for the resulting partial function. Bouchard (1995) makes a similar point.

11. Similar points are made in Sesardic (1993b).

12. Tooby and Cosmides in effect assume that all groups have been exposed to identical selectional pressures. They admit the theoretical possibility of "selectively driven quantitative deviations between populations" in "mean arousal, or threshold for anger" (1990: 48), but largely ignore it.

13. The *Essay*, I, 1, 5. In fairness, Locke was concerned with "ideas" rather than behavior, a distinction no longer as sharp as it was thought to be in his day.

14. Bouchard (1990) emphasizes how large trait-relevant environmental correlation due to selective placement would have to be to induce even modest correlations between monozygotic twins. Thus, if r_E = the correlation between environments induced by placement is .5, and r = the correlation between environment and phenotypic IQ is .6, the correlation between monozygotic twins reared apart would be $r_E \times r^2 = .18$. Empirically, r_E tends to fall in the .1–.2 range, and the contribution of placement to MZ correlations < .01. Bouchard replies to other deconstructions of twin studies in (1983, 1984, and 1987).

15. The density function for the absolute difference $|x - y|$ for two independent normally distributed variables x and y is $\iint |x - y| f(x) f(y) dx dy$, f the normal density function; the mean of the distribution of $|x - y|$ works out to $(\sigma_x + \sigma_y)/\sqrt{\pi}$. When $x = $ this becomes $2\sigma_x/\sqrt{\pi}$; when x is normalized, so that $\sigma_x = 1$, the mean is $2/\sqrt{\pi} = 1.12$. See Kendall (1960: 241–242); also see Plomin and DeFries (1980: 22).

16. That $h^2 = $ co-twin/co-twin r follows from the remarks after Tables 3.2 and 3.. Let r_{xy} be the correlation between x and y for P. (a) h^2 for phenotype P is $(r_{G/P})^2$. (b) If is the only factor on which both x and y load, $r_{xy} = r_{xz} \times r_{yz}$. (c) The only factor common to identical twins reared apart is their genotype G. So (d) $r_{P/P}$, the correlation between the phenotypes of pairs of separated twins, is $r_{G/P} \times r_{G/P}$ (e) Clearly, the product $r_{G/P} \times r_{G/P}$ is $(r_{G/P})^2$, and the square of r_{xy} is the proportion of variance in x explained by y. Thus $r_{P/P} = h^2$.

17. Plomin (1990b) and Falconer (1989: 111–184) explain these methods in more detail. Another important survey is Bouchard and McGue (1981).

18. To repeat, these numbers are not correlations, but estimates of h^2 inferred from correlations; see McGue, Bouchard, Iacono, and Lykken (1992, Table 1), for a survey of familial IQ correlations. Goldberger and Manski (1995) criticize as a "misconception" that "taints" Herrnstein and Murray (1994) the idea that heritability is a measure of parent-child resemblance in IQ. Of course it is not, and Herrnstein and Murray never say it is, although they do (properly) predict parent-child correlations from estimates of h^2. Goldberger and Manski are also troubled by the "classical biometrical" assumption—which they accuse Herrnstein and Murray of ignoring (see 764–765, 770)—that a phenotype is the additive sum of genetic and uncorrelated environmental influences. Herrnstein and Murray do indeed make this assumption, but since the environments of monozygotes reared apart *are* uncorrelated, it is satisfied. In any case, I and others emphasize gene/environment correlation primarily to point out that it may cause underestimates of h^2.

19. The influence of genes on environment is inferred from twin studies; ratings of home environment by identical twins reared apart are more alike than ratings of home environment by reared-apart fraternal twins.

20. The unweighted mean of the black and white populations is 92.5, with an SD of 7.5. $(5/7.5)^2 = .44$.

21. Unless excluding one factor somehow raises the probability of involvement of each remaining environmental factor.

22. See Rushton (1995b: 113–133) for an extended discussion of this material.

23. Gould (1978) accused Morton of underestimating the size of Negroid crania from an unconscious desire to prove white superiority. Michael (1988) has vindicated Morton's measurements, and shown that such errors as he made underestimate Caucasoid/non-Caucasoid differences. Gould admits (1981: 66) to unconsciously underestimating the size of the Caucasoid crania in Morton's sample, presumably from a desire to prove racial equality.

24. With a constant of proportionality 8×10^7, by eq. 3.15 of Jerison (1973: 79).

25. Tobias (1970), often cited as having refuted race differences in brain size, does no such thing. Tobias cites control failures for body size, cerebro-spinal fluid (CS), and other variables in the race $\times$ brain size studies h153e surveys, but he presents no evidence that Negroid and Caucasoid brains are equally massive when these factors are held constant. In fact, even the studies he faults are meta-analytically significant. Distortions like body size and CS fluid, being unrelated to race, should cancel out. The odds are small that differences would persist when sources of error vary randomly.

26. Weinberg, Scarr, and Waldman largely attribute the uniform decline in IQ scores from 1975 to 1986 to "test revision and test norm obsolescence," citing Flynn's studies documenting an average decline of seven points in scores from the WAIS to the WAIS-revised. The race gap has remained fixed across test revision, and test revision is controlled for in the Scarr-Weinberg study. The significance of the Flynn effect is discussed in section 4.18.

27. Levin (1994) defines the unit of d as the weighted pooling of the SDs of all four cohorts. Other possible units are the square roots of the means of the weighted or unweighted variances. One can apply t-tests (such as those in Cohen 1983) directly only to d's expressed in variances; I follow Weinberg, Scarr, and Waldman in using a computing formula based on the (pairwise) mean of the SDs.

28. Perinatal privation not caused by whites will not sustain the moral claim that whites have damaged blacks. Yet the earlier the onset of an environment impediment to black IQ, the less likely it is to be due to whites.

29. The hypothesis that many blacks pursue entertainment as a way out of poverty similarly fails to explain the distinctive characteristics of black comedy and popular music. Earlier in the century Jews flooded popular entertainment, but their music, comedy, and films were unlike blacks'.

30. Colleges are often accused of "exploiting" blacks by luring them with scholarships and feeding hopes of a professional career, when in reality blacks often drop out, and relatively few make the pros. In truth, most people would appreciate that sort of exploitation for their children. Black athletes are given scholarships to colleges to which they would otherwise have no hope of admission.

31. The sum of the sample sizes for the ten studies for which size is given is 5352.

32. Much criticism of Lynn's work has focused ad hominem on its appearance in *The Mankind Quarterly*. However, Lynn (1978) was brought out by Academic Press, a most reputable scientific publisher.

33. A black Zimbabwean, resolving the issue of tester bias.

34. Since the scores and distribution of errors for the 15-year-old Ethiopians resembled those for 9-year-old Israelis, Kaniel and Fisherman interpret the gap as "a developmental delay, and not a different cognitive style," and express confidence that the Ethiopians can " 'catch up.' " Since the Ethiopians are near or at the end of the period of intellectual development, this optimism is curious.

35. While the mean IQ of Asians exceeds that of whites, the Asian variance may be smaller (see Vining 1983), leading W. Block (1993) to suggest that there may be fewer Asian geniuses of the sort responsible for scientific breakthroughs. Sociological explanations of Caucasians' superior inventiveness—for example, that Asian societies are conformist—beg the question of why Asian societies exhibit traits hostile to innovation.

36. Draper explains these family patterns as a response to a perceived abundance of food and space in Africa. Curiously—for her paper was published in a journal entitled *Ethnology and Sociobiology*—she does not consider that this abundance might, over time, select for such responses.

37. Hart's and Asimov's rankings are two men's opinions, but their reader will probably find himself concurring. Historical importance is discussed in chapter 6.

38. Baker also presents craniological evidence that Egyptians were not sub-Saharan (1974: 517–519).

39. Japan also did not contribute to the development of science, but since encountering science in the nineteenth century it has become a world leader in technology.

40. Some African problems may be due to the socialism instituted after decolonization. Jews have very high IQs, but the quasi-socialist Israeli economy is in shambles. The former Soviet empire and China have talented populations, but the Eastern Bloc was near collapse when it abandoned Communism, and China remains poor. At the same time, a free market only flourishes under democracy, and the inability of Africans to sustain democracy, perhaps connected to a strong individual dominance drive, may make command economies inevitable in nations with black populations. In that case the effects of a command economy would be genetic correlates.

41. This figure cannot be taken uncritically. Reed (1995) points out that fewer than 1% of the functional gene loci in man have been identified and used in population surveys.

42. Where σ is the coefficient of selection and p its initial probably, Loehlin, Lindzey, and Spuhler (1975: 270) approximate the number of generations required t change p from .01 to .99 as $2[-\log_e(p/(1-p)/\sigma)]$.

43. I assume that the m in the second offset equation on 270 has become a in the equation $g = ah2$.

44. The basic idea is that the phenotypic expressions of genotypes that differ in being aa and AA at one locus are normally distributed with means $2/15 = .13$ S apart. Selection of intensity i ensures that only some upper segment of both curves will reproduce. The ratio of these surviving segments is directly proportional to i and the mean difference of .13. But this ratio is just s. See Falconer (1989: 201–202); I ha

given his a the simplifying value 1 and normalized his σ_p to 1.

45. Ethan Akin has pointed out that the variances for IQ become too small, in the 4–5 range. These variances can be enlarged by increasing the number of loci and assigning q extreme values. Relaxing linearity induces still more plausible values. One hopes better-fitting models will be proposed.

46. Let e be the proportion of an individual's deviation from the mean due to environment; if A lies at kz and B at $(k + 1)z$, environment explains ek and $e(k + 1)$ respectively, hence $e\%$ of their separation.

47. For x and y independent normally distributed variables with SDs σ_x and σ_y, the mean σ_z for the distribution $z = x - y$ is $\sqrt{\sigma_x{}^2 + \sigma_y{}^2}$, so, when y is x, σ_z takes the value $\sqrt{2}\sigma_x$ (Mosteller, Rourke and Thomas 1970: 345–348). When x is normalized, as here, $\sigma_z = \sqrt{2}$. Ninety percent of all pairs of environments are within $1.79/\sqrt{2} = 1.265$ of each other. I do not use the distribution $|x - y|$, since the question is not how different the white and black environments are, but how much *worse* the black environment is.

48. As σ for the distribution of pair differences is $\sqrt{2}$, the position of the point pair $<x_1, x_2>$ is $(x_1 - x_2)/\sqrt{2}$. 95% of the points on any normal curve lie to the left of $z = 1.64$. $1.64 = (x_1 - x_2)/\sqrt{2}$ implies $x_1 - x_2 = 2.318$.

49. DeFries' (1972) quite different formula for computing H^2 from h^2 requires the groups being compared to occupy the same range of environments. The question relevant to race is how different the black and white environments must be for $H^2 = 0$.

50. Note that $1.85/\sqrt{2} = 1.3z$ in the difference distribution, $2.4/\sqrt{2} = 1.69z$ and $3.2/\sqrt{2} = 2.26z$.

51. To anticipate the topic of chapter 8, there is no obvious *injustice* in more money being spent on white children than blacks, so long as the money is spent by whites, because whites are wealthier than blacks. The 12.8% of Connecticut's population that is black does not pay 12.8% of the taxes supporting its schools. That anything like the same money is spent on black as white children should be described as white generosity.

52. After 30 years of affirmative action, blacks constitute about 1% of the profession, mostly pursuing marginal topics related to race.

53. Chapter 5 considers the charge that explaining culture via biological variables is "reductive."

54. Hacker's remark that "Certainly, compared with other continents, Africa remains most like its primeval self" sought to neutralize the bearing of African achievement on hereditarianism.

55. Klineberg testified to this effect in *Brown*.

56. Black health problems are usually attributed to anger and stress over racism; see, for example, Goleman (1990). It is, however, known that blood pressure varies inversely with Caucasian ancestry (Maclean et al. 1974), suggesting a nonsocial cause for race differences in circulatory complaints.

57. Beals, Smith, and Dodd (1984) find thermoregulation the cause of the climate/head size (and brain size) nexus: rounder crania maximize the volume/surface area ratio, hence conserve heat better and are more adaptive in the cold. This theory does not, however, explain the difference in dizygotic twinning.

5

Determinism, Reductionism, Reification, Racism

Explaining human behavior via genetic factors commonly provokes charges of "genetic determinism" or "biodeterminism," and is said to be reductive (Lewontin, Rose, and Kamin 1984). Biology is supposedly the wrong place to look for the causes of social phenomena. Mention of biological race differences in particular is apt to be called "racist." Since much of what follows hinges on a biological approach to social behavior, these charges must be considered.

5.1. "BIOLOGICAL DETERMINISM"

Determinism is the thesis that everything has a cause, or equivalently, that everything can be explained, or again, that every event falls under some natural law. Were God to return the universe to precisely the state it occupied at the Big Bang, determinists hold, all would unfold just as before. Put literally, determinism maintains that, for every variable x, there are variables $x_1, \ldots, x_n$ such that the value of x at time t is a function of the values at some earlier time t' of $x_1, \ldots, x_n$.[1]

Determinism is less restrictive than it might at first appear. For one thing, it in no way restricts the variables $x_1, \ldots, x_n$ on which a given variable x may depend. A determinist may regard social effects, in particular, as determined by any combination of factors whatever—biological, environmental, familial—as long as there are some factors which determine these effects. Determinism is not inherently "biological." For another, calling x a function of $x_1, \ldots, x_n$ does not deny that other factors are at work in the world. What determinism says is that, once values for suitable x_i are fixed, only one value for x is possible regardless of the values of any other variable. Thus, the claim that genes and physical environment determine phenotypic intelligence no more repudiates the social environment than it repudiates quasars.

Most scientists outside the field of quantum mechanics, and reflective nonscientists, are determinists. We all presume there are reasons for everything, including human behavior. Environmentalist accounts of behavior are as deterministic as hereditarian, for they assume, as do hereditarian accounts, that identical initial conditions yield identical results. The difference between environmentalists, social environmentalists and hereditarians is that environmentalists take the conditions relevant to behavior to be exclusively environmental, social environmentalists take only social factors to be relevant, while hereditarians include biological factors as well. If being deterministic is a flaw, it mars all accounts of social phenomena.

An easier target is "biological" determinism, the thesis that social phenomena are functions of biological variables only, so that, in particular, race differences depend on biology alone, to the exclusion of environment. The boldest form of biological determinism, namely "genetic" determinism, holds that, given an ensemble of genotypes, only one society can emerge from them whatever the environment.

Biological and genetic determinism are indeed implausible, but they are not implied by the view that genes contribute significantly to race differences. They are not even implied by setting $h^2 = 1$ or $H^2 = 1$. So long as genes interact with environment, social phenomena, including those involving race, depend on both genotypic and environmental variables. No sensible position—including the one adumbrated in chapter 4 that the character of a society is determined by the expressions of the genotypes of its members in a given physical environment—holds that outcomes flow from genotypes alone; to explain a society, the (physical) environment in which the genes of its members develop must also be specified, a tenet that might be called "bio-environmental determinism." By way of comparison, the principle that gravity controls all motion near the Earth does not mean the final velocity of a falling rock follows from the law of gravity alone. The rock's final velocity also depends on the height from which it was released, a factor independent of gravity itself. A physical determinist need not be a "gravitational" determinist.

"Genetic" and "biological" determinism are thus straw men. The traits of a society are "genetically determined" only when they are functions of genotypes with flat reaction ranges expressed as phenotypes whose heritabilities are 1. If everyone in group G_1 in Figure 5.1 has the same genotype for intelligence, and this genotype expresses itself as IQ 80 in all environments, genes alone dictate a mean group IQ of 80. Contrast G_1 with genetically uniform group G_2, in which the heritability of intelligence is also 1, but whose reaction range stretches from 90 to 100. Although h^2 is also 1 for G_2, its mean phenotypic IQ depends on its mean environment. Likewise, genetic determinism can be false even when genetic variation explains all group differences in a phenotype. Suppose H^2 for groups G_2 and G_3 is 1 but the reaction range of the genotype of G_3 is 110–140. In both environments E and E' the mean $G_2 - G_3$ difference is due wholly to genes, but the size of the difference varies with environment. $G_3(E) - G_2(E) \neq G_3(E') - G_2(E')$. For a group difference to be "genetically determined," h^2 and H^2 must both be unity and the reaction ranges of the genotypes of both groups must be congruent, as are those of G_3 and G_4; here $G_3(E) - G_4(E) = G_3(E') - G_4(E')$ (again see Figure 5.1).

Although broader than biodeterminism, bio-environmental determinism is narrower than determinism per se. Holding the characteristics of a society t depend on the genotypes of its members and the physical environment excludes culture itself as an explanatory variable, a ban prompted by the patent circularity of using cultural variation to explain variation in cultures. Baker illustrates this circle in connection with numeracy. The Arunta Australids, according to an anthropologist he cites, did not develop counting because "The Arunta . . . has no need of a system of numbers comparable to ours. He possesses nothing that he must necessarily count, no domestic animals, no merchandise, no money." "But *why*," asks Baker, "had they nothing to count? *Why* were they content with this situation?" (1974: 527). When in the same vein it is asked why the Greeks invented science and philosophy, bio-environmental determinism rejects "Their culture encouraged speculation" as a non-answer. Viewing culture as a phenotype concedes the impact of abiological contingencies like disease and encounters with other groups, taken as features of the environment in which a population develops. But bio-environmental determinism denies that culture itself is an abiological contingency, and holds open the likelihood that group responses to historical accidents—for instance, which of two warring groups will dominate the other—reflect evolved presocial dispositions.

Figure   5.1
Genetic   Determinism   Almost   Always   False

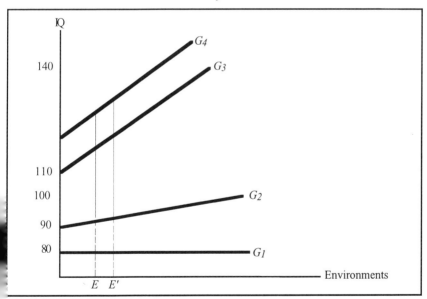

Gene/environment accounts of culture are noncircular because physical environment—climate and natural resources—are independently explainable, by geophysics and evolution respectively. These factors may be used to explain culture because they are known to exist and have causes other than culture, just as the egg may be said to precede the chicken when there is an account of eggs independent of chickens—for instance that the first chicken egg came from a protochicken. Had science been bestowed on mankind by extraterrestrials á la *2001*—a seemingly nonbiological source—we would still want to know why the extraterrestrials but not humans were able to develop science on their own. The answer would likely be an endogenous difference between us and them. One weak form of purely biological determinism may be tenable, however. Look again at Figure 5.1. While the phenotypic values of G_2 and G_3 and the value of $G_2 - G_3$ both depend on the environment, $G_3 > G_2$ throughout the indicated range. In such a case the environmental variable does cancel out, and we can say that $G_3 > G_2$ is determined by genes alone.[2]

5.2. WHY GENETIC EXPLANATIONS ARE DISTURBING

A puzzling aspect of the race issue is the depth of feeling provoked by genetic explanations. People "obsess about nature versus nurture," chide Herrnstein and Murray (1994: 131), a compulsion they trace to the erroneous equation of "genetic" with "unalterable." This diagnosis is surely superficial; practical worries about the difficulty of raising black intelligence cannot explain the rage and disdain mere mention of genes can provoke. The diagnosis that talk of genes evokes Hitler does more justice to the emotional tone of the debate, but it does not explain why ordinarily acute individuals—capable, one would think, of seeing that hereditarianism is one thing, the Third Reich another—lose all perspective when genes are mentioned.

The true cause, I suggest, is a felt connection between genes and personal identity. Most informed persons in the late twentieth century conceive genes as part of a person's essence, what makes him what he is. Flaws due to genes are consequently considered constitutional in a way that phenotypically indistinguishable flaws due to environment are not. Some such thought must be imputed to anyone who calls talk of genetically lower intelligence an insult to blacks yet insists that blacks do poorly in school because deprivation stunts their mental growth, since the second hypothesis as much as the first implies that blacks are less intelligent than whites and is therefore no less negative. It is not simply a case of people ignoring unwelcome implications of their own words; in denouncing genetic explanations while embracing environmental ones, people are responding to a difference between them. The difference is this: if a child, neonate, or fetus suffers brain damage from an environmental cause like malnutrition or trauma, it makes sense to say that *he* might have been more intelligent, and that *he* would have been more intelligent under other circumstances. But if his neurological condition was in his genes, *he* could not have been more intelligent, even with different genes, since different genes would not have produced *him*. Genes are of the essence of personal identity, so talk of a genetic black intelligence deficit seems to ascribe a defect to the black *essence*. Most people are reluctant to draw so harsh-sounding a conclusion.

This fear could be allayed by decoupling genes from identity, but the connection is tenacious.[3] The average person would almost certainly agree that *he* could have been born with different arms or legs. He would probably agree that, if the very sperm and egg that produced him had met in a different place and time, and the person they produced went on to lead a life wholly unlike th life he has actually led, this person would still have been *him*. Finally, the average person would probably agree that, had the sperm and egg that actually produced him never fused, but another zygote in his mother's womb had grown into an adult with his appearance and personality, that person would *not* have been him. Judgments of identity track genotypes.

Kripke (1973) compares the relation of a person to his genes to that between a physical object and its constituent atoms. Once your kitchen table has been destroyed, a table made of different atoms, no matter how similar, cannot be that very table redivivus. Judgments of physical identity track atomic constituents. As to why identity works this way, Kripke (1973) considers these relations "metaphysically necessary," while empiricists suspect they are by-products of linguistic convention (Ayer 1982, Levin 1987a). A conventionalist analysis of identity intuitions might draw some of the sting from genetic differences in highly valued traits, but as things stand the widespread albeit inarticulate belief that genetic traits (hence genetic flaws) are essential is one more disturbing aspect of race.

5.3. REDUCTIONISM I

The question now shifts from the existence of laws connecting (individual) genotypes to phenotypes, to the possibility of explaining social phenomena via the genetically determined phenotypes of individuals. The case against this "reductionist" goal rests on the idea that social phenomena are autonomous, explanatorily and ontologically. Thus Natalie Bluestone:

All biological explanations of culture suffer from the impossibility of reducing an independently valid domain with valid explanatory principles of its own to what are considered more basic principles. But just as it does not make much sense to "explain" that the table represented in Cezanne's painting "The Card Players" is "really" only a system of electrically charged particles, it does not make sense to assert that women do not paint great pictures for biological reasons. (1987: 189)

This claim is immediately suspect, since it seems perfectly sensible to say that, for biological reasons, no fat man has ever won a marathon—an event in the "domain" of culture—or that, for biological reasons, chickens do not paint great pictures.[4] So far as I am aware, though, the anti-reductionist literature does not respond directly to concrete examples of successful reduction. Rather, it presses two abstract arguments.

The first (Winch 1958) points out, correctly, that social phenomena are specified by *rules* and *norms*. What counts as a crime, for instance, is defined by legislative decision; there is no physical characteristic common to all crimes. A hand movement may constitute forgery in one social context, yet a physically similar movement in another context—as that of an actor in a historical drama writing "John Hancock"—may be licit. Conversely, any hand movement whatever could be wrongful in some context or other (see Fodor 1974). But only physical traits can be programmed genetically, the argument continues, so "gene-based" and "social" categorize behavior in disparate ways. This mismatch between social and biological categories prevents biology from explaining social behavior like crime. There can't be a gene for criminal hand movements because genes cannot determine or anticipate external norms.

An analogy may clarify this argument. Multiplying is an abstract mathematical operation defined by the rules of arithmetic, not a physical process. You can't learn it by watching what happens inside a computer. Trying to understand society by studying genes, say anti-reductionists, is like trying to learn to multiply by looking inside a computer.

As might be expected, this first argument is too good to be true. In the first place, while in principle "crime" does not coincide with any one type of bodily movement, *in practice* certain movements are almost always criminalized, and many criminal acts consist of specific movements. The physical act of striking another human being with one's fist is criminalized in every society with a formal legal code, and informally sanctioned in all others. Readiness to punish such behavior is apparently found even among nonhuman primates (Ellis 1987a). Striking another human being is acceptable on occasion, as a boxing match, but a disposition to throw punches eventually violates the norms of any group. And, since genes can program behaviors like punching, there can be genetic explanations of crime consisting of such behavior. Thus, while it may make no sense to speak of a "crime gene" or to call blacks genetically more crime-prone than whites, it does make sense to say that blacks are genetically more prone to behavior that is in fact criminalized in virtually all societies.

In terms of our analogy: while "multiplying" cannot be defined in terms of computer hardware events, certain hardware events characteristically *amount to* multiplying. Once you understand what is going on inside a computer, you can see how it embodies multiplication (how, in particular, some series of events in the computer parallels the steps in the abstract multiplication program). In this way, a seemingly mathematical phenomenon, computer multiplication, can be "reduced" to physics.[5]

In the second place, the genes of more developed organisms program learning as well as specific behavior. A pitcher plant's genes simply tell it to secrete digestive fluids when its cilia are touched, but a squirrel is told to return to where it has previously found acorns. Human genes set even more sophisticated learning tasks, such as determining what skills impress local females. Since human genes can program learning, there is nothing to keep them from programming an impulse to identify, and refrain from, locally proscribed behaviors.[6] And if there can be innate impulses to identify and follow rules, there can be biological explanations of individual and group differences in the strength of those impulses. It therefore becomes meaningful to conjecture that an innate tendency to follow certain rules, or rules in general, is stronger in one group than another. This offers a precedent for reducing race differences in social behavior to genetic differences.

5.4. REDUCTIONISM II

The second major anti-reductionist argument runs that social facts concern relations between individuals that transcend their individual traits, and since genes program only individual traits, they cannot explain social facts. A standard example: the behavior of a mob emerges from the interaction of its members and cannot be predicted from the traits of those individuals alone. Indeed (the argument goes) social roles shape their occupants: people are the way they are because of their places in society, not vice versa, leaving the heritability of phenotypes doubly irrelevant to social behavior.

Although this argument does not appear especially relevant to race, its application turns out to be quite far-reaching.

Once again, the factual basis of the argument is incontestable. Collective behavior does transcend and constrain individuals; an isolated Robinson Crusoe cannot vote or be rude, and a normally well-behaved man caught up in a riot may smash a window because of his proximity to other rioters. Moreover, social groups like mobs have lives of their own, persisting despite the loss of individual members. Yet, while a mob's existence is independent of any one of its members, it is not independent of all of them. Once all the rioters disperse the mob perishes. In contrast, the existence of an individual does not depend on any of the groups he belongs to, or all of them together; a mob's erstwhile members survive its dissolution, and all men could live, albeit unhappily, as isolated Robinson Crusoes. So social phenomena depend ontologically on individuals in a way that individuals do not depend on social phenomena. And social phenomena depend on their individual constituents in a further sense: if all the individuals in one society behave exactly like their counterparts in another, the two societies being isomorphs, the same social facts will be true in both.[7] Group properties are thus "reducible" to the properties of individuals in the strong sense that the traits of a group are functions of the traits of its members.

Furthermore, because the traits of a group depend on the character of the individuals composing it, not just any individuals can generate just any social fact. No matter how many penguins congregate on an ice floe, or for how long, they will never form a constitutional monarchy. *It therefore makes perfect sense to seek to explain a group phenomenon via the characteristics of the group's constituent individuals, and equally good sense to suppose that some of those characteristics are genetically influenced.* There can therefore be two-step genetic explanations of group phenomena: from genes to individual traits, and then from individual traits to group traits. This is the structure of genetic explanations of cultures, and cross-cultural differences—in particular, differences between cultures composed of different races.

The belief that society is "emergent" confuses two kinds of trait an individual may possess: what he is like when he is alone, and how he relates to others. This confusion plays itself out in the following way. Emergentists and reductionists agree that society consists of people standing in social relations (see May 1987); their disagreement, or apparent disagreement, concerns the basis of these relations, why they hold. Were reductionism the claim that social facts follow from individual traits of the first sort—that the nature of society is predictable from the properties people exhibit when alone—it would obviously be false. How someone reacts to others cannot be inferred from how he acts in isolation. But social facts may still follow from an inventory of individual traits that includes relational ones. An individual's behavior in a mob *can* be inferred from his individual properties, when those properties include his tendencies to react to others. If one of Smith's relational traits is that he starts throwing things when others around him do, his throwing rocks during a riot is a consequence of (one of) his individual traits. It is this sort of reduction of social to individual behavior that sophisticated reductionists aspire to.

Sophisticated reductionism recognizes "intrinsically social" phenomena. It can accept Greenwood's suggestion that these are phenomena "constituted by or constructed out of—and maintained and sustained by renewed and fresh commitments to—arrangements, conventions and agreements" (1994: 95), and that "intrinsically social" psychological explanations are those that refer to the "recognition and acceptance" of such conventions. Reductionists insist, only, that recognition of a convention or an agreement is an individual (relational) psychological state, and that such states explain the tenure of conventions and agreements.[10]

It would be cheating to classify reactive tendencies as traits of individuals rather than as traits which "emerge," were that distinction well drawn. But it isn't (see Hempel 1965a, Levin 1989c). Just as a sugar cube in splendid isolation on the Sahara is disposed to dissolve in water, a mute castaway on a desert island may be disposed to chat in company. That disposition is as intrinsic to him as his height, needing others to trigger it but not to exist. Whether social behavior "manifests latent individual traits" or "emerges in relations" collapses into a verbal issue.

The arbitrariness of the property/relation line goes to the heart of biosocial explanation. No one doubts that genes influence the traits of individuals, so once it is understood that these traits include tendencies to form social relations, it becomes clear that social relations too can be influenced by genes. The point, although perhaps not this terminology, is doubtless already familiar to the reader. He has probably heard of Konrad Lorenz's week-old ducklings programmed to "imprint" on the most salient figure in their environment (in the wild usually their mothers), forming thereby a life-long attachment. In other words, ducklings are genetically disposed to form a rudimentary social relation. Genetically programmed dispositions to form human social relations become an obvious possibility that cannot be ruled out by a priori arguments.

5.5. REIFICATION

Since biological reductionism accommodates social relations, it denies nothing that plainly exists. But anti-reductionism, which treats society as autonomous, often affirms what plainly does not. "Society" is a collective name for interacting individuals and the upshot of their actions, perfectly harmless so long as it is not taken to *explain* those actions. But taking "society" to name a thing in its own right yields pseudo-explanations of a sort that seems particularly common when the topic is race.

Take this sentence: "The country has watched passively for more than a generation as its urban cores have devoured the people who live there" (Deparle 1992). Its author is treating the place where blacks reside as an entity with causal powers (which, it is implied, whites have negligently failed to restrain). Yet, speaking literally, urban cores devour no-one. When different people lived in those places, there was no "devouring"; blacks there now are being victimized by other predatory blacks. Or consider the oft–heard remark that black children misbehave in school because they "bring the streets with them." Black children don't literally bring pavement into the schoolroom, nor the values taught by pavement. They bring the values of the children who live on those streets; which is to say, their own. The equally familiar diagnosis of black teenage crime and destructiveness as a product of "peer pressure" is also reification, disguising the very phenomenon that needs to be explained—the source of the malign "peers," who are, after all, merely other black youths. Peer groups are not things in their own right. Accounts attributing the disadvantages of blacks to life amidst "crime," "poverty" or "drugs" are equally obscurantist. "The Carter family is being stalked here by what the clan's 54-year-old matriarch, Regina, calls a monster—crack cocaine. She has watched it swallow her daughter, and now she is fighting it for her grandson's soul" (Terry 1995: A12) is so luridly dishonest as to need no further comment.

The essential emptiness of these reifying explanations is distilled in the following brief description of Chicago's Robert Taylor Homes, a virtually all black public project of 60,000:

Garbage, human waste and graffiti have reduced the project to a high-rise hovel. For the Taylor residents who are trying to provide proper homes for their children, the deterioration of the buildings is both maddening and frustrating. "We complain," one mother notes. "Complain to whom?" (Ruth, 1989)

Use of the active voice, abetted by focus on nondestructive residents, represents filth as a natural force. This force wreaks havoc, such wording suggests, because the Chicago Housing Authority is not stopping it.[11] In reality, of course, the filth is produced by Taylor residents who urinate and defecate in its hallways—not all the residents, to be sure, but sufficiently many to create the conditions described. No doubt these conditions disinhibit further destructive impulses, establishing the feedback loop characteristic of genotype/environment correlation.[12] Still, the ambience of public housing cannot explain the behavior of Taylor residents, since the behavior explains the ambience.

Scholars as well as journalists make similar mistakes. Fischer et al. (1996) propose the vacuous "community conditions" (18) as a way to explain poor black outcomes. Like a great many other writers, they assert that segregation "reduced living standards and housing quality" (180), without explaining how the sheer proportion of black people in a region can effect housing quality unless black people either individually or collectively tend to squalor. Most absurdly, they explain the educational difficulties experienced by blacks in all-black neighborhoods in terms of greater exposure to crime and lower levels of safety (see 196), as if crime were a force independent of blacks themselves.

Reification also obscures the causes of impoverishment in areas left by whites, with concomitant job growth in white suburbs, often ascribed to whites "taking jobs with them." One writer describes the feelings of "the minority community" in Connecticut this way:

Some complain that they were bypassed by the wealth that surged through Fairfield County in the 1980's, giving rise to sprawling office parks and roadways crowded with luxury cars. . . . For tenants of Carlton Court, a privately owned project that is almost entirely black and Hispanic, a telling sign of that disparity is the waterfront district. . . . "When you live in a place like Carlton Court, there is a lot of wealth and opportunity around you, but you can't partake of it," said Robert Burgess, executive director of Norwalk Economic Opportunity Now. (Levy 1993)

"Jobs," "wealth," and "opportunities" are here treated as things that surge through neighborhoods, build roads, and magically produce luxury cars, and that unnamed (but easily identified) forces prevent blacks from "partaking of."

This is a muddle. The occupational tasks characteristic of a group are determined by its members' abilities and preferences, and can no more exist apart from it, or be "left behind" when the group leaves, than can the social structure of a wolf pack stay in place when the wolves migrate. Certain abilities allow performance of certain highly valued tasks. When an ability is so widespread that a group comes to rely on its being exercised, and those with the ability come to rely on its exercise being rewarded, the resultant feedback creates an occupational role. The "job" of whittling darts presumably took hold among the Yekuana of South America because enough Yekuana could whittle darts for other Yekuana to hunt monkeys with blowguns, and for each generation of craftsmen to train the next. The Yekuana could "leave" this skill by teaching it to any group that replaces them, should their replacements have sufficiently many craftsmen and a taste for monkey meat. But dart-making does not exist in the jungle apart from the Yekuana, and is not, like a bar of gold, physically transferable. If the Yekuana are replaced by fishermen, they will not have "taken" anything they could have left. It would be absurd for a community of fishermen near the Yekuana to complain that hunting was "surging" through the Amazon Basin but bypassing them.

White flight "takes" jobs in the sense that tasks characteristic of whites cease being performed in the absence of whites. There is no reason to expect these tasks to be performed, in light of the differences between black abilities, preferences, and levels of persistence and those of the group that created them. When colonial whites left Africa, the roads they built began to disintegrate in the absence of engineers to maintain them. Detroit fell into similar disrepair when whites left (see Chafetz 1990). The reified entity "Detroit" ceased manufacturing automobiles because too few of the blacks who remained were able or inclined to manufacture cars on their own.

5.6. "RACISM"

"Racism" is a Janus word (see section 3.8) whose evaluative face predominates; calling someone or something "racist" automatically condemns him or it. In fact, the fierce emotions accompanying "racism" suggest that its core meaning is "grossly improper race consciousness." Yet at the same time "racism" is freely used of an enormous range of beliefs, attitudes and practices, many of which seem in no way grossly improper, or improper at all. That is why the word serves only to obscure.

The chief problem the word creates is that of begged questions. Precisely because things racist are bad by definition, it is tempting to try to force condemnation of an attitude or practice by labeling it "racist," when in point of logic the attitude or practice in question must *first* be shown to be bad by some independent standard *before* it can be so labeled. In legal language, "racism" is conclusory, and cannot be used as a premise. Yet, because incessant denunciations of "racism" has made the epithet unchallengeable, that is often just how it is used.

Natural Janus-words, which inherit their evaluative force from a social consensus about the value of their referents, do not lend themselves to this kind of abuse. Since everyone agrees and is known to agree about what sort of butter is fit to eat, no one would try to condemn perfectly fresh butter by calling it "rancid." But "racism" as currently used did not inherit its negative force from a universal dislike of its referent. It might once have denoted Hitlerian racial beliefs while also encoding rejection of those beliefs—and when it did, less egregious racial offenses were called "bigotry" or "prejudice." In that usage, racism was a systematic *theory*; this theory entailed certain attitudes and modes of behavior, but those attitudes and behaviors did not by themselves constitute racism. By contrast, today's "racism" was coined *for the purpose* of condemnation (in part by summoning up emotions evoked by the old word), and for the condemnation of anything belonging to almost any category. I trust the reader will agree that all he can conclude when he hears "racist" employed today is that its referent is something to do with race that the speaker dislikes. "Racism" is not so much uttered as shouted; its conversational function is to shut conversation down. This torrent of unpleasantness saturates whatever the word is attached to, however arbitrary the attachment. What has created an aversion to "racism" is less disapproval of what the word denotes than a wish to avoid anti-racist wrath.

Calling claims of genetic race differences "racist," in particular, begs not one but four questions: (1) Are race differences in themselves bad? (2) Is believing in race differences bad? (3) Is saying there are race differences bad? (4) Is studying race differences bad? Once it is realized that an affirmative answer to each of these questions must be established *before* the charge of racism can be made to stick, the charge itself collapses.

Consider question (1) first. Race differences, as facts of nature, have no moral dimension. They either exist or do not exist. Reality may frustrate our wishes, but it is not in itself bad or good. Since a thing must be bad to be racist, race differences, if they exist, are not racist.

The only challenge I can think of to this seeming truism is the contention that facts—about race, or society, or the world in general—do not exist apart from belief in them, but rather are "socially constructed," that is, exist only insofar as we believe in and value them. (Devotees of this view have lately insisted that the medium of construction is discourse.) The quickest way with such Berkeleyan idealism is to observe that if *all* facts about race are constructed, so must be the fact that anyone believes in race differences.[13] On the constructionist theory, there are hereditarians only because there are people who believe in and talk as if there were hereditarians. Were this so, constructionists could end belief in race differences merely by ceasing to believe, and convincing others to cease believing, that anyone believes in race differences. Their constant inveighing against "racism" shows that not even constructionists themselves take their theory seriously.

Turning to (2), factual beliefs in themselves are merely true or false, not good or evil. Therefore, belief in race differences cannot, in itself, be racist. The motives for holding such beliefs may be bad—some people may believe whites more intelligent than blacks from a desire to find blacks inferior—but there is nothing wrong with believing in race differences from estimable motives, such as intellectual persuasion. Precisely the same assessment applies to assertions, which, like beliefs, are merely true or false, not good or bad. A maliciously motivated *act* of asserting that blacks are less intelligent than whites may be bad, and possibly racist, but I have repeatedly pointed out good reasons for making this assertion, such as a desire for justice. Calling attention to race differences, considered apart from the motives for doing so, is not inherently racist.

Finally, research into race differences is bad, and potentially racist, only if driven by bad motives, such as active enjoyment of humiliating blacks. But, once again, such research can just as easily be driven by good motives, such as curiosity or a desire for justice. So research into race differences is not in itself racist.

Many people grudgingly admit that discussing race differences may not be inherently bad, but warn that it risks bad consequences, including distress to blacks and encouragement of hate. The question is then whether blindness to these possibilities is so negligent and "insensitive" as to be "racist." This is a matter of judgment, of course, but a strong case can be made against the "racism" verdict. After all, frank discussion of almost any important issue is bound to offend somebody. Talking about evolution bothers religious fundamentalists, yet is morally permitted. And how sensitive must one be to avoid being a "racist"? To judge from their blanket condemnation of all talk of race differences, many people accept only silence as adequate, turning "sensitivity" into a demand for self-censorship. All told, "racism" is so strong a word that only active malevolence can merit it, not unintentional disregard of black feelings.

That empirical beliefs cannot be criticized on moral grounds may have eluded Jaynes and Williams (1989):

For some people, racism means any form of race recognition, especially instances in which members of privileged groups act in a manner injurious to a disadvantaged group. Others, however, reserve the term for patterns of belief and related actions that overtly embrace the notion of genetic or biological differences between groups. Still others use the term to designate feelings of cultural superiority. . . . We use the term racism to denote biological racism, as in the second interpretation above. Societal racism, borrowing from Frederickson (1971), is used to denote negative racial attitudes or outcomes that lack a clear basis in a belief in inherent racial inferiority [*sic*]. Mere recognition of social groups based on "racial" characteristics is not treated here as a form of racism, but as being "race conscious." Cultural preferences that do not include systematic ranking of social groups and clear hostility toward out-groups is termed ethnocentrism. The concept of racism, however qualified and defined, involves a value judgment. Racism of whatever variety is undesirable; racist outcomes are wrong; and people who advocate racist ideas are typically viewed as being morally deficient, if not dangerous. (556; there is no further analysis of "racism," or index entry for "racism")

Since they identify "racism" with "the notion of genetic or biological differences between groups," an ostensibly factual belief, Jaynes and Williams have at first glance so defined it as to avoid a "value judgment." But they immediately add that racism does involve a value judgment, and is, moreover, undesirable. They do not, it is true, infer the biological identity of all groups from the undesirable racist character of believing otherwise; rather, they bury the factual issue beneath an avalanche of pejoratives—"wrong," "morally deficient," "dangerous."

Jaynes and Williams' foray into semantics illustrates the conflict between the negative connotation of "racism" and its chosen referent. The repeated occurrence of this clash within commonly offered definitions suggests that no coherent definition of "racism" is possible and the word consequently unusable.

Many people would adopt the third definition mentioned in the citation from Jaynes and Williams, namely that "racism" is belief in racial superiority. The first point to make about this proposal is that it spares the belief that whites are more intelligent and self-restrained than blacks, for intelligence and time preference are empirically defined traits whose ascription implies no value judgment (again see section 3.8). One may well value intelligence and self-restraint, and anyone who does so, while also believing that whites typically possess these traits to a greater degree than blacks, is committed to believing that whites typically possess more of something valuable. But the belief that whites are more intelligent than blacks does not *by itself* imply this, so is not "racist" in the sense now under consideration.

In any event the definition is unsatisfactory. It takes belief in the superiority of one race to another to be inherently bad, whereas, or so I argue in chapter 7, such beliefs are typically "operationalized" as factual beliefs. As factual beliefs are not good or bad per se, "operationalized" judgments of racial superiority are not good or bad either. Moreover, as we will also see, these beliefs are supported by the evidence (and probably accepted by the reader), so since there is nothing wrong with accepting reasonable beliefs, do not deserve to be stigmatized. Indeed, there seems nothing inherently wrong with thinking one's own group better in a frankly cheer-leading sense. Most individuals experience this impulse, and, when black-white comparisons are not at issue, it is considered healthy. I am not talking of venomous emotions like hatred, but the feeling that "British is best," the vicarious satisfaction many Jews take in the number of Jewish Nobel laureates, and other forms of ethnic pride. Racists "are typically viewed as being morally deficient, if not dangerous," say Jaynes and Williams, yet an American who thinks the Japanese are crazy to risk their lives eating *fugu* is not dangerous, nor is a German calmly citing Beethoven to prove the superiority of German culture morally deficient. It is hard to see why these attitudes are bad enough to be "racist."

Now the fact is that, whether they should or not, most people do value intelligence, so the statement that blacks are typically less intelligent than whites, no matter how intended, is usually perceived as derogatory. ("Intelligence is valuable, and blacks are inferior to whites in respect of this valuable trait" is an "operationalized" value judgment.) And, while unflattering beliefs about groups held on good grounds are acceptable, what of "stereotypers" who hold such beliefs because they want to, and use evidence selectively to confirm what they want to believe?[14] Defining "racism" as the *desire* that blacks lack highly-valued traits[15] captures how such people think, and, as this desire is unattractive, it appears consistent with the negative force of "racism."

Yet this definition, too, is unsatisfactory. For one thing, while racism is commonly said to pervade American society, there is relatively little racism in this sense. Once bad motives are required, negative beliefs stemming from observation rather than animus are excluded. "Unconscious" and "institutional" racism become contradictions in terms. Finally, and less apt to be noticed, there are many noninvidious—hence non-"racist"—reasons for wanting blacks to have disvalued traits. Whites may hope for a genetic black IQ deficit to absolve them of guilt for black failure, just as any defendant in a liability suit hopes the plaintiff's case will unravel. Or else, whites may be indulging the understandable impulse to resist those who willfully personalize every disagreement. It is human nature to oppose whatever the office blowhard is for, and whites, weary of blame for every black failure, naturally want the blame-mongers shown up by the existence of innate black shortcomings. These impulse are not bad, let alone "racist."

Similar problems face the broader definition of "racism" as hatred of racial groups other than one's own. Without denying the existence or corrosiveness of race hatred, this attitude must be distinguished from numerous milder ones. Hatred implies a desire to destroy, yet, just as it is possible to sneer at pro wrestling fans or readers of supermarket tabloids without hating them, it is possible to dislike black music without wanting it banned, or dislike the black personal style and wish to avoid blacks altogether without wanting to see blacks suffer. One may avoid black neighborhoods for fear of crime while sympathizing with black victims of black criminals. None of these feelings are "racist" if racism implies hatred—yet anyone who confesses distaste for black music or admits wishing to minimize contact with blacks will surely be branded "racist." Those who bandy "racism" in this way use it too broadly to mean "race hatred."

"Racism" is stretched to its snapping point when defined, as it often is, as "treatment of individuals on the basis of their race." More so than any other, this definition is too sweeping to support the negative overtones of "racism," and condemns legitimate practices virtually by word magic. There is nothing obviously wrong with sending Asian policemen to infiltrate Asian gangs, or casting a white woman as Desdemona, or preferring to marry someone of one's own race, all counted as "racist" by this definition

A few writers have tried to salvage the word by neutralizing it, defining it as race-consciousness of any sort while dropping the axiom that all forms of racism are bad. D'Sousza's (1995) term "rational racism" is a stab at neutrality. Such attempts are unlikely to succeed, in my view, for the pejorative connotations of "racism" have become too deeply entrenched. In fact, contemporary discourse combines the worst of both worlds, using the neutral and the pejorative senses of "racism," often in the same breath. The unfortunate result is guilt by equivocation. Something is labeled "racist" in the neutral sense; there is a switch to the pejorative sense; and what is innocent stands condemned. One victim of this gambit was Dale Lick, a candidate for the presidency of Michigan State University. Years prior to his candidacy he had said, a propos the recruitment of basketball players, "A black athlete can actually outjump a white athlete, so they're better at the game. All you need to do is turn to the N.C.A.A. playoffs in basketball to see that the bulk of the players on those outstanding teams are black." Once made public, this remark and Lick himself were instantly called "racist" and Lick was denied the job (*New York Times* 1993b, 1993c). His statement had indeed been "racist" in the neutral sense—it was a generalization about race—but not "racist" in the bad sense, since there seems nothing wrong with citing well-known facts, about basketball or anything else, in aid of a conclusion they support. Yet Lick's statement was treated as "racist" in the bad sense.

Jaynes and Williams write: "Many people question any scholarly use of this concept because it is so manifestly value laden" (1989: 556), a question very much in order. The elasticity of "racism" permits denunciation of virtually anything, including whites doing on Tuesday what was demanded of them on Monday. Ignoring the racial impact of higher academic standards for college athletes is racist, but so (as Dale Lick discovered) is noting the racial composition of college basketball teams. It is racist to treat black executives as if they are special, racist to forget their isolation. Media coverage of slum crime is racist, yet the media turning a blind eye to slum crime is racist indifference to black victims. Too few police in black slums is also indifference—while more police, when white, are an occupying army, or, when black, are Uncle Toms. It is racist to overlook the deprived upbringing of black criminals, racist to insult the spiritual richness of black life by pointing out these deprivations. Failure to act decisively against drugs lets them ravage the black community; treating drug dealers harshly is an assault on black males. Movie and TV depictions of blacks as jive-talking pimps are racist, but so are sugar-coated depictions of blacks as middle-class. (Gates [1989] makes both complaints; DeMott [1996] calls media representation of equality repellent.) Racist colonial powers exploited tribal rivalries in Africa, racist colonial powers imposed the European concept of nationhood on rivalrous tribes. It is racist to say that black children differ cognitively from white, and racist to ignore the black learning style.[16]

Jettisoning "racism" will not by itself end such nonsense, but nonsense is easier to spot when shorn of verbal camouflage. There would remain the problem of labeling the position that race differences exist and are important determinants of human affairs; Christopher Brand (1996) has suggested "racial realism." The important obligation is removal of verbal obstacles to clarity.

NOTES

1. More concisely, if $x(t)$ is the value of x at time t, and v ranges over n-tuples of variables, determinism says that $(\forall x)(\exists < f, v >)(\forall t)x(t) = f(v(t'(t)))$. For refinements see Berofsky (1971: 268–269). Russell (1929: Lecture 8) rejects $t' < t$, and R. Montague (1974: 337–338) rejects "periodicity." Should the notion of a cause for the Big Bang

itself seem problematic, determinism may be construed as a thesis about every event *in* the universe, with the cause of the universe itself left open.

2. We can easily have $f(x, y)$ and $f'(x, y)$ both dependent on two variables, while some functional $F(f, f')$ depends on x alone. When $f = x^2 + x + y$, $f' = 2x^2 + 3x - y$, and $F(g, h) = g - h$, $F(f, f') = 3x^2 + 2x$.

3. Traditional criteria for personal identity are notoriously inadequate. Bodily identity won't do, since science fiction stories in which someone switches bodies are readily understood. Continuity of consciousness won't do, since, apart from the fact that people survive dreamless sleep, it is circular to define a person's experiences as those he remembers happening to *him*. Green and Winkler (1980) identify persons with the actual physical basis of a series of connected experiences, namely the brain. If identity involves the brain, by a natural extension it involves the genetic factors that produce that brain.

4. There can of course be mediate cultural explanations of cultural phenomena, for instance that Cambridge is so named because it bridges a river called the "Cam." There can be noncausal explanations of cultural phenomena to which biology is irrelevant, for instance that it is improper to belch when meeting the queen of England, since belching is discourteous and one must be courteous to the queen. The issue here is whether the cultural realm is *completely* self-contained.

5. The reduction is a mapping from hardware events to actions in the abstract multiplication program.

6. Wittgenstein's influential (1953) seems to deny the reality of rules; see Kripke (1980).

7. See the notion of "full coverage" in Quine (1981). The converse does not hold; identical social facts can be true in different societies where different individuals play isomorphic roles.

8. At some risk of question-begging these might be called "personal" and "relational" traits. Logicians speak of monadic and polyadic predicates denoting unary and n-ary relations.

9. Provably so. The decision problem for the monadic predicate calculus but not the n-adic calculus is solvable, so statements about relations are inequivalent to monadic statements.

10. For instance, to call wearing ties a convention is to say, roughly, that men wear ties because they believe others expect them to do so. My recognition of this convention is my belief that others expect me to: wear a tie because I believe they expect me to do so. An analysis along this line was first proposed by Grice (1957).

11. The federal Department of Housing and Urban Development has since assumed control of Taylor.

12. There may also be gene/environment interaction, as exogenous public assistance cancels the negative consequences of destructive behavior, further disinhibiting destructive impulses.

13. Constructionists such as R. Schwartz (1986) try to minimize the nihilism of their view by pointing to artifacts like eyeglasses, which are manmade yet objective. However, no one supposes that biological race differences—to say nothing of electrons and stars—are manmade.

14. "[Nothing] will . . . deflect individuals like Arthur Jensen and William Shockley from sifting through research reports for evidence of racial superiority. . . . [E]ven Nobel Prize winners can end up seeing not what is actually there, but what they

want to see. . . . Even individuals who have never taken a science course feel free to cite some studies as authoritative, if they agree with their findings. Thus racism has always been able to come up with a scientific veneer" (Hacker 1992: 27-28). Weizmann et al. say that Rushton "scavenges whatever materials lay at hand, whether ecology, anthropology, psychology or paleontology. His tendentious borrowing of materials, often themselves tainted by racism, is quite unscholarly. Libraries are full of so-called data which can be used to support almost any point of view about the causes of differences among people" (1991: 49). Neither Hacker nor Weizmann et al. indicate what supporting evidence environmentalists may find by sifting or scavenging.

15. "A racist, I am inclined to say, is someone who wants there to be racial differences along dimensions of value, and who wants these differences to go in a certain direction" (Nozick 1982: 325).

16. Taylor (1992) catalogues further double binds.

Part II. Values

6

Race and Values

This chapter has two aims: to explain why group differences do not imply that any group is better than any other, and how the norms of different groups can diverge widely without any being deviant or pathological. Chapter 7 has the complementary aim of explaining how, despite the absence of any absolute standpoint, one may still reach normative conclusions about racial disputes. I should emphasize at once that when I deny that intelligence or self-restraint are inherently good I am not advising the reader to be indifferent to such qualities, nor am I suggesting that I am. My point is the purely logical one that in themselves race differences are facts of nature without moral significance.

These aims are connected by a theory of the origin and differentiation of values laid out in sections 6.1–6.4. Inspiring this theory is commitment to naturalism, the position that human values can be explained solely in terms of preferences, reinforcement, and selection for preferences and reinforcement, without the assumption that anything in the universe is actually good or right. The possibility of a value-free social science is one corollary of naturalism.) The interest of a plausible naturalism is that, by banishing value from our world picture,[1] it also banishes absolute and comparative judgments of the value of groups. More intelligent and cooperative groups are seen to be just that—more intelligent and cooperative. There is no cosmic perspective from which greater intelligence and cooperativeness are "better," just as there is no cosmic perspective from which wings are "better" than fins. The races simply differ, in abilities, behavior, and standards of evaluation. For the naturalist, different abilities and intragroup norms, like different organic structures, are adaptive responses to divergent selectional pressures.

The more philosophical aspects of the present discussion are confined to section 6.5, which can be skipped by readers already convinced that factual and normative issues are distinct.

6.1. VALUING

The first order of business is to ask what it is to value something, to find it valuable.

Valuing, whatever its subjective accompaniments, is not a state of mind. Valuing X cannot be defined as believing that X is good, since the very notion that needs to be explained is "believing X good." Nor can valuing be defined as a kind of approval, as many philosophers have thought, since, according to the dictionary, to "approve" of something is "to think favorably" of it, and "to think favorably" is, once again the very notion in need of clarifying. Accounts of valuing in terms of other beliefs and attitudes run in the same circle.

This circle can be broken by appeal to the notion of *reinforcement*.[2] As the reader may recall, a stimulus is said to reinforce an organism when it strengthens any behavior (the "response") with which it is associated. The taste of sugar reinforces those coffee drinkers more apt to drink coffee with sugar than without. Calling a stimulus[3] a reinforcer for an individual may seem little more than a convoluted way of saying that he likes it, but, by giving the cash value of liking, the behaviorist idiom has the virtue of blocking pseudo-explanations. "Smith reads mysteries because he likes them," we say, although the cash value of "Smith likes mysteries" is simply that Smith reads them when he can. Talk of reinforcement makes this clear.[4]

From a behavioral standpoint, to value something is to be reinforced by it and to explain someone's values is to explain why those stimuli that reinforce him do in fact do so. Evaluations are preferences as revealed in behavior: you value Wedgwood china insofar as you strive to see, touch, and own it. You value honesty insofar as you strive to make people honest, that is, reinforce their honesty. You value being honest yourself insofar as you try to keep yourself honest, and experience the aversive emotions of distress and conflict when you lie. Acquisition of a valued object reinforces you; valued behavior not only reinforces you, you seek to (i.e., you are reinforced when you) reinforce it.

"Taste" is ordinarily reserved for trivial preferences and "values" for important ones, but beneath this dichotomy runs a spectrum of preferences of varying strength. At one end lie "tastes," in dining ware for instance, which control only small stretches of behavior, concern only oneself, and are easily overridden; people who like Wedgwood seldom care about—that is, are only weakly reinforced by—Wedgwood's reinforcing anyone else. At the other end are "moral" values, which I take to be behaviors one is always, categorically ready to reinforce, and whose reinforcement is always, categorically reinforcing. You value honesty "morally," rather than prudentially or as a matter of form, when you want everyone to be honest, and want everyone want everyone to be honest. Valuing honesty "morally" entails preparedness punish all dishonesty everywhere, at least to the extent of directing anger at It will limit verbiage to use "reinforcement tendency" toward a behavior mean a tendency to be reinforced by that behavior, to reinforce it, to reinforced by its reinforcement, and to reinforce its reinforcement. Mo approval, then, is a categorical reinforcement tendency,[5] and the morality of group are its shared categorical reinforcement tendencies—the rules its memb want everyone to follow, and want others to want everyone to follow.[6]

It is natural to ask why some reinforcement tendencies become categorical. It is also natural to observe that one can imagine categorical reinforcement tendencies having nothing to do with what is ordinarily considered morality: the desire to get everyone to jog and reinforce jogging would be less a moral conviction than an obsession. The two problems are connected, and jointly solved, by the link between categorical reinforcement and the characteristic subject matter of morality. Categorical reinforcement and everyday morality both focus on practices that tend to benefit everyone when but only when followed by everyone (see Baier 1958; Gert 1970; Gauthier 1967, 1984). Concern with honesty is "moral," whereas concern with jogging isn't, because the benefits of jogging for one individual are independent of how many other joggers there are, whereas honesty benefits an individual only to the extent that the reciprocal honesty of others lets him rely on their words. There is no point in being truthful and some point in lying when everyone else is a liar. (There are also advantages to lying when everyone else is honest, which is why honesty must be reinforced.) This is the intuitive difference between a moral imperative like "Tell the truth!" and a maxim of prudence like "Jog!" "Moral" rules are distinguished by sign-change and multiplier effects: the value of adherence to them tends to become positive when the number of other adherents passes a minimum, and then increase rapidly with number of adherents.[7]

Now, everyone has an interest in getting others to do what tends to benefit everyone when done by everyone, and in getting others to get others to follow suit. Hence one can expect to find categorical reinforcement of precisely those rules the benefits of following which increase with the number of followers. No one has much to gain from getting everyone to jog, but everyone has something to gain from getting everyone (himself included) to be honest, since honesty benefits each if, but only if, others are honest as well. These incentives align the universalistic form of everyday morality with that large part of the content of everyday morality that concerns reciprocity and equity. Form and content may diverge; some rules of sexual behavior, such as the injunction to premarital chastity, lack any obvious multiplier effect yet are intuitively 'moral" because those who subscribe to them do so categorically. So we may say that morality consists mainly of rules that are categorically reinforced because they tend to benefit all when obeyed by all.[8]

This brief characterization of morality already foreshadows group differences herein, and some possible consequences.

(1) Following a given rule may not be equally beneficial in all environments. Cooperation, for instance, may be more individually advantageous where food must be hunted rather than gathered, so cooperation may be deemed more important by groups in hunting niches than groups in gathering niches. One crucial "environmental" variable affecting the benefits to an individual of following a rule is the readiness of the other members of his group to follow that rule, or rules in general, along with him. The readier one's short is to reciprocate honesty, the more individually advantageous honesty will be.[9] When the increase in benefit-per-recruit of following a rule in one environment exceeds the increase in benefit-per-recruit of following that rule in another environment, members of a society situated in the first have a stronger

incentive to enforce the rule than have members of a society situated in the second.

It is thus possible that different rules, and different degrees of overall concern with rule-following and reciprocity, might develop in different environments. Assuming the disposition to follow advantageous rules is adaptive, a genetic factor enters variation in the content of group morality and the intensity with which a group's morality is reinforced. We should therefore not be surprised to find groups differing not just in intelligence and time preference, but in readiness to conform to moral considerations. Should honesty, cooperativeness, and reciprocity have been differentially advantageous in various environments, groups evolving in those environments will have developed varying levels of these traits. Some groups might be on average less concerned than others to think along categorical lines. Since universality and reciprocity are built into the concept of morality, members of some groups might be more inclined than others to embrace moral values, which means, from a behavioral point of view, to be reinforced by moral thinking.

(2) It is important to stress, given the possibility of genetic divergence in moralities and concern for morality itself, that on the present analysis, moral values are not *better* than nonmoral ones. The moral/nonmoral line distinguishes kinds of rules and kinds of attitudes toward rules; it does not say which kind is preferable. That by definition morality concerns categorical reinforcement tendencies is merely a fact about language. "Moral" is the name conventionally given to rules that benefit everyone when generally followed, and "moral conviction" is the name conventionally given to readiness to reinforce categorically. These conventions neither imply nor suggest that rules benefiting everyone when followed generally are obligatory, that categorical reinforcement tendencies are superior to more limited ones, or that people who care whether their maxims would benefit everyone if followed by everyone are better than those who do not. Nonhuman animals are entirely amoral, but there would seem to be no cosmic scale by which a universe with people in it is better than one with only nonhuman animals.[10]

(3) The proposed analysis explains how moral judgments can influence behavior as effectively as more tangible reinforcers. Calling a behavior "right" expresses the speaker's categorical readiness to be reinforced by it, and secondarily, reports that that behavior reinforces other group members.[11] Social creatures like man enjoy[12] pleasing other members of their species. They are reinforced by reinforcing. So doing what others call "right" is doing what others have signaled they find reinforcing, which is itself reinforcing. Conversely, "wrong" expresses a categorical readiness to be reinforced negatively, so doing what others call "wrong" is experienced as aversive.[13]

But the interplay of eagerness to accommodate others with divergence in morality may lead to mismatches. When two intermingling groups differ in their values and their threshold for expressing angry disapproval, the more placid of the two may find itself constantly caught off guard by the belligerence of the more volatile one. Members of the more placid group will be bewildered, wondering what they have done to cause outrage. I suggest section 6.8 that this dynamic explains some puzzling features of black/white interaction.

6.2. PSYCHOSOCIAL ACCOUNTS OF MORALITY

The most obvious mechanism for instilling moral values is socialization. Children learn to obey rules by associating obedience with parental affection and disobedience with parental anger. A socialization account is incomplete, however, for explaining individual adherence to certain rules by a learned association between adherence and favorable outcomes does not explain why a society creates that association. Socialization theory does not reveal why a society trains its children to follow some rules rather than others, or why different societies socialize their children to different rules.

For answers to these deeper questions one initially looks to the *function* of morality, where the "function" of a social practice are those of its effects that explain its perpetuation.[14] Since a rule, having no life of its own, exists only when people follow it, the functions of a rule are those effects of adhering to it that ensure a steady supply of new adherents. Rules that endure do so, therefore, because heeding them, training one's offspring to heed them, and encouraging one's cohort to heed them, enhance the fitness of cohort members. It follows that the function of a group's morality is group survival, and it is easy to see how typical moral rules discharge that function. Injunctions to within-group honesty, cooperativeness, and nonaggression facilitate trade, construction projects, organizing for battle, and other activities helpful to all. Such rules become entrenched because the behavior they prescribe is adaptive. And since obedience to such rules enhances group fitness, so does the reinforcement of obedience.[15]

However, while a functional account explains why rules persist once they start being followed, it does not explain the origin of rules, how they come to be followed in the first place. Finishing the story evidently requires appeal to biology.

6.3. SELECTION FOR MORALITY

No biological account of morality was available before Darwin, despite speculations by Hume, Smith, and Hutcheson about "moral sentiments" built into "human nature." Even after Darwin, biological accounts of morality seemed precluded by the all-too-familiar conflict between moral behavior and self-interest. Robust phenotypes must enhance or at least not diminish fitness, and moral conduct—altruism, putting one's children first, doing what would be good if everyone did it even when others do not—often requires self-sacrifice and risk. Having neighbors willing to defend me benefits me in the long run, and my helping defend my neighbors will, in the long run, encourage them to defend me, but my short run bravery on their behalf may cost me my life. What mechanism could favor genes that coded for it? As Kant might have asked, how is morality biologically possible? Given the depth of the problem, is it no wonder that mankind has long acquiesced in intuitionistic explanations of moral behavior ("The Good attracts the soul"), theological explanations ("God tells us what is right"), or the nonexplanations of common sense ("Of course we sacrifice for our children—they're our children, after all"). But in recent years a powerful evolutionary account of the selection of moral reinforcement tendencies has been developed by Hamilton (1963, 1964, 1975), Trivers (1971), E. Wilson (1975), Dawkins (1976), Axelrod and Hamilton (1981), Axelrod (1984), and others. (An excellent nontechnical review is Sesardic [1996].)

That morality is under genetic control, and therefore probably related to fitness, is shown by the universality and irreversibility of the sequencing of moral development in children (Piaget 1948; Kohlberg 1982; see J. Gould [1982] on irreversibility and universality among conspecifics as criteria of innateness). Now think of morality not as saintly self-sacrifice, but, as here, readiness to help those ready to help back, or "reciprocal altruism." Under certain conditions, genes expressed as aid toward others who reciprocate aid will be fitter than genes expressed as selfishness. The main such condition (Axelrod 1984: Chapter 3, and Appendix B, Prop. 6) is a sufficient initial density of cooperative genes. A lone cooperator amid a population of noncooperators will be taken advantage of unto extinction—a reciprocal altruist must initiate cooperation to see what kind of individual he is dealing with, so is cheated at least once by every nonreciprocator he meets—but a critical proportion of cooperators allows them to interact with each other often enough to accumulate fitness benefits.

Fitness in this theory is of course conceived inclusively: a cooperator among cooperators is more apt to leave partial as well as total copies of his genes. And this suggests a source for the requisite initial clustering. Fitness-enhancing altruism toward kin is common in nature—experience shows that people favor siblings and offspring over strangers, the same being true of less developed organisms (see Pfennig and Sherman 1995)—so it is plausible that the first cooperators were mutated kin-altruists. On this theory, generalized human altruism evolved when mankind lived in small tribes, where a tendency to individually risky altruism was likely to increase inclusive fitness by benefiting relatives. When enough mutant cooperators appeared, more or less simultaneously, cooperation spread at a rate determined by the adaptiveness of cooperation in the local environment, "environment" again including the dispositions of the other individuals with whom cooperators interacted. Recall the hypothesis that conditions in northern Eurasia strongly favored cooperation. Groups of men hunting large game had to trust each other to be where they were supposed to be, to respond to shouted instructions, and to agree on the division of meat. Humans genetically disposed to cooperate with other group members and to reinforce cooperation would have been more likely to hunt successfully, survive in that environment, and transmit this disposition. So cooperation would have spread especially quickly among hunters.[16]

The adaptiveness of cooperation transposes the benefits of categorical reinforcement into a genetic key. Programming a desire to jog enhances genetic fitness, but there is no fitness advantage in programming a desire that others jog. However, since honesty and cooperativeness benefit each individual to the extent that others are respond in kind, there *is* a fitness advantage in programing a desire that others be honest and cooperative. Genes that invest in the reinforcement of jogging waste resources, while genes that invest in the reinforcement of honesty earn the fitness-enhancing dividend of the honesty of others.[17]

Since naturalism turns on this biosocial theory of morality, a bit more should be said of its explanatory power. We may begin with the startling consequence of game theory (the analysis of strategic interaction) that cooperation cannot emerge from the individually rational pursuit of self-interest. Cooperation is rational once it becomes widespread, but the first interactant to extend a helping hand may be taken advantage of. Although everyone would like everyone to cooperate, taking the first cooperative step is therefore too risky. So, if everyone is rational, no one initiates cooperation, and, paradoxically, everyone ends up worse off than he and everyone else could could be and see they could be (see Luce and Raiffa 1957: 94–113). This stalemate is just the problem of the emergence of cooperation, now facing conscious interactants instead of genes.

On the present sociobiological[18] account, evolution cut this Gordian knot by blindly bringing forth enough mutant cooperators to make cooperation rational. The theory does not show that initiating cooperation is rational—nothing can do that—but it does show how cooperation can arise despite being irrational. What is worrisome is that the irrationality of cooperating seems to leave cooperation in danger of breaking down the moment people come to their senses. Sociobiology tells us to stop worrying. We can't help being reciprocators—or, more precisely, we are so strongly disposed to reciprocate benefits that we can be depended upon to do so even when game-theoretical reason counsels otherwise.

Figure 6.1
The Prisoner's Dilemma [19]

		B	
		Cooperate	Don't Cooperate
A	Cooperate	1, 1	–1, 2
	Don't Cooperate	2, –1	0, 0

Interpersonal cooperation and the emergence of cooperative phenotypes both resent Prisoner's Dilemmas. Interactants A and B are wondering whether to cooperate when: both get $1 if both cooperate; both get nothing if neither cooperates; and the initiator loses $1 while defector gets $2 if one initiates cooperation while the other withholds it (see Figure 6.1). Noncooperation dominates for both, so neither cooperates, yet the joint noncooperative outcome <0,0> is inferior for both to the joint cooperative outcome <1,1>. Communication is no help, for both players know that noncooperation is the best strategy against cooperation, so neither trusts the other's promise to cooperate and they still end up at <0,0>.) The dilemma can be partially solved by having A and B play repeatedly, for then one of them can signal intention to cooperate in the future by initiating cooperation early on, while being prepared to follow up unrequited overtures with a return to noncooperation. Here again, though, reason inhibits the process by revealing the risk that the first signal will go unreciprocated, and incur a dead loss. The sociobiological hypothesis has it that nature brought forth so dense a burst of cooperators that their initial blind-faith gestures were reciprocated. Because mankind for most of its history has lived in small communities where people interact frequently—setting up iterated Prisoner's Dilemmas—it was adaptive in the long run to reply cooperatively to signals of cooperation, instead of taking the quick cheater's payoff in hopes of never again meeting the party one duped.

Inverse to the problem of initiating cooperation is the temptation to cheat. For millennia, the sad fact that injustice sometimes pays has sent moralists searching for self-interested reasons for being just—a foredoomed search, since there is no way to show that justice pays when it does not. Again we find able thinkers trying to square a logical circle because of apprehension, this time that, once word gets out that injustice sometimes pays, people will cease being just. And, again, while evolution cannot do the impossible—show that justice always pays—it makes the impossible unnecessary. That fair dealing became adaptive when there was a high enough likelihood that the average person encountered would reciprocate the treatment he received, implies selection for a *standing readiness* to deal fairly that is insensitive to local fluctuations in the cost of so doing. Those prone to take their opportunities when they saw 'em were selected out, leaving a filtrate strongly disposed to behave justly even when there are short-term gains in cheating. We have evolved a tendency to treat everyone, strangers included, as if we might meet them tomorrow, a tendency that withstands intellectual awareness that there is little chance of doing so.[20]

6.4. SELFISH GENES, UNSELFISH PEOPLE

Phrases like "gene strategy" and contentious book titles like *The Selfish Gene* (Dawkins 1976) open sociobiology to charges of nothing-buttery—of calling men nothing but tools of their genes, of calling altruism nothing but disguised selfishness (Stove 1992a, 1992b; also see Symons 1979: 38–41 Levin 1992a, 1993a). The flashpoint is sociobiology's claim that the function of altruism is to enhance genetic fitness. Since morality is supposed to be unselfish, sociobiology seems to deny that moral behavior as ordinarily conceived is possible (see, e.g., Baumrind 1991).

This criticism rests on three confusions: of selfishness with reinforcement seeking, of the function of a desire with its object, and of epiphenomena with mediating causes. To begin with the first, reinforcement-seeking cannot be enough to make behavior selfish, or all purposive behavior, innate or learned would be selfish almost by definition. The selfishness of behavior determined, rather, by the *kind* of reinforcer sought. Altruists are reinforced by reinforcing others—in everyday language, they want others to be happy. An unselfish father prefers playing with his children to watching TV; the selfish father prefers TV. Nor is the unselfish father's secret aim to get pleasure pleasing his children, which would be selfish were it possible, [21] but his children's pleasure, which pleases him. As the selfishness of a behavior is determined by the reinforcer it seeks, to which questions of origin are irrelevant, the *innateness* of a behavior in no way implies selfishness. An innate desire for someone else's happiness is no more secretly selfish than a conditioned desire for that same end.

Calling moral behavior (innately) reinforcing does seem to contravene Kant's dictum that moral action must be done from a sense of duty, not desire. Yet acting "from [or for the sake of] duty" is ambiguous (see Dietrichson [1967] for a scholarly treatment). It may mean, and sometimes Kant seems to mean, that an act gets moral credit only when done from *the desire to do one's duty,* rather than some other desire. Sociobiology recognizes moral action in this sense, since it allows—indeed insists—that men desire to be what is ordinarily called moral. However, the dictum may also mean, and at other times Kant seems to mean, that an act is dutiful only when done from no desire whatever, but rather flows from a purely intellectual "respect for the concept of law." A naturalistic outlook, which considers all behavior reinforcement-seeking, must deny moral action in this sense. But so does common sense, although perhaps unsteadily. Common sense certainly reserves a special kind of praise for the man who does what he thinks right despite not wanting to, yet it also holds that such a man does what he does because he *wants* to act rightly: the conflict of inclination with duty is interpreted as a clash between two desires, an immediate, usually self-regarding one with a longer-range one with a more impersonal object. Common sense, taking a *desire* to act rightly as sufficient for "acting from duty," has no quarrel with sociobiology.

To understand how unselfish motives can perform a selfish function, consider a lazy inventor designing a conscious android housekeeper. The best way for him to get his android to work hard is, clearly, by programming it to *want* to work. The *function* of the android's custom-made desire for tidiness, the reason this desire exists, is to indulge his master's laziness, but the *object* of the desire, what the android itself wants, is a clean house. The android genuinely likes work, although it likes work because its creator dislikes work. The cause of a preference may thus be quite distinct from its object. And, just as lazy creators can make industrious creatures, selfish genes can make unselfish people. The *adaptive function* of unselfishness and altruism, the reason they exist, is the replication of genes that program these impulses, but the organisms created by those genes genuinely and unselfishly care about others. A man may want to please his children (or a complete stranger) because his genes make him want to, and his genes may make him want to because programming that desire enhances (or once enhanced) inclusive genetic fitness, but the desire thus created is unselfish. Dawkins' should have called his book *Why Selfish Genes Make Unselfish Organisms,* although even that title is inaccurate. Genes do not literally have desires, selfish or unselfish; they are chemicals which it is often convenient to describe teleologically. Dawkins' book might best have been named *Why Genes, Which May for Heuristic Reasons Be Called Selfish, Make Unselfish People.*

The third source of confusion about sociobiology is a tendency to collapse causal chains—to think that, when X causes Y and Y causes Z, the intermediate factor Y drops out, leaving X the real cause of Z. Collapsing the chain from genotype to phenotypic preferences to behavior makes the genotype appear the true cause of the behavior, with the preferences superfluous. But this view of causality is mistaken. Speaking literally, Y is superfluous in the sequence $X \rightarrow Y \rightarrow Z$ if, given X, Z will occur even without Y. In that sense, intermediate factors are typically not superfluous. When a gale topples a tree, which then blocks traffic, it is not "really" the gale that blocked traffic, where this means that traffic would have stopped anyway, whether or not the tree had fallen. Nor is it "really" the father's genes rather his desires that cause him to read to his children, where this means that he would have read to his children whether he had wanted to or not. Had the tree not fallen there would have been no traffic jam, and the father would not have read had he not wanted to. What causes confusion is that, given X (the wind, genes), mediator Y (the fallen tree, preferences) is *inevitable*, so, since Y also makes Z (the traffic jam, behavior) inevitable, Z is inevitable given X. However, the inevitability of Z given X does not make Y superfluous when Y is *how* X makes Z inevitable. The causation of behavior by genes does not mean behavior would occur absent preferences. The answer to the question "What would you get if you had genes without desires?" seems to be "behavior," leaving desires superfluous. But the correct response is, "You *can't* have genes without desires; once you have the genes you have the behavior *because* the genes produce the behavior-producing desires." Conative drives are how genes bring behavior about. They are not epiphenomena, but necessary intermediate mechanisms.[22]

In attributing behavior to factors beyond individual control, sociobiology is (bio-environmentally) deterministic, and behavior is often thought to be free only when the agent himself has chosen its cause, the cause of its cause, and so back to its ultimate origination. And since freedom is necessary for moral responsibility, we seem to have another reason for thinking sociobiology wars with morality. As we saw in chapter 5, however, bio-environmental sociobiology is no more deterministic than environmentalism or socio-environmentalism. Sociobiology may make us puppets of genes, but environmentalism makes us puppets of society. The underlying worry is not *genetic* determinism, then, but determinism per se. The whole topic of free will is taken up in chapter 9; until then it may lessen anxiety to reflect that most behavior ordinarily considered free and responsible is caused by factors beyond the agent's control. Nobody decides to fall in love or get hungry, yet marrying and eating lunch are paradigms of voluntary behavior. The everyday test of the voluntariness of a behavior is not that the behavior be uncaused, but that it be *alterable by reinforcement*. The burglar planning a getaway, guided by fear of arrest, is held more responsible than the compulsive voyeur who lurks in the bushes despite dread of discovery. The burglar's getaway plan shows he would desist were he faced with imminent arrest, while the voyeur cannot stop himself from acting in ways he knows he will regret. Bio-environmental determinism is consistent with the shaping of behavior by reinforcement; the very existence of unconditioned reinforcers, together with a capacity to learn to associate unconditioned reinforcers with novel stimuli, implies the plasticity of behavior. Biological sociology permits conditioning and learning, and therewith all the freedom common sense demands.

6.5. VALUE SUPERFLUOUS

One can agree than an individual's values are his reinforcers, yet still insist that values are objective. After all, what people *do* value is commonly distinguished from what they should, what is valu*able*. But a biosocial account of valuing challenges this distinction, by excluding value as explanatorily superfluous. It seems reasonably clear that no natural phenomenon needs value to explain it, and, on the present account, no aspect of human behavior does either. Moral conviction and action are consequences of natural selection, rather than apprehension of a platonic Good. As an economist has put it succinctly, "Normative statements are not part of any scientific explanatory model" (Robin 1977: also see G. Harman 1977: 6–9). Once people's actual values have been specified, further questions about the value of their values are empty.

Value is not the sort of thing that can be observed,[23] and, generally speaking, belief in what is not observed must be warranted by its indispensability in explaining what is. Scientists believe in units of electrical charge because, so far as we know, oil droplets would not move as they do in a magnetic field unless unit charges (called electrons) exist. Were all observations explicable without the postulation of electrons, there would be no reason to believe in them; should all observations prove to be explicable without the postulation of value, there is no reason to believe in value. Agnosticism about value would in that case also be untenable, since like many theoretical posits, value is identified by its putative relations to the data. The substance phlogiston was once supposed to be what is given off during combustion; after Lavoisier found that nothing is given off during combustion, it would have been absurd to be a phlogiston agnostic. Value has traditionally been supposed to be what is apprehended in moral reflection and what, when apprehended, guides moral choice; if moral reflection and action are not matters of apprehending anything, value does not exist.

This is where the biosocial analysis kicks in, for it implies that moral convictions and choices—categorical reinforcement tendencies and their manifestations[24]—can indeed be explained without mention of value. Take Harry Truman's decision to use the atom bomb. Any explanation of this event would seem to involve value, since it must cite (among other things) Truman's view of his duty to American soldiers and the cost in innocent lives. Yet duties and costs explain nothing. Truman's decision was caused by his *weighings* of those factors, weighings caused in turn (on the biosocial view) by his upbringing and, ultimately, biological heritage. The rightness or wrongness of those weightings were irrelevant to Truman's decision,[25] and predict nothing about Truman's other behavior.[26] Whatever the merits of his convictions, Truman would have made the same choice had he had the same convictions, leaving the merits of his convictions idle.[27] The point of the example can obviously be generalized: nothing about the observable universe remains puzzling when intrinsic value is left out of our world picture.

It is natural to object that everyone, skeptical naturalists included, endorses his *own* values, and in endorsing them shows that he believes them to be in some sense correct. Why would anyone be honest unless he thought honesty *better* than lying? Why would you buy a car instead of investing your money unless you thought the car was a more worthwhile use of your money? At the moment of choice a chooser must believe the alternative he is choosing is best. Did he not think this, he would be unable to explain why he was making one choice rather than another. Someone who explains his own present choice naturalistically, without reference to its superiority to alternatives, regards what he is doing as happening to him, hence not a choice at all. Should he say he is going to study law because his parents drummed a legal career into his head, or because "that's just the way I am," he views his attending law school as beyond his control. Since each of us makes choices, each of us must believe that some choices, some values, are better than others.

So we must. It is a fact—exploited in the next chapter—that everyone believes his present values correct. But it does not follow that his values or anyone else's *are* correct, since unbelievability is not the same as falsehood. You cannot believe you are dreamlessly asleep, but that doesn't mean you are always awake. Nobody can believe he is kidding himself—the moment you realize that your belief that you will get a raise is a pipe dream, you stop believing you will get a raise. Yet people kid themselves all the time; the only thing they can't do is kid themselves *knowingly*. Sorensen (1988) calls unbelievable truths "blindspots;" we have blindspots about being asleep when asleep, and permanent blind-spots about self-deception. The naturalist suggests that each of us has a blindspot about the causes of his choice at the moment of choosing. The nature of choice forces an agent to think his choice the right one, and that he is making the choice he is making because it is right, just as the nature of belief forces you to believe that your current beliefs are warranted and that you accept them because of their warrant But that no more guarantees that your choices *are* right than it guarantees that all your beliefs are warranted.[28]

An interesting question is just what it is that someone believes when, at the moment of choice, he believes his preference "best." The skeptical naturalism defended here implies that the idea of objective value has no determinate content; how then can anyone mistakenly embrace it when there is no "it" to embrace? Lacking anything constructive to say about this puzzle, which would in any case lead us far from the topic of race, I will leave it here.

A logical problem created by skeptical naturalism is treated in Appendix C.

6.6. MORALITY AS A PHENOTYPE

Smudging the line between innate and socially inculcated behavior challenges the ancient conception of norms, or "culture," as Reason's governor on "natural" impulses. Plato and Kant thought pure reason revealed noumenal realm beyond man's empirical character; Hobbes thought instrumental reason delivered man from his natural state of war; romantics like Mark Twain invert the relation, opposing the innocence of nature to the corrupt artifices of society. Against all such views bio-environmental determinism maintains that socialization expresses genes just as surely as do presocial drives like sex or hunger. Nurture is natural.

An innate tendency to follow a norm can be expected to correlate with an innate tendency to train the young to follow it, and an innate tendency in the young to respond positively to the training, for all these tendencies are apt to be adaptive when any one of them is. Where cooperation (say) increases inclusive fitness, so does training one's children to cooperate, and so does the responsiveness of one's children to training in cooperation. Training and its positive uptake enhance the cooperativeness, hence the inclusive fitness, of one's offspring, and what enhances one's offspring's inclusive fitness enhances one's own. In an environment favoring cooperation, that genotype is likeliest to spread half-copies of itself which programs cooperation, the training of offspring in cooperation, and offspring who cooperate when trained. Instinctual drives emerge as passive genetic correlates of socialization.[29] Training is no epiphenomenon; it does modify instinctual drives. But it is a collateral effect of the genes that produce the drives it modifies, with the causal relations posited as in Figure 6.2.

Figure 6.2
Socialization as a Passive Gene/Environment Correlation

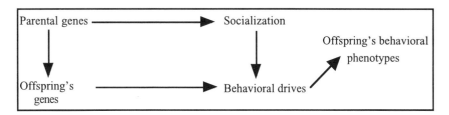

The idea that morality transcends nature comes of asking what unsocialized humans would be like, a question that assumes there could *be* unsocialized humans. This assumption is invited by the occasional feral child, who indeed acts like an animal, and by the ways in which differences in socialization contribute to between-culture differences. Since training obviously counts for something it is natural to extrapolate, to ask what would happen absent all training—if, in effect, every child were feral. The wholly asocial behavior then imagined, usually something out of *Lord of the Flies*, becomes the baseline for what is "natural," and the gap between this baseline and actual human conduct is interpreted as the contribution of "culture." If training is a correlate of innate drives, however, socialization of some sort must emerge in all human groups. The evolutionary pressures that selected wildness in children also selected for adult impulses to curb it. The best evidence for this is the fact that socialization to codes of conduct *has* emerged in every human society without the intervention of any outside agency. The (unnaturally all male) children in *Lord of the Flies* will grow up to resemble the adults who rescue them, or they won't grow up at all.

The hypothesis of a hypothetical question must be consistent with the laws of nature. The need for this "cotenability" condition (Goodman 1966) is clear a propos such a question as, "How fast could a 50-foot-tall man run?" On one level the question is intelligible: a normally proportioned, ambulatory giant violates no rules of logic and is easy to visualize. But according to biomechanics, the science of the movement of creatures of various shapes and sizes, a giant cannot exist. He would collapse under his own weight. So the question "How fast could a man run were he 50 feet tall?" is ill-defined after all, suspending as it does the very laws needed to answer it. Likewise, the question, "What would a world of unsocialized human beings be like?" suspends the very evolutionary laws that produced human beings, and govern what human beings would do under hypothetical circumstances. As the development of socializing mechanisms is part of man's nature, according to these very laws, there is nothing unsocialized men would be like.

6.7. BETWEEN-GROUP MORAL DIVERGENCE

The hypothesis that group norms are evolved reinforcement tendencies predicts that groups developing where divergent reinforcement tendencies are adaptive will develop different moralities, or different levels of concern for morality. A tendency to reinforce honesty, cooperation, and within-group altruism is probably always adaptive to some extent, and so to be found among the norms of every group, but groups developing in different environments may assign different weights to these core tendencies.[30] Again idealizing, imagine hunters in a cold climate, exploiting dangerous animals, storing and sharing food, and gatherers in a more forgiving climate. The hunters must cooperate or die, whereas cooperation is less important for the gatherers, each of whom can more nearly support himself and his (or her) offspring by individual effort. Since each hunter has more to gain from the cooperation of others, there is a greater return in fitness on the efforts of each to get other hunters to cooperate. Hunters can therefore be expected to develop stronger cooperative tendencies, to reinforce cooperation more strongly, and to respond more positively to the reinforcement of cooperation. The gatherer ethos will not be absolutely uncooperative, but it will be less cooperative than the hunters'. Hunters will deem a higher level of cooperation morally obligatory, and each group will probably disdain the minimum degree of cooperation expected within the other. There may be differences in sexual norms as well. Since it is easier for female gatherers in a warm climate to support their offspring, there will be less intense selection for females who prefer fidelity in their mates, hence, by what Darwin called sexual selection, less intense selection for males disposed to conform to this female demand. Hunters may regard gatherer sexual morality as loose, while gatherers regard hunter sexual morality as inhibited.

Since an individual's environment includes other group members, one factor determining the adaptiveness of cooperation for any single individual is the mean time preference of the individuals he interacts with. Greater concern with the present means greater likelihood of defecting in Prisoner's Dilemmas, in disregard of the long-run benefits of reciprocity. Environments favoring cooperation produce individuals with low time preferences, which in turn makes cooperation adaptive for others seeking to enter the population, allowing the most adaptive level of cooperation to stabilize at some high value. Conversely, seeking to cooperate is maladaptive when the mean time preference of cohorts is high. No time preference, low or high, is intrinsically irrational. High time preferences and relative uncooperativeness are individually rational, and maximize inclusive genetic fitness, when the mean time preference of the cohort is high. Gatherer time horizons are not aberrant, and will be stable until some environmental change or improbable mass mutation favors a different equilibrium. (If time discounting is exponential, the advantageousness of cooperation can be shown to vary inversely with the minimum concern for the future necessary for cooperation.)

Conceiving group differences in value as expressions of biological differences makes plain the circle in explaining group differences via differences in values. Supposing (for instance) the phenomenon to be explained is the failure of members of group A to create as many institutions dependent on honesty as have members of group B, it is no explanation to say that honesty is less important for A than for B. What needs explaining is the institutional manifestation of this very difference. A group's values cannot be explained by intergenerational transmission for the same reason, since transmission presupposes the presence of those values without explaining how they first arose. The theory that values are correlates of genes recognizes the reinforcement of one generation by a previous one, but insists that this mechanism cannot explain which values are reinforced.

We saw in chapters 3 and 4 that the races differ in a number of empirical characteristics, in great measure, so the evidence suggests, for genetic reasons. Of these, time preferences in particular are continuous with what are ordinarily called values. But the upshot of skeptical naturalism is that this is no basis for ranking groups. We cannot say that the time preferences or values of one race are better than those of another, because such comparisons explain nothing. Similarly, the higher intelligence of whites and Asians no more makes them superior to blacks than a cheetah's greater speed makes it superior to a horse. Such comparisons lack empirical content. Like the cheetah's supple spine and the horse's hoof, the levels of intelligence of the different races were responses to environmental pressures—as were the values embraced by different groups. Known race differences in intelligence, impulsivity and aggressiveness, and such race differences as there prove to be in cooperativeness and rule following, together with the divergent evolutionary histories of the races, suggest that the races have also evolved divergent *evaluations* of cooperativeness, aggression, rule-following, and concern with the future. We will see further evidence of this in chapter 7. However, should whites have evolved preferences and modes of thought closer to those conventionally called "moral," we must remember that there is no Archimedean point from which to judge "moral" thinking better than nonmoral thinking.

6.8. THREE COROLLARIES

The theory that group differences in norms reflect evolved adaptations has three noteworthy consequences. They may be viewed as testable predictions of the theory, or, if the phenomena in question are granted to be real, as confirmation.

(1) The response of a group to externally imposed socialization is unlikely to resemble the response, to the same socialization, of the group in which the socialization evolved. If the pressures that select a socializing regimen also select responses to it, children in groups that evolved the regimen have been shaped to internalize it more readily than children of other groups.

Consider the suggestion that making school seem as important to black children as it is made to seem to Asian children will improve black academic performance. This suggestion is often made by circumlocution, to avoid implied criticism of black culture as anti-intellectual, yet however broached it assumes that presenting high grades and the esteem of teachers in a good light will benefit black children as much as it benefits children of other races. But such representations will make little difference to black performance or attitudes if valuing academic success is a correlate, not a cause, of educability. Low time preferences are necessary both for achieving high grades and for wanting to achieve them, so impulsive individuals will not tolerate the delay necessary to reap the benefits of studying, no matter how fulsomely they hear studying praised. Should the factors causing Asian parents to extol academic excellence also be the cause of Asian children caring to excel, encomia to schoolwork will bounce off a higher proportion of black than Asian children. Should genotype/environment feedback make the encouragement given children in part an effect of children's responses to encouragement, the norms emerging from the Asian socialization process will have even less effect on performance when imposed from without.

(2) Violence will increase when a group acquires killing technologies it did not evolve. Stable groups that manufacture devices such as firearms must also have evolved the self-restraint not to employ them to maladaptive excess. A group developing the intelligence to invent weapons ahead of sufficiently strong inhibitions against their indiscriminate use will tend to destroy itself. New technologies themselves exert selectional pressure toward ability to control their use. (Whether any population is restrained enough to survive the discovery of atomic energy is not yet clear.) Therefore, a group that acquires a weapon from another group, rather than having developed it, is less likely to have evolved the appropriate inhibitions. Groups that do not discover fermentation have no reason to develop tolerance for alcohol, and when introduced to it often fall prey to drinking problems; access to exogenous killing techniques may be expected to yield parallel results.

There has been a sharp increase in gunshot homicides among black males in the last half century. In 1943 there were 44 handgun homicides in New York City; in New York in 1992, 1500 black males died of gunshot wounds inflicted by other black males. Since 92% of the 2200 murders in New York that year were committed by blacks, black males must also have killed several hundred other nonblacks with firearms. In 1992, 23,760 murders were committed in the United States, for a murder rate of 9.3 per 100,000 (9.8 in 1991, 9.3 in 1990; FBI 1993: 13), far greater than that of France (4.5), Germany (3.9) or Austria (2.3) (1990 European figures, Rushton 1994). The involvement of firearms in 70% of these homicides (FBI 1993: 17) supports the widespread view of the United States as a violent country. But disaggregating homicides by race, following Taylor (1994), presents a markedly different picture. In 1992, 55% of offenders whose race was known were black, and 43% white (FBI 1993: 17). Since blacks are 12.1% of the population, this means that the murder rate for blacks in the United States was 44.9/100,000, while that for whites 4.78—much closer than 9.3 to the figures characterizing European whites. Detroit is perceived as having become extremely violent in recent decades, its homicide rate among juveniles having grown from 8/100,000 to 30/100,000 during the 1980s (Ropp et al. 1992). However, the homicide rate for white juveniles in Detroit remained constant during this period; the upward trend was due to black-on-black murder, and that increase "could be attributed almost entirely to firearm homicides" (2907). Much of the June 12, 1996, issue of the *Journal of the American Medical Association* is devoted to firearms and violence. One California study reported that the hospital discharge rate for gunshot wounds for blacks between the ages of 15 and 24 was 439 per 100,000, while those for non-Hispanic whites in that age group was 28 per 100,000. (The rates in the 25–34 age group were 235 and 17, respectively.) Assaults accounted for 74% of all firearm-related crimes. Assault rates for males aged 15 to 24 years were 598 per 100,000 population for blacks versus 27 per 100,000 for whites.

Gary Kleck (1991: 282, 311, Table 7.6) has found that even controlling for gun availability, black males between the ages of 15 and 24 have almost ten times as many fatal gun accidents as whites of that age; the ratio falls to 5:1 for the 25–39 age group. Most of these accidents involve display of weapons, mock fighting, or mock threats. Kleck speculates that "accidental and intentional killers may share some underlying personality traits, such as poor aggression control, impulsiveness, . . . willingness to take risks, and sensation seeking" (1991: 286). These are traits which distinguish blacks from whites (see chapters 3 and 4).

The race difference in gunshot homicides is largely due, I suggest, to the failure of blacks to evolve inhibitions adapted to firearms like those evolved by whites and Asians during their long march to the discovery of explosives. The black gunshot homicide rate is not explained by the sheer accessibility of guns, since guns have been available to whites and Asians—who invented and now manufacture them—for centuries. Every Swiss male citizen owns a gun, yet the annual adult homicide rate in Switzerland is 3.2 per 100,000. Guns have become more available to whites as well as blacks in recent decades, but the acceleration of the black homicide rate far exceeds the white. Apparently, the race gap in gunshot homicides is due to an increased availability of guns *to blacks*. In many cities a large proportion of young black males, far higher than he proportion of young white males, carry firearms. Blacks are evidently more drawn to guns and readier to use them, especially in disputes whites would ind trivial.[33] Possibly, thresholds of aggressive reaction and levels of empathy hat "work" for African weapons are too low to regulate more efficient European weapons. Death is relatively unlikely when sudden anger must be ented through is a spear or assegai or one's fists, which are hard to wield, asy to fend off or flee, far more likely with guns at hand.

(3) Divergent evolution can scramble moral signals.

In any cooperative group, an inclusively adaptive response to another's injury is sympathy, aid, a desire to suppress the cause of injury, and, when the cause of injury is perceived to be oneself, guilt feelings. To select these responses, evolution must find a way for organisms to signal such subjective accompaniments of injury as anguish, a problem it solves by associating subjective states with spontaneous outward expressions. Human beings raise their voices when angry, smile when pleased. These natural signs also tend to elicit responses, as a shout of anger will often cause others to attempt calming measures. We may call shouting a *natural indicator* of anger and a *natural elicitor* of efforts to calm. In the case of aversive subjective states the feedback between utterance and response is generally negative, as the response diminishes the signal by soothing the injury.

So suppose (to put the point abstractly at first) that R is the fittest response to an injury of severity I in population W. Suppose as well that, initially, injury I has two natural indicators among the W, utterances U and U'. If for any reason, one of these utterances—U, say—becomes more likely than the other either to indicate I or elicit response R, there will be selection for uttering U on sustaining I and for responding to that utterance with R. A bandwagon effect will lead to U's becoming the sole natural indicator of I and elicitor of R. We may suppose that this has happened in W, and that utterance U' has come to indicate the more serious injury I' and elicit the more galvanic response R'.

Figure 6.3
Crossed Signals

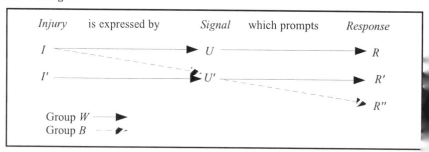

The associations between I, U, and R are contingent. Internal states need not have the indicators they do have; overt utterances need not signal the internal states or elicit the responses they do. Smiles express one feeling for humans, another for chimpanzees. So suppose that, in another group B, U' has been selected to indicate injury I and elicit response R'', in B the fittest response to I. Perhaps U' is more strident than U because Bs are less empathetic, and stronger signals are needed to compel attention. The signals cross, as in Figure 6.3. Here U', the indicator of a serious injury for group W signals a less serious injury for group B, and elicits different responses from members of the two groups.

Should these populations interact, the crossing of signals will mislead W

about expressions of injury uttered by Bs. A W will construe a protest from a B to indicate the injury it would indicate if uttered by another W, and respond to this protest as if it had been uttered by a member of his own group. He will interpret U' when uttered by a B as meaning serious injury I', instead of the less serious injury I that it does mean, and he will emit stronger response R'. One result of this misunderstanding is that Ws will overestimate the seriousness of anger displayed by Bs. Another result will be failure of the W response to extinguish B anger, since this response will not soothe the injury B is actually expressing. A W will respond to U' with R', while in B utterance U' is supposed to elicit R''. The Ws will be left bewildered by B's intransigence, as unsatisfying responses amplify misunderstood anger.

I conjecture that given levels of anger signal a less serious injury among blacks than whites, and that a positive feedback between black expressions of anger and white efforts to assuage it explains some otherwise puzzling aspects of race relations. White guilt and black intransigence may in part be effects of white misinterpretation of black reactivity and an overall mismatch between black emotional cues and white responses.[34]

That this sort of misunderstanding regularly takes place can hardly be denied. According to Kochman, "Whites *invariably* interpret black anger and verbal aggressiveness as more provocative and threatening than do blacks" (1983: 44; also 31, 58). A natural hypothesis is that white expectations about anger have been shaped by dealings with other whites, and that a show of belligerence which *among whites* is normally caused by injury (and predicts retaliation) is, among blacks, often caused by what to whites are trifles. More precisely, any given level of anger is apt to signal more serious injury when reached by a white than when reached by a black. Not having evolved to interpret black displays, whites tend to interpret black anger, including anger directed at them, as indicating the more serious injury such anger would signal from whites. As guilt and solicitude are evolved white responses to perceived injury, whites tend to blame themselves for black rage and seek to ameliorate it. Again this seems confirmed in experience. "Whites . . . wonder what horrendous offense the other person could have committed to deserve such an emotionally powerful response" (Kochman 1983: 123).[35] However, as white reactions are geared to injuries that cause anger in whites, they do not address the causes of black anger. The reduction in display that usually follows a placatory response does not ensue, leaving whites feeling they have left something undone.

(A further stage in this dynamic is black perception of solicitous white responses as rewards, which reinforces black anger. "It seems plausible," writes Trivers, "that the emotion of guilt has been selected for in humans partly in order to motivate the cheater to compensate his misdeeds and to behave reciprocally" (1971: 50); at the same time, the existence of guilt may select for 'Sham moralistic aggression when no real cheating appears [to] induce reparative altruism [and a] feeling [of] genuine moralistic aggression even when one has not been wronged if so doing leads another to reparative altruism." The races may not have interacted long enough for blacks to have evolved race-specific sham moralistic aggression, but white guilt may elicit a generalized readiness to display it.)

Consider by analogy how evolution has shaped infants to cry to express needs, and adults to respond to infant crying by meeting those needs. Crying gets babies fed. But nature has made the mediating emotions less than entirely pleasant; a baby's crying provokes irritation as well as concern, and often parents exert themselves "just to get some peace." Now imagine a hypercolicky baby whose signals are nonstandard. Cries that normally express hunger do not express hunger for him, so attempts at feeding do not quiet him; the least wetness produces "diaper rash" crying not stilled by a change of diaper. Twenty-four hours of dealing with such a baby would leave the typical adult unstrung. I am proposing that black anger, white failure to read it, and the consequent failure of white solicitude to reduce it leave whites unstrung.

One point favoring the crossed-signals hypothesis is its agreement with everyday observation. As Kochman confirms, the black communicative style is more confrontational. Black loudness is a "stereotype," and eavesdroppers will be struck by the manner in which black conversation may suddenly jump to a level of intensity that among whites would prefigure a fight, and as quickly subside. Such conversations probably agitate white listeners more than the black participants. A second point in favor of it is that the usual explanation of white guilt, namely unblinking recognition of white misdeeds, is inadequate. Most whites know that no living white has any direct connection with slavery, and few remain who were associated with Jim Crow. Only 20% of antebellum Southern families owned slaves, so few living Southern whites are likely to be connected to slavery even by descent. Blacks today enjoy civil rights, legal privileges under affirmative action, and full access to a system of public schools supported primarily by whites. Black-on-white homicides, which are relatively common, are seldom publicized, while white attacks on blacks are treated as national scandals. Many of the whites who are most solicitous of black feelings have done nothing personal to atone for. Actual mistreatment of blacks, then, seems insufficient for white guilt.

Consider in the light of this hypothesis a number of incidents that have received media coverage in the United States in the past few years.

• A father wrote a short letter to the *New York Times* describing the theft of his son's bicycle by blacks, as a consequence of which his son "no longer wants to be around 'black kids' " (Blake and Bray 1993). A black asked by the editorial board to respond, wrote

The young men who jumped your son and took his bike saw a little white boy with a bike whose parents can afford to buy him another one. The synergy of race and class privilege allowed them to act. . . . In the end, it is your example that will best help him heal. The black friends who come to your home, whom your son sees you socialize with, will reassure him. . . . He's been introduced to something painful about race . . . by accident of birth, he is white and male and economically secure in a city where many others are none of those things. (Blake and Bray 1993)

Apart from the dubious proposition that the thieves asked themselves whether the bike was replaceable, it is suggested that they were entitled to the bicycle and above all, that responsibility for any ill-will that may result lies with the boy's parents. It is not suggested that any black should make any gesture toward the boy.

• When black Washington Mayor Marion Barry was arrested for possession of crack cocaine, prominent blacks complained that whites were out to "get" black officials. Reporters skeptical of the charge itself nonetheless judged it understandable that blacks would believe this, and the Barry case came to symbolize the difficulties facing successful blacks.

• To escape her stepfather's wrath at her partying, a 15-year-old black girl named Tawana Brawley covered herself with dog feces, wrote "Nigger et [*sic*] shit" on her chest, and claimed to have been raped by white policemen. She immediately came to represent the suffering of black women. When her story was found to be fraudulent, its plausibility to blacks was cited as evidence of white police brutality and the sexual advantage taken by white men of black women.

• Edmund Perry was a Harlem teenager recruited on full scholarship by Exeter, an exclusive New England preparatory school. Perry appears to have been performing academically at about the white mean, far below the level normally required for admission to an elite institution. A few weeks after graduating he was killed by a plainclothes policeman he was attempting to mug. Angry demonstrations ensued, and Perry's act was attributed to Exeter's alienating atmosphere. *New York* magazine found it significant that his white classmates preferred the Rolling Stones and slighted Perry's interest in soul music. His death was taken to indicate the many steps whites still had to take to make blacks feel comfortable. (Anson 1988 details the Perry affair.)

• In 1989 eight black teenagers beat, raped, and left for dead a jogger in New York's Central Park. They explained that they raped the woman because "it was fun" and described their activity as "wildin,' " which proved to be a black colloquialism for banding together to commit random violence. Many prominent blacks criticized the media for calling attention to the boys' attitude and language.

• In Milwaukee, black alderman Michael McGee announced the formation of a black militia for the shooting of white people if his district was denied more money. "Our militia will be about violence," he said. "I'm talking actual fighting, bloodshed and urban guerrilla warfare" (Wilkerson 1990: A12). While McGee drew some muted criticism, a white alderman commented, "He is certainly pricking our conscience about the real world. It's a legitimate question. Where are our priorities when you spend millions for downtown amenities and dribbles for the neighborhood?" The *New York Times* referred to McGee's threats, as well as increasingly frequent black looting, as "a cry of despair."

• At the University of North Carolina at Greensboro, where a black student center was already housed in a campus building, blacks demanded a free-standing, separate building of their own. After numerous demonstrations, the administration reluctantly granted their wish. But when it was decided that the building would be placed across the street from the main campus rather than on it, further demonstrations, backed by rap music, were staged. When asked to explain their purpose, one black student said, "I start from the fact that my great-grandfather was a slave" (Sanoff et al. 1993). Another black student told a group of whites, "I don't think it's the responsibility of black people to step outside of who we are" (Sanoff et al. 1993).

• After a black youth was witnessed shooting to death a district attorney actively prosecuting gangs in Boston, the first, and angry, demand of black spokesmen was that black teenagers not be rounded up in "sweeps." The Boston Police Department complied, indicating in public statements that its first priority was not apprehending the murderer, but ensuring that "racism" would not be indulged.

In each case, black misbehavior was represented as victimization, and proof of how much more whites must do to earn black trust. In each case, blacks adopting aggressive postures leveled accusations that whites, for the most part, accepted. Warrants issued for the arrest of members of the Brawley family involved in the hoax did not result in arrests. Looking beyond these particular incidents, there have been some appeals by high-profile blacks to end black-on-black crime, but to my knowledge no apology for black-on-white crime, nor any demand from whites for one. A biological component in the dialectic of black anger and white bafflement seems probable.

6.9. "SICK" AND "DYSFUNCTIONAL" VALUE SYSTEMS

Many people disturbed by black behavior but reluctant to sound judgmental describe black values as a "pathology" produced by oppression and "the ghetto." Just as pneumonia is bad yet nonculpable, "pathological" norms can be described as bad yet not blameworthy, and perhaps as distortion of something more agreeable.

For their part, critics of the medical analogy (led by Szasz 1970) see it as a veneer for moralism. Sickness is somatic, they say, so behavior cannot be "sick"; criticism of behavior should be stated frankly, without pretense of scientific neutrality.

Both views are partly right. Behaviors and value systems as well as organic structures can be unhealthy, contrary to critics of the medical analogy. At the same time, the radically divergent, conflicting value systems of groups that have evolved apart may all be healthy, and behavior that is pathological when exhibited by members of one group may be healthy when exhibited by members of another. Black deviation from white norms does not by itself imply anything wrong with black norms or behavior.

The key issue is the meaning of "health." Calling a condition "unhealthy" makes no reference to any particular cause, since people recognized sickness before knowing what caused them. Nor is "sickness" a name for conditions people dislike (see Englehardt 1986), since ugly people are not considered ill. Also, health covers more than the absence of disease, since a broken leg, although undiseased, requires medical attention. We call such conditions "abnormalities," and, in fact, abnormality includes disease as a special case, diseased structures being in an abnormal condition but not (as the broken leg case shows) vice versa. The fundamental opposition is that between health = normality and abnormality.

The normal/abnormal distinction in turn rests on the notion of "function." A structure or behavioral drive is in its normal state when performing its function: healthy hearts do what hearts are for. Now, in its core meaning, as applied to artifacts, a thing's function is what it is intended to do (Woodfield 1976), so when the world was thought to be God's creation, the function of a thing was thought to be what God intended it for. This does not leave "function" tied to theology, however. It has a more general meaning that includes purpose, human and divine, as a special case: the function of a structure or drive are those of its effects that explain its existence (Wright 1973, 1976; Symons 1979: 10–14, 34–38; Levin 1984b, 1997). To say that blood circulation is the function of the heart is to say that hearts exist because they pump blood.[36] To call eating the function of the hunger drive is to say that organisms have a hunger drive because it leads them to eat.

In the familiar case of artifacts, an effect explains its cause via the effect's having been intended, the intention being the cause. Emery boards are for filing nails because someone wanted nails filed, saw that emery boards would do the job, and made emery boards because they would. (The effects of divine artifacts explain their causes via God's expectations and intentions.) But another way in which effects can explain why their causes *per*sist, if not exist, is natural selection. Once hearts appeared, their ability to circulate blood enhanced the fitness of genes coding for hearts; those genes replicated, producing more hearts. Pumping blood is the heart's function—and a heart pumping blood is functioning normally—because blood pumping got hearts selected in. A normal appetite gets animals to eat what nourished their ancestors in the wild, perpetuating the hunger drive by perpetuating the animal.

The upshot of this analysis is that the normality of a behavior depends on its evolutionary history. The same behavior can be normal for one organism, abnormal for another. Hunger for hay is normal for horses but not lions, since eating hay was adaptive for horses but not for lions.

"Normal" is evaluatively neutral: that behavior was adaptive in ancestral environments does not automatically make it good or right. Still, it carries positive connotations, primarily because of the positive correlates of normality. Using body parts for what they are for is pleasant. It is more invigorating to use the muscles evolution gave us than to atrophy in front of a television.[37] And normal behavior feels good because environmental pressures that select for a behavior, making it normal, also select for its being reinforcing. Enjoyable behavior is emitted more frequently, so enjoyment of adaptive behavior is itself adaptive; selectional pressures, we may assume, have made meat taste good to lions.[38] This lemma, that normality feels good, has the diagnostic corollary that only behavior accompanied by negative affect—conflict, guilt, compulsivity, dissatisfaction—is presumptively abnormal. Absent strong evidence to the contrary, behavior free of these accompaniments is normal.

Suppose the conflicting values of hunters and gatherers evolved in very different environments. A degree of helpfulness considered obligatory by hunters is considered foolish by gatherers, whereas hunters might regard gatherers as selfish. Each may think "something is wrong" with—and dislike[39]—the other. Yet, as these incompatible norms were each adaptive in the hunter and gatherer ancestral environments, there is nothing "wrong" with either. Hunters and gatherers who conform to their own respective values each act normally. In fact, a reinforcement tendency normal in one group may be abnormal in the other. There may be nothing wrong with a gatherer disinclined to help others, but a good deal wrong with an equally indifferent hunter. Symptomatically, hunters might experience guilt at failing to cooperate while equally uncooperative gatherers are emotionally placid.

Here is a model that allows black values to be normal. The separate evolution of blacks and whites, which appears to have produced cognitive and temperamental differences, makes it possible, indeed likely, that behaviors and norms pathological for whites are not pathological for blacks, and that identical behaviors and norms have different functional significance for the two races. It must be shown rather than assumed that behavior that is abnormal when displayed by whites is abnormal when displayed by blacks. Fighting and stealing often symptomatize personality disorders in white youths, for instance, so it is tempting to conclude that black youths, who fight and steal more frequently, are more disturbed. Should such behavior occasion weaker feelings of guilt among blacks than among whites, however, it may simply be that levels of aggression normal for blacks are higher than those normal for whites, and normal levels of concern for others lower. There is certainly evidence that, for blacks, violent behavior is accompanied by fewer negative feelings. The emotions of black teenage murderers have been described this way: "Their initial reaction after the shootings was not sorrow or remorse, but survival: How do I get out of this?" (Treaster 1992). This represents a minority of young black male murderers, to be sure, but there is no comparable white subpopulation. A survey by *USA Today* found that 35% of the looters arrested after the Los Angeles riots reported participating in the looting "because it was fun." Among many blacks, shoplifting is an accepted method of procuring stylish clothing and jail is a rite of passage. Nothing here indicates guilt or distress.

Commentators should bear in mind the possibility that the readiness for violence displayed by American blacks was adaptive in ancestral environments—or, more precisely, that the genotype expressed as violence in contemporary white society might once have been expressed as some, possibly distinct, adaptive phenotype. In that event, black violence would neither need nor be amenable to cure.

Likewise, such other putative black pathologies as drug use, reliance on public assistance, illegitimacy, and (what whites see as) lack of ambition, whatever they signify when displayed by whites, may express a genotype fit in ancestral environments. If Caucasoids and Mongoloids evolved longer time horizons because of the exigencies of survival in Eurasia, the level of concern for the future implied by unwillingness to work at boring, poorly paid jobs, while abnormally low for Caucasoids and Mongoloids, is normal for blacks. Indeed, identical activities will be perceived as differentially "boring" and "not worth the effort." Assuming that more rapid sexual maturation, greater expenditure of energy in mating than parenting, more intense male competition for females, and lower levels of male engagement in child-rearing were differentially adaptive in black ancestral environments, present black levels of illegitimacy and abandonment are also normal. One would always expect the fittest child-rearing practices to be selected for in any population. The most adaptive treatment of lion cubs is that administered by lionesses, rough as it is by human standards. The low academic achievement of black children adopted by whites (see chapter 4) suggests that black parenting patterns, although negligent and harsh[40] by white standards, are also a biological adaptation.

Think again of the spike in black illegitimacy and crime over the last 50 years (Jaynes and Williams 1989: 518; Hacker 1992: 197; *Encyclopaedia Britannica* 1914; Herrnstein and Wilson 1985: 465; Murray 1984: 116–120), probably facilitated by such environmental factors as the lower costs of illegitimacy and crime. I have conjectured that this change, together with white failure to respond similarly to these same factors, illustrates the gene/environment interaction stressed by Jencks and Lewontin: the black genotypes controlling violence and sexual behavior express themselves very differently in the pre- and post-1965 environments, with black and white genotypes diverging more in the post-1965 environment. On the present analysis of "normal" there would then be nothing abnormal about black levels of illegitimacy and crime *before or after* 1960. Both behaviors in both eras express the same once-adaptive genotypes. Most people deplore crime and illegitimacy, but "sick" used of dispreferred behavior without evidence of deviation from a past functional role—evidence including the behavior's likely maladaptiveness in ancestral environments, its present maladaptiveness, or its present association with negative affect—really is a disapprobative under a (patronizing) medical cover.

The point has been appreciated by some psychologists, although it is sometimes conflated with issues of bias (see Hare 1985). Gynther cautions that high MMPI scores predicting maladjustment among whites may not predict maladjustment among blacks, especially within the black community:

[O]ne cannot assume that a profile or code type has the same behavioral referents for blacks and whites. The attribution of maladjustment to blacks' MMPIs on the basis of white norms appears unwarranted and highly undesirable. (1972: 391; also see Dahlstrom 1986)

Gynther is right that nonwhite scores on diagnostic instruments may indicate good adjustment for blacks among other blacks, but it does not follow that black scores mean nothing. They do reliably register race differences in temperament and behavioral propensities; for both blacks and whites, higher Psychoticism scores indicate a propensity to behavior that violates white norms (Elion and Megargee 1975). The point is that a propensity to violate white norms need not be disordered or dysfunctional for all groups.

"Antisocial" is another group-relative term. Behavior works against a society when it frays social bonds, so pro- and anti-sociality may vary among societies held together by different bonds. The relations uniting group X may be impervious to or strengthened by behavior that weakens the relations uniting group Y, in which case anti-Y behavior will be neutral or pro-X. To the extent that white and black social groups are maintained by different relations, black behavior that is anti-white society may be pro-black society. One instance might be the swaggering black walk noted by many commentators (Wolfe 1987; Argyle 1988; L. Thomas 1992). The male gang appears to be an important element of black society, making aggressive physical display, which helps determine rank in loose male hierarchies, black-prosocial. The same oppositional/defiant body language is disruptive of white society, in whose crowded cities constant physical challenge is intolerable, and where hierarchical status is determined by more symbolic displays of dominance-aggression.

Calling black crime, drug use, illegitimacy and academic failure "pathologies" may just be a way of saying that these behaviors are maladaptive in the contemporary American environment. However, behavior is literally maladaptive only if it retards reproductive success, and the total reproductive rate of 2.2 for American blacks (Wattenberg 1987: 77) exceeds the population replacement rate of 2.1. In fact, despite a lower life expectancy and higher infant mortality, black fertility is greater than white. Brody (1992: 278), following Vining's somewhat different analysis, puts the white fertility rate at 1.46 and the black at 1.94. Hacker estimates that black women have 1.3 babies for every baby born to a white woman (Hacker 1992: 71; also see Jaynes and Williams 1989: 513–514); the fertility rate for the United States population as a whole is 1.8 (Wattenberg 1987: 142), also yielding a white rate below 1.5. Moreover, it is hard to see how black fertility could be higher if illegitimacy, drug use, and crime were lower. Thirty-three percent of all black children (and their mothers) are now supported almost entirely by the resources of genetically unrelated whites in the form of public assistance, rather than by their biological parents. Black success at inducing whites to divert resources from their own children to the children of unrelated blacks is successful exploitation of the environment rarely matched in nature.

Black behavioral norms might eventually *become* maladaptive. On the standard theory, *r*-populations are subject to collapse when environments turn unfavorable, for instance by the exhaustion of resources. Specific scenarios are hard to project, but one may speculate about what might happen should the tax base for welfare erode at the same time that white and Asian food retailers withdrew from black neighborhoods. Nonetheless, at the moment black norms are highly adaptive.

6.10. TOLERANCE, RELATIVISM, CROSS-CULTURAL COMPARISONS

A view of values as biological phenomena must not be allowed to add to the confusions surrounding "moral relativism" and "multiculturalism."

First of all, the incommensurability of group norms does not oblige anyone to do in Rome as the Romans do, or, when in Rome, to keep to the morés of his own culture, or in Rome or at home to tolerate the ways of others. Adopting local customs is one norm, defying them another, refusing to judge a third; naturalism accords them no special status. Naturalism does not tell American blacks to conform to white majority standards, or that they should, or are entitled to, defy them. It does not tell whites to maintain white values, or, come the demographic day when they are a minority, to adopt nonwhite ones. Nor does it counsel either race to tolerate the other, or to seek dominance. Having repudiated objective morality, the naturalist has no advice to give about what anyone ought to do.

A view of group norms as products of evolution must also be distinguished from "moral relativism," the doctrine, as it is usually expressed, that each group's values are right *for them*, and that each group should be judged by its own standards (see, e.g., Ladd 1973, and the more equivocal Wong [1984]). This popular view is not only not entailed by naturalism, it is, despite its surface plausibility and seeming broad-mindedness ("Judge whites by white standards, blacks by black"), unintelligible.

When relativists declare that the moral quality of an act depends on the society in which it is done, they are trying to say that moral terms are meaningless in isolation, and make sense only in such contexts as "right-for-[or wrong-for-]social-group-G." The very same act can be wrong for A and permitted, even obligatory, for B, and is such when forbidden by A's group but permitted or commanded by B's: as Christians but not Moslems accept eating pork, Christians but not Moslems may eat pork. Straightforward as this idea seems, however, it is on a par with saying that Mt. Everest is 29,000 feet tall for Christians but not for Moslems. Just as Mt. Everest either is 29,000 feet tall or it isn't, period, so too, once actions are granted moral qualities of any sort, eating pork is either permissible or impermissible, period.

The relativist replies that height differs from moral qualities precisely in being nonrelational. "Eating pork is wrong," he suggests, is better compared to "Mt. Everest is 10,000 miles from here," whose truth varies as "here" does. To see why this analogy only leads to more trouble, let us press the relativist by asking whether it is wrong, period, for someone to do what is wrong-for-him. Given that eating pork is wrong-for-Moslems, is it wrong for a Moslem to eat pork?

Suppose the relativist agrees that it *is* wrong for a Moslem to eat pork—that, for any group G, members of G should not do what is wrong-for-G. Then that initial "should not," the first "wrong" in "It is wrong for Moslems to do what is wrong-for-Moslems," is *not* relative to any system of norms. In agreeing that people should (not) do what their society tells them (not) to, the relativist has endorsed an absolute value, conformity to the norms of one's own group, and ceased to be a relativist.[41] The distance analogy sheds light on how this happened. "Mt. Everest is 10,000 miles from here" is indeed relative to "here," but once a point of origin is fixed the statement becomes absolute. When "here" is New York, what is being asserted is that Mt. Everest is 10,000 miles from New York, which is true independently of any point of origin; it is just as true in Perth as in New York that Mt. Everest is 10,000 miles from New York. Comparing morality to judgments of distance from a variable origin leaves morality absolute once a group is fixed.

The relativist can avoid this embarrassment only by declining to judge whether it is wrong, period, for Moslems to eat pork. That judgment too is relative, he must say: a Moslem's eating pork is wrong-for-those-who-think-people-need-conform-to-the-standards-of-their-group, while allowed-for-those-who-deny-that-people-should-conform-to-the-standards-of-their-group, and he must take the same prolix line on judging people by their own standards. The trouble facing the relativism now is that his "wrong-for- . . ." has begun to sound like the sophomore iconoclast's "true-for" in the pronouncement that "God's existence is true for believers, false for atheists." Sophomoric types wish to be relativists about truth, but all that their tendentious "true-for" idiom amounts to is that some people believe in God and others do not. This truism is not new, or interesting, or especially relativistic. The relativist's parallel "wrong-for-*G*" has degenerated down to "believed by group *G* to be wrong," so that when he reports that a practice may be wrong-for one group but permitted-for another, he is saying only that various groups disagree about right and wrong. This truism is not new, or interesting, or especially relativistic.

There is a connection of sorts between naturalism and tolerance. While in point of logic the one does not imply the other, in point of psychological effect a naturalistic outlook does tend to make people less judgmental. For one thing, it strongly suggests that some groups may be *unable* to conform fully to the values of another. Since no one is obligated to do what he cannot do, deviant groups are relieved of blame for falling short of societal standards they cannot live up to, and normative groups are relieved of the disagreeable chore of passing negative judgments. It is as pointless to insist that individuals with innately high time preferences and low empathy should follow white rules of cooperation as to insist that Japan field world-class basketball teams. Second, realizing that there is no right answer to a dispute often leads the disputants to disengage. Nobody would criticize Martians—who sense only gamma rays—for indifference to Holbein, nor would Martians criticize humans for indifference to their short-wave masterpieces. Shakespeare enjoys the highest prestige among Caucasoids, but biologically based differences in aesthetic response might lead others to find his work less compelling (see Smith 1988), and his language simply baffles people whose IQ falls below some minimum. White musical tastes favor harmony, complexity, and rhythmic regularity while blacks favor improvisation and syncopation. People debate the merits of jazz and classical music, with increasing rancor as the subject has acquired ideological baggage, but naturalists see the issue as irresolvable. How could an observer, whose neural wiring disposed him to enjoy neither, decide between saving the last score of a Bach Passacaglia and the last recorded Louis Armstrong solo? Once one agrees that there is no way, the jazz-versus-classical debate mercifully loses its point, and heat.

Yet interracial conflicts are unlikely to abate so easily. Living and letting live is easy when it comes to art, but what of crime, public deportment, claims of redress? Does sociobiology offer anything beyond power struggles?

Despair is premature. It is a commonplace of moral philosophy that many ostensibly normative disputes are really about facts, very often the most efficient means to an accepted end. Battles about which route the family "ought" to take boil down to which highway will get everyone to his destination most quickly, and, once tempers subside, can be resolved by studying a road map. Many cross-cultural disagreements are also at bottom instrumental. One society may urge its farmers to pray to the rain god while another recommends experimenting with fertilizer, but they agree in wanting greater crop yield; where they disagree is how to get it. All such disputes can be settled by empirical investigation.

Even when intergroup conflicts rest on ground-floor divergence in values, as some unquestionably do, there is reason not to despair. The practical consequences of abandoning the search for a God's-eye point of view are less ominous than might be thought.

That a world without a vantage point should seem fearful is easy to understand. It is a fact that people dislike departures from their own values, and those who do the departing; values being reinforcement tendencies, this hostility is almost a matter of definition. Insofar as moral variance has a genetic basis, hostility toward those whose values differ, and who therefore are likely to carry different genes, enhances inclusive fitness. A certain degree of moral intolerance may have been selected for. It is also a fact that hostility impedes the peaceful resolution of conflict. The parties to a dispute cannot agree to disagree when just one claim can prevail, and, should compromise prove impossible, the only recourse is fighting. Belief that there is an objective right and wrong in a conflict gives hope that both sides will recognize them, but denying that any side is right seems to concede as inevitable, perhaps to sanction, the use of force.[42]

The saving consideration, so to speak, is that impasses are just as likely if there *is* a perspective from which to discern objective values. Disputants will come to blows even if one side is objectively in the right when the side in the wrong won't see things that way. Strife is rooted in human intransigence, not theories of value. Conversely, if the parties to a dispute can come to an agreement, peace will break out whether or not their agreement can be validated externally. Once disputants adopt a common perspective, they do not need God's.

Despite the superficial resemblance of naturalism to relativism and the loose connection between naturalism and tolerance, there is no affinity whatever between moral relativism and political liberalism. D'Sousza (1995) argues that white America tolerates black behavior because followers of the anthropologist Franz Boas popularized the view that no group's norms are better or worse than any other, hence that black behavior cannot be judged "worse" than white. This conviction led whites to abandon reinforcement of their own norms, and D'Sousza hopes that blacks can be persuaded to aspire to white norms once the grip of relativism is broken.

Historically this account is far-fetched, since the Boas school had little influence outside academic circles. It is doubtful that any of the editorialists, activists, lawyers, or legislators who championed civil rights had more than the vaguest idea of or interest in anthropologists' views of morality (although Margaret Mead's fabulations about Samoa did convince many people that sexual norms are arbitrary). The most influential thinker about race of the period was in fact Gunnar Myrdal, who maintained (Myrdal 1944) that blacks were just like whites, and behaved otherwise only because of discrimination. Far from thinking that black norms *differ* from those of whites but finding himself intellectually unable to criticism them, Myrdal accepted white norms as a benchmark, believed blacks at bottom shared them, and expected this commonality to emerge once race prejudice ended. White liberals followed Myrdal's lead, not Boas'. Conceptually, the sheer existence of the present chapter is its own refutation of D'Sousza. He would certainly classify naturalism as a form of relativism, yet it is integral to a position he could scarcely count as "liberal."

What naturalism does need is some pragmatic point of view for adjudicating racial conflicts; one is offered in the next chapter. Before turning to that, however, some words are in order about "multiculturalism."

6.11. "MULTICULTURALISM" AND "AFROCENTRISM"

If critics (D'Sousza 1995; Lindsay 1996; Mack 1996) of the increasingly pervasive trend toward "multiculturalism" and "Afrocentrism" associate it with "relativism," its friends associate it with "tolerance." Who is right?

Standard "Afrocentric" works—Van Sertima (1983), Hunter Adams (1988), James (1952), C. Williams (1974), Diop (1974)—are too meretricious to shed much light on this question or repay extended discussion for any other reason. For instance, part of James' case that the Greeks stole philosophy from Africa is that Alexander's conquest of Egypt gave Aristotle access to the library of Alexandria (1992: 17). Alexander of course *founded* Alexandria, there having been no library there previously. Incompetence of this order leaves the commentator little choice but to reconstruct the debate in his own terms.

Taken most charitably, and as usually presented, "multiculturalism" holds that all cultures deserve understanding and respect. "Afrocentrism" adds that the cultures and achievements of Africa and black America are unduly neglected in the contemporary United States, hence deserve a larger role than the marginal one they now receive in the education of black and white children. Hearing more about blacks will raise black self-esteem, damaged by racism, while white children, raised to think their culture best, will be made more modest. Black children do not need to learn any more about white culture, as they are already inundated by it.

So understood, multiculturalism is neither nihilistic nor relativistic. Far from forbidding cross-cultural comparisons, it insists that black culture is at least as good as white, and that it is a good thing, period, to play down European culture. Multiculturalists think *everybody* should know more about black achievements.

Some critics (e.g., Bloom 1988, Henry 1994) reply to this official or exoteric version of multiculturalism that the European classics deserve pride of place because of their intrinsic excellence and universality. However, the notion of "intrinsic excellence" is ill-defined from a biosocial perspective, nor is it clear that European culture is esteemed equally highly by blacks and whites. E Hirsch's (1989) more empirical argument, that the United States is a European country which for better or worse must maintain contact with its European origins to comprehend itself, invites the reply that such conservatism *is* for the worse, an excuse to perpetuate oppression.

Bloom, Henry, and Hirsch miss the underlying issue because they treat multiculturalism and Afrocentrism as *normative* claims about the *value* of black culture and its proper place in education. In fact, multiculturalism and Afrocentrism are claims of *historical fact* about the attainments of Africans and American blacks. To be sure, these claims are made part of a case against the majority society. Grant that black achievement is central to history, that science and mathematics are black creations stolen by the Greeks, and that this great truth is scarcely known because whites have suppressed it: it follows at once that whites are villainous. The claims themselves, though, are empirical in nature, not about educational policy in any controversial sense. Everyone would instantly accede to the demands of multiculturalists were the facts as they say.

Assessment of Afrocentrism by the usual tests of historical accuracy is impeded by the bellicosity of Afrocentrists, which keeps pushing the issue back into the realm of partisanship. Thus, after claiming in *The Destruction of Black Civilization* (1974) that Africans invented, among other things, democracy, Chancellor Williams concludes:

The white man is their Bitter Enemy. For this is not the ranting of wild-eyed militancy, but the calm and unmistakable verdict of several thousand years of documented history. [There must be] a mental revolution out of which Black America faces up to the stark reality that white America as a whole is its enemy. . . . (310–322)

This rage runs on until the reader forgets that it is fueled by the factual claim that Africans invented democracy.

It is especially frustrating to pursue the issue though the minefield of charges of "racism," detonated by almost everything anyone says, but the resulting contradictions are inseparable from "multiculturalism" itself. We have already seen how Hacker classifies belief in mental differences between the races as "bigotry" (1992: 27), yet considers it "inherently racial" malice to ask black children to adapt to a "white" school setting in view of their distinctive "learning style" (171). Hacker goes on:

black children should be given more opportunities for expressive talking, since black culture gives as much attention to style as to the substance of speech. Here, too, it has been found [no references given] that black youngsters apply themselves more readily to lessons involving actual people than to more abstract situations. Similarly, teachers should be tolerant of more casual approaches to syntax, time, and measurement. . . . [W]hite children tend to tell stories in a "linear" fashion, while black children are more apt to employ a "spiral" style. . . . Black children are also more attuned to their bodies and physical needs. So, some educators argue, they should be allowed more leeway for moving around the classroom. (1992: 171–2)

Hacker suggests that blacks would be happier being educated away from whites, although he would surely denounce the idea that white children would be happier away from blacks who cannot speak grammatically, sit still, or keep to the point.

The standard resolution of this inconsistency is to say that black children differ from whites because of discrimination, and for that reason need exposure to their own cultural heritage—which is anyway more attractive than "a world dominated by technology, administration and corporate priorities" (Hacker 1992: 175). Furthermore, to convince black children that they are the intellectual equals of whites and Asians, they should be taught that their ancestors produced cultures comparable to those of Europe and Asia. This is how Hacker comes to urge teachers to tell black children that Africans invented vaccination and carbon steel blast furnaces, and beat Columbus to the New World (172–174), although he carefully avoids saying these claims are true. By this pedagogical route does genetic egalitarianism inevitably lead to Afrocentrism.

Only resolute focus on the empirical question, the actual historical contribution of blacks, can keep discussion from decaying into irrelevance and emotionalism. (Hacker also recommends that black students adopt African names, and he falls into the language of "mental genocide"). The issue is not white villainy or the needs of black children, but whether blacks have been as important to American and world civilization as Afrocentrists say. When it is asserted in "African and African-American Contributions to Science and Technology," a "Baseline Essay" around which Portland, Oregon elementary schools design their science curriculum, that Africans have been "the wellspring of creativity and knowledge on which the foundation of all science, technology and engineering rests," the single pertinent question is whether this is in fact so.[43]

Some might object to this question on the grounds that historical importance is itself a normative and selective notion, hence deformable by (white) bias. There is little to be said in support of this familiar claim. Importance has an objective measure: impact, the difference an event, individual, or group has made. Columbus was more important that Cortez because the gap between our world and the world that would have existed had Columbus never been born is wider than the gap between our world and the world that would have existed had Cortez never been born. Classical Greece was more important than fifteenth century Australia because the present would have been about the same without Australian aborigines, but unrecognizable without the Greeks. Every group's history matters to the group itself, but not every group has mattered equally, made an equal difference, to mankind as a whole. Admittedly, what humans counts as a difference reflects their concerns (Nozick 1989: 172–173), but once what is to count as a difference has been specified, historical importance—the difference made—becomes a question of fact.

As everyone knows but has become reluctant to say,[44] the world as a whole would hardly have noticed had sub-Saharan Africa not existed or never been contacted by Europeans and Asians. No important discovery, invention, or world leader emerged from Africa. The art, music, architecture, literature, and political history of Eurasia owe virtually nothing to Africans. Trade with black Africa (as opposed to European exploitation of the mineral wealth of the African continent) has always been negligible. Afrocentrists point to or, as in Bernal (1991) exaggerate, the contributions of Egypt,[45] but in any event Egyptians were not black (Baker 1974).

Blacks in the United States formed a small, rural population until well into the twentieth century, played no role in the construction of the major American cities, and had little contact with the nineteenth century immigrants who influenced the country's character. The Civil War, although it revolved around blacks, was conducted by whites. Democracy and liberty, the basic ideals of American public life, are European (see Schlesinger 1992). The relatively small cadre of black entrepreneurs, scientists, and doctors has not appreciably influenced American business or medicine. Some inventions by blacks, such a the traffic light and railroad coupling, while significant, cannot compare in impact with the airplane or the telephone (although they now are often given more prominence in history texts); other widely cited attributions, such as Benjamin Banneker's design of the District of Columbia, seem to be fiction (see Bedini 1972: 109). While the patent office keeps no racial statistics, i seems likely that blacks hold disproportionately few patents. Of the formative works of American literature, only *Huckleberry Finn* is concerned significantly with race: the topic is scarcely mentioned by Hawthorne, Melville, Poe, Whitman, Dreiser, James, Wharton, Sinclair, Hemingway, Salinger, Heller, or Roth. Few blacks have achieved eminence in areas other than sports, entertainment, the demand for rights, or writing about race itself.

Afrocentrists accuse conventional white authorities of inventing these facts. But white historians, who freely acknowledge the attainments of Asians, have no reason to lie about blacks. The truth is that, until recently, most whites gave blacks relatively little thought, and would not have cared enough about blacks one way or the other to invest energy in obscuring their achievements. Blacks simply have not mattered as much to the white world as Afrocentists implicitly claim they did—and perhaps wish they did, for having powerful enemies is considerably more flattering than being ignored. The limited interest in blacks shown by conventional historians is best explained by the belief that further interest was unwarranted.

Probably because of its weakness under objective scrutiny, Afrocentrism has adopted the social constructionist denial that objectivity of any sort is possible. This denial defeats itself, as philosophers have recognized since Socrates took on Protagoras. "[M]ost scholars are willing to admit that there can be no such thing as an 'objective' history or 'neutral' social science," says Hacker, without citations (1992: 170); "All depictions bolster some interests; all interpretations support some present view." But the denial of objective facts implies that it is not an objective fact that there is no objective history, nor an objective fact that most scholars are willing to admit there is no objective history. Hacker's claims entail that they themselves are merely interpretations he has imposed to bolster some interests. And if conventional sources can't be trusted because of their motives, neither can Africanist sources, who have more to gain in prestige and power, from fabrication. A single standard must be used: the evidence offered by conventional white authorities cannot be dismissed as doctored while the evidence offered by unconventional black authorities, such as it is, is taken at face value. Africanists cite scattered references to blacks in Herodotus to support a Nubian origin for Greek religion, but ignore Aristotle's silence about Africa. Why should Aristotle have lied, but not Herodotus?

Running through the entire "stolen heritage" thesis is the idea that knowledge is a tangible good that, akin to a stolen car, can be lost to its original possessor. Knowledge is not like that. After Marco Polo "stole" the preparation of noodles from the court of Kubla Khan, the Chinese remained able to make noodles. The secret *spread*. Afrocentrists speak of a "raid into Africa" in Aristotle's time to steal African science, but ignorance of the Pythagorean Theorem on the part of modern Africans cannot be explained by the Greeks having made off with it. The currency among blacks of the "stolen heritage" doctrine, with its curious conception of knowledge, is perhaps more evidence of race differences in cognitive functioning.

At the risk of lapsing into the moralism I have sought to avoid, I would make some suggestions about "multicultural" education.

The principal point is that, whatever the United States may eventually become, its schools at present are white institutions—created and financed by whites, based on white pedagogical ideals, which include white criteria for adequate performance. It seems reasonable to expect anyone demanding access to these (or any other) institutions to conform to their extant standards. Blacks who think it important for their children to learn about black culture can teach it to their children after school hours; no reason comes to mind for forcing white children to learn this material in public schools. Blacks to whom the curriculum in white schools is intolerable—who wish to spend as much time studying Africa as Europe, and pursue literature courses which "balance" white writers with black ones—are and should be free to create their own schools. That is the meaning of tolerance and diversity. After all, black universities exist without anyone insisting that they make whites comfortable. Blacks without the resources to create the schools they wish must accept attending the schools others have created, abiding by the standards these others consider appropriate.

An obvious parallel is religious instruction. Catholics would like their children to learn about Catholicism, but they do not insist that the public schools teach it. They either send their children to Catholic schools, or, if their children attend public schools, send them for religious instruction after. They do not insist that all children, Catholic and non-Catholic, learn religious dogmas. Insistence that someone else—the government, in the case of public schools—do a job that black cultural institutions should be doing breeds resentment, and confirms the idea that blacks on their own cannot sustain an educational system.

No one has forced blacks to attend white schools; blacks demanded admission to them. Surely it is unseemly to demand access to an institution in the name of equal opportunity—opportunity, that is, to compete *by the rules the institution has established*—and then insist that its rules be changed. What would we think of a short man who asks for only a "fair chance" to play basketball, to be judged by his ability rather than his height, and then, once on the court, insists that the basket be lowered or the ball changed? In being allowed to play by the going rules, he got a chance that was fair by his own stipulation. In asking for more he complains of rules he has implicitly agreed to accept. That is what blacks do when, after entering schools know to be of European design, they complain of a "Eurocentric" curriculum they have difficulty mastering.

NOTES

1. The reality of value is often explicated as the claim that some things are truly c actually right, or that there are moral facts or moral properties, but these explication are empty. A thing is "truly" or "actually" so-and-so when it is so-and-so (see Lev 1984d, 1991; Horwich 1991), and facts and properties are metaphysically superfluo (see Levin 1979: I.3–I. 5).

2. What follows owes much to Hocutt (1977); also see R. Perry (1926).

3. When a stimulus (such as the taste of sugar) has a unique distal cause (such sugar), the distal cause may be called a reinforcer; compare the discussion "environment" in section 4.1.

4. It has been denied that any characteristic pleasure-experience accompanies reinforcement. After all, people "like" to read mysteries and eat ice cream although the experiences accompanying these two activities differ. Furthermore, pleasure is inseparable from its putative cause. The taste of ice cream is what is enjoyed, not the pleasure of having the taste. Still, the psychologist's terms "feeling tone" and "positive affect" seem to name something—perhaps awareness of reinforcement.

5. Contra Brink (1989: 26, 78; also see Gibbard 1982: 39), a behavioral analysis accommodates a readiness to judge historical and fictional figures beyond one's influence. Preparedness to condemn Macbeth or Caligula is a residue of stimulus generalization, the process that leads experienced drivers riding in the passenger seat of a car to hit a phantom brake in heavy traffic. A disposition to brake when the car one controls is going too fast can be partially triggered when a car one is riding in goes too fast. Likewise, a disposition to condemn murder can be partially triggered by tales of murder. Partially triggered—edgy passengers seldom grab the wheel, and playgoers don't call the police when Macbeth stabs Duncan.

6. Perhaps "conscience" names an introspective awareness of categorical reinforcement tendencies—what it feels like to be categorically disposed to reinforce a behavior, especially a behavior one has omitted.

7. If unbeknownst to science cardiovascular systems resonate, so that everyone's health increases with the number of joggers, "Jog!" would still not feel like a moral imperative. This suggests that a "moral" rule should be defined as one that is reinforced because it is *believed* to benefit each if and only followed by all. People urging each other to jog after they discovered the resonance probably *would* make a characteristically moral appeal: "You're benefiting from the jogging of others; do your bit."

8. One might formalize the notion of a moral rule as follows. Letting $b_r(m,e,n)$ be the mean benefit to individuals of following rule r in environment e when m others in a population of size n also follow r, r is "moral" only if $\partial b_r / \partial m > 0$. Should $\partial b_r / \partial m = a$—so that the rate at which each new recruit enhances the benefit of following r be constant—a natural solution is $b_r = a(r,e)[m - n/k(r,e)]$, where $a(r,e)$ and $k(r,e)$ are parameters and $k < n$; obeying r is more advantageous in environment e than in environment e' if $a(r,e) > a(r,e')$. Here $1/k$ is the proportion of the population that must follow r for following r to be advantageous. Adherence is advantageous when $m > n/k$, indifferent when $m = n/k$, disadvantageous when $m < n/k$. If $\partial b_r / \partial m = am$, so that the rate at which the benefits of following r change is proportional to the number of adherents, a natural solution is $b_r = a[m^2 - n^2/2k^2]$.

9. "[W]hen [the probability of interaction] is sufficiently great, there is no single best strategy regardless of the behavior of the others in the population" (Axelrod and Hamilton 1981: 1392).

10. So talk of the "morality" of wolves is misguided. Within the pack wolves show a concern for others that resembles morally motivated altruism, but they don't generalize beyond that circle. Wolves protect their own pups but are happy to eat moose calves. There is no such thing as "lycine" morality; wolves are amoral.

11. M. Hocutt, in correspondence.

12. Here and elsewhere I use mentalistic language for brevity.

13. Brink asks "*why* anyone should be so keen on getting others to share his attitude, or on recommending a course of action, unless he thought the attitude he was expressing or the course of action he as recommending was correct or valuable" (1989: 8). Answer: because humans find agreement pleasant.

14. "Function" is discussed more fully in section 6.10.

15. Here is an instance in which talk of "group selection" is well-defined, and convenient. However, these remarks can be recast in terms of individuals. reinforcement of morality, because individually fitness-enhancing, sustain groups of

morality-reinforcers, allowing group fitness to piggyback on the fitness of group members.

16. Marinoff (1990) objects that a pure defection strategy is evolutionarily stable, meaning that, by definition, mutant cooperators would always be wiped out. This objection overlooks the point that change in the genetic relations between individuals changes the payoff matrix under which defection is stable; see Axelrod and Hamilton (1981: 1394). More important, stability refers to the impossibility of any *single* variant gene penetrating a gene pool, while in the Trivers-Axelrod-Hamilton account a cluster of (interacting) cooperators enters the cheater gene pool simultaneously.

17. There are explanations of genuine saintliness less cynical than E. Wilson's of Mother Theresa (1978: 165), that she wins the security of her church. Perhaps 1 person in 10^6 is a genuine saint. Following Ellis (1991: 7), let a 10-loci polygene control altruism, with qs between 0 and 1 leading to a normal distribution of genotypes with saints in one tail, the dominant alleles always coding for selfishness. If the population is in equilibrium when q = .2, 1 in every $2^{10} \approx 10^6$ genotypes will be homozygous recessive, that is to say saintly.

18. Paralleling "biopsychology," the biological approach to mind, a biological approach to society might better be called "biosociology" than "sociobiology." Precedent aside, the adjective "biosocial" is more graceful than "sociobiological."

19. Further constraints must be imposed on the matrix to rule out "mixed," probabilistic strategies.

20. To avoid infiltration by deceptive cheaters, cooperators must have evolved cheater detectors. There is speculation that these detectors were so good that an interactant could convince cooperators that he would reciprocate only by *genuinely intending* to reciprocate. That may be why interactants can see the advantages of cheating yet be unable to bring themselves to cheat. It has been argued that Hobbes believed such bootstrapping is what inclined men to "perform their covenants made."

21. Butler observed that one cannot find pleasure in an activity he does not like for itself. If one liked *only* the pleasure the activity yielded as some sort of consequence, the activity would yield no pleasure of any sort. Unless the father finds reading intrinsically pleasant, he could not get pleasure from it. Also see Feinberg (1989a).

22. Some sociobiologists do talk as if choice does not matter; often their critics gratuitously impute this error to them. Not all sociobiologists are nothing-butters; here is Symons (1979) on jealousy: "it would be inaccurate to infer that a furious cuckolded husband only imagines himself to be angry at his wife's sexual peccadilloes when, in some more profound sense, what he is 'really' doing is promoting the survival of his genes."

23. It has been held that value is detected by a quasi-perceptual faculty of moral intuition, but this theory does not explain why moral intuitions conflict. It won't do to say that some people are morally blind or suffer from moral illusions, since many optical illusions are cured by a closer look, whereas closer looks seldom resolve moral disagreement. Some optical illusions like the Müller-Lyer survive close inspection, but perceivers can be brought to realize intellectually that the lines are equal—whereas nobody ever says "I know X is wrong, but I just can't help thinking it is right." Color blindness is revealed by inability to discriminate, but someone who sees no moral difference between abortion and infanticide can still in a literal sense tell these practices apart. At least one of two societies in fundamental moral disagreement must be suffering from an ongoing moral illusion, on the moral intuition theory, whereas who societies do not suffer from ongoing optical illusions.

24. That what needs explaining are behavioral tendencies, not beliefs with propositional content, meets an objection from Lycan (1986: 87–90). He argues that an explanation of moral judgment must refer to value because its explanandum is not merely the making of these judgments, but their *contents*. Just as scientific theories

about observations explain not only scientists' beliefs about what they see ("I thought I saw red litmus paper") but what they see (red litmus paper), a theory of morality must explain not only why lying is thought wrong, but why it *is* wrong. In Lycan's terms, the "intuited" as well as the "intuition" must be explained.

A partial reply is that scientific theories must explain the content as well as the occurrence of observation reports because the occurrence of most scientific observation reports are caused by what is reported. When a scientist thinks he sees red litmus paper, it is usually because the paper *is* red. Hence, explaining his belief about what he saw involves explaining what he saw. Were observation reports explicable without the assumption that scientists see what they think they see, the reports themselves *would* be the sole explananda. What distinguishes moral intuition is that, unlike observation, its deliverances seem explicable without their truth being assumed.

But the basic trouble with Lycan's argument is that, by treating moral intuitions as beliefs, it begs the principal question. Since all beliefs have propositional content, moral beliefs would have to have content also, which, on the analogy with science, must (just as Lycan says) be taken at face value by an explanation of morality. But reinforcement tendencies have no propositional content, to take at any value. On the view defended here, moral convictions are dispositions; the conviction that lying is wrong is a disposition to punish lying and reward probity. Hence, all that needs explaining is this disposition. There are no propositional contents, no "intuiteds," to explain. (On "non-doxastic" analyses of morality, see Papineau 1993: 199-201).

Brink's use of "recognition of what is morally required" (1989: 48, 225) as a primitive is similarly biased toward realism, since it suggests that moral judgments have uneliminable propositional content.

25. Sturgeon (1984: 442–444.; also see 1986; also Sayre-McCord 1988; Railton 1986) argues that moral facts do help explain beliefs and behavior. We think Hitler evil, according to Sturgeon, because he *was* evil. We think him evil because of what he did, and only an evil man would do what he did. It is not that Hitler's wickedness *as opposed to* his hatred of Jews caused the Holocaust, but that, since such hatred is necessarily wicked, Hitler would not, indeed could not have done what he did without being wicked. One may reply that this counterfactual, even if true, is not explanatory; Hitler's behavior was caused by motives on which, as Sturgeon and others (see Kim 1990, 1993) put it, evil supervenes. Well, Sturgeon would presumably ask, what is so wrong with that? Given a necessary connection between a supervenience base and supervenient properties, why not attribute what the base causes to the supervenor? Because allowing supervenors to be explanatory rules out epiphenomenalism in philosophy of mind and elsewhere. For suppose, as epiphenomenalists do, that neural state S causes both mental state M and behavior B, M being causally inert. As M is not an intervening mechanism by which S produces B, M is an epiphenomenon if anything is. Yet B would nonetheless not happen unless M did, for B would not happen without S, and, since S is sufficient for M, B cannot happen without M. Hence, if supervenient value explains actions, M explains B. Epiphenomenalism may be false, but it should not be falsified so easily.

Brink the realist (1989: 186) cheats a bit on the explanation issue. Since skeptics say values make no explanatory difference, Brink correctly claims the right to assume the existence of values for the purpose of proving that they would make a difference if they existed. But Brink thereupon permits himself the much stronger assumption that the wrongness of cruelty *enters into the explanation of the belief that cruelty is wrong*. That is more than he is allowed: it is what he must *show* from the weaker assumption that cruelty is wrong.

In discussing the explanation issue, Nagel (1986: 144–146) first concedes that "you don't *need* to explain normative judgments in terms of normative truths." But then he draws back. "[A] severe headache . . . seems to me to be not merely unpleasant, but a bad thing . . . and the true explanation of my impression may be the simplest one, that headaches are bad, and not just unwelcome." According to Nagel, a headache is bad, not merely unwelcome, because the sufferer "has a reason to get rid of it." No doubt the sufferer does have a reason to get rid of it, but that reason is surely the pain, not the pain's badness.

26. Following Sturgeon, Brink argues from examples that normative statements predict: "Zenobia is a good person" jointly with "Good people keep their promises" predicts that Zenobia will keep her promises (1989: 137). However, Brink's criterion for predictiveness is far too loose. "Zenobia is zorful," jointly with "Zorful people invest," predicts that Zenobia will invest—yet statements about zorfulness are cognitively empty. (See Berlin 1939; Hempel 1952; a counterexample-proof criterion of "predictive hypothesis" remains elusive.) Explanation and prediction being partially symmetrical, the trouble with moral prediction resembles the trouble with moral explanation. What predicts Zenobia's fidelity is a naturalistic trait, her consistency, which is (at most) the supervenience base of her goodness; Brink replies that the supervenience base *is* her goodness; see n. 28.

27. Railton (1986) offers a two-stage argument for the explanatory efficacy of value. He first explicates "X is intrinsically good for A," naturalistically, as "A would want himself to want X if he knew everything" (173–174). He then points out (178) that someone's health might be explained in terms of the intrinsic goodness of his diet, as when we say "He knows what's good for him." Social phenomena, he goes on, can also be explained in terms of value, as when denial of the objective good of large numbers of people causes unrest (192, 199). Even the collapse of a roof can be explained by saying that it *ought* to have had 2×8 rafters instead of 2×6s (185).

One may first query whether it is not the supervenience base that is doing the explanatory work—that Smith is healthy because he knows what will make him healthy in the long run, not because long-run health is intrinsically good for him. In addition, Railton's explication of objective values is hardly clear. What would me +(his term for myself omniscient) want me to want? If me $^+$'s desires are not mine, me $^+$ might think I should want what he wants. But then the question becomes what I would want were I infinitely knowledgeable, a question I see no way to answer. Alternatively, we might assume that me wants what I do, or, at any rate, that when me $^+$ wonders what I should want, he frames that question in terms of my wants, not his. In that case, thought, my "intrinsic good" comes down to satisfaction of my wants, or possibly my long-run wants. We are back to wants, not value, doing the explanatory work.

28. Some philosophers (Boyd 1988; Gibbard 1982—but see Gibbard's qualifications on 43) regard naturalistic explanations of preference as identifying, not eliminating, value. Just as chemists discovered that water is H_2O, they argue, sociobiology may have discovered that goodness is reciprocal altruism. The synthetic character of this identity is said to overcome Moore's meaning-based open question argument against naturalistic definitions (Brink 1989: chapter 6), and to allow moral properties to be explanatory, since they *are* the natural properties that cause behavior (Brink 1989: 19). But comparing value to water is initially suspect because the existence of water itself is not in doubt, whereas the existence of value is the very question at issue. Identifying goodness with a natural property also breaks the seemingly analytic link between valuing and behavior. It becomes possible for someone—Brink's "strong amoralist"—to consider an act right without being inclined to do it. If this attitude were possible, so would be others that seem plainly *im*possible. Consider "cool," applied by teenagers to clothing styles and pop stars. The same thing may be cool at one time and uncool at another, in just the way (according to Brink) the supervenience bases of moral qualities are disjunctive; there are many ways a distribution can be just and, according to Brink, the same distribution may be just in one context but not in

another. Why, then, is coolness not a disjunctive, multiply realizable property of dress and pop stars, with the strong amoralist matched by the "strong nerd," the science whiz who knows what is cool but just doesn't care about it? Coolness-realism is surely less plausible than coolness-emotivism, according to which "cool" expresses approbation, strong nerds are impossible, and ordinary nerds either wish but are unable to be cool, or think (although not in those words) that science is cool.

29. Lionesses teach their cubs to hunt, but the whole process by which lionesses are prompted to teach and their cubs to learn is innate (again see J. Gould 1982). Direct expression of the genes as the phenotypes of Figure 6.2 may seem a more efficient strategy, but presumably nature rejected it as unfeasible. Prey may be so multifarious and encountered so haphazardly, for instance, that innate tendencies to catch each type in the most suitable way confer no advantage. A gene programming the ability to learn the habits of prey and teach what is learned might be fitter.

30. Among the three conditions Trivers (1971) lists for the emergence of reciprocal altruism is "degree of mutual dependence."

31. "Educational psychologists should take the lead in studying the Asian work ethic to ascertain its key motivational behaviors and to see whether they are applicable to non-Asian students" (Yee 1992: 117).

32. Thus, black culture may be praised for not instilling dubious traits, like "skills needed for managing administrative systems," necessary for success in white society because of "willingness [on the the part of white youths] to adopt a structure of success set up by adult society" (Hacker 1992: 144–145).

33. Sixteen-year-old Brian White emptied the magazine of an automatic into 12-year-old Quentin Carter because White had felt dissed by the manner in which Carter had asked him for a quarter (C. Goldberg 1995).

34. The situation is more complicated, since whites who seek to placate blacks generally do so at the expense of other whites. Supporters of racial preferences do not resign their own positions for blacks; the white intellectuals, lawyers and government officials who defend and administer preferences usually escape its consequences. So whites do not literally disadvantage themselves for blacks; one group of whites disadvantages another. There may be further biological dimensions here, since the two white groups are competing for resources.

35. He adds: "White sympathies will typically be with the person on the receiving end of such an exchange." On the present hypothesis, whites—including whites on the receiving end—tend to sympathize with the ostensibly injured party, the black.

36. Wright's analysis has provoked a large literature, some references to which are given in Levin (1997). The most challenging counterexample is Boorse's (1975): an accidental leak in a pipe releases deadly gas, preventing repairmen from closing the leak. The persistence of the leak is explained by its release of gas, but the leak is not for releasing gas. However, the leak differs from the heart in that the cause of the leak is not itself explained by what it does. Leaks are caused by corroded gaskets, but gaskets do not corrode because corroded gaskets leak. By contrast, hearts not only exist because of what they do, but the physiological mechanisms that produce hearts also exist because of what they do, namely produce hearts, and the DNA that codes for these mechanisms exists because of what it does, and so on. So we may extend Wright's analysis by saying that F is a function of X when the X-F sequence ends a series of causes each of which is explained by its effect; see Levin (1997).

37. People watch too much TV now because resting when possible was once adaptive, and is not yet maladaptive.

38. No trait should be more maladaptive and anhedonic than homosexuality, yet homosexuality appears to have a major genetic component (Levay 1991). See Levin (1996) for a discussion of how homosexuality might have been selected, and whether it normal.

39. "Agreeing on what is to be a community's 'normal' discount rate is a highly charged social process. Individuals or groups who accept a higher discount rate than the consensual one tend to be ostracized as shiftless, but they are apt to seem less threatening to other people than are individuals or groups who have achieved a lower rate, who often are accused of being misers and are persecuted, as in the pattern of Western anti-Semitism, or the attacks on Indians in Africa or the Chinese in Southeast Asia" (Ainslie 1992: 233, n. 3).

40. See Moore (1986) on the greater negativity of black as compared to white mothers of black adoptive children in teaching situations.

41. It might be argued Moslems should not eat pork because, although eating pork is neither right nor wrong, Moslems think it is wrong, and one should never do what one thinks is wrong. But this reply assumes the absolute judgment that one should never do what he thinks is wrong.

42. Leibniz called naturalism "extremely dangerous." Marlowe's Machevel in *The Jew of Malta* says, "Many will talk of title to a crown: What right had Caesar to the empery? Might first made kings, and laws were then most sure When, like the Draco's, they were writ in blood." Naturalism seems to encourage this attitude.

43. Africanists when challenged often decline to defend the truth of their assertions, pleading instead that they stimulate discussion, are optional, or do no harm because no one believes them. "The essays have played a role in framing a new and improved debate," according to Portland's Coordinator for Multicultural Education (cited in Travis 1993: 1122). Of the claim in the baseline essay that Africans invented the airplane, she says "I as a classroom teacher would not teach that as fact. . . . We have not said to people, you must read it, believe it, and teach it." Sometimes Afrocentrist excess is blamed on whites: "It's the story of something that is not very good that went into a vacuum" (an official of the American Association for the Advancement of Science cited in Travis 1993: 1121).

44. "One reason for the silence may be that educational groups feel it's tough to criticize Afrocentric material without being called racist. 'Everyone's so damn cautious about stepping on non-white toes' says Eugenie Scott, executive director of the National Center for Science Education. To some extent, William Aldridge, executive director of [the National Science Teacher's Association] agrees: 'Any criticism at all is going to be criticized as racist. It's a no-win situation' "(Travis 1993: 1122).

45. Surviving Egyptian religious writings lack the dialectical character of Greek philosophy. Although the Egyptians knew special cases of the Pythagorean theorem and the volume of certain solids, their mathematics lacked the characteristic Greek rigor and generality (see Dunham 1990). Lefkowitz (1996) is a scholarly debunking of claims that Greece "stole" important elements of their culture from Africa.

7

A Vantage Point

Value may not exist, but everyone has values; human beings, after all, have to make decisions, and decisions involve the ranking of alternatives. Specifically, while there may be no definitive comparison of different groups or their norms, a vantage point of some sort is needed when genetically dissimilar groups interact. We must decide what counts as justice, a fair share, a legitimate means to control disorder. The perspective adopted here is the Caucasoid value system, a seemingly self-serving choice defended in sections 7.1–7.4 along with some general remarks about the nature of moral reasoning. The balance of the chapter outlines what I take Caucasoid values to be.

7.1. WHOSE NORMS?

As moral principles cannot be derived from nonmoral premises—one way of putting the conclusion of chapter 6—all moral discussion must take some norms for granted. In most cases, the values assumed by interlocutors are those they share. When disagreement arises, however, the most effective appeal is often to the values of one's interlocutor rather than one's own. You may be able to sway him by showing him that your position is implied by principles or precedents he already accepts. Much everyday moral reasoning proceeds this way. You persuade someone to heed the speed limit on an unpoliced stretch of highway by first getting him to agree that others should not be judges in their own case as to when it is o.k. to break the law, and then getting him to admit that that is exactly what he would be doing by speeding. This ad hominem element in moral argument does not always produce agreement, but it is ineliminable: no one is persuaded by principles he does not accept.

Moral philosophy also rests on accepted norms. Whatever the classical thinkers may have thought they were doing, they were in fact seeking to codify what Rawls calls "fixed points" of moral intuition. Extending the metaphor, moral philosophy is a curve fitting exercise, with ordinary reflective judgments the data points and ethical principles the curves fitted to them.[1] A moral theory is typically considered plausible when it coheres with ordinary judgments, implausible when it violates them. Aristotle's doctrine of the mean reads like a generalization of a pattern of thinking he detected in the actual deliberations of his fellow Greeks. (Aristotle seems more interested in habituation to virtue than in the identity of virtuous acts, which he evidently finds obvious.) Kant stresses the congruence of the categorical imperative with "ordinary rational knowledge of morality." Mill frankly admits that the only proof he can offer of utilitarianism, the doctrine that pleasure alone is desirable, is that (so he thinks) pleasure alone is actually desired. The standard criticisms of utilitarianism and Kantianism is that they flout common sense. Thus, utilitarianism is accused of demanding, contrary to "moral intuition," that a woman forced to choose between saving her baby and saving ten strangers from a fire should save the strangers. "Act" utilitarians who accept this counterintuitive conclusion and call their critics sentimentalists convince no one. More sophisticated utilitarians disown the conclusion, explaining that they favor *rules*—like putting one's own children first—which maximize happiness in the long run when generally followed. The lesson to draw from these exchanges is not who gets the better of them, but what is counted as getting the better, namely consistency with common sense.

Contrary to their reputation, moral revisionists rarely challenge the values of their society wholesale. Apparent moral innovation is often a reorientation of old values around new facts, or purported facts. Prophets base religious reforms on news about God's will—all the while assuming, and assuming others agree, that God should be obeyed. Environmentalists urge pollution control to avert consequences that everyone would regard as disastrous. Or, innovators may argue that their new norms are implicit in society's old ones. Nietzsche's promise to "transvalue all values" relied on his readers to deplore self-deception, which he thought permeated Christianity. Civil rights rhetoric called on the United States to live up to its long-standing ideals of equality. Animal rights, a voguish notion seemingly at odds with everyday convictions, presents itself as entailed by those convictions; its advocates appeal to the capacity of animals to suffer and reason, banking on the truisms that pain is bad and that Reason confers moral status. To be sure, there are zealots openly hostile to all the norms of their society, but their very readiness to resort to violence shows their disinterest in persuasion. On the whole, revisionists realize that a new value just announced out of the blue, such as a right to hear a good joke every day, would be greeted by puzzlement.[2] Moral argument must rest on accepted values. This is why the values to be deployed here are the Caucasoid ones: they are the ones the reader, my interlocutor, most probably accepts.

Since everyone necessarily thinks his values are correct—he wouldn't hold them otherwise—the ad hominem component in moral reasoning does not leave it less than fully persuasive. That universal blindspot, our inability to see our own values as does the gimlet-eyed naturalist, prevents anyone from regarding appeal to his own values as a weakness. Within an individual's moral framework—the perspective he cannot help taking—a deduction from his own standards is proof, and inconsistency with his own standards is refutation.

But one catch keeps moral discussion from becoming routinized, and gives disagreement its occasional intractability: strains in a set of moral convictions can be relieved in many ways. When an attractive principle is shown to yield an unattractive verdict in concrete cases, one can regretfully drop the principle, gamely accept the verdict, or modify both until they rest in "reflective equilibrium."[3] Naturalism despairs of any observational test for a "correct" revision. This slack in what to change has the familiar real-world consequence that favored convictions can be retained come way may; the driver who wishes to speed can declare that, on reflection, it sometimes is all right to be the judge in one's own case.[4] (This point is brought to bear on the affirmative action debate in section 7.3.) Still, most of us want our values to be consistent, however they are made so, and this desire sustains moral suasion.

7.2. CRITERIA

Because norms are not discovered in the way facts are, and moral reasoning knows no standard beyond consistency, the role of facts in ethics is often underplayed. In reality, most overtly "moral" arguments concern matters of fact, a point particularly pertinent to discussions of race.

Peter Singer's purported illustration of the irrelevance of facts inadvertently highlights this point. In adopting the principle of "equal consideration for all," he says (1993), we decide that the empirical traits of individuals and groups have no bearing on how anyone should be treated. Without facts to guide its application at every step, however, Singer's principle is plainly blind. Paying two employees the same wages accords them equal consideration only if they are equally productive, and whether they are equally productive is a factual matter. Higher arrest rates for blacks constitutes unequal treatment only if, in fact, blacks commit no more crimes than whites. The equal consideration principle relies on empirical facts all down the line.

The most familiar way facts mesh with norms is in the determination of instrumental value, the best means to agreed-on ends (see section 4.12). As the equal-pay case shows, however, facts can also determine rights and duties. A certain empirical trait serves to specify "equality," and two individuals are taken to deserve treatment that is identical in some respect when they display that trait to the same extent. Such traits may be called *criteria*, in this case for equality. To vary examples, the norm that responsibility for a fight creates liability to reprisal needs criteria for responsibility before it can pin liability on anyone. The usual criterion is initiation, which in different conflicts may amount to throwing the first punch, being first across a border, or issuing the first threat. The norm of compensation that will occupy us in chapter 8, that an agent who causes injury must make up the loss, requires criteria for injury and loss.[5] Even so humble an evaluation as distinguishing "prime" from "choice" beef requires empirical criteria—the meat's marbling, among others.

Normative principles have the following schematic form: "If criterion C, then assessment S."[6] Two equally *productive* people deserve the same *pay*; whoever *throws the first punch* may be *struck in return*; should *A break B's pot*, then *A* must *compensate B*. So it is easy to see why, when a body of principles is widely accepted, disputes about particular entitlements usually hinge on satisfaction of the relevant empirical criteria. Whether country X may mobilize against country Y depends on whether Y overflew X; whether the mover owes the dancer damages depends on whether he dropped the piano on her foot. The factual issues to be decided are not always clear-cut, since criteria may be vague and multipronged. There are many ways to start a fight; new situations may require new criteria, as when innovations in communication technology create new ways to threaten. These complications aside, most moral inquiry is undertaken by disputants who agree on the connection between fact and assessment, and decide on assessments by finding the facts.

Those who stress the is-ought gap are right to insist that criteria are chosen, not discovered. It is not written in the heavens that productivity is relevant to pay, for remuneration could be based on effort, or need, or shoe size. And this freedom to choose criteria is another reminder that facts, although in practice usually decisive, can never force anyone to abandon a verdict his heart is set on. If extant criteria don't yield the favored verdict, he can always search for, or stipulate, one that does. A schoolyard authority bent on blaming a particular boy for a fight can simply seize on something he did to call an infraction. He can say Tommy started things by looking at Billy provocatively.[7] This approach to argument, inventing new principles to get the conclusion one wants, is both logically impeccable and exasperating—which brings up the topic of affirmative action.

7.3. UNDERSTANDING THE AFFIRMATIVE ACTION DEBATE

That in practice facts are dispositive while in theory commitment is unshakeable—one can dig in his heels by endlessly shifting grounds—helps explain the wayward course of the affirmative action debate. Advocates of preference have practiced just this kind of footwork.

Lyndon Johnson introduced affirmative action in 1965 with an appeal to compensation: "You do not take a person who for years has been hobbled by chains," he said in his famous Howard University address, "and liberate him, bring him up to the starting line, and then say, 'You are free to compete with all the others.'" Martin Luther King asked for "compensatory preferences," and Jimmy Carter called preferences "compensatory discrimination." The president of the Urban League criticized a 1989 Supreme Court decision as opposing "reasonable affirmative action decrees that remedy past discrimination" (*New York Times* 1989); Bill Clinton defended affirmative action as "redress." One might reasonably conclude from the words of three Presidents that compensation is what the debate is all about. Yet, as the compensation argument has tottered—mainly with growing awareness that the beneficiaries of affirmative action have never been discriminated against, and that its white victims have never discriminated—there has been a migration to new grounds few of which were heard of in 1965: fighting discrimination, fighting stereotypes, justice for groups, changing demographics, diversity, role models, a representative workforce, black self-esteem, a black stake in society, administrative convenience (since no one knows which blacks have suffered, see Nickel 1975; Ezorsky 1991; Harwood 1993), a need to make blacks feel loved (West 1997). Bergmann (1996) urges virtually all of these fallbacks. Chapter 8 looks at some of them more closely, but some preliminary remark are in order about the cost to quota advocates of their proliferation.

First, some of these new arguments are so post facto as to be patent rationalizations, the efforts of people who have forgotten, or never knew, why they supported racial preference in the first place. Shopping around for reasons after making up one's mind is not an attitude to inspire confidence.

Second, the nature of a claim changes as the grounds for it do. Should Smith demand back a sweater that Jones stole, Smith wants restitution; should Smith demand the sweater because he is cold, he wants charity. Should he say Jones owes him the sweater for both reasons, his claim is ambiguous, if not incoherent. Real-world equivocations of this sort tend to be overlooked because people are often more interested in who is on their side than why they are. Opposing a war as unwinnable is not the same as opposing it on moral grounds. Wanting to preserve endangered species useful to man is not the same as valuing biological diversity in itself; an ecological utilitarian won't miss a useless species, while an ecological absolutist will. Vegetarians for health reasons need not oppose animal experimentation, while animal-rights vegetarians oppose it vehemently. Critics of capital punishment who deny that it deters are no friends of abolitionists who call it "barbaric" and would oppose it whether it deters or not. Consequently, and despite appearances, there is no such thing as pacifism per se, or *the* environmentalist position or *the* case against the death penalty; there are as many positions as there are reasons for them. So too, the constant ground-shifting by preference advocates has blurred their position. Is preference meant to remedy past harm or preempt future bigotry? Is it charity or therapy ("self-esteem"), good for blacks only or for everyone? Will it continue indefinitely, or end at some specified time when blacks are again whole? That the various rationales for affirmative action imply contrary answers to these questions suggests that the positions these rationales support are in fact very different.

The more grounds there are for a claim, the more they vary in force. Saying preference for blacks is necessary because blacks need jobs is less compelling than calling preference a matter of justice. Newly minted values like "group rights" leave the unconverted cold. Adding to the confusion, when arguments of varying strength are treated as interchangeable, is the almost unconscious pull toward the strongest. Since justice is commonly deemed more urgent than utility, charity, therapy, or convenience, most officially noncompensatory arguments may be expected to rely tacitly on compensation, and this is what, in chapter 8, we will find.

7.4. ADOPTING A VANTAGE POINT

Because of the the ad hominem nature of moral reasoning, the values most conducive to agreement on racial questions are those of one's audience, in the present case my reader. I henceforth assume his values are those characteristic of Europe and European America. Thus, when I call a position "justified," or speak of "ordinary standards," I will mean "in accordance with standards the reader of this book probably hold." It is pertinent to ask what Caucasoid values imply about race differences not because that point of view is God's, but because it is yours.

Some readers may well be embarrassed to have Caucasoid values attributed to them. Whites have been ridiculed so often, by whites[8] as well as blacks (e.g. McCall 1994),[9] that "white middle class" has become a term of derision signifying blandness, conformity and hypocrisy. Yet what ideals does this phrase summon? Hard work, certainly, and self-reliance, self-control, modesty, honesty, punctuality, politeness, sportsmanship, and considerateness. A Caucasoid may not always live up to these ideals, but he thinks he should—and I doubt that, on reflection, the reader can easily dismiss them as ridiculous.

More than a list is needed, however, and the next section offers a fuller account.

7.5. CAUCASOID VALUES

Fully recognizing the risk of oversimplification, I would summarize the primary Caucasoid value, the core of Caucasoid morality, as the golden rule—what Locke called "that most unshaken rule of morality and foundation of all social virtue."

The best way to determine someone's values is not to examine his sociopolitical views. With political judgments especially it is difficult to separate normative commitment from empirical beliefs about means-ends relations, and political preferences are behaviorally shallow, often amounting to no more than an occasional vote. An individual's deepest values are revealed, rather, in his private behavior (see Sommers 1984): the company he keeps, the way he raises his children, the actions that occasion guilt, conflict or satisfaction—in a phrase, his values are revealed by his criteria for "good person."

The ordinary criteria for being a good person do not emphasize being happy or making others happy. Happiness plays a role in everyday moral thinking, but it is not basic, and is valued only when thought to be deserved. The happiness of a child molester is less important than that of a productive citizen (and not merely because the child molester's happiness is outweighed by his victim's misery). On the whole, the average person thinks less in terms of happiness than of rights and duties. There is no need to make heavy weather of these concepts;[10] a right, as ordinarily understood, is a freedom that should be allowed, and a duty correlative to a right is an act whose omission interferes with that freedom. You have a right to speak when others should not (i.e., have a duty not to) silence you, where forbearing from silencing is the act whose omission is an interference. You have a proprietary right to a thing when others should (i.e., have a duty to) let you do with it as you please. Noncorrelated duties (e.g., to be tactful) are obligatory acts whose omission violates no right. Enforceable rights are freedoms that others may protect by force; strongly enforceable rights are freedoms that others should protect by force.

The golden rule enters as a higher-order principle about the distribution of rights: it decrees that *basic rights must belong to everyone*. This rule goes by many other names: reciprocity, the categorical imperative, what i sauce for the goose is sauce for the gander, walking a mile in the other guy's shoes. Hobbes advises that "a man . . . be contented with so much liberty against other men, as he would allow other men against himself." Rawls' "first principle of justice" assigns "[e]ach person . . . an equal right to the most extensive liberty compatible with a like liberty for all." Restricted rights must flow from the exercise of universal ones: although I alone own my shirt, I acquired it by trade, an activity in which everyone may engage. The restricted duties recognized by common sense, such as my obligation to expend more care on my children than the children of strangers, are also regarded as flowing from universal ones: everyone has an obligation to care for his own children more than on the children of strangers.[11]

Despite its vagueness—who is "everyone"?; does "compatibility" in the like-liberty test mean satisfiability at the same time or something more abstract—the golden rule clearly distinguishes one class of rights, namely "negative" rights (Berlin 1969) to to be free from interference. Included are a right against theft, assault and harassment, a right to speak freely, and a right to associate with anyone willing to associate back. Everyone can be free of coercion at the same time: my not interfering with your speech requires no interference with mine; my allowing you to keep the company you please (in both senses) is consistent with my being allowed to keep the company I please. The enforcement of these rights is also universalizable: stopping A from stifling B's speech is consistent with stopping C from stifling A's speech. Everyday morality certainly recognizes these "negative" rights.

The like-liberty test finds "positive" rights, such as a right to someone else's company, more problematic. If Jones has a right to associate with Smith, however Smith may feel about it, Smith cannot have the reciprocal right to associate with Jones however Jones may feel about it. Jones' freedom to associate or not with Smith at will forbids Smith to refuse the association once Jones has demanded it. Everyday morality agrees in rejecting unilateral rights to association. Nobody has a right to force himself on others.

Basic positive rights to goods also violate the golden rule. A right to a particular thing, like a shirt, is unshareable. A general right to the necessities of life—food, clothing and shelter on demand—imposes a duty on anyone who must work to provide those goods that prevents him from having that same right. Were a provider also unconditionally entitled to the necessities, he would have the right to cease work and demand food, clothing and shelter for himself, wrongfully leaving others without the goods they are entitled to. Ordinary morality again agrees that the world owes no one a living.

The situation is more complex with respect to the conditional right to support *if* one is unable to care for himself *and* there are others productive enough to support themselves and him (see Levin 1984; Melnyk 1989; Levin 1989a, 1989b). What seems clear is that, while everyone could be *said* to have this right, not everyone could *claim* it. It is logically impossible for everyone to be needy while there are people around who can support themselves and others. A society already supporting as many paupers as it can must tell the next one who presents himself that he is out of luck—a catch-22 cloaked by the prosperity of welfare states to date.[12]

To be sure, the average person does think himself obligated to help those in need, since he recognizes that in the same situation he would want help. At the same time, he regards this duty as uncorrelated with any *right* to his assistance. Moreover, he sees the urgency of his duty to help others as waning with their distance from him, as he moves from children to friends to countrymen, and, again by the golden rule, he understands that those distant from himself feel stronger obligations to their own offspring than to his own.

Some limits on negative freedom are universalizable. Civil rights laws are of this sort: it is possible for Jones to have the right to associate with Smith so long as Smith's only reason for not associating with Jones is Jones' race, *and* for Smith to have the reciprocal right to associate with Jones so long as Jones' only reason for not associating with Smith is Smith's race. It is also possible to *will*[13] a world in which no one refuses associations for reasons of race—indeed, the American public thought it was willing that with the 1964 Civil Rights Act. Nevertheless, most people would agree with Hobbes and Rawls that, all else equal, freedom should be maximized. No one likes having his own universalizable freedom limited, so, by the golden rule, he cannot sanction limits on anyone else's. When, as he would, the typical heir of the European tradition agrees that my freedom to swing my fist ends at your nose, he will quickly add "but it does extend right up to it." Where more liberty is possible, the party of less liberty carries a heavy burden of proof.[14]

Now, reciprocity, nonaggression, honesty, and helpfulness can be promoted in two ways. Aggression, for instance, can be forbidden by the absolute personal prohibition "Never attack nonaggressors," what philosophers call a deontological rule, or by the goal-directed, teleological rule "Do what you can to prevent attacks on nonaggressors." The two differ in that the first forbids aggression under all circumstances while the second permits and possibly commands aggression that reduces aggression overall. Should a terrorist hide an A-bomb that will blow up Manhattan, the teleological rule allows torturing his son if that is the only way to make him reveal the bomb's location, while the deontological rule forbids it. Many dilemmas, including some about race and crime considered in chapter 9, are conflicts between deontological rules and the values they serve. The question, then, is whether ordinary morality is goal-directed or absolute.

Morality *feels* absolute. Parents tell children to be honest, not to maximize honesty. Praise, blame, and culpability are allotted as if what counts is adherence to deontological rules. The responsible officials might lose sleep for lacking the will to torture the terrorist's son, but the terrorist, not the officials, will be punished in case the bomb goes off. But most people are also theoretical maximizers. Make the numbers large enough in hypothetical cases, and they will accept the torture of an innocent child to prevent the murder of millions. In such cases we start to think teleologically, showing a willingness to overstep the moral line a little to keep others from overstepping it a lot.

This ambivalence has often been remarked on, with many writers taking i to show that moral justification proceeds in two stages: particular acts are justified by strict conformity to deontological rules, while deontological rules are justified teleologically, by the good consequences of following them. This analysis has much to recommend it, but I would rephrase it naturalistically there is an uneasy truce between the goal of morality and the means by which this goal is achieved. Moral rules, I suggested in chapter 6, exist because they promote adaptive social behavior like honesty, so their function is teleological. Yet, oddly, absolute rules promote such behavior more effectively than rules explicitly commanding that this behavior be promoted; instructing people to maximize honesty where they can yields less honesty than instructing them not to lie and otherwise mind their own business. This is so because the typical individual, unlike a leader whose acts have vast consequences, deals only with a small immediate circle. Since honesty within his circle is unlikely to cause lying elsewhere, the lying he foregoes is his net contribution to honesty. Thus, the average person best maximizes honesty simply by not lying, regardless of what others do. On the other hand, a rule that lets the average person lie to serve some further end, even overall honesty, offers him a ready excuse for lying whenever temptation presents itself. Because promulgation of teleological rules thus produces less adaptive behavior than promulgation of absolute rules, absolute morality has been selected for. This normally hidden strain between the form and function of morality becomes visible in extreme cases when adherence to absolute rules threatens to frustrate the function they serve.

As to whether people *should* think teleologically or deontologically, the naturalistic answer is that there is no answer. Apparently, morality is absolutist in everyday circumstances, teleological in emergencies, and in each context blind to the other alternative. This blindness itself is probably adaptive: rules couldn't do their everyday job of maximizing adaptive behavior if people thought of them as less than absolute, yet it would also be maladaptive to cling to rules on those rare occasions when doing so would be catastrophic. We are of two minds, and there the matter ends. About all that remains invariant across the divide is the golden rule, acknowledged even in emergencies. If a situation permits or requires you to break a rule, anyone else in the same situation may break the rule, even with you on the receiving end.

7.6. KANTIANISM

As the foremost theorist of universalization in ethics is Kant, let us call an individual *kantian* to the extent that he conforms himself to the golden rule.[15]

Kantianism—the tendency to inhibit oneself from doing what he would not want others to do, especially to him—is a descriptive trait of behavior that permits a rough, qualitative scaling of moral goodness. To say that everyday morality flows from the golden rule is to claim a strong connection between kantianism and everyday criteria of personal goodness, a connection that does appear to hold. A kantian can be expected to see things from a variety of perspectives. He will follow general rules, not constantly seek to make an exception of himself. He knows that other people take their own ends as seriously as he takes his, so he does not treat others as mere resources. Nobody wants his own preferences overridden for the sake of someone else's, so a kantian will not selfishly override the preferences of others. A kantian who wishes others to serve his own ends attempts to recruit them as he would wish to be recruited, by persuasion or bargaining rather than threat, coercion, or deception. Kantians are aware that they sometimes need help, so they are inclined to help others. Since a kantian like everyone else wants to be able to rely on promises, he is trustworthy. The similarity between ideal kantians and the ideal Boy Scout is not coincidental, since the Boy Scout code also encapsulates Caucasoid morality.

A number of valued traits not especially kantian on their face, such as intelligence, industriousness, patience, self-sufficiency, and development of one's talents also appear to associate with kantianism. Brand (1987) reports a correlation between altruism and IQ. Chapter 3 mentioned the correlation found by Cattell and Kohlberg between IQ and "moral maturity." In fact, what Kohlberg means by "moral maturity" is attainment of the highest stages of his sequencing of moral development, where principles display "logical comprehensiveness, universality, and consistency . . . (the Golden Rule, the categorical imperative)" (1981: 382–383). The cognitive demands of kantian thinking may explain its connection to IQ. An egoist ponders the benefits of his actions for himself, an act-utilitarian the benefits of his actions for everyone, and a rule-utilitarian the hypothetical benefits for everyone of everyone acting as he does. A full-fledged kantian (see n. 13) ponders the logical consistency of everyone acting as he does, the highest level of abstraction. That many bad people have been highly intelligent shows that intelligence is not *sufficient* for goodness, but it does appear necessary. To paraphrase Kohlberg, you can be smart without being good, but there are limits to how good you can be without being smart. The higher the mean IQ of a group, consequently, the more kantian its morality is likely to be.

The reader may value kantianism under rubrics like "integrity" or "*menschlichkeit*." He tries to judge others and himself by a single standard. He is irked when, remonstrating with a larcenous coworker that no office supplies would be left if everyone stole them, he is told "Don't worry, no one else will." When he asks his children why they should not steal, he wants to hear "Because stealing is wrong," not "Because you might get caught." Among Caucasoids, calling someone "morally good" almost always means, and will certainly be taken to mean, that he is principled, is aware of the desires of others, and does not think he is special. Rawls defines a good man as a good citizen, which in turn means a man with whom you would wish to design a society (1971: 435–438). Since, according to Rawls, you would want your fellow society-designers to be disposed to follow basic rules adopted by equals, good citizenship is kantianism.

Concern with rules is currently derided in some quarters (see Calhoun 1988; Gilligan 1983) as overly "masculine," since women evidently value consistency less, and empathy more, than men do.[16] Caucasoid women nevertheless value kantianism highly, as it includes, in addition to rigorous consistency, a willingness to see other people as subjects of experience like oneself rather than as resources or obstacles. Looking toward racial comparisons, black males on average score lower than white males on the "femininity" scale of the MMPI and higher on the "masculinity" scale, so black males are likely to be farther from the typical Caucasoid female's idea of a "good person" than is the typical Caucasoid male.

7.7. RACE DIFFERENCES IN PERSONAL GOODNESS

Turning explicitly to racial comparison, earlier chapters reviewed evidence that blacks are typically less intelligent and inclined to follow rules, and more aggressive, self-assertive, and impulsive than whites. Blacks are more likely than whites to agree that "It is not hard for me to ask help from my friends even thought I cannot return the favor." These differences may be summarized by saying that blacks are typically less kantian. Since kantianism is the principal Caucasoid measure of personal worth, it follows that, by ordinary Caucasoid standards, the average white is a better person than the average black. Assuming the composite trait of kantianism distributes roughly normally in both populations, a greater proportion of black than white behavior also falls below the ordinary thresholds of decency, and of tolerability.

These statements sound monstrous, but they follow from data difficult to gainsay. Since intelligence correlates modestly but significantly with moral maturity and altruism, the race difference in intelligence by itself suggests a race difference in moral reasoning. Conceivably the IQ/kantianism correlation holds within but not between races, but no author I have cited gives any indication that this is so. Likewise, when Herrnstein and Murray (1994) report a moderate correlation between IQ and what they call the "Middle Class Value Index," they mention no interaction with race. In any event a moral maturity/IQ correlation among whites but not blacks would be a significant race difference in its own right.

Next consider the vast race difference in crime, discussed in more detail in chapter 9. It is a virtual truism that nobody, black or white, wishes to be a *victim* of the sorts of acts that are usually criminalized. Therefore, the fact that blacks are many times more likely than whites to commit violent crimes and non-violent felonies shows a marked difference in willingness to do what nobody wants done to him. It is in the context of crime that Banfield (1974: 182–199) explicitly invokes Kohlberg's classification of the stages of morality. Lower-class (and disproportionately black) preconventional morality defines "right" in terms of what can be gotten away with instead of "some universal (or very general) principle that [is] considered worthy of choice"; it associates with willingness to inflict injury, the absence of "conscience" and short time horizons. "Most stealing," writes Banfield, "is done by persons who want small amounts *now*."

Other patterns of black behavior illustrate a difference in moral orientation. 'Trash talk," the stream of arrogant banter by which black athletes seek to intimidate and humiliate opponents, is alien to white ideals of sportsmanship see Kochman 1983: 72, 140–148). Numerous fights among black males result from "dissing," the pursuit of dominance by shows of disrespect—that is, by behavior that, directed toward oneself, would be found intolerable. Anderson 1994) and Baumeister, Smart, and Boden (1996) report that in black street angs "prestige and respect are gained by depriving others of them." One item n an IQ test for children that has drawn accusations of bias asks the subject vhat he would do if he threw a baseball through a neighbor's window. The nswer scored as correct is offering to pay for the window, whereas, it is said, a "Sorry, man," would suffice in a black neighborhood. Assuming this so, it indicates looser attachment to property rights and norms of compensation. Kochman (1983) vividly describes race differences in attitude toward various rule-governed social interactions. In formal negotiations, he finds, whites are more interested in following "the rules of negotiating" and "the negotiating procedure," whereas as blacks are more driven by their emotions and see conformity to these rules as defeat (37–42). In turn-taking situations such as the classroom, "the white classroom rule is to raise your hand, be recognized by the instructor, and take a turn in the order in which you are recognized. . . . The black rule, on the other hand, is to come in when you can. . . . Within the black conception, the decision to enter the debate and assert oneself is self-determined, regulated entirely by individuals' own assessment of what they have to say" (24–28). My experience debating blacks in public has certainly been consistent with this description.

Kochman addresses race differences in kantianism most directly in a section pregnantly entitled "Doing Unto Others" (119–127), which concerns balancing the impulse to express emotion (as in "talking back" to the screen during a movie) against the sensibilities of others (the rest of the audience).

[W]hites have been taught that to act on behalf of their own feelings is unjustified if someone else's sensibilities might become offended as a result. . . . Whites are not simply being altruistic in placing the sensibilities of others before their own feelings. Whites, after all, are "others" in other people's consideration just as other people are "others" in theirs. . . . The black social interaction rule . . . grants individuals the right to claim consideration from others for their feelings. [W]hile sensibilities also have a moral claim on other people's consideration, feelings are seen to have a preemptive claim . . . individuals must place their own feelings first . . . even if other people's sensibilities might become offended in the process. . . . With the shift in focus from *doing unto others* to *doing for oneself*, blacks can also act as their feelings direct without subsequent guilt. (121–126)

Here the race difference in impulsiveness is immediately expressed as a difference in kantianism. Baumeister, Smart, and Boden (1996) summarize the findings of several urban sociologists that attainment of status in "modern youth gangs" is "essentially a matter of acting as if oneself is above the rules that apply to others" (22).

Since some tendency toward kantianism is necessary for the survival of any society, there has doubtless been widespread selection for appreciating it However, as kantianism itself seems not to have been as strongly selected for among blacks, one expects that *valuing* kantianism has also been selected for less strongly. This conclusion is consistent with the high self-esteem felt by blacks despite display of traits considered shameful by whites, such as regular brushes with the law. If blacks do value kantianism less than whites, then while in some respects whites may be perceived as "better" by black as well a white standards—a black, for instance, may trust whites more than other blacks—the overall ranking of the races by black standards may differ markedly from the white. There is certainly anecdotal evidence reinforced b media accounts that young black males, who prize toughness and dominance within male hierarchies, hold less aggressive whites in contempt. Whites ma also be considered sexually "up tight."[17] There is no reason to assume black criteria for personal goodness to coincide with white.

This diversity is a reminder that any racial ranking is internal to a value system, not a consequence of the nature of things. But it must also be remembered that, since the criterion for "superiority" used by Caucasoids is the descriptive trait of kantianism, group comparisons *in this operational sense* are factual issues, empirically decidable. In Robert Nozick's language, there are "dimensions of value" along which groups may differ (1982: 325).[18] On the evidence, whites exceed blacks along dimensions that matter to whites. The average white differs from the average black in ways whites care about, and—although possibly not to the same extent—blacks care about as well.

The plural "dimensions" is also a reminder that, contrary to Gould (1981: 24–25, 159), rankings are seldom defined by a "unilinear scale" in a "single series." Rankings are more commonly determined by a number of partially overlapping scales assigned various weights. On the other hand, summary judgments are possible when one person or group outscores another on all or most scales. There is no linear ordering of value by "one number," as Gould prefers to put it, but, since whites outscore blacks on the major dimension of kantianism and the secondary factors of intelligence, self-restraint, and time preference, whites are superior overall by white standards.

Since racism is bad per se, and believing well-supported generalizations is not bad, this conclusion is not racist. There may be people who think whites superior to blacks absolutely, and that belief may be racist, but it is not what most people have in mind when (perhaps only to themselves) they indulge in racial rankings. When asked what they mean, whites cite specific disvalued traits they believe blacks exhibit more frequently than whites, such as "pilfering, careless work, reneging on debts and promises" (Miller and Dreger 1968: 35). The belief that blacks steal and break promises more frequently than whites is a claim of fact which is not racist if true and believed on the basis of evidence. Should honesty and reliability be white criteria of personal moral stature, the claim that blacks are morally inferior to whites becomes operationalized, empirical and, if supported by the evidence, not racist.

Many people say blacks cannot be "racist," no matter how strident their demands or their vilification of whites. The oddity is that this proposition is intuitively *correct*: "black racism" seems oxymoronic. Yet if demands for special treatment on the basis of one's own race and phillipics against other races do not count as racism, what does? The usual explanation for this double standard, that whites have the power (Hacker 1992), is unconvincing. None of the definitions of "racism" reviewed in chapter 5—belief in biological differences, belief in racial superiority, race-based judgment of individuals—include the ability to enforce one's views. To exist, says Hacker, racism must have "an impact on the real world" (Hacker 1992: 29), which not only implies that Hitler was not anti-Semitic until he became chancellor, but is circular. An idea cannot include its enforceement, since since there is then no way to specify *what* idea is to be enforced. (Even Hacker admits problems here.) It is also doubtful that running through people's minds is the idea that, since whites have power, all white criticism of blacks must be driven by the bad (hence "racist") desire to oppress, whereas all black criticism of whites is driven by a benign desire for equality. Many black tirades are far from benign.

I suggest that those, particularly whites, who deny that blacks can be racist do so for a quite different, forbidden reason. Such individuals *do* think of "racism" in its one of its conventional senses, namely belief in racial superiority, and they find it inconceivable that anyone, black or white, can really believe blacks are superior to whites. And they find this inconceivable because they are using Caucasoid criteria for personal worth, tacitly operationalizing the claim blacks are "better" as the *factual* claim that blacks are more productive, cooperative, honest, and law-abiding. This factual claim seems too preposterous for anyone to maintain seriously. Since this reason for denying black racism is deeply taboo, a more acceptable black/white asymmetry must be substituted, and the black/white power asymmetry is handy. The simplest explanation of reluctance to call any black "racist" is that most people operationalize judgments of personal worth.

The empirical character of racial comparison emerges, finally, in defenses of affirmative action like James Nickel's (1975), who justifies preference for *all* blacks as an "administrative convenience." He admits that discrimination has not harmed all blacks, but argues that, since any black is "highly likely to have been victimized," there is a "high correlation" between being black and deserving preference. Race can therefore go proxy for victimhood without incurring unacceptable costs in injustice. But Nickel worries, momentarily, that "racists" will claim they discriminate against blacks because of a high correlation between being black and being "lazy or untrustworthy." Nickel allows that laziness and untrustworthiness are good reasons to doubt someone will make a good employee, but replies that the preference advocate differs from the racist because the racist "has to make claims which can be proven to be erroneous about the correlation between being black and having some relevant defect such as being lazy or untrustworthy, while the defender of racially administered preferential policies can make a plausible case without using erroneous premises." Nickel, then, hinges the distinction between preference and "racism" on the claim that blacks are as industrious and honest as whites. Independently of whether this is actually so, it is, as Nickel explicitly admits, an empirical claim. Nickels is discussing good employees rather than good persons, but this more limited context nonetheless illustrates how readily questions of worth and the correlation of worth with group membership reduce to questions about empirical criteria. Nickel freely allows that some forms of discrimination against blacks would be justified (hence not "racist") if the "racist" claim is empirically correct.

In light of chapter 6, the race differences cataloged here are not most accurately described as differences in morality. By definition, morality is the domain of universal rules. Groups have different moralities when they universalize different rules; differential interest in rules per se is not a difference *in* morality, but a difference in concern with morality itself. Blacks may therefore be said to be on average less interested than whites in morality —not more immoral, but more amoral. Blacks, like whites, have *values*, preferences revealed in behavior, but preference for conformity to the golden rule is not as strong an impulse for blacks as for whites. This does not make either whites or blacks better in any absolute sense. From a biological point of view, mora thinking, putting a premium on universalizable patterns of behavior, is an adaption that is not better (or worse) than amorality. But from the Caucasoid point of view moral thinking *is* better than amoral thinking. For that reason, whites who accept empirical race differences will find racial comparisons difficult to avoid.

7.8. EFFORTS TO AVOID RACIAL RANKING

Many commentators seek to disrupt the racial ranking implicit in white values by minimizing or denying the differences between black and white attitudes. Of a group of black teen-age murderers, the *New York Times* wrote: "The most shocking thing about them is their ordinariness. Like so many young teenagers, they are insecure, materialistic, impressionable, not always in command of their anger and aggressiveness" (Dugger 1993: A1). A *Newsweek* story about illegitimate black children begins: "Devonna Ruckett is 7, and, at least outwardly, she has all the trappings a kid could need. There's the hot bike. The cool clothes. What Devonna lacks is a live-in father. Born to a 16-year-old unwed mother without the financial and emotional resources to take care of her . . . [h]ers is a family of women" (Ingrassia 1993: 25). The question this story does not ask is how Devonna's mother could afford that bicycle. As Devonna and her mother live on welfare, the answer is: with other people's money.

Hacker (1992) compares black unmarried mothers on welfare to widows receiving social security, and argues that welfare but not social security is criticized because "those receiving these Social Security benefits are overwhelmingly white." This is not so. Critics of welfare would reply that the late husbands of surviving widows had contributed to the Social Security system, and moreover were helped to work and be productive by their wives' support in the home. The typical unmarried welfare mother has put nothing into the public treasury from which her welfare payments are drawn. Hacker also forgets that Social Security has numerous white critics who worry that current claimants are receiving more from the system than they paid in, and regard the possible inequitable burdening of future generations as a moral and fiducial crisis. Blacks appear entirely untroubled by their disproportionate receipt of public funds.

Jencks goes farther in attempting to assimilate disparate value systems. He admits the willingness of "most welfare recipients to lie and cheat":

Yet welfare mothers operate on the same moral principles as most other Americans. They believe that their first obligation is to care for their children, and they assume this means providing food, shelter, heat, electricity, furniture, clothing and an occasional treat. Since the welfare system seldom gives mothers who follow its rules enough money to pay for such necessities, they feel entitled to break the rules. (1992: 208)

This caricatures the "moral principles" of "other Americans." White mothers believe their first obligation is to care for their children *through their own efforts and those of their husbands*. White norms forbid a woman to have children until she has married a man able to support them. Whites find public assistance shameful. The average white would not be comfortable buying her children "occasional treats" with someone else's money, let alone feel entitled to, and would regard it as profoundly wrong to break rules she has agreed to obey to get money taken from others by a third party. This difference in attitudes probably illustrates the interaction of motivational and cognitive factors in morality. One reason an unmarried 17-year-old black mother whose IQ is 80 does not worry about the ethics of welfare is that she cannot comprehend where her welfare check comes from or the workings of the political system that provides it.

The version of the equivalency thesis implicit in calling black teen-age murderers "materialistic . . . like so many teenagers" is that black values merely exaggerate those borrowed from white society. According to Joan Countryman (1994),

the enthusiasm of poor black children both for conspicuous consumption and for violence suggests that inner-city residents aspire to embrace all the trappings of mainstream America. . . . Chauntey Patterson, one of the most intelligent and attractive children [in a study under discussion] became a drug dealer and later went to jail, "largely because of the grinding memories and the delusions of psychic relief that his American upbringing has offered to him." . . . To compensate for the hurt of growing up poor, jobless and outcast, they have sought these essentially American dreams.

Nike, a manufacturer of running shoes popular with black teenagers, has been accused of inciting blacks to steal to get them by using the slogan "just do it" in its advertising.

It hardly needs to be said that drug trafficking, killing, and stealing running shoes are not "mainstream America" or "essentially American dreams." White teenagers are exposed to the same advertising and material goods, yet do not as frequently resort to violence when unable to afford them.[19] Some advertisers do cater to black values, and perhaps—by white standards—they should not. Still, these merchants are merely associating their products with stimuli that already appeal to blacks, capitalizing on existing preferences rather than creating new ones. Nor is this spectacle the free market in all its vulgar glory, since, insofar as black resources are derived from welfare, such advertising is not market-driven.

In part, white values and black—or "slum," or "lower class"—values may appear more similar than they are because in a nonbehavioral sense the races agree. Blacks surveyed will say they deplore crime and filth, and that they want jobs. Criminologist James Hagedorn reports that two-thirds of drug dealers he interviewed "would like to settle down with their families in homes with white picket fences. [They] know that they are poisoning the community and believe it is morally wrong but don't feel they have an option" (cited in Gest 1995: 53). These words mean little, for values are revealed in behavior. Anyone who "believes it is morally wrong" to sell drugs does not sell drugs unless compelled to by a motive weightier than lack of "options." Whatever their words, blacks and whites differ in behavior relevant to morality.[20]

These differences are also obscured by the mistaken belief that people consciously set out to be "good" or "bad." In real life nobody follows the pronouncement of Richard III that "I am determined to be a villain." Moral judgments are passed externally on actions and motives that seldom have explicit moral content, and may, to the subject himself, seem legitimate. The neo-realist movies made by and about young blacks, such as *New Jack City, Juice, Menace II Society, Boyz 'n' the Hood, Clockers, Dead Presidents, Set it Off*, and *New Jersey Drive*, usually portray their protagonists sympathetically, as trying to overcome or falling victim to "the ghetto" or racism. To themselves, to the directors of the movies, and to the black audiences cheering them, these kids are not evil. Yet the mayhem they commit is, by white standards, evil.[21]

A simple experiment involving the Prisoner's Dilemma (see section 4.3) would go some way toward testing for race differences in morality. Although noncooperation is rational in a single Prisoner's Dilemma encounter, as we saw, cooperation is rational in the early stages of a long series of Prisoner's Dilemma plays as a signal from one player to the other that he will cooperate if the other one does.[22] However, players whose short-term desire for the defector's payoff overwhelms their desire for a future series of cooperative payoffs will nonetheless defect if offered cooperation. Short-sighted persons will also miss the advantages of future co-operation, hence not initiate a cooperative exchange. Here we see the intimate relation between time preferences and intuitively moral behavior. The hypothesis of race differences in morality presented here predicts race differences in the play of iterated Prisoner's Dilemmas. Whites, it is predicted, will be more apt than blacks to initiate cooperation, to respond cooperatively to cooperation, and occasionally to meet noncooperation with cooperation. Economists and social scientists experienced in running Prisoner's Dilemma tournaments should find it easy to design a test of these predictions.

7.9. EGALITARIANISM [23]

Many writers[24] claim that Caucasoid values are egalitarian, assigning equal intrinsic worth to all men, although not of course all deeds. Obviously, a sweeping renunciation of all interpersonal comparisons removes racial comparisons with it. While my main concern is what Caucasoid values *are*, not what anyone thinks they *should* be, egalitarianism seems wrong descriptively as well as prescriptively.

Let us begin with Rosenfeld's "postulate of equality . . . the normative proposition that all individuals are morally equal qua individuals." He continues: "This proposition is counterfactual in the sense that it does not depend for its validity on any empirical proof of the existence of particular descriptive equalities" (1991: 20). Is this postulate actually embraced by anyone?

It might be accepted in words by those who believe that God gave everyone a soul of infinite worth, although this belief would have to be distinguished from the assertion in the Declaration of Independence that "all men are created equal, endowed by their Creator with certain unalienable rights," which affirms equality of *rights*, not worth. But once again we must look beyond verbal assent to behavior, and, to judge by people's actions, no one, not even egalitarians, values all men equally. Whatever anyone professes, he prefers certain kinds of people to others. No one is indifferent to the company he keeps. No egalitarian would say his children turned out equally well if one grew up to be like Socrates and another like Charles Manson. Nor would anyone have much trouble saying what descriptive traits make Manson the less attractive.

Given the difficulty of separating equal worth from "descriptive equalities," a natural and popular fallback position is that intrinsic worth rests on admittedly empirical traits, but traits that all men share. The traditional equality-making criterion has been reason; more recent candidates include the ability to be moral, the ability to claim rights, the need for self-respect, and the capacity to conceive oneself as a purposive agent. As Pojman (1991a, 1991b) points out, however, worth is ordinarily thought to vary with the degree to which these traits are present, not their presence per se. Reflective persons are prized more highly than less thoughtful ones, and the unswervingly honest man more than the frequent liar. Nouveau standards like the capacity to conceive oneself as an agent are too abstruse to inform ordinary thinking, and in any case do not yield the desired universal equality. Thus, it is quite true that an agent must regard his own purposes as worth pursuing, and feel entitled to ask others to respect them. He must also recognize that others regard themselves and their purposes in exactly the same way. But none of this implies that either he or anyone else is *right* to do so. Some agents may be entitled to think well of themselves while others should be ashamed. People aren't valued for simply having purposes, but for the nature of their purposes.

Right after purportedly divorcing normative from descriptive equalities, Rosenfeld acknowledges, perhaps inadvertently, the role of empirical criteria in determining worth.

[W]hat is crucial in the context of liberal political thought are not descriptive equalities of race, physical strength and intelligence, and the like, but rather, as [Amy] Gutmann indicates, a broadly conceived equal 'ability to abide by the law and choose a reasonable plan of life,' as well as an equal capacity for self-respect and human dignity. (1991: 20)

Yet law-abidingness is a descriptive trait, and to say two individuals share it equally is to ascribe a descriptive equality. It is a plain matter of fact that some people are less law-abiding than others, and the racial disparity in crime rates implies that the races exhibit this trait to different degrees on average. "Ability to choose a rational life plan" and "capacity for dignity," although very vague, are also descriptive and evidently vary among individuals. Prostitutes who exchange sexual favors for drugs presumably are relatively unconcerned about their dignity. Chapter 3 mentioned a reality/aspiration gap among blacks: many black teenagers in interviews glibly report planning to be lawyers or surgeons. According to documentaries like *Hoop Dreams* and books like *Last Shot* (Frey 1994), basketball is virtually the only interest of a large number of young black males sustained by the fantasy of a professional basketball career. The police officer who arrested the eight rapists in the Central Park Jogger case, when asked what they wanted to do with their lives replied "Kids like that don't know what they'll be doing a half hour from now." Blacks seem on average less able to plan their lives rationally. Given Rosenfeld's own criteria, then, blacks and whites on average display unequal personal worth.

A major motive for defending equal human worth is a desire to protect equality of *rights*, the promise of the Declaration of Independence. Like many thinkers, Tom Regan runs these quite separate issues together:

[W]e must believe that all who have inherent value have it equally, regardless of their sex, race, religion, birthplace, and so on. Similarly to be discarded as irrelevant are one's talents or skills, intelligence and wealth, personality or pathology, whether one is loved or hated—or despised and loathed. The genius and the retarded child, the prince and the pauper, the brain surgeon and the fruit vendor, Mother Teresa and the most unscrupulous used car salesman—all have inherent value, all have it equally, and all have an equal right to be treated with respect, to be treated in ways that do not reduce them to the status of things. (1985: 21)

"Used car salesman" is a rather blinkered notion of the nadir of depravity; one wonders whether Regan equates Mother Theresa with a child molester. Regan's main error, however, is that of taking "the right not to be reduced to the status of a thing" to depend on "inherent value." Were there such a dependency, differences in worth would imply differences in rights, perhaps contrary to the golden rule. But rights can be separated from worth, as in fact they are in modern liberal societies. Nothing in the golden rule denies equal rights to people of very different "inherent value." It harms no one to let everyone, whatever his worth, associate, worship and speak freely. Negative rights can expand forever, and, fully expanded, yield benefits in the way of free inquiry and social harmony.[25]

At the same time, negative rights are not (Regan's fervor notwithstanding) completely independent of personal worth. Felons are deemed so worthless by others as no longer to deserve a full array of rights; incarcerated thieves can no longer associate freely, and condemned murderers have lost their right to life. Even a man on death row cannot of course be used in medical experiments against his will; no one in modern Western society is ever reduced "to the status of a thing" by loss of all immunities. Nonetheless, common sense apportions rights by worth when behavior becomes bad enough. (It is useless to argue that felons are [or may be] deprived of rights but still are [or should be] regarded as equal in value to the next man, for by this point "equal worth" has ceased doing any work.) Deprivation of wrongdoer's rights on golden-rule grounds is not difficult to understand. I as a kantian must refrain from aggressive acts because I wouldn't want anyone aggressing against me. But may I (or my deputies) aggressively incarcerate or kill an aggressor? Well, I recognize that if *I* aggressed against someone, I could not, consistently with the golden rule, complain if he retaliated aggressively. In other words, I recognize that I would be willing to have others retaliate against me if I aggressed against them. Hence, when I let the justice system retaliate against aggressors in my name, I do nothing I wouldn't be willing to have done unto me. The golden rule accords everyone a right to equal treatment until he violates someone else's rights. Then a number of bets are off.

Many egalitarians urging a universal right to respect cite Kant's endorsement of a duty to "respect all rational nature." The right they have in mind is fictitious, and not sponsored by Kant. "Respect" as colloquially understood is esteem, which is thought to be *earned* by accomplishment or character. Equal respect for everyone in this sense is no part of anyone's moral code; the hardy mountaineer is respected more than the layabout. As for the "right to respect" Kant deduces from the categorical imperative, it amounts to freedom from coercion, deception, manipulation and like circumvention of one's will—the familiar negative rights others should respect.

To be sure, Kant sometimes[26] expresses this universal immunity by saying that all "rational wills" have "absolute worth," but there he departs from common usage. We have little respect and little use for wills that rationally chose to lie, cheat and parasitize.[27] That respect for freedom constrains us from coercing murderers and thieves for reasons unrelated to their transgressions is hardly to esteem them equal in value to everyone else. Yet at other times[28] Kant recognizes something higher than a rational will, namely a *good* will, one that seeks to do its duty—what we would call good character and what he considers "valuable without qualification." In this latter sense of "inherent value," Kant does not think everyone has equal inherent value.

The ranking of individuals and groups goes uncontested in nonracial contexts. Let the reader imagine himself forced to choose between spending a month with highly kantian *A* or far less kantian *B*. *A* is more intelligent, cooperative and restrained, less self-assertive, less inclined to respond angrily to disagreements and to exempt himself from rules he wants you to follow. Kantian *A*'s style is dispassionate, impersonal, and nonchallenging, while *B*'s is confrontational. All else equal, the reader (no doubt in the company of Rosenfeld, Gutmann, and Regan) would almost certainly prefer to spend the month with *A*: he would regard himself as justified in so choosing, and by ordinary standards he would *be* justified. In this behavioral sense he regards *A* as better than *B*. This preference does not commit him to concluding that *B* deserves fewer rights than *A*. He simply prefers *A*'s company.

It is also indisputable in nonracial contexts that groups as well as individuals can differ with respect to valued traits. Because of empirical mean group differences, the reader can expect to value randomly chosen members of some groups more than those of others. Few egalitarians would have the effrontery to deny that the average minister has more qualities he admires than the average murderer. Generalize the thought experiment of the last paragraph to two villages *A* and *B* differing in mean kantianism. There is less violence in *A*, more predictable behavior, more trust, and fewer confrontational personal interactions than in *B*. *A*'s mean IQ is higher. The reader forced to choose between spending a month in village *A* or *B* would almost certainly choose *A* as I assume Rosenfeld, Gutmann, and Regan would also. Once again, this preference just assigns a higher value to *A*, not a wish to restrict *B*'s rights.

White values are not superior to black values sub specie aeternitatis. There is no warrant in the nature of things for caring about the personal attributes Caucasoids care about to the extent Caucasoids care about them. But some outlook must be assumed as we turn to substantive issues, and that outlook will be the Caucasoid.

NOTES

1. The data tend to be judgments considered obvious by a philosopher's society. Having internalized his society's training and sharing genes with its other members, he can access these data by introspecting his own reinforcement tendencies. Linguists access their mother tongues similarly.

2. An interesting case is Kagan (1989), who defends, against the "moderate" view of ordinary morality, the "extreme" view that everyone at all times should strive to maximize the world's total good, there being no "option" to do otherwise when suffering could be relieved. (Kagan thinks the government should redistribute income for this purpose until taxation reduces the incentive to work; see 395.) Yet the attentive reader will find that Kagan's main arguments are attempts to show that an "extreme" commitment to maximizing the good is already part of ordinary morality, or at least one strand of it. "I believe that the moderate is in fact committed to the existence of a pro tanto reason to promote the good," he says (46); "my claim is that the existence of a pro tanto reason to promote the good provides the best explanation of a number of judgments that the moderate wants to make" (also see 48, 49, and 50 on this point). On the existence of an obligation to try to overcome one's partiality for self, friends and family, he remarks: "This argument [that agents have a pro tanto reason to overcome bias] is an attempt to spell out part of what the moderate himself must be committed to" (322; also see 352–353, 361, 369, 399).

3. There is no reason to view stable commitments as better (see Rawls 1971; Daniels 1979; Brink 1989), since the equation of stabler with better explains nothing.

4. There is also some free play in conforming scientific theories to data, as Quine (1951) emphasizes, but in the end a scientific theory must answer to observation. Observation statements *entail* no statements about physical reality, but this Cartesian point allows an evidential role for observation in science via inverse inference.

5. Empirical criteria do not *define* moral terms, just as operational "definitions" in science do not define theoretical terms. In fact, there are extensive similarities between criteria and operational definitions. "Force" as it occurs in the law of inertia $F = ma$ just abbreviates "ma," leaving the law a tautology. The law acquires predictive yield only when F is operationalized as gravitational attraction, which connects any two masses m_1 and m_2 in proportion to $m_1 m_2/d^2$, d their distance apart. But the law of gravity is equally empty without the force law. For let a_{ij} be the acceleration induced in m_j by m_i when $d = 1$ ($i, j = 1$–3), and suppose $a_{12} = 2a_{23}$. Neither inertia alone nor gravity alone imply anything about the motion of m_1 relative to m_2, but together they imply $a_{32} = 2a_{23}$. The law $F = ma$ is analogous to the moral principle that people may do what they wish with their property, which is tautological, laying down no prescriptions absent criteria for "property." One such criterion is Locke's: a man owns what he mixes his labor with. The principle of property alone does not give a baker title to a pie he bakes because it does not say the pie is his, and Locke's criterion alone, which does say the pie is his, does not say what rights follow. But principle and criteria together say the baker can do what he wants with his pie. Note that the gravitational and Lockean criteria supply sufficient but not necessary conditions. There are forces beside gravity and legitimate modes of acquisition beside manufacture.

6. Sometimes the "if" can be strengthened to "if and only if."

7. This completes the analogy between moral and legal reasoning. By joining a penalty to a precisely defined act, legislators supply a criterion for guilt. Judgments of guilt or liability consequently hinge on whether some act of the defendant fits this legal criterion. That factual question is what judges and juries try; the bench's instructions to the jury always concern the finding of fact on which its verdict must rest. Legal criteria may be vague; a jury that must decide whether a man forged a signature may have to be told what counts as a "signature." Innovations in artistic techniques, like computer-

assisted design, may prompt the legislature to add new criteria for forgery. Particular findings of fact are usually sufficient but not necessary for a verdict, there being many ways to forge. Finally, lawmakers are unconstrained, able in theory to penalize any act they choose to. Lawmakers bent on punishing someone can always find something he does to criminalize.

8. "Whites are said to be multi-cultural: There's the gun culture and the car culture and the boat culture and the monster-truck and pro-wrestling cultures, to name just a few. . . . Think of the most obnoxious White Guy you know. Guess what? He's going to croak" (Carrey,1993). Carrey is white.

9. The dominant emotion of McCall's memoir (1994) is anger at whites. He unapologetically recounts ganging up with teenage friends to beat whites who entered his neighborhood, but fumes that his sentence for armed robbery exceeded that for attempted murder (because, he says, his intended victim was black).

10. I have been influenced here by Hohfeld (1923).

11. If A should care for A's children, the universalization of A's duty is neither $(\forall X)(X$ should care for A's children) nor $(\forall X)(A$ should care for X's children), but $(\forall X)(X$ should care for X's children). The universal duty associated with A's obligation to do $O(A)$ is $(\forall X)(X$ should do $O(X))$, not $(\forall X)(X$ should do $O(A))$. Many "refutations" of Kant founder on this confusion. Morris Raphael Cohen argued that if someone thinks he ought to earn his living as a baker, the categorical imperative requires him to think that everyone should be a baker, which of course is an absurdity. In fact, the baker's maxim is probably something like "I should earn my living at a task for which I am well-suited," whose universalization, "Everyone should earn his living at whatever he is suited for," can easily be willed.

12. Baird (1996) worries that the epithet "negative" is too negative, and suggests distinguishing "non-rivalrous" from "rivalrous" rights. This terminology nicely captures the logic of the situation, and deserves to become current, but the negative/positive distinction is so well entrenched I use it here.

13. There are, following Kant, two ways a policy can violate the golden rule. Attempts to universalize it may self-destruct; there cannot *be* a world in which everyone takes advantage by cheating, since cheating is advantageous only when everyone else follows the rules. Or, a policy that could without logical contradiction be followed by everyone may be such as no one could *want* everyone to follow. There could be a world in no one helps anyone else, but no one would want to live there.

14. MacCallum (1967) disputes the negative/positive distinction on the grounds that any positive liberty can be construed as negative by adjusting criteria for "interference." Jones can say Smith's refusal to associate with him interferes with his associating with Smith. MacCallum's proposal abuses language, if nothing else, for it also allows Jones to call Smith's refusal to teach him algebra interference with his education. Usage aside, there remains a substantive difference between freedom from coercion and freedom from ignorance. There cannot be a world in which everyone may coerce everyone else: if Jones may restrain Smith at will, and Robinson may restrain Jones at will, Robinson is both allowed and forbidden to stop Jones from restraining Smith. There can, however, be a world in which no one is obliged to relieve ignorance.

15. Kant has been faulted for overemphasizing motives at the expense of actions; some acts, like cheating, seem wrong apart from the reasons they are done. Kant could reply that some actions (such as cheating) seem intrinsically bad only because they could never be done from conscientious motives. The mirror-image problem dog utilitarians, who hold that motives are irrelevant—only action count—and attribute everyday condemnation of certain motives to their invariably bad consequences.

16. As with the case of wolves discussed in chapter 4, this means that women are more amoral than men, not that there are "male" and "female" moralities.

17. It is expected that *r*-strategists able to conceptualize behavior would favor loose

sexual norms.

18. Nozick is a moral realist.

19. Acceptance of black "materialism" illustrates white abandonment of white norms once blacks begin to transgress them. For generations America has been criticized as too materialistic. William James famously deplored "the bitch goddess, Success." Yet as prominent black athletes and entertainers have become increasingly open about their desire for money and celebrity, this criticism has become muted (except when filthy white lucre is condemned for seducing blacks).

20. "Slum persons generally are apathetic toward the employment of self-help on a community basis, they are socially isolated, and most sense their powerlessness. This does not mean that they are satisfied with their way of life or do not want a better way to live; it is simply that slum apathy tends to inhibit individuals from putting forth sufficient efforts to change the local community. They may protest and they may blame the slum entirely on the outside world, but at the same time they remain apathetic about what they could themselves do to change their world" (Banfield 1974: 71–2).

21. "Every wicked man," remarks Aristotle in the *Ethics*, "is in a state of ignorance as to what he ought to do and what he should refrain from doing, and it is due to this kind of error that men become unjust and, in general, immoral. . . . Ignorance in moral choice does not make an act involuntary—it makes it wicked."

22. Cooperation is plainly irrational on what is known to be the last play of an iterated Prisoner's Dilemma, and if cooperation is known to be irrational at the nth play it is irrational at the $(n-1)$st, from which it apparently follows that cooperation is never rational in an iterated Prisoner's Dilemma. However, cooperation remains rational if neither player knows when the iterated game will end, the situation in the small groups in which mankind evolved. Cheating in small groups with open-ended interaction is unwise (and will be selected against) because one can never be sure he will never again encounter a man he has cheated.

23. My discussion here has been influenced by Pojman (1991a, 1991b, 1992d), and Sher (1987: 142–144).

24. Rosenfeld (1991), Dworkin (1977b), Gutmann (1980), Nielsen (1985), Nagel (1970), Gewirth (1978), and Regan (1985), among others.

25. Personal merit *is* felt to matter to the ascription of enforceable positive rights. Many people resent subsidizing women with illegitimate children more than they resent subsidizing farmers, because farmers, unlike unwed mothers, are felt to be hard-working contributors to society. Those who deplore such comparisons can make them unnecessary by rejecting enforceable positive rights.

26. Section Two of the *Groundwork of the Metaphysic of Morals.*

27. When Kant speaks of the absolute worth of all "rational wills," he uses the term in such a way that a rational will *cannot* deviate from duty, the dictate of "practical reason." It then becomes a matter of definition that rational wills have maximum worth, but an open empirical question whether any human wills are rational.

28. Section One of the *Groundwork.*

Part III. Implications

8

Justice

8.1. THE ARGUMENT FROM THE INTERACTION OF GENES WITH ENVIRONMENT

Many writers deny that a genetic basis for race differences, even if granted, would have any moral or practical significance.

This verdict is sometimes based on a caricature of what race differences might require, as in Lewontin's suggestion that they would mean that all black women should be domestics. At other times race differences are said not to matter because people should be treated on the basis of their "individual" traits—which assumes (wrongly, I argue in chapter 10) that race is not a trait of individuals. A negative verdict may also be rendered on the ground that deserved equality of treatment is a "moral postulate" rather than a statement of empirical fact, hence irrefutable by empirical race differences. This ground is infirm, for, as the last chapter made clear, what treatment counts as "equal" depends on the descriptive traits of individuals and groups.

The most popular reason for discounting genotypic differences, however, is genotype/environment interaction. The idea is that genetic race differences would be "politically neutral" (Jencks 1992: 94, 109) because of the variability with which genes can express themselves, an argument also given by Layzer 1973), Lewontin (1974), Block and Dworkin (1976a: 483–485), Hirsch (1970: 3–94), McGuire and Hirsch (1977: 68–70), Feldman and Lewontin (1975), Loot (1993: 70–72),[1] Block (1995), and Goldberger and Manski (1995: 64–765). Bodmer and Cavalli-Sforza (1970) mention interaction as well, linking it to environmentalism.

This argument begins by insisting that (a) the same gene can express itself in different ways in different environments and (b) two genotypes that express themselves differently in one environment may express themselves identically in another. As the reader is by now aware, these points are correct, indeed truistic. Lewontin is entirely right that "no statistical methodology exists that will enable us to predict the range of phenotypic possibilities that are inherent in any genotype" (although the factual accuracy of Lewontin's examples of interaction—the height of the Achillea plant and larval viability in fruit flies—has been challenged by Bouchard [1995]). From this it is inferred, again correctly, that environmental manipulation might reduce the discrepant effects of any race differences in the intelligence polygene(s). The high heritability of IQ relative to known environments (see section 4.5) does not rule out undiscovered environments in which black and white genotypes express themselves as equal phenotypic IQ. Such new environments, were they realized, would simply reduce genetic variance and with it heritability. These possibilities are taken to imply, finally, that the persistence of the race gap is a social decision, not a decree of nature. We can, says Lewontin, boost IQ "as much or as little as our social values may eventually demand" (1970: 112). In the words of one of his students, "The decision as to how strenuous environmental intervention should be must rest on ethical considerations of social justice and compassion, and not on inferences about malleability drawn from a heritability index" (Baumrind 1991: 386). According to Block, "we should *ignore heritability* . . . and simply *try out improved environments"* (1995: 124–125).

Jencks illustrates the prospects for intervention with hair length. Genetic sex predicts this trait, making it highly heritable in contemporary America, but only because hair styles reflect sex differences in fashion. The heritability of hair length is an artifact of environmental differentiation triggered by the genetic sex difference, which society could seek to overcome by teaching boys and girls to adopt the same coiffeurs. Similarly, Jencks speculates (1992: 99, 107), race differences in criminality and IQ resulting from differences in treatment triggered by genetically determined variations in appearance would also be artifacts spuriously raising the heritability of criminality and IQ. Creating new environments might raise black IQ and reduce black crime, incidentally reducing h^2 for IQ and criminality. (Jencks' point is precisely the one Block makes with the "clubbing redheads" fable; see section 4.8)

An obvious objection to this argument is that it very nearly treats what is possibly possible as if it were actual. The reaction ranges of the genes controlling intelligence in humans *might* be as wide as those controlling hair length, so that race differences in IQ *might* be as malleable as sex differences in hair style. Indeed, there *are* environments in which blacks and whites coincide in intelligence: the IQs of black and white infants left exposed on the Moon would quickly converge to 0. But the sheer *possibility* of environments in which the races agree in (nonzero) intelligence does not show that such environments actually exist, or that, if they do, they could sustain a human society. The reaction ranges for some genes, like those controlling eye color, is quite narrow; IQ may be equally constrained. There might be selection for narrowness of range of highly adaptive traits like intelligence to ensure their presence in all environments. (In the language of geneticists, highly adaptive traits "go to fixity.") Or else, as I stressed in chapter 4, the reaction ranges of the black and white IQ genotypes might be wide but covary with environment, fixing the race *difference* across all environments. Proponents of the interactic argument offer no evidence that their conjectures are more than that—evidence, in other words, that the reaction range(s) for the (allelic variants of the intelligence polygene is as wide as Jencks assumes the range of the hair-length gene to be. Indeed, evidence mentioned in chapters 3 and 4 suggests that feasible environmental interventions will not significantly raise intelligence, hence are unlikely to shrink the race gap. This failure of special stimulation to boost the IQs of black children further weakens the interaction argument: every new experiment that leaves the race gap intact reduces the likelihood that some gap-reducing environment waits over the horizon. Indeed, sex differences in sexual cues, causing males and females to call attention to their gender, might make large reductions in the heritability of even hair length more difficult than Jencks supposes. Sex-typing of display may be an environmental correlate of genes.

In short, even if all "political" questions look forward to what can be done, gene/environment interaction shows at most that biology *might* be "neutral," not that it *is*. Should the reaction range for the IQ gene be narrow—and the mere possibility of its being wide proves nothing to the contrary—biology may forbid what "social justice" demands.

But this criticism of the interaction argument is superficial. The deep failing of the argument is that it examines "political" issues from the wrong perspective. Looking forward is often less important than looking back, since what should be done about a situation often hinges on how it came about. Despite the fact that I am comfortable while you are chilly, the question most relevant to who should get the extra sweater in my closet is not who would benefit most from having it or how costly a transfer would be, but how I came to possess the sweater in the first place. If I took it from you, I should, and others may make me, return it. If I knitted it myself, it is more difficult to maintain that you should get it, and much more difficult to justify a forcible transfer. The same is true for the outcome of competitions. That a year's training for *A* can close a genetic difference in running speed between himself and *B* does not bear on whether *B* beat *A* fairly in today's race. Nozick (1975) calls the idea that the way things came about limits what may be done about them the "historical theory of justice," but it is less a theory than a commonsense truism—so much so that, by ignoring it, the argument from interaction puts itself almost completely beside the point.

Retrospection is crucial when the topic is racial justice. Whether or not there are environments in which the attainments of the races would be equal, if *in fact*, in the environment(s) in which blacks and whites have *actually* functioned, the extant race difference in attainment was caused by genetic factors rather than white misdeeds, this difference is not an injury, hence not an injury for which whites are to blame, hence not a condition whites are obliged to remedy. Given that race differences express genetic variation in the environments that have actually existed, how these same genetic factors would respond to other environments is irrelevant to questions of fault, just as, given that I *did* not steal the sweater from you, it is absurd to argue that you deserve because it might have been yours. It is absurd to accuse *B* of cheating *A*, and give *A* a share of the prize to make up for it, because *A* might have run faster.

From this perspective, the significance of genetic race differences is what they *dis*prove, namely, every claim denying race differences or attributing them to a racist environment. A genetic explanation of black failure refutes these accusations against whites and undermines the policies premised on them. That is why genetic race differences are anything but "politically neutral."

Ironically, Herrnstein and Murray's (1994) *objections* to intervention programs and affirmative action are subject to the same criticism. The ineffectiveness of intervention to date persuades these authors that continued investment in slow learners can never repay its costs. In fact, they say, the *basic* point about the race gap is the difficulty of closing it, whatever its origin.

[M]ost important, it matters little whether the genes are involved at all. . . . If tomorrow you knew beyond a shadow of a doubt that all the cognitive differences between races were 100 percent genetic in origin, nothing of any significance should change. The knowledge would give you no reason to treat individuals differently than if ethnic differences were 100 percent environmental. By the same token, knowing that the differences are 100 percent environmental in origin would not suggest a single program or policy that is not already being tried. (312–315)

This follows only on a forward-looking view. For suppose that the "100 percent environmental" cause of the race difference in IQ is past and present racism. Then, while trying to atone for these sins by boosting black IQ might still be futile, special efforts for blacks of some sort would be far more defensible than they would be were the IQ gap caused by genes or environmental factors unrelated to white misdeeds.

A forward gaze is not entirely misdirected. There may be reasons of utility or distributive justice unrelated to the past for trying to reduce the race gap, a possibility explored in sections 8.13–8.15. However, as white responsibility for black failure is the dominant theme of discussions of race—more so, we will see, than most discussants realize—the key issues are harm, responsibility and rectification.

8.2. DEFINING PREFERENCE AND AFFIRMATIVE ACTION

I henceforth use "preference" and "affirmative action" interchangeably, to mean favoring blacks in competitive situations in which ability has heretofor been, at least ideally, the selection criterion.[2] In this context, "ability," "standard" and related notions are task-relative. A task is specified—curing th ill, say—and the associated ability is that cluster of traits needed t accomplish it. "Qualification" is sometimes meant more broadly to cove predictors of ability as well, such as certification by a medical board. Race occasionally a qualification; a white singer cannot portray Joe in *Showboat*.

The forms of preference are protean, including quotas, set-asides, lowe standards for blacks, training reserved for blacks, efforts to recruit blacks n matched by efforts to recruit whites, race-norming of tests, banding, a recentering. The first five are self-explanatory, but the latter three, bein relatively recent, call for some elaboration.

A test is *race-normed* when a testee's score is compared only to the scores other testees of his race. For example, the race-normed percentile rank of black on a civil service exam is the proportion of black competitors he outscores, while the race-normed rank of a white is the proportion of whites outscored. Since scores on any instrument that registers mental ability distribute like IQ, race-norming a test taken mostly by whites in effect boosts each black 1 SD: a black at the 50th percentile overall will outscore 84% of all other blacks, outranking a white who scores at the 80th percentile overall.

Banding is the identical treatment of all candidates who score in a specified range on some predictor. A college "bands," for example, when it fills the 700 places in its freshmen class by randomly choosing 700 applicants whose combined SAT score exceeds 1250, instead of the 700 highest scorers. This procedure favors blacks because black scores, being lower on average, tend to cluster toward the low end of any band (see section 8.9). Proportionately more blacks who score over 1250 on the SAT score *just* over it, hence are likelier to be selected if a 1250 is considered as good as a 1350.[3]

Recentering, better termed "compression," is the renorming of a test on a less able reference group without a corresponding rise in the maximum score. Its effect is to squeeze poor performers closer to good performers, as can be seen in the 1995 recentering of the SAT.

The SAT was originally scaled from 200 to 800 with a mean of 500 when normed in 1941 on a virtually all white and predominantly male population. Over the last three decades the mean SAT Verbal (SAT-V) score has fallen to 424 and the mean SAT Math (SAT-M) score to 478, in large part, according to the College Board, because the test is now taken by "more women, minorities and economically disadvantaged students," inducing the board to recenter to "reflect more accurately the diversity of students now taking the test" (cited in Krauthammer 1994). In so doing it set the current mean performance at 500,[4] but kept the top score at 800 instead of letting it climb, as the highest IQs climb when IQ is renormed on a less able population (see section 3.9). As a result, raw scores above the current SAT-V mean, which in 1994 were spread over 800 – 424 = 376 points, are now spread over just 800 – 500 = 300 points. With fewer pigeonholes, the new scale must conflate some raw scores that were previously distinguished. Although the conversion fluctuates from year to year, a raw score of 4 wrong will now typically earn an 800, whereas 800 previously denoted a perfect score while 4 wrong was a 760. This sounds like a boon for higher (but imperfect) scorers, but the overall result is faster growth for low scores. A raw score previously scaled at 424, hence 760 – 424 = 336 points below a raw score that formerly earned 760, is now 800 – 500 = 300 points below the former 760. Thus, the recomputation aids lower, disproportionately black, scorers. Consider that in 1990 the black SAT-M mean was 352 and the Asian mean 535. Recentered, 352 becomes 402—a gain of 50 points—while 35 becomes 555, a gain of 20 points, "reducing" the black/Asian gap by 7%. The College Board gives its subscribers conversion tables for calculating old scores from new (Herszenhorn 1995)—belying the argument that the new scores are more helpful—but, since a recentered 800 covers the old interval from 760 to 800, which includes proportionately more whites and Asians than blacks, it becomes more difficult for college admissions officers to interpret the (high) scores of whites and Asians than the lower scores of blacks.

Affirmative action "goals" are sometimes said to avoid the discrimination against whites that quotas unarguably involve, but the distinction is illusory. Suppose nine out of ten engineers in a firm are white, and management adopts the goal of increasing the proportion of black engineers. A new position opens for which a black and a better-qualified white apply. Should the firm ignore race and hire on merit it will select the white, allowing its goal to become empty words. Should the firm take its goal seriously enough to hire the less-qualified black, race will have overridden merit and the goal will have become discriminatory, in effect attaching a racial requirement to the open position. Defenders of goals protest that they favor hiring the qualified: after all, they say, the firm will recruit a black engineer, not any black off the street. In the words of Justice Powell's pivotal opinion in *University of California Regents v. Bakke* (1978), race is to be used simply as a "plus factor." However, any black whose qualifications are inferior to those of a white by less than the "plus factor" enjoys a decisive advantage solely because of his race, as clear a case of racial discrimination as could be wished. Think of the problem in legal terms. Should the firm hold a government contract under Office of Federal Contract Compliance Programs rules requiring "good-faith efforts" to hire more black engineers, or have been ordered by a court to do so, hiring the white will create a liability that hiring an identically qualified black will not—which is to say that such regulations and orders legally enforce discrimination against whites. Defenders of the goal/quota distinction have never satisfactorily explained what employers in the position of our hypothetical firm should do, or what quotas demand of it that goals do not. Goals are simply quotas in slow motion.

Four points should be noted about defining preference solely in terms of favoring blacks.

First: As in the engineer example, it is usually pertinent to distinguish private from legally mandated affirmative action, since private but not mandatory preferences may be viewed as an exercise of freedom of association (a point we eventually consider; also see Levin 1984a; 1987: 124–126). Nonetheless, the basic issue is whether blacks deserve preference, not whether their title to preference is strong enough to be enforceable. A definition of preference should therefore downplay the public/private distinction, pivotal though it is in other contexts.

Second: Many critics of quotas say they support affirmative action in its original sense, which was, according to them, the taking of pains to find fully qualified blacks otherwise apt to be overlooked. However, including what such supporters of "affirmative action" explicitly reject does not distort the present definition, for it covers the laws and policies that have fallen under that rubric since the Nixon administration's Philadelphia Plan in 1971. The substantive question is the legitimacy of these preferential policies by whatever name.

Third: Some (e.g., J. Richards 1980: 112) seek to reconcile preference with merit criteria by inserting a racial component into task descriptions. They say, for instance, that redefining a "doctor" as someone who cures the ill in a way that leads people to think of blacks as competent makes race a legitimate qualification for being a doctor—a curious argument perhaps echoing the more familiar idea that institutions like medical schools have changed (or should change) their goal from that of producing doctors to serving wider social needs

The catch, of course, is that a "doctor" can also be redefined as someone who cures the ill after graduating from a race-blind medical school, or as someone who cures the ill in a way that leads people to admire whites. It must then be explained why the first but not the second or third redefinition is appropriate, and that is the original problem of justifying affirmative action all over again. So preference advocates might as well not tug at their definitional bootstraps, and understand "qualification" as it is usually understood.

Fourth: In deference to the many grounds offered for affirmative action, my definition avoids reference to its purpose.[5] While I go on to argue that a compensatory purpose is central, defining it as central would beg that question and stipulate away other possible rationales. At the same time, I should stress the range of competent authorities on both sides of the issue who do stipulate a compensatory aim. Hacker puts "reparation" at the center: "Some have argued that black Americans deserve reparations for their centuries of bondage and subjection. That is what the argument about affirmative action is essentially about" (1992: 219). Pojman, a critic of preference, writes:

Affirmative Action is the effort to rectify the injustice of the past by special policies. Put this way, it is Janus-faced or ambiguous, having both a backward-looking and a forward-looking feature. The backward-looking feature is its attempt to correct and compensate for past injustice. This aspect of Affirmative Action is strictly deontological. The forward-looking feature is its implicit ideal of a society free from prejudice; that is both deontological and utilitarian. (1992: 183)

Pojman's characterization is not ideally clear, since the first sentence implies that both the backward- and forward-looking orientations of affirmative action are meant to rectify past injustice, yet freeing society from prejudice *henceforth*, as specified in the final sentence, is not rectificatory. (Perhaps Pojman is unconsciously shifting from rectification to fairness in some broader sense.) Despite this unclarity, Pojman plainly conceives preference as reparation.

Rosenfeld, a strong proponent, defines affirmative action as "the preferential hiring, promotion, laying off . . . preferential admission . . . or preferential selection of businesses . . . for purposes of remedying a wrong or of increasing the proportion of minorities" (1992: 47). I take "remedy" here as synonymous with redress. Moreover, since "increasing the proportion of minorities" simply redescribes racial preference, only the first disjunct actually specifies a goal. So Rosenfeld's definition too is compensatory.[6]

Edwin Hettinger also offers a disjunctive definition:

By "reverse discrimination" or "affirmative action" I shall mean hiring or admitting a slightly less qualified woman or black, rather than a slightly more qualified white male, for the purpose of helping to eradicate sexual and/or racial inequality, or for the purpose of compensating women and blacks for the burdens and injustices they have suffered due to past and ongoing sexism and racism. (1997: 305)

Here Hettinger seems to contrast equality with compensation, but he makes clear that the inequality affirmative action is intended to end is inequality of opportunity, and that aim turns out to be a version of compensation (see 1997:

312, cited more fully in section 8.6).

The most recent textbook discussion known to me, Shaw (1996), offers "compensatory justice" as the sole argument for affirmative action (321–322).[7]

8.3. COMPENSATORY JUSTICE

Compensatory justice requires the restoration of damage caused by injury, where "damage" is taken to include the loss of abilities as well as the loss of goods, and "injury" includes force, fraud, and, lately, discrimination.[8] The moral basis of compensation is that injured parties should be where they would have been but for the wrong they suffered, or, more generally, that the world should be as it would have been had that wrong not been done. "Where the injured party should be" might be thought to mean any position as good as the one he should have been in—in economist's terms, any point on the indifference curve he should have occupied—justice demanding only that he be as well off as he would have been absent the original wrong. Proponents of compensatory preference seldom regard equivalent outcomes as a sufficient remedy for discrimination, however; for reasons discussed in section 8.7, they ask instead for outcomes as close as possible to the very ones nondiscriminatory dealings would have produced.

The ideal of restoration is embodied in three principles of increasing generality that almost, but not quite, nest. The narrowest of these is *pure compensation*:

• An injured party deserves what he lost, to be restored by the tort-feasor.

However, a broader principle is needed for wrongs affecting third parties. After A steals B's car, kills B, gives the car to his own son, and then dies, B's survivors are entitled to reclaim the car from A's survivors. The principle covering such cases is that of *illicit advantage*:

• Those worse off because of a wrong deserve to be as well off as they would have been but for that wrong, and those better off because of it must surrender, to the worse off, their ill-gotten gains.

The illicit advantage principle coincides with pure compensation when a wrongdoer improves his own position, for then the wrongdoer's victim is worse off and, by the illicit advantage principle, deserves to have his losses restored from the beneficiary-wrongdoer's gains. But the principles diverge when a wrongdoer worsens his own position as well as his victim's, as when a car thief injures himself while wrecking the vehicle he has stolen. In those cases the pure principle asks the wrongdoer to repair his victim's loss, but, as no one is better off because of the wrong, the illicit advantage principle demands nothing from anyone. To resolve this conflict, confine the pure compensation principle to wrongdoers and the illicit advantage principle to indirect beneficiaries and victims; so restricted, the two principles cover everyone consistently, unequivocally requiring worse-off, reckless car thieves to compensate their worse-off victims.

The illicit advantage principle is too strong as stated, for not every wrongful act leaves its innocent beneficiaries liable. The advantage on an examination enjoyed by intelligent Jones over Smith, the latter feeble-minded because his mother negligently dropped him on his head when he was a baby, is perfectly legitimate. The intuitive reason is that Jones' relatively superior performance is unconnected to the wrong that created Smith's inferior one.[9] In legal language, an advantage created by a wrong is illicit only when linked to the wrong by an *appropriate causal nexus*. The nexus paradigmatically appropriate for redress is transmission of a material object: after Jones steals Smith's car, whoever comes to possess it must return it to Smith or his heirs. Causal relations between wrongs and benefits involving consanguinity, as when a swindler's children inherit his money, are also considered appropriate. The existence of a nexus is distinct from both liability and negligence, incidentally. Agents are compensatorily liable for the foreseeable consequences of their acts (which may be construed quite broadly, as in the classic *Palsgraff v. LIRR*), but innocent beneficiaries may retain advantages foreseeably consequent to a wrong. Jones need not take his examination next to a jackhammer even if Mrs. Smith should have foreseen, did foresee, or even intended that her derelict handling of her son would impair him.

Further complications are created by the death of a wrongdoer whose actions have harmed everyone. In that event there are no ill-gotten gains to surrender to victim and family since there are no gains of any sort, and the element of guilt that demands the pound of flesh from worse-off wrongdoers is not to be found in worse-off innocent bystanders. A pedestrian slightly bruised when Jones kills himself wrecking Smith's car did not himself harm Smith, and should not suffer further loss in the form of payment to Smith just because Smith is yet worse off. This problem is (or should be) of concern to defenders of compensatory preferences, since some historians (e.g. Genovese 1967: 26-31) believe slavery harmed whites as well as blacks by preventing investment in industrialization and otherwise retarding progress.

There is a third principle to cover these cases, that of *net illicit advantage*:

- If the gap in advantage between A and B is increased by a wrong, and the causal nexus is appropriate, A must surrender to B his net illicit gain relative to B.

The scope of this principle is limited. As in the case of the reckless car thief, A need not always make B whole when A's own position as well as B's has been worsened absolutely, but A comes out relatively further ahead. However, this requirement does seem reasonable when A and B are obliged to compete. Consider: a kidnapper orders his two hostages Jones and Smith to box each other, promising to release the winner, and then immobilizes Smith's right hand. Many people would judge that Jones enjoys an unfair advantage, and ought to be handicapped in some way, despite the fact that he (like Smith) should not have to fight for his freedom in the first place. While there should never have been a fight, we think, any fight that does take place should at least be held under equitable conditions. Since the setting for affirmative action is a competition between some whites and some blacks, the net illicit advantage principle kicks in if slavery and discrimination harmed whites as well as blacks, but harmed blacks more.

8.4. AFFIRMATIVE ACTION AND COMPENSATORY JUSTICE

The application of compensatory justice to race is immediate and familiar. Black underrepresentation in the more prestigious and remunerative fields is taken to be an effect of the wrongs of slavery, discrimination public and private, and an atmosphere of contempt. Today's blacks may enjoy full legal rights, but they lack the credentials and abilities they would have had had their ancestors competed on a level playing field. Preference seeks to annul these handicaps by putting blacks "in the position the minority would have enjoyed if it had not been the victim of discrimination" (*Rios v. Enterprise Association Steamfitters Local 698* [1984]). Or again: "the remedy is necessarily designed, as all remedies are, to restore the victims of discriminatory conduct to the positions they would have occupied in the absence of such conduct" (*Milliken v. Bradley* [1974]). Whites must forego the "fruit of the forbidden tree" of discrimination, their wrongfully acquired competitive advantage: "Compensatory justice requires (even innocent) beneficiaries of racism to give up their unearned competitive edge over innocent victims of racism" (Harwood 1991: 68) because, in general, "innocent beneficiaries of injustice should compensate victims of that injustice" (Harwood 1993: 80). Boxill epitomizes the argument in two syllogisms:

Black people have been and are being harmed by racist attitudes and practices. Those wrongs deserve compensation. Therefore, black people deserve compensation. Preferential treatment for black people is an appropriate form of compensation for black people. Therefore black people deserve preferential treatment. (1984: 148)

We find a black lawyer arguing:

All colleges have an obligation to deal with compensatory education. We can show that we were harmed by racial policies of the past that persist today. Once you document it, then you have a substantial ground for compensatory education. (Cited in Ridgely 1992: 17)

In addition to these sources and those cited in sections 7.3 and 8.2, the reader may also consult Sher (1975), Rosenfeld (1991, esp. 290–291 and 307–308), Boxill (1978), Thomson (1977) and Wasserstrom (1985a).

None of the authors cited questions that sameness of race of wrongdoer and beneficiary is an appropriate nexus. As for how "compensatory justice" is to be interpreted in these passages, it can mean the illicit advantage principle as well as the principle of pure compensation, and Harwood clearly does mean this. Faced with evidence that whites as well as blacks are worse off because of slavery, compensation theorists could and presumably would invoke the net illicit advantage principle, asking whites to step aside for blacks because success in competition is determined by comparative qualification, an whatever else slavery did (they would argue), it widened the gap between black and white qualifications.

Racial preference cannot be justified as punishment because only wrongdoers deserve to be punished, and the few racial wrong-doers still alive are not especially likely to be touched by affirmative action. Preferences are to assesses compensatory, not punitive, damages. Yet an unmistakably punitive edge emerges whenever, by quirk of circumstances, pursuit of the stated goal of a preferential policy stands to benefit whites. Racial preference in the allotment of public housing is officially intended to encourage integration, but the courts have forbidden reservation of housing for whites intended to forestall white flight. It is considered permissible for black students in mostly white schools to associate with each other for purposes of morale, but scandalous—and illegal by the courts—for white children to remain together in mostly black schools (E. Harrison 1993, *New York Times* 1993a). Pleasure at white discomfort leaps from Hacker (1992) (recall his gratuitous jibe at "white kids in suburban schools" being inferior to Asians in mathematical ability). Much of the emotional force behind the compensation argument come from a feeling that whites have got discrimination coming. The "crossed signals" theory of white guilt presented in chapter 4 may help explain why many whites concur.

8.5. VARIANTS OF THE COMPENSATION ARGUMENT

The compensation argument is central not merely because claims of justice are compelling, but because a number of superficially dissimilar arguments, some already mentioned, reduce to it.

Transition to the Illicit Advantage and Net Illicit Advantage Arguments

While preparing this chapter I was told several times that sophisticated proponents of preference now reject the compensation argument. Since compensation would seem to be exactly what everyone had and still has in mind, this disclaimer was puzzling. But the puzzle vanishes as soon as one learns why the compensation argument has been abandoned, and what usually replaces it. Compensation is an unsuitable rationale, it is agreed, because contemporary whites who lose under affirmative action are not the ones who harmed blacks. However, in its place one finds the claim that today's whites benefit from the damage done to today's blacks—and that claim is merely a shift from one compensatory principle to another. Preference advocates sophisticated enough to recognize that the typical white is not better off than the typical black because of something that *particular* white did almost always insist that that particular white should still forego his advantage because he is better off as a result of something some other whites did.

A good example of this transition is Janet Richards' version of the role-model argument (1980: 109–118), which she applies to women but would apply equally in the case of blacks. Richards begins by seeming to reject the compensation argument precisely on the grounds that women apt to benefit from preference are not the ones who have been harmed by discrimination. Rather, she says, "positive discrimination" for reasonably well qualified women will "make people in general more used to seeing women in former male preserves." There remains, however, the troublesome better-qualified man passed over in the process who is in no way responsible for the present sexual division of labor. But this man

is not necessarily deprived of anything he was entitled to. A fair scheme of positive discrimination . . . would give these positions to . . . women who would have had them anyway, in a fair competition with men, if women had had the same advantages men had in the past. But the only men excluded on this sort of principle would be the ones who, as far as we could tell, would not have succeeded anyway if the situation had been fair. (118)

Yet this assurance contradicts the earlier one that favoring women today has nothing to do with benefiting women passed over previously, for now preferences are to give women the positions "women would have had anyway . . . if the situation had been fair." Indeed, if preference meant to change expectations does not work "substantial injustice" (118) because it gives women the jobs they should have gotten in the first place, that alone can be seen as justifying preference without a detour through expectations. (Prior [1982] develops this point.)

Although the illicit advantage and net illicit advantage principles are distinct from that of pure compensation, their similarities are patent. Appeal to all three is driven by the conviction that blacks must be put where they would have been but for past wrongs. Appeal to all three takes this restoration to require that whites competing against blacks yield some of their competitive advantage—whether because those very whites harmed those very blacks, or because those whites benefited from harm to those blacks, or because injuries done to both harmed those whites less than those blacks. Rosenfeld (1991) is content to call the following statement, which clearly concerns illicit advantage, an appeal to "compensatory justice":

Provided an affirmative action plan is precisely tailored to redress the losses in prospect of success attributable to racism or sexism, it only deprives innocent white males of the corresponding undeserved increases in their prospects of success. Thus, insofar as affirmative action brings the prospects of success of all competitors (and potential competitors) to where they would have been absent racism and sexism, it merely places all competitors in the position in which they would have been if the competition had always been conducted in strict compliance with equal opportunity rights. (307–308)

This passage can also be read as an appeal to the *net* illicit advantage principle when "undeserved increase" is construed as "undeserved relative increase," or "increase in the difference in prospects of success" (an interpretation Rosenfeld does not explicitly consider).

In one way or another all three arguments trace race differences in prospects to white misdeeds, interpret these misdeeds as creating a black title to preference, and collapse together if black prospects are not inferior because of white misdeeds. All three are variants of a single idea.

Group Rights

Consider next the argument that groups, like individuals, have rights, and that blacks as a group, having been injured, deserve to be made whole. Indeed, affirmative action has been hailed as a breakthrough for calling attention to this hitherto unrecognized class of rights.

In fact, although the reification of groups is so striking a doctrine and its difficulties so evident that it attracts attention, it plays almost no logical role in the group rights defense of preference. As is obvious on reflection, the existence of groups and their endowment with rights does not by itself determine what rights which groups possess. Different groups presumably have different rights, which depend on more than sheer group existence. And the only plausible reason for ascribing a preferential right to the group Black—and the only reason ever cited in the group-rights literature—is the harm inflicted on it by the group White. Yet, if White owes Black preference because of past harm, then, while the entities who owe and deserve preference are collective instead of individual, the basis for preference remains compensatory. In fact, the basis remains compensatory in the narrowest sense, since the very entity said to have benefited from the harm, the group White, is the one said to owe restitution, and the very entity said to have been harmed, the group Black, is the one said to deserve it. Furthermore, since, according to group-rights theory (see May 1987: 135, 138) groups are benefited and harmed through benefits and harm to their members—for how else can groups be affected?—damage to Black's ability to compete can be remedied only by compensatorily preferring individual blacks over individual whites.

In the end, collective entities function only as media through which to diffuse harm to blacks ostensibly never harmed by discrimination, and advantages to whites who ostensibly never benefited from discrimination (see May 1987: 141–145, and Rosenfeld 1991: 85–86). Some such agency is needed to lift preferences over that great stumbling block, the innocent unadvantaged white, who can now be said to have gained simply by belonging to the group White. The existence of groups is merely a lemma toward a theorem about compensation applying to every white and every black.

Role Models

Another supposed alternative to compensation is appeal to role models: the sight of more blacks in high status fields, it is said, will raise the aspirations of other blacks. It is hard to believe any black remains unaware that he can become a doctor or lawyer, or that a great many people stand ready to help him do so, and I know of no evidence that the perception of blacks in high-status jobs inspires emulation. But let us assume the role-model effect real.

The difficulty here is that a comparable effect must also be assumed to work for whites, leading one to ask why whites are not entitled to the regnant number of white role models to inspire them. Why deny white youths the marginal inspiration of an additional white doctor so that black youths may gain the (possibly greater) marginal inspiration of an additional black one? Why not let role models sort themselves out race neutrally? A number of standard replies—that higher aspirations among black youths will bring more blacks into the professions, for instance—beg the question, which is whether anything is wrong with the extant distribution. A more pertinent reply is that, because of oppression, there are fewer black role models than there should have been, and black aspirations are consequently lower than they should be. Lack of opportunity has made black role models not merely scarce but *too* scarce. Clearly, this reply turns the model argument into an appeal to compensatory justice, which now asks that blacks be given the ambitions they would have developed absent past wrongs.

The seemingly noncompensatory, forward-looking argument that role models are *needed* to inspire blacks also passes through compensation territory. After all, a reasonable estimate of one's life chances, however pessimistic, is not in itself bad. There is poignancy but no injustice in a boy's distress when, taking stock of his athletic abilities, he realizes he will never be a major leaguer. It is a contemporary shibboleth that all children should aspire to be anything they wish, but in everyday life, in dealing with concrete individual problems, realism is considered paramount. Nobody would encourage his own children in false hopes. By that standard there is nothing inherently wrong with a black youth concluding from the scarcity of black doctors that his own chance for a medical career is poor. The reply, of course, is that this inference is proper only if the small number of blacks in medicine reflects black ability, not factors tending to obscure black ability. And the only reason that suggests itself as to why black doctors *should* be more conspicuous than they are is that blacks have been kept from being doctors. In that case, a black who doubts his chances is less like the boy who knows he is not good enough for the majors than a boy who doesn't know how good he is because he and his friends have never been allowed to swing a bat. But if that is the argument, preferential selection of black role models is once again intended to restore something, namely aspirations, that blacks would have had in a fairer world. We are back to compensation.

One finds many "individualists," who urge people to base aspirations on their own talents rather than preconceptions about the groups they belong to, also endorsing the role model argument. Individualists thus at odds with their own principles generally explain that group numbers are important now because a black can tell from current numbers that he *won't* be judged by his own talents. Individualism would be appropriate for blacks in an unprejudiced world, but now they need a sign that the world is becoming unprejudiced enough for them to stop thinking of themselves as members of their race. "If a country wants to vouchsafe that it has overcome discrimination and prejudice," says Hacker, "visible evidence is necessary" (1992: 122). This last argument is not exactly compensatory, but it does assume that black doubts about the value of effort is an effect of past wrongs it is the responsibility of whites to overcome.

Dworkin

The compensatory character of role models is often left obscure by the absence of any positive attempt to justify them. Ronald Dworkin's "utilitarian and ideal" defense of preferences (1977b: 223–239) is a case in point; his best known argument is in fact not an argument *for* preference at all, but a criticism of a criticism of preference that becomes an argument on the pro side only when fortified by compensatory considerations.

Dworkin's chief concern is the apparent symmetry between favoring blacks and favoring whites. Why should only one kind of favoritism be allowed? The difference, he says, is that favoring blacks is meant to satisfy their "personal" preference for self-respect, while favoring whites caters only to the "external" preference of whites for black failure. As Dworkin defines these terms, personal preferences concern "[one's] own enjoyment of some goods or opportunities," while external preferences concern "the assignment of goods and opportunities to others." Favoring blacks gives blacks something, namely self-respect, that does them some good, whereas favoring whites gives whites nothing that does them any good.

Like group rights, the personal/external distinction is at once so eye-catching and so dubious as to seem central to the argument, yet under analysis it proves to be peripheral. Dubious it certainly is. To begin with, the two kinds of preference do not exclude each other, as Dworkin plainly means them to. A man's desire that his child's murderer suffer is an external preference, but also personal if he has a right to the murderer's suffering. More important, the distinction, even if exclusive, does not show why discrimination against blacks is worse than discrimination against whites. Dworkin insists that personal preferences (such as those allegedly served by affirmative action) are superior to external preferences (such as those alleged to drive white discriminators), but, like other utilitarian rankings of desires, this one conflicts with utilitarianism itself. If satisfaction of desire is intrinsically good, as utilitarians hold, the *object* of a desire has no logical bearing on whether it should be satisfied. When the satisfaction of an external preference and that of a personal preference promise equal long-run consequences for happiness, utilitarians must be indifferent between them (see Ely 1983; Alexander 1992; Rosenfeld 1991: 107).[10]

Dworkin (1977b) claims to find a reason to downgrade external preferences in a conflict between external preferences and preference-maximizing:

Suppose many citizens, who are not themselves sick, are racist in political theory, and therefore prefer that scarce medicine be given to a white man who needs it rather than a black man who needs it more. If utilitarianism counts these political preferences at face value, then it will be, from the standpoint of personal preferences, self-defeating, because the distribution of medicine will then not be, from that standpoint, utilitarian at all. (235)

I return in chapter 10 to this caricature of the motives of "racists," left as opaque as Iago's by the unexplained allusion to "political theory."[11] But even granted Dworkin's understanding of "racism," his reasoning is plainly circular. Of course counting external preferences is not utilitarian "from the standpoint of personal preferences"! Catering to external preferences has not been shown to be *self*-defeating; catering to external preferences conflicts with maximizing utility only because utility is conceived from the start to exclude them.

Dworkin senses he is running in place, for his next sentence begins "In any case, self-defeating or not, the distribution [that takes racist preferences into account] will not be egalitarian," and concludes that under this distribution "the egalitarian character of [utilitarianism] is corrupted." This sounds like replacement of the self-defeat argument with a new one about equality. But the new one turns out to be the old one thinly disguised, for the sense in which utilitarianism is committed to equality, Dworkin explains, is that "the chance that anyone's preferences have to succeed" should depend only on the demands they make on "scarce resources," not "on the respect or affection [others] have for him or his way of life." Surely "the feelings of others about preferences" is just another name for external preferences—which are, again, being disqualified by stipulation. Counting external preferences is corrupt because it counts external preferences. Dworkin has yet to advance an inch.

Dworkin's gears refuse to engage because he has mistaken who is being inegalitarian when external preferences are treated like personal ones. Utilitarians are being consistently egalitarian when they consider all preferences, including external ones, equally important; it is the holder of external preferences who is the inegalitarian. Should Smith want to graduate college while Jones wants Smith to flunk out just so he can feel superior, *Jones* is being inegalitarian, but it is not inegalitarian to try to satisfy obnoxious Jones. A utilitarian fairy godmother would almost certainly grant Smith's wish instead of Jones', since in point of actual consequences Smith graduating will almost certainly do more good than Jones gloating. But she must take Jones' wish as seriously as Smith's, and she does not grant it only because she cannot. Calling her an enemy of equality is like calling her an enemy of classical music because she tries equally diligently to satisfy the preferences of classical and rock devotees. Utilitarianism is not pro- or anti- any musical genre, merely pro-satisfaction.

Dworkin's argument, then, amounts to stipulating that utilitarians should heed only personal preferences and deducing therefrom that utilitarians should ignore preferences of other kinds, which hardly marks a significant moral asymmetry. But even allowing that the personal/external distinction does distinguish anti-black from anti-white discrimination, it would not *justify* the latter. That discriminating against whites is not as bad as discriminating against blacks does not make discriminating against whites good, or constitute a reason in its favor. Granted that a desire for one's own self-respect is better than a desire for someone else's failure, why should black self-respect be enhanced at white expense? Why not let self-respect be determined race neutrally, by everyone's perception of his own talents, his race's success, or whatever else influences him? Dworkin says that increasing black self-respect will make society "more equal," which he equates with "more just," but the internal/external distinction in no way explains why black self-respect is a matter of justice.

The answer Dworkin finally gives (1977a: 139) is:

increasing the number of blacks who are at work in the professions will, in the long run reduce the sense of frustration and injustice and racial self-consciousness in the black community. . . . [Affirmative action] tries to provide "role models" for future black doctors, not because it is desirable for a black boy or girl to find adult models only among blacks, but because our history has made them so conscious of their race that the success of whites, for now, is likely to mean little or nothing to them.

I take it Dworkin is not saying that the black sense of injustice should be relieved whether it is warranted or not, but that this feeling *is* warranted, by "history" (by which he presumably means slavery and discrimination). Ultimately, then, Dworkin's argument for preference is that blacks need role models because "history" has kept them from finding encouragement in the success of nonblacks. And this need to offset a current negative consequence—alienation—of racism resembles a, rather tortured, appeal to compensation. I am not altogether comfortable calling this argument compensatory, since Dworkin's label for it, "utilitarian and ideal," leaves unclear whether he thinks greater self-esteem is owed to blacks, is a demand of justice in some broader sense, or is simply a good that society should promote. Still, the argument has the crucial element of seeking to annul an alleged effect of past wrongs.

I suspect Dworkin misperceives his defense against an attack on preference as an argument for preference, and is vague about the reasons for enhancing black self-esteem, because he assumes that opponents of preference carry the burden of proof. Finding it obvious that what he assumes are effects of history should be annulled, he feels no pressure to argue *for* preferences once he has, to his own satisfaction, met arguments against them.

Dworkin's whole discussion illustrates the risks of ignoring empirical facts. For one thing, he takes for granted that black self-esteem *is* low, contrary to data cited in chapter 3. More seriously, his disregard of race differences numbs him to the double edge of even his own oddly shaped sword. Still granting that only personal preferences deserve respect, it is clear by now that there are unarguably personal preferences that discrimination against blacks might serve; Dworkin rests so much on the personal/external distinction, I suspect, because he literally cannot conceive this possibility. The desire of white children (and their parents) for orderly, stimulating classrooms is personal; the evidence that black children are on average less intelligent than white children is overwhelming, and there is considerable evidence that black children are more disruptive, especially in integrated settings (Shuey 1966: 120; Humphries 1991). The scanty data that exists from the Minnesota Transracial Adoption Study (see chapter 4) suggests that close association with black children may affect white children in undesirable ways. Although Coleman et al. do not emphasize this fact, they found the academic achievement of whites (as well as blacks) to vary inversely with the proportion of blacks in their schools (Coleman et al. 1966: 307). One might also wonder about the effect on white children's self-esteem of knowing they must attend school with less able, more disruptive black children for the good of the blacks, not their own. As school integration appears not to benefit blacks (Scott 1993; Herrnstein and Murray 1994), a utilitarian argument for school segregation from Dworkin's standpoint would be sound. A second utilitarian argument sound from Dworkin's standpoint concerns rape: black males rape almost as many white females as black females, and rape white females at twice the rate white males do (see chapter 9). Given the limited opportunities for unplanned interracial contact allowed by residence patterns, a preference on the part of black rapists for white victims seems indicated. If keeping black men away from white women promised to reduce the victimization of white women without increasing the victimization of black women, the assuredly personal preference of white women for avoiding rape would justify shielding them from black men.

Nepotism

Proponents of affirmative action often point out *tu quoque* that colleges once preferentially admitted sons of alumni, and ask, rhetorically, what difference there is between favoring blacks and favoring sons of alumni. Actually, that question is easily answered. As Sowell observes (1993: 131), the scale of race preferences is vastly greater. Also, admitting underqualified children of alumni increases alumni support, financial and otherwise, whereas admitting underqualified blacks brings no comparable benefit. Finally, the gap in credentials between children of alumni admitted to a college and their peers is far narrower than that between typical admittees and blacks admitted by quota (Zelnick 1996: 147). These differences are so obvious, in fact, that the the appeal of the nepotism argument must be found elsewhere.

That appeal may lie in a sort of caricature of compensation, whereby preference is seen as turning the tables on white males, who once benefited from discrimination in college admissions and are now deservedly its victims. Taking pleasure in the reversal makes no logical sense, since the competitors over whom the sons of alumni were once chosen consisted entirely of whites, precisely the competitors over whom blacks are chosen via racial preference. But this reversal is satisfying at the emotional level at which whites are viewed as one undifferentiated entity. Turnabout is fair play. Let's see how The White Male likes being on the receiving end. Taunts about "nepotism" are a vindictive version of the compensation argument.

8.6. NONJURIDICAL ARGUMENTS FOR PREFERENCE

While there is no surveying every conceivable rationale that might be proposed for preference—such was the lesson of section 7.2—a few that do not invoke justice are instructive. They show why empirical facts about race must be heeded, and, by their weakness, confirm the importance of the redress argument.

Preference is Needed to Prevent Discrimination [12]

On its face this is a non sequitur. The appropriate way to stop an action is to forbid the action itself, not forestall a suppositious consequence; you don't prevent theft by paying people who might be robbed. The overt reason this illogical rationale is favored anyway is that whites are such inveterate discriminators that merely forbidding discrimination will not stop them. Only pre-emptive favoritism toward their victims will. But quota advocates relying on this defense must also assume that disproportional outcomes are tantamount to bias, or they would not be so sure that proportional outcomes do no more than foil bias. When for instance Burd (1993) says that the failure of blacks to win National Institute of Health (NIH) grants at the white rate—a tally that excludes the 1.5% of the total NIH budget reserved for grants to blacks—requires "enhancement" of programs for blacks, he is taking it as unquestioned that blacks win grants at a lower rate than whites for reasons other than inferior proposals. The bias-fighting rationale, while not compensatory, shares the assumption of rough interracial parity in talent.

In some cases this or related assumptions can be tested. Advocates of the bias-fighting rationale frequently cite studies showing race differences in mortgage approval rates. Across all income levels, 37.6% of black applicants for home loans were rejected in 1991, as against 17.3% of white applicants (Brimelow and Spencer 1993a: 48; the rejection rate for Hispanics was 26.6%). Controlling for criteria of creditworthiness reduces the black rejection rate to 17%, but also reduces the white rate to 11%, a 6% discrepancy. A more recent survey of mortgage loans in the Boston area (Munnell et al. 1996) that controlled for 38 variables affecting creditworthiness reduced a raw racial gap of 18%—10% rejections for whites vs. 28% rejections for blacks and Hispanics—to 7%: the predicted rejection rate for whites exactly like blacks was 21% (47). In effect applying the bias test of section 3.13, Brimelow and Spencer argue that the 1991 discrepancy shows bias only if blacks who do get mortgages default less frequently than whites who do, and would presumably argue similarly about the Munnell et al. data. Empirically, black repayment is not overpredicted; blacks and whites deemed equally creditworthy default at equal rates, indicating that blacks are not being held to higher standards.

Munnell et al. reply (1996: 44–45) that equal prediction of default rates "tells us nothing. . . . Because the average minority applicant has weaker financial characteristics, it is possible that the marginal minority loan that qualifies for a mortgage is of a higher quality than the marginal white loan, while the average creditworthiness of the accepted minorities is lower than that of accepted white." The hypothesis Munnell et al. seem to have in mind is this. Suppose 1000 blacks and 1000 whites have already received loans, the repayment rate having been 50% for blacks and 52% for whites. (The default rates must be discrepant, or the assumed predictors of creditworthiness would be invalid.) As a result, lenders grant mortgages to the next 300 blacks only if their probability of repayment is (say) .58, while granting mortgages to the next 300 whites whose probability of repayment is .52. The overall black and white default rates are both 48%, yet the last 300 blacks have been discriminated against. Empirically, this hypothesis can be tested against race × default rate correlations with creditworthiness held constant, in effect stratifying loan applicants to isolate effects at the margin; Munnell et al. cite some unpublished papers reporting equivocal findings on this score. Conceptually, however, the hypothesis appears to assume prior discrimination favoring blacks, since the base default rate for blacks (50%) could not have exceeded the white (48%) unless black applicants had been held to lower standards, or uniform standards applied across races overpredicted black reliability. As a reason to suspect bias, discrimination at the margin defeats itself.

Munnell et al.'s conclusion that no economic factor (including unknown causal variables for which race is a proxy) can explain the residual 7% discrepancy in loan approvals is seemingly supported by data from "testers," whites and blacks provided with similar credentials pretending to shop for housing, jobs, or loans. In some (a minority of) cases the blacks are treated somewhat less well. Lawyers for plaintiffs in discrimination suits brought in the name of testers have told me privately that the credentials provided blacks are often inferior to those of the control whites, but even if credentials are assumed to be perfectly matched a higher rejection rate for the black testers need not be bias. A loan officer's principal fiduciary obligation is to protect investor's money, and if in his experience blacks have been more apt to default than whites they resemble on paper, it is reasonable for him to be more dubious of black applicants. Differential treatment of testers is thought to reveal bias because there is no concrete reason for trusting the white tester more than the black one, as the credentials shown to loan officers, real estate agents, and personnel managers have been contrived to remove suspicion. But a loan officer's evidence is not limited to credentials. It can include both previous experience with blacks and the demeanor of the individual before him; he may have a visceral sense of something fishy.[13] Since his suspicions are correct—the black tester *is* deceiving him—it is hard to call them groundless. The only evidence relevant to creditworthiness the loan officer can point to supports the black tester, but beliefs can be justified by more than the evidence available to conscious articulation. According to recent "reliabilist" analyses in the theory of knowledge (Goldman 1967; Dretske 1971; Nozick 1982; Levin 1993), a belief is justified when produced by a process that in fact produces true belief sufficiently often, whether or not the process is known to be reliable, and whether or not the subject is aware of the process at work.[14] On this analysis, the loan officer need not be able to say why he is dubious for his suspicion to be justified. He is justified if suspicions of the sort he harbors, however inchoate, are usually correct. Assuming it is past dealings with blacks that have caused him to believe, correctly, that there is something phony about the tester, he is right to believe this whether or not he can say why he does. This state of mind, the presence of strong but hard-to-articulate reservations, is probably common in interracial dealings and interpersonal dealings in general.

Two final remarks about the issue of lending bias remain to be made. First, the fact that already-identified variables eliminate more than 60% of the race gap in mortgage acceptance (7/18 = .39), leaves, so to speak, 60% of the original charge of bias unfounded. Bias becomes a hypothesis to explain the residual variation, not a patent fact. Second, the minuteness of the discrepancy in absolute terms bears out Taylor's remark (1992) that the disappearance o real discrimination in the United States has forced "racism" spotters to magnify any discrepant treatment of blacks. (Another recent discovery is "environmental racism," the alleged practice of municipalities of building waste processing facilities in poorer areas.) Discrimination on so small a scal —if that is what the 6–7% represents—cannot account for the large rac differences in prosperity.

Diversity Promotes Values other than Compensatory Justice

Many advocates find the value of quotas in nothing so negative as righting past wrongs or preventing present ones, but in serving positively desirable ends. Quasi-utilitarian arguments of this sort tend to be extremely vague, especially since preference advocates seldom specify just what concrete effects preferences are to have.[15] Their vagueness makes it useful and necessary to treat these arguments as a class from a somewhat abstract point of view.

The first point that needs clarifying is just what extrinsic benefits quotas are supposed to confer. Sometimes it is said to be diversity itself: more blacks in different walks of life. Now, whether or not diversity is desirable, saying so is not an *argument* for affirmative action, since affirmative action is just another name for bringing more blacks into different walks of life. Praising affirmative action as good for promoting diversity is like praising carnivory for promoting consumption of meat.[16] Affirmative action must promote more than diversity, that is, affirmative action, to be justified by its effects.

The "benefits" argument is unclear in a more fundamental way, however. To ascribe positive consequences to diversity implies that there are valuable traits distinctive to blacks, since changing the black/white ratio in government, the workplace or the university changes nothing else if blacks and whites are fungible. One pressing question that then arises is whether these distinctive black virtues are unique to blacks or merely more prevalent among blacks. I defer this issue to chapter 9, raising here a second question: in whichever way black virtues are distinctive, where did they come from?

Only three by-now-familiar answers are available: biology, black culture, or white oppression. In addition to being circular, appeal to culture invites the thought that if culture is the source of distinctive black virtues, it might also be the source of black failings. Should hiring black executives be desirable because they will bring a culturally acquired spontaneity to management, perhaps this same spontaneity helps explain the lower black savings rate. Such an admission undercuts most other arguments for preference, for it is hard to justify white sacrifices to offset a byproduct of black culture. In the same way, a biological explanation of black virtues suggests that black failings may also be biologically based, undercutting other arguments for preference.

So it is understandable that diversity advocates generally attribute black virtues to oppression. Blacks are said to make better policemen for the slums because they remember the slums, more scrupulous judges because they have known injustice, more compassionate economic planners because they have known poverty. Apart from a lack of evidence that oppression ennobles, however, this idea contradicts the claim, otherwise insisted on, that oppression produces a poor self-image, a sense of grievance, disdain for authority, erratic work habits, and a readiness for violence. The preference advocate who attributes black virtues to oppression must either say that oppression has actually helped blacks, after all, or that truculent, suspicious individuals improve the workplace.

There is one contribution blacks alone are said to be capable of making that does not require special virtues, namely destruction of stereotypes. Since the idea is that seeing more blacks will convince whites of black capabilities (Greenawalt 1983: 64), blacks hired to achieve this effect need not be exceptional and in fact, the more like whites they are the better.

Even allowing that the destruction of stereotypes is an important good, this argument requires that preferences do have the intended effect. What little research exists on the taboo topic of white attitudes (Lynch 1984, 1989; Beer 1987; Lynch and Beer 1990) suggests, to the contrary, that whites resent affirmative action blacks. In an attitude survey, Sniderman and Piazza found whites significantly "more likely to perceive [blacks] as irresponsible and lazy merely in consequence of the issue of affirmative action being brought up" (1993: 129). When whites are asked about blacks in contexts in which affirmative action is not mentioned, the number of negative characterizations decreases (109). My own experience in the academic world is that the white professorate supports hiring quotas in the abstract, but scorns Black Studies departments and other concrete manifestations.

Coate and Loury (1992) have developed a mathematical model of quotas that admits of two possible effects. Temporarily rewarding blacks more than whites for the same output might convince blacks that developing their "human capital" is worthwhile, leading blacks to enhance their productivity, thereby disconfirming stereotypes. On the other hand, preferential rewards might convince blacks that they need not work harder, thereby suppressing black effort and confirming stereotypes. Which result will occur depends on the values of certain parameters which the model provides no way to determine. One indirect test of the model is black productivity, which would have risen since the late 1960s if affirmative action encourages effort. Empirically, however, black productivity was about 2/3 that of whites when affirmative action began, and was no greater in 1983 when Leonard (1983) investigated the matter. Later Census Bureau figures pegged black wages at about 57% of white (Jaynes and Williams, 1989: 274), also implying black productivity just below 2/3 that of whites. Since that time the ratio has fluctuated, but not as to indicate a surge of black productivity. Common sense would find it perverse of anyone to infer from being held to lower standards that he should work harder to meet higher ones, and there is no sign that blacks have done so.

White attitudes toward blacks have changed markedly during the twentieth century. Contemporary whites view blacks more accurately, and more favorably, than they once did. It is now recognized that a wide range of ability is present in the black population, if not necessarily in the same proportions as in the white. Sweeping judgments about all members of any group are much rarer than in less sophisticated times. But these changes have come mainly from contact with blacks who are proficient at race-neutrally defined tasks. If blacks are in fact as competent as whites, affirmative action may shift white attitudes further in a positive direction, despite any accompanying resentment. On the other hand, preferences may work against a positive attitudinal change should they bring whites into contact with an increasing number of black incompetents. This latter appears to be what is happening. Hartigan and Wigdor (1989: 188) concede that race-norming the General Aptitude Test Battery, used for referral by the Labor Department, has placed some blacks in jobs for which they would not normally be considered qualified (219). Black teachers fail competency exams at more than twice the white rate (Herrnstein and Murray 1994: 493). Blacks drop out of college at much higher rates than whites.

While it is reasonable to expect that putting blacks where they are not suited to be will lead to more negative judgments than might be held otherwise, the evidence about white beliefs remains ambiguous. Do Sniderman and Piazza's respondents think less well of blacks because of their experience with affirmative action hires, or do their negative attitudes merely express anger at injustice? The point to stress is that, while the destruction-of-stereotypes defense of quotas is certainly not compensatory, it shares an empirical premise with the compensation argument. In assuming stereotypes false, it assumes that blacks are on average as able as whites.

8.7. FLAWED REPLIES TO THE COMPENSATION ARGUMENT

Compensation theorists must construe the black attainment deficit as evidence of, not constitutive of, injury. Defects, however severe, are not injuries unless caused in a certain way, namely by malfeasance or negligence, so the attainment gap must be caused by white misdeeds to be compensable. Since the heart of the compensation argument is thus a theory of the cause of the race gap, a more plausible account of the gap in terms of factors other than white misdeeds would annul black rights to a remedy. Such an account can acknowledge past wrongs; it need only deny that current black limitations are among their effects. Innate race differences in intelligence and time preference offer such an alternative.

Taking the origin of the race gap as the heart of the matter explains why criticisms of preference that do not reach that issue, however forceful they may be on their own terms, are dismissed as picky, legalistic, and irrelevant.

Knowledge Problems

Of all the challenges that concede lingering effects to racism, most worrisome to compensation theorists, as we have seen, is the unknowability of who was damaged and by how much. The contemporary victims and beneficiaries of past discrimination cannot be precisely identified, nor, if they could be, could their losses and gains be measured; so vague an identification of victim, tort-feasor, and injury would not normally sustain a finding of tortious liability.[17] Consider *Weber* (1979) and *Bakke* (1978), two central "reverse discrimination" cases. In the first, Brian Weber's employer, Kaiser Aluminum, chose less senior blacks over him for an in-house training program. (The Supreme Court held that favoring blacks was a permissible exercise of Kaiser's freedom of association, without explaining why favoring whites does not equally permissibly exercise this freedom, or why the Civil Rights Act does not impermissibly curtail it.) In *Bakke*, the medical school of the University of California at Davis rejected Alan Bakke, whose grade point average was 3.8 out of a possible 4, while admitting blacks whose mean average was about 2.4. Now, granting arguendo that Weber might have enjoyed less seniority and the blacks placed ahead of him more in a world without slavery or Jim Crow, passing over him now annuls an illicit advantage just in case *those very blacks* would have been senior *to Weber* absent slavery and Jim Crow. How can anyone possibly know that? How can anyone know that the very blacks admitted instead of Bakke would have had higher grades than Bakke in a world without discrimination? The mean SATs of blacks admitted to American universities remain about 200 points below those of whites (Herrnstein and Murray 1994: 452; Zelnick 1996: 129); can anyone say for sure that, had blacks come to America as freemen, black SATs today would be 200 points higher?

But these uncertainties are swept aside by the rhetoric of oppression. Perhaps it cannot be proven with certainty that today's blacks suffer the consequences of past mistreatment, the compensation theorist replies, but who can doubt it? Why else the continuing black educational failure, the high crime rates, the poverty? Perhaps (he may concede) whites are not responsible for some of the immediate causes of black failure, like illegitimacy and drug use, but aren't these self-destructive patterns themselves effects of racism? Rosenfeld issues essentially this challenge:

[I]t seems reasonable to presume that the measure of the reduction in the prospects of success of blacks is roughly equivalent to the difference between the ratio of blacks with desirable employment to the total number of blacks in the workforce and that of whites with such desirable jobs to the total number of whites in the workforce. . . . While it may be rational to assume that the disparity between the proportion of whites and the proportion of blacks who hold certain desirable jobs is attributable to the effects of first-order racial discrimination, it is of course possible that at least part of this disparity is ultimately due to some other cause. This mere possibility certainly should not foreclose the constitutionality of using affirmative action as a compensatory device. Indeed, in the case of long-standing, massive, and systematic wrongs such as those perpetrated against American blacks, it seems warranted to adopt a rebuttable presumption that existing discrepancies in prospects of success are the result of first-order discrimination. (1991: 290–291)

Harwood's language is crisper:

Some would place on blacks the burden of proving that they are victims of racism. But racism against blacks has been so thoroughgoing, systematic and pervasive throughout America's history, that history more than establishes a strong presumption of any American black's victimization. Since any black is presumably a victim, and since proving racism is usually hard, the fair and efficient procedure would be to put the burden of proof of non-discrimination on those who oppose AA. . . . When will it all end? When will AA stop compensating blacks? As soon as the unfair advantage is gone. The elimination of unfair advantage can be determined by showing that the percentage of blacks hired and admitted at least roughly equaled the percentage of blacks in the population. (1993: 78–82)

Actually, the presumption of innocence *does* forbid compensatory penalties until causes of black disadvantage other than racism are ruled out. And Harwood might have reflected that the difficulty of proving racism is evidence against its existence. But these niceties are no match for the rhetoric of "long standing, massive and systematic wrongs" resulting from "thoroughgoing systematic and pervasive racism." The critic of preference must attack the presumption of black victimization lest he resemble a lawyer grasping a technicalities, exploiting the rules of evidence to evade what he cannot deny.

Inappropriateness of the Nexus.

A criticism that grants both the efficacy of racism and precise knowledge of its harm concerns the causal nexus. Whites may be at an advantage because of past wrongs to blacks, but, as we saw, an advantage caused by a wrong is illicit only if the two are appropriately connected. The paradigm nexus for redress is transmission of a material object, but (black) ability is not an identifiable object which whites have transferred to each other across generations and can be returned to blacks. Other nexuses coalesce around intent, and it is conceivable although unlikely that slave owners and segregationists intended to advantage late twentieth century whites. Still, the nexus created by such intentions seems insufficient for liability. When, unbeknownst to Robinson, Jones hobbles Smith so that Robinson beats Smith in the race, Robinson hardly seems obliged to surrender his medal.[18] What of the nexus of consanguinity? Hacker (1992: 122) likens races to families, but the metaphor does not hold up. The nineteenth century owner of the great-great-grandfather of a black I am competing against is no literal relative of mine. The races of Jones and Smith in the foregoing example was surely irrelevant. All that links past racial wrongdoers to their alleged contemporary beneficiaries is sameness of race, a link too tenuous to anchor the illicit advantage principle.

To the compensation theorist all this sounds like pettifogging. "Look," he says, "systematic wrongs were done, with terrible consequences for blacks today. Do we just say 'too bad?' Do we let millions languish in poverty when all that keeps them from realizing their fullest potential is refusal to lift the dead hand of the past? Doing something is every white's responsibility." This chain of association must be at work in Hacker's mind, lest the last sentence in the following passage be a non sequitur:

So in allocating responsibility the response should be clear. It is white America that has made being black so disconsolate an estate. . . . [O]f course, life can be unfair. We cannot vouchsafe that every infant will be born with sight or hearing or the full use of his or her limbs. However, not all disabilities derive from nature; many are contrived by society. Some have argued that black Americans deserve reparations for their centuries of bondage and subjection. (1992: 218–219)

Complaints about spurious nexuses cannot match "centuries of bondage and subjection."

Affirmative Action is Post Facto

A third criticism inspired by legal logic is that affirmative action is post facto. Slavery, segregation, and private discrimination were lawful in their time, and to annul actions thought lawful and just when done is unfair.

The standard defense of post facto rulings is that people may place reliance only on *legitimate* laws and customs, and, so the compensation theorists will say, white gains are tainted. Just as Germany could not hope to keep art treasures it seized during World War II by pointing to the legality of the seizures under the Reich, whites cannot point to the legality of Jim Crow. According to Harwood, "there is no *reasonable* reliance or *legitimate* expectation to keep ill-gotten gains due to America's *notoriously* deep and widespread racism. Expecting or relying on the unfair advantages from past racism to be perpetuated is unreasonable and illegitimate" (1993: 81). The post facto reply to compensation depends on the advantages of today's whites being more like money acquired by financial tactics subsequently banned rather than the spoils of aggressive war. Once the advantages of whites are attributed to "*notoriously* deep and widespread racism," the war analogy is hard to rebut.

Personal Identity

The who-was-harmed and genetic causation issues expand into a general problem about personal identity (Levin 1979, Morris 1988; also see section 5.2). A finding of compensatory entitlement requires that the complainant would have existed had the wrong he suffered not occurred, for his award is meant to put him where *he* would have been but for that wrong. The ballerina who asks $1 million in damages from a piano mover is claiming that *she* would have earned another $1 million with an intact leg. It follows that a complainant cannot have been made worse off by a wrong responsible for his existence. Yet contemporary American blacks said to deserve restitution for slavery would not have existed but for slavery, since, without it, their ancestors would have remained in separate parts of Africa. The typical American black cannot be put where *he* would have been absent slavery, since absent slavery there would have been no him, just as a Jew conceived in a concentration camp, who would not have existed but for his parents' internment, has no "wrongful conception" claim against Germany.

The failure of this argument to seize the popular imagination is not due entirely to its abstruseness. It is straightforward once stated. Rather, people to whom it is explained tend to fall back on the Nazi analogy anyway. "All right," they say, "Today's blacks can't deplore where slavery left them, because without slavery there wouldn't be any 'them.' But that does not cancel the oppression actual living blacks suffered. Perhaps Jews conceived in Auschwitz should not complain about being born there, but they can complain about what happened to them afterwards. So too with blacks." Problems of personal identity may force compensation theorists to reformulate their claim, but so long as the Nazi analogy is handy—so long as the plight of each black is attributed to what whites did to him once he was born—they will feel no pressure to withdraw it, or moderate their indignation.

Inefficiency

A common objection to preferences is the inefficiency of placing blacks in positions for which they are not the best qualified. Compensatory remedies must normally be feasible; a ballerina who loses her foot in an accident demands what the accident cost her, not the right to be a crippled Princess Aurora. Brimelow and Spencer estimate that, all told, preference costs the impressive sum of $350 billion annually (1993b: 82). Paying every American black a reparative lump sum of $100,000 would be cheaper than a decade of affirmative action.

Yet compensation theorists rarely consider monetary reparation, and when they do it is to supplement, not replace, preferential placement. To some extent this disdain for cost may be due to ignorance ("The winds of utility blow in many directions," says Sher 1975), but it is rooted in the substance of the oppression theory. By denying blacks the opportunity to develop their abilities, whites denied blacks the satisfaction of exercising those abilities. The self-respect thus lost, according to many writers (e.g., Rawls 1971: 440), is a basic moral good. Hence

> the wrongs done . . . make jobs the best . . . form of compensation. [B]lacks and women were denied . . . full membership in the community; and nothing can more appropriately make amends . . . than precisely what will make them feel they now finally have it. And that means jobs. Financial compensation . . . slips through the fingers;[19] having a job, and discovering you do it well, yield—perhaps better than anything else—that very self-respect which blacks and women have had to do without. (Thomson 1977, p. 373)[20]

Once one agrees (as Thomson clearly does) that black self-respect is low, that paid employment is as important a source of self-respect for blacks as for whites, and that self-respect is as vital as Rawls says it is, restoring blacks to the right indifference curve ceases to be enough. Only the right point, the full measure of self-respect associated with employment, will do. The right level of self-respect becomes a singularity, incommensurable with any other mix of goods. Discrimination is not just another remediable harm; it is a blot on the national soul that must be completely expunged. No wonder fuss about efficiency seems obtuse.

Preferences Harm Blacks

A feeble evasion of the justice issue remarkable for its popularity is the worry that preference demoralizes blacks (see e.g. Steele 1990). Proponents of this claim cite no empirical evidence, and sometimes seem to believe that preference has a demoralizing effect simply because it should,[21] but let us—as so often—assume arguendo that this is so. Surely, asking blacks to forego something rightfully theirs because it might be bad for them is about as persuasive as asking the injured dancer to forego a large settlement lest it blunt her zeal for rehabilitation. It is for the injured party, not those in his debt, to say whether a compensatory remedy will harm him. The implication of preference that blacks can't make it on their own weakens the destruction-of-stereotype argument, but does not faze the compensation theorist. "Of course blacks can't make it on their own," he replies; "they can't because of what whites have done to them. That is why they deserve preference, and why they will deserve preference until they can." The primary question is whether blacks deserve something. Whether getting it is bad for them is secondary.

Disparate Outcomes Are A Fact of Life

Yet another evasion is the argument that race differences in the United States are unsurprising, since ethnic subgroups vary in interests, habits, and wealth in all societies.[22] This generalization is correct, but what is said to warrant affirmative action is not mere variation in outcomes for blacks and other groups, but the systematically *poorer* showing of blacks on all indicators of success. Boxill easily disposes of the argument:

> A more serious attempt to explain the group differences in qualifications between blacks and whites is [that] blacks are simply not as interested as whites in positions of affluence and prestige. . . . [But] it isn't as if blacks are underrepresented in the public school system or in law or in banking or in the professions. It is that they are underrepresented in all of these. . . . Unless we assume that some cultures have no interest in any of the traditional areas, we cannot explain a group's general underrepresentation in all desirable positions in society by citing cultural differences. (1978: 163–164; also Hacker 1992: 122)

Just so: to speak as if blacks were choosing different avenues of success instead of failing across the board avoids the real issue.

A related technical objection to preferences raised by Sowell is that any statistical test for discrimination yields false positives (also see Epstein 1992: 207–211; Herrnstein and Murray 1994: 481–483). Thus the presumption of the Equal Employment Opportunity Commission that a standard discriminates when its pass rate for blacks is less than 80% of the pass rate for whites will find about one in twenty moderate size nondiscriminating firms to be discriminators just by chance. Sowell's statistical point is correct: in any long series of sets of 25 tosses of a fair coin, about one set out of 20 will contain 14 or more heads; that is, the "pass rate" for heads will be more than 80% that for tails. Nonetheless, once one concedes that blacks are as able as whites one must also concede that statistical tests can discover discrimination, and accept false positives as a side effect of an otherwise reasonable procedure. They can always be reduced by raising the minimum number of employees for firms the 80% rule covers, or lowering the cut-off to, say, 70%. The issue becomes one of balancing the harm of erroneously charging (and legally punishing) discrimination against the benefits of detecting it. Once an acceptable limit on the number of false positives is decided on, the statistical parameters of an effects test can be adjusted so that this limit is met. Sowell's argument leaves unchallenged the moral basis of the EEOC regulation.

Affirmative Action Changes the Rules

Finally, critics (e.g., Yates 1991: 14) often claim affirmative action substitutes "equality of outcome" for "equality of opportunity": instead o giving everyone a right to strive with no guarantee of success, affirmativ action gives blacks the right to succeed.

It should be clear by now that compensation theorists countenance no suc shift. The whole problem, they claim, is that blacks lacked equal opportunit for centuries, and imposing equal outcomes now simply yields wha opportunities equal in the usual sense would have produced. If prospector Jone has more nuggets than prospector Smith because Smith was kept from panning in the stream, asking Jones to back away from the stream for a while to let Smith pan by himself is not a new kind of equality. By traditional criteria Smith deserves the time he should have had earlier. Thus Hettinger:

Affirmative action takes away the greater than equal opportunity white males generally have, and thus it brings us closer to a situation in which all members of society have an equal chance of succeeding through the use of their talents. . . . It is not the purpose of affirmative action . . . to disadvantage white males in order to take away the advantage a sexist and racist society gives to them. [23] But noticing that this occurs is sufficient to dispel the illusion that affirmative action undermines the equality of opportunity for white males. (1997: 311)

Compensation theorists do not see themselves as changing the rules. If blacks are innately as able as whites, unequal outcomes indicate unequal opportunity. "When will it all end? . . . As soon as the unfair advantage is gone, A[ffirmative] A[ction] will stop. The elimination of unfair advantage can be determined by showing that the percentage of blacks hired and admitted at least roughly equaled the percentage of blacks in the population" (Harwood 1993: 82). As justices thunder from the bench, "Asked and answered!" by the racism hypothesis.

8.8. RACISM, RACE DIFFERENCES AND THE ATTAINMENT GAP

Let us review the considerable evidence against the amorphous claim that "racism" explains the attainment gap.[24]

The gap in attainment between blacks and whites in Africa is wider than the American racial gap, yet blacks were not enslaved in Africa and colonialism, which began in the 19th century, ended after World War II. Extensive economic investment in black Africa has not been accompanied by economic growth, nor prevented Africa from reverting to pretechnological conditions. Occam's razor, paring away multiple explanations of the economic failure of similar populations, implies that oppression is not a significant cause of the attainment gap in the United States.

The sheer passage of time makes reference to slavery progressively less plausible. Slavery in the United States ended almost a century and a half ago, yet the attainment gap, while narrower than decades ago, has persisted, and other mistreated populations, conspicuously Jews and Asians, have come farther in less time. The Asian comparison is particularly telling, since Asians, like blacks, are visually identifiable. Sheer identifiability cannot be the factor distinguishing blacks from more successful ethnic groups.

The allegedly unique American black experience of slavery and Jim Crow is not unusual in global perspective. Enslavement and subjugation have been common occurrences throughout history, and judged wrong only by the morality developed recently in western Europe. About 400,000 blacks were shipped to the United States during the eighteenth century, the height of the slave trade (Reed 1969), and about 750,000 in all. The usual estimate for the death rate among transshipped slaves is one in five, a rate three orders of magnitude lower than that of the Nazi extermination of the Jews. The archives at Tuskegee University contain records of 4709 lynchings of blacks between 1886 and 1966 (Applebome 1992); the highest estimate I have seen for the number of lynchings of blacks is 5000, an unremarkable level of violence, historically speaking. 1300 whites were also lynched in this period (Taylor 1992: 92), so lynching was not the war of whites against blacks it is usually represented to have been.

The Jim Crow era began with *Plessy v. Ferguson* in 1896, following great federal efforts to aid blacks during Reconstruction, and ended with school integration in 1954. Private discrimination against blacks became illegal in 1964; preferences favoring blacks were in place by 1970. It is difficult to recall another instance in which a dominant group voluntarily restricted itself in this way. Since 1970 blacks have received set-asides, special programs, and the sympathetic attention of academics, the media, and business. I mentioned earlier the $2 billion spent by the National Science Foundation and NIH since 1972 to stimulate the intellectual development of blacks. By now hundreds of billions have been spent on Head Start. Publications such as *Directory of Financial Aid to Minorities* (Schlacter and Goldstein n.d.) and *Directory of Special Programs for Minority Group Members* (Garrett n.d.) list literally thousands of training programs, scholarships, fellowships, internships, and awards reserved for blacks (and, usually, non-European Hispanics, along with women).[25] Blacks at Pennsylvania State University are given $500 for every grade over *C* (Taylor 1992). Although their high school records are much worse, blacks are admitted to college at much higher rates than whites; in 1995, for instance, Rice took 25% of white applicants as against 52% of black applicants, while Amherst took 19% of white applicants and 51% of black (Zelnick 1996: 133–134). The mean Law School Aptitude Test Score of first-year black law students is 1.5 SD below the mean white score (Herrnstein and Murray 1994: 455). White elites strive to put blacks in a favorable light: producers of movies and television programs go out of their way to portray blacks as able, high-status figures; the black computer whiz has become a staple of action movies, and advertisers scrupulously include blacks in group photos. A willingness on the part of whites to give blacks a chance would seem to be shown by the election in recent decades of black mayors in New York, Los Angeles, Trenton, Philadelphia, Chicago, Cleveland, and Denver and a black governor in Virginia. The Joint Chiefs of Staff has had a black chairman. Most of the budget of Howard University, the best-known black college, is supplied by the federal government. For many decades the legal staffs of the National Association for the Advancement of Colored People and the (now unaffiliated) NAACP Legal Defense Fund have been mostly white.

The acid test of what someone thinks of a situation is what he does about it, and blacks clearly prefer the United States to other countries, including those in which blacks are the majority. Thousands of Haitians risk the trip from Hispaniola to the American mainland in rickety boats, and the immigration quotas of African nations are always filled. At the same time American blacks have shown no interest in emigrating to Haiti, the Ivory Coast, Botswana, or Rwanda. This is not the behavior of a group whose prospects are limited by oppression.

All told, as W. J. Wilson (1987: 11) has remarked, it is difficult to maintain that racism has increased in the United States since World War II.[26] Objectively speaking, a black born after 1965 has experienced, not oppression, but unparalleled privilege—yet by many standards, including crime, marital stability and illegitimacy, blacks are worse off.

Indeed, American blacks at the end of World War II may usefully be compared to the Germans and Japanese of that period, whose countries lay in ruins under the occupation of foreign powers. Japan had suffered two atomic attacks, a collective psychic trauma (if one believes in such things) as severe as any in black history. Contemporary black father absence is often traced to the disruption of families during slavery; by way of comparison, 25% of the adult male population of Germany had been killed in the war, and hundreds of thousands of German women raped by Russian soldiers. By any measure, Japanese and Germans were worse off than the American black population in 1945. Yet by 1950 Germany had attained its 1936 production level (Spanier 1962: 43), and within three decades Germany and Japan had become world economic leaders. Family stability is not a problem in either country. It seems plausible to attribute the different trajectories of their three cultures to the traits of Germans, Japanese and blacks.

Postwar European prosperity is often traced to the Marshall Plan, now almost a metaphor for what blacks need, as in "For today's inner cities, perhaps only a domestic Marshall Plan will do the job" (Popenoe 1992). This comparison only highlights the investment that has already been made in the "inner cities." Between 1948 and 1952 the United States spent $12 billion rebuilding Europe (Spanier 1962), or $204 billion in 1990 dollars.[27] In 1990 the total outlay for Aid for Families with Dependent children, food stamps, housing and other subsidies for the poor—what is colloquially called "welfare"—was $215 billion. Blacks received 41.3% of this (Rector 1992b), or about $88 billion. The black/white income ratio stabilized at 57% in the early 1970's (Jaynes and Williams 1989: 287; Currie and Thomas 1995: Table 2); as federal and state income taxes are progressive, blacks may be assumed to pay about 50¢ in taxes for every dollar whites do. So blacks, at 12% of the population, collectively pay about 6% of the cost of welfare, or roughly $13 billion, for a net annual white-to-black transfer of roughly $75 billion. This is in effect a Marshal Plan for the "inner cities" every three years. Another way to conceive the actual situation is that the black population, constituting 40% of welfare dependency and paying 6% of its costs, receives about 41/6 = $6.83 for every $1 it pays in welfare, while the rest of the population pays 94% of the cost of welfare to receive .59/.94 = $.63 for every $1 paid. The net black rate of return for spending on public assistance is thus roughly ten times the white rate (a ratio which increases when non-European Hispanics are excluded from the white population).

According to Hacker (1992), "Even today, America imposes a stigma on every black child at birth" and he implies that the casual killing of blacks is a common occurrence. (He remarks, in mock surprise, that "the odds are that [a black] traveling across the heart of white America will reach his destination alive.") Extravagant claims of this sort are commonplace. Yet virtually the only specific evidence still cited for the prevalence of racism consists of white mistrust of blacks. The grievance literature tells story after story of white women clutching their purses when black men approach, of taxis passing blacks by, of clerks watching black customers, of policemen suspicious of blacks (Hacker 1992: 48; Harper 1994).[28] A case can be made—I make it in the next chapter—that black crime justifies this mistrust, but, warranted or not, slights of the sort recounted seem insufficient to explain black failure. They are trivial in themselves, and other groups have responded to disdain with intensified efforts to attain higher status.

Pressed for evidence of continued oppression, Hacker cites despair at injustice, expressed in riots and paranoid beliefs—for instance in the conviction held by two-thirds of blacks that the government encourages black drug use, and the conviction held by one-third that AIDS may be a creation of scientists to annihilate blacks (Hacker 1992: 49). Yet feelings of injustice, whatever their evidentiary value would be, do not seem to be what actuates the black behavior it is said to. Many if not most racial disturbances nominally staged as protests end in stealing and looting, whereas people seeking to make a moral point do not use the occasion for theft. The men who savaged Reginald Denny during the 1993 Los Angeles riots stole his wallet. If their beating was a gesture of defiance, they would have left his wallet alone.

There is a test of sorts that Hacker (1992), Fischer et al. (1996) and perhaps others (for Fischer et al. credit to "an economist," although they may be thinking of Hacker) propose for measuring the disadvantage borne by American blacks. How much, they ask, would the reader want to be paid to become black while otherwise retaining his present personality and abilities. Hacker, a university instructor, reports that his students put the figure at $1 million. Perhaps so, but the rational white acquainted with the facts would willingly *pay* for the suggested exchange. A moderately bright white child scoring slightly above the white mean on standardized tests would, were he black, be eligible for a program called "Prep for Prep," which pays all costs of attendance at elite private elementary schools. Value: at least $30,000. He would then get to go to an elite private preparatory school for four years, again at no cost. Value: at least $60,000. College scholarships available to him that are unavailable to similarly able whites may conservatively be estimated to be worth another $50,000. Even before entering the job market, where affirmative action kicks in, being black is, at a conservative estimate, worth $130,000. Speaking purely behaviorally, blalcks in current American society are more highly valued than whites.

In short, the historical record alone suggests that the attainment gap is due to factors other than white oppression. Since there is also overwhelming evidence that races differ in the achievement-relevant traits of intelligence and time preference, these traits present themselves as alternatives. Occam's razor cuts away other causes.

Some compensation theorists have registered awareness of this possibility but their proceeding undaunted does not mean that they have met it, or tried very hard to. In fact, there has been a noticeable declension toward slovenliness in their treatment of the subject. Recall that, in an essay first published in 1974, Block and Dworkin recognized that some arguments for quotas do rely on assumptions concerning the causes of phenotypic differences. Compensatory arguments assume that some proportion of the phenotypic differences between groups is due to past unjust treatment. . . . To the extent that arguments are advanced for proportionate-to-population quotas which rely on assumptions about the distribution of genotypic abilities, it becomes relevant to assess the validity of such assumptions. (1976: 512)

Their main topic was not preference and they say no more about it, but this passage contains the whole issue in a nutshell. Writing in 1978, Boxill found it necessary to defend some "general assumptions" of the illicit advantage argument:

The most obvious is that the black and white groups are roughly equal in native talent and intelligence. If they are not, then unless differences in native ability between the groups are remediable and justice requires that they be removed [see section 9.9 below], it is not at all clear that the lower qualifications of blacks are any indication that they have been wronged. Fortunately, this difficulty can be avoided. The weight of informed opinion is against Jensen, but even if it is ultimately shown to have some merit, his theory that as a group blacks have less native intellectual talent than whites is, for now, extremely controversial. Jensen himself, though regrettably not chary enough in proposing policies based on this theory, is tentative enough in stating it. Consequently, given its present uncertainty and the great injustice that would be wreaked on a people if it proved false and educational policies were based on it, I submit that we are not warranted now in basing any policies on it. (1978: 165)

Boxill cites no reference for his claim about "informed opinion"; his sudden allusion to education pulls the argument off course; he fails to consider that environmentalism might also promote injustice; and, unlike Block and Dworkin, he seems to think genes explain all racial variation or none. Still, the key idea is present.

But few subsequent defenses of affirmative action have referred in any way to the challenge of race differences in intelligence. Greenawalt (1983) does not mention the topic. Wasserstrom (1985a) denies in a footnote that the races differ in any socially important characteristics. By recent standards the passage cited in chapter 4 from 24–27 of Hacker (1992) is candid, and the passage from Rosenfeld (1991) cited in section 8.7, for all its vagueness about the "other causes," concessive. But Rosenfeld says no more about alternative hypotheses, and eventually stipulates that anyone admitted to "the dialogical process designed to determine the legitimacy of affirmative action" must "accept the proposition that the substantial underrepresentation of blacks in education, employment and business is due to past and on-going first-order discrimination against them" (309).

So the state of play is this. Some compensation theorists realize they must deal with genetic variation in intelligence as an explanation of the race gap in achievement, but they have made no serious effort to do so, and are increasingly complacent about needing to. My guess is that in the time since the appearance of the papers of Block and Dworkin and Boxill, hasty and inadequate as they were, it has become less acceptable to mention intelligence differences even for purposes of repudiating them.

8.9. INTELLIGENCE AND RACE DIFFERENCES IN ATTAINMENT

Chapter 3 mentioned that the correlation between the mean IQ of incumbents in an occupation and its prestige rating is .9, and the correlation between individual IQ and occupational status is .3–.5. The r between IQ and social status runs between .3 and .4. Although I have seen no studies with unrestricted range correlating IQ with income, income is correlated with schooling, which is correlated with IQ. In prestigious occupations, those above the 90th percentile in IQ outearn those below by about 30%, a finding replicated for those in less prestigious occupations (Herrnstein and Murray 1994: 98, 685). IQ correlates with on-the-job success, the tie strengthening as the IQ demands of jobs increase. True r between academic attainment and IQ exceeds .6. The achievement correlates of IQ makes it unlikely that they are bias artifacts, and the variety of its nonsocial correlates across as well as within races makes it unlikely that IQ/achievement correlations are socialization artifacts. Intelligence is an attainment-relevant trait that blacks on average have less of than whites which promises to explain lower average black attainment.

The question is how much of the gap is explained by differences in intelligence (and how much of the remainder by differences in time preference). The strongest bias hypothesis, that IQ explains none of it, predicts that black representation in a field remains constant when ability is controlled. This seems to be the view of compensation theorists,[29] and, given that black IQ *is* lower than white IQ, it is demonstrably wrong. Compensation theorists could consistently hold that racism explains some but not all of the achievement gap, but it is understandable why they seldom do: "Take from whites their ill-gotten gains" is a more galvanic slogan than "Take from whites that proportion of their gains corresponding to the variance in attainment left unexplained by cognitive and temperamental variables." Still, in point of logic blacks might deserve compensation for the (large) remainder if IQ explains only a small proportion of the variance.

The proportion of the attainment gap explained by IQ can be determined by examining the hypothesis that IQ explains it all; the more closely this hypothesis fits the data, the more IQ explains. Assuming the mean IQ of whites to be 1 SD above the black mean (for a black mean of $-1z$), and equal variance in both populations, 16% of the black population is at least as intelligent as 50% of the white population. There are then $50/16 = 3.12$ whites for every black with an IQ over 100, population size held constant, implying that, absent discrimination,[30] there should be about 3 whites for every black on jobs requiring an IQ of 100. Since there are about 7 whites for every black in the United States, whites should outnumber blacks in those jobs by about 21 to 1.

However, because the normal distribution is curved, the race ratios on jobs predicted by a 1 SD difference in mean IQ vary with the IQ range from which job incumbents are recruited. Notice how in Figure 8.1 the IQ curve falls faster at point x_1 than at x_0. Also bear in mind that the number of individuals at or to the right of a point on the x axis is represented by the area under the curve to the right of that point. Next, Figure 8.2 superimposes the two bell curves

Figure 8.1
Change in Slopes

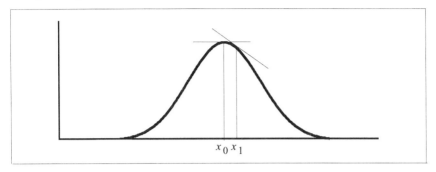

$$x_0 \; x_1$$

W(hite) and *B*(lack), whose variances are identical and whose means are 1 SD apart. Here, x_0 is the mean for B but -1 SD for W; x_1 is the mean for *W* but $+1$ SD for *B*; and x_2 is $+1$ SD for *W* but $+2$ SD for *B*. As before, 50% of the *W* population and 16% of the *B* population lie to the right of x_1. However, because *B* is falling faster than *W* at x_1, there are relatively fewer *B*'s than *W*'s to the right of x_2. This effect intensifies further to the right. As x_2 is 1 SD to the right of x_1 but 2 SD to the right of x_0, 16% of the W population but only 2.3% of the B population lie to the right of x_2, yielding a ratio of the populations to the right of x_2 of 16/2.3, or about 7:1. Above 2z from the *W* mean, which is 3z from the *B* mean, the ratio is 17.5:1. It is possible to compute the ratio of the proportions beyond a given point of two normally distributed populations with unequal SDs so long as their mean difference and the ratio of their SDs are known. If for instance the SD for *B* (call it SD$_B$) = 81 SD$_W$ and x_2 = .743 SD$_W$ = 2.28SD$_B$, the *W*/*B* ratio to the right of x_2 is about 20:1. Small differences in variance have large tail effects.

The higher the minimum intelligence for a job, the larger the predicted ratio

Figure 8.2
Relative Cutoffs

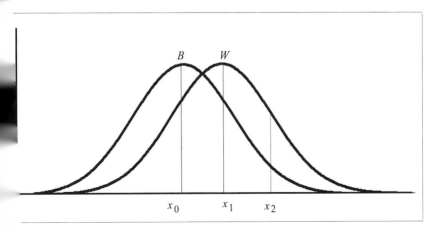

$$x_0 \qquad x_1 \qquad x_2$$

Table 8.1
B/W Ratios of Male Employees versus B/W Ratios Predicted by IQ

1. Job	2. IQ range	3. % B above	4. % W above	5. % B in range	6. % W in range
Physician	114+	1.1	23	1.1	23
Engineer	114 +	1.1	23	1.1	23
Secondary teacher	108-134	3.3	35.2	3.3	32.7
Real estate sales	108-134	3.3	35.2	3.3	32.7
Fireman	91-117	28.4	74.5	27.8	56.9
Policeman	91-117	28.4	74.5	27.8	56.9
Electrician	91-117	28.4	74.5	27.8	56.9
Truck driver	86-112	42.5	83.1	40.8	56.4
Meat cutter	86-112	42.5	83.1	40.8	56.4

of white to black incumbents, as derived from the black and white IQ distributions and the range of IQs from which job incumbents are recruited. The proportion of a population suitable for a job is the area under the segment of the curve between the high and low ends of its IQ range. After that proportion is derived from z tables, the black-white ratio is calculated as before. The fit between these ratios and the actual ratios for various jobs measures the proportion of variance in vocational achievement explained by the race difference in IQ. This fit has been studied by Gottfredson (1986, 1987, 1988) Table 8.1 is adapted from Table 2 of Gottfredson 1986.

Take the rows marked Physicians and Engineers. As determined by the General Intelligence Scale of the General Ability Test Battery, used by the U.S Department of Labor in studies of job performance, column (2) indicates that male physicians and engineers are recruited from the population whose IQ is 114 or above, a range with no ceiling. Columns (3) and (4) give th proportions of the black and white populations above that cutoff, using a whit IQ mean of 101.8, a white SD of 16.4, a black mean of 83.4 and a black SI of 13.4, derived from Hitchcock (1976). Observe that SD_B = .81 SD_W. 114 i .743 SD_W above the white mean and 2.28 SD_B above the black mean; 23% c the white population and 1.1% of the black population lie above that poin Columns (5) and (6) indicate the proportions of the populations within th recruitment range for physicians and engineers, which, since this range has r ceiling, are again 1.1 and 23. The ratio of blacks to whites in that range then 1.1/23, rounded to .05 in column (7); this is the number of blac physicians there would be for every white physician, holding population si constant, were physicians recruited solely by intelligence. Using Census Bure data, columns (8) and (9) display the actual ratios of black to white physicia in 1970 and 1980, again with populations constant.

The actual ratio of black to white physicians in 1970, shown in column (8 was .23, or 23% of the proportion required by racial parity, an coincidentally, the "absurdly small" number Dworkin cites as showing t

Table 8.1 cont'd.

7. B/W ratio in range	8. B/W ratio in 1970	9. B/W ratio in 1980	10. Expected B/W ratio, B recruitment range .5 SD lower
0.05	0.23	0.3	0.22
0.05	0.12	0.25	0.22
0.1	0.59	0.59	0.35
0.1	0.18	0.23	0.35
0.49	0.27	0.65	0.87
0.49	0.69	0.87	0.87
0.49	0.33	0.5	0.87
0.72	1.59	1.48	1.07
0.72	0.98	0.98	1.07

Source: Adapted from Table 2 in Gottfredson (1986)

need for affirmative action (1977a: 140–141). Yet 23 is more than four times greater than the proportion expected were the recruitment of blacks based solely on intelligence. The same effect is seen elsewhere. Comparing columns (7) and (8) to (9), in 16 of 18 cases blacks are *more* numerous than they would be if recruitment were based solely on intelligence.

This overprediction is clarified in column (10), the expected black/white ratios when blacks are recruited for jobs from an IQ range whose cutoff is .5 SD_W below the cutoff of the white range for those same jobs. Take the "Physician" row again, once more holding population size constant; there are about 100 whites whose IQs are 114 or over for every 22 blacks whose IQs are 107 or over. So, if white doctors were recruited from the IQ range 114+, and black doctors from the IQ range 107+, there would, according to column (10), be about .22 black doctors for every one white doctor. That is near the actual 1970 ratio.

The desire to "keep blacks down" should intensify with job status. Since IQ demand correlates positively with job status (a measure included in Gottfredson's original study) the discrimination hypothesis predicts that the gap between the numbers in columns (8) and (9) and those in (7) will increase as one ascends the columns toward higher status jobs. This is not found.

Controlling for intelligence explains all the discrepancy and more between the black/white ratio in the general population and in a wide assortment of occupations. The data fit the hypothesis that race differences in representation are due to the intelligence gap, and strongly suggest that blacks are recruited from an IQ range lower than that from which whites are drawn. IQ overpredicts black presence only among firemen and electricians in 1970, but the discrepancy is too small for bias to have much to explain. In the most discrepant case, firemen in 1970, proportionality predicts 3.7 times as many black firemen as there were, 12% instead of the observed 3.2% (populations varying). But this prediction is halved, to 6%, when IQ is fixed. Even in this case IQ explains 2/3 of the discrepancy between proportion of population black

Table 8.2
Race Differences When IQ Is Controlled For

	%B	%W	IQ	%B at IQ	%W at IQ	%B/W Gap Reduced
High school graduation rate	73	84	103	93	89	136
Holding bachelor degree	11	27	114	68	50	200
Being in a high-IQ occupation	3	5	117	26	10	800
In poverty	26	7	100	11	6	74
Unemployed	21	10	100	15	11	64

Source: Herrnstein and Murray (1994)

and occupational representation.

Herrnstein and Murray (1994) have also compared black to white performance on a number of educational and economic variables before and after controlling for IQ. Their results are summarized in Table 8.2. Like Gottfredson, Herrnstein and Murray show that the mean IQs of blacks in various job categories are systematically lower than those of whites, which means that blacks are more numerous in various categories than they would be if recruited solely on the basis of IQ. For instance, although 84% of whites have graduated high school against 73% of blacks, 93% of blacks whose IQ is 103 have graduated high school, but only 89% of whites at that IQ have. Although the average annual black income is lower than the white (they give $20,954 as against $27,372), the income gap shrinks to $25,001 versus $25,546 when IQ is fixed at 100. Indeed, Herrnstein and Murray find discrepancies in job representation larger than Gottfredson's, as shown in Table 8.3. When these discrepancies are recomputed using the IQ and jobs of individual blacks and whites rather than group means, they find that a black is over 1.5 times more likely than a white with the same IQ to be employed in the professions, a technical occupation, or a clerical job (1994: 489–492).

There were fewer black doctors and policemen in 1900 than IQ would have predicted, (redundant) evidence that in 1900 able blacks were not reaching desirable positions. But according to the data assembled by Gottfredson and Herrnstein and Murray, discrimination explained a minute portion at most of black vocational failure in 1970, and ceased altogether to be a factor by 1980. Ability and representation apparently intersected some time after World War I but before the start of affirmative action, by which time the effects of slavery and discrimination had been attenuated to nonexistence. (Herrnstein and Murray estimate that employment opportunities equalized in the professional fields in 1963, and clerical jobs in 1967; see 1994: 490.) It is reasonable to conclude that there are now at least as many black doctors and other professionals as there would have been absent slavery and discrimination. *per impossibile* the blacks selected over Alan Bakke had existed absent slavery, they would probably have gotten roughly the grades they did get.

Table 8.3
Mean Race Differences in IQ in Various Occupations

Job Category	B/W Difference in SD
Professions	1.3
Managerial	1.1
Technical	1.5
Sales	1.4
Clerical	1.1
Protective services	1.4
Other service jobs	1.4
Craft	1.1
Low-skill labor	1.1

Source: Herrnstein and Murray (1994)

There is also striking agreement between the present analysis and data from the upper tails of the IQ curves. In addition to the failure of the National Science Foundation's multibillion dollar preference program to increase the number of blacks in science, the Johns Hopkins Center for Talented Youth, which identifies the brightest 12-year-olds across the United States on the basis of SAT scores, reports "concern about the low number of minority children (mostly from inner-city schools)" that meet CTY criteria (Mills 1992: 189). Jensen (1991a, 184; also see 1980: 112) gives 130 as the IQ predictive of success in an academic career, operationalized as earning a Ph.D., which covers 2.3% of the white population and .135% of the black. As there are about 420,000 black youths in any one-year cohort, fewer than 600 blacks are predicted to be capable of earning a Ph.D. in any one year. The anticipated ratio of black to white Ph.D.'s is .135/2.3 × .12 = .007. In fact, blacks earned 838 doctorates in 1990 and whites earned 35,199 (De Palma 1992: A18), a real-world ratio of .023, three times greater than predicted. (In 1987, 1988 and 1989, respectively, blacks earned 767, 813 and 811 doctorates.) This discrepancy may indicate lower standards for blacks, a cutoff for Ph.D.'s below 130, a concentration of blacks in the less intellectually demanding areas, or some combination of these factors, but the discrepancy is not in the direction it would be if able blacks were being denied degrees.

Another attainment discrepancy conventionally blamed on bias is the absence of blacks from mathematics, which in turn limits black participation in science. In a typical year 3 or 4 blacks earn the Ph.D. in a mathematical science; sometimes as many as seven do, sometimes none. The mean IQ for mathematicians is 143 (Jensen 1980: 342). Using the Gottfredson-Hitchcock estimates for the black and white IQ distributions, this is 4.59 SD above the black mean, corresponding to about .00001% of the black population. Assuming the black SD to be as large as the white reduces z to 4 and increases the proportion of the black population with an IQ of 143 to .00317, which predicts about 13 blacks in any one-year cohort as intelligent as the average mathematician. Considering the small numbers involved, this prediction fits the data remarkably well.

One may reflect that as the below-replacement birthrate for black females attending college and the higher birthrate of low-IQ black women lowers the mean black IQ further, more aggressive preferences will be needed to sustain the present representation of blacks in high-status jobs. Furthermore, although birthrates vary inversely with education for white women as well, the proportion of all black women who are both low-IQ and having children is greater than that for the corresponding white female cohort. This discrepancy may widen the race gap in IQ in coming decades, so that maintaining current levels of preference will require the recruitment of ever less competent blacks. Environmentalists will presumably interpret any decrease in black representation as a recrudescence of the racism that proves the need for stricter quotas.

The race difference in mental ability also explains most or all black overrepresentation in low-status areas. Rosenthal and Jacobson's well-known *Pygmalion in the Classroom* (1968: see esp. 53–54) argued that normally able black children are placed in remedial classes because of low teacher expectation, where they internalize these expectations and perform accordingly. This theory of "learned failure" has become a dogma for many educators, although the "Pygmalion effect" has not been replicated (Fleming and Anttonen 1971). But Jensen (1980: 91–92) cites a striking example of the IQ discrepancy predicting the precise racial contours of remedial placement. In the early 1970s, blacks made up 28.5% of the population of San Francisco, and 66% of the classes for the educable mentally retarded (EMR). EMR classes recruit from an IQ range of 75 and below, where 4.8% of the white population and 25.4% of the black population fall, a ratio of about 1:5. (Including Asians in the nonblack population reduces the ratio further.) Since the proportion of whites to blacks in San Francisco is about 2.5:1, there should be fewer than one white child for every two black children in EMR classes, a proportion slightly smaller than observed. The IQ gap predicts the race discrepancy in EMR classes with reasonable accuracy.

The curvilinear character of the distribution of mental ability explains two further features of the attainment gap commonly attributed to discrimination. One, the race difference in selection error rates on apparently unbiased predictors, was discussed in chapter 3. The second is the average black/white attainment difference within job and education categories, for example, the failure of black scientists to win peer-reviewed grants in proportion to their numbers. This phenomenon is quite general; on average blacks earn less than whites with the same number of years of schooling (Jaynes and William 1989: 288, 301), and less than whites in the same occupation. As credentials seem to have been controlled for in these cases, bias is a natural suspect. But looking again at Figure 8.1, one sees that at points x where the bell curve is falling rapidly, the mean IQ of the population beyond x will be fairly close to x, since almost everyone above x clusters near it. For a point x at which the curve is falling more slowly, the mean value of the IQs above x is considerably higher than x, since many members of the population lie well to its right. This effect is strongest between $-2z$ and $+2z$, which includes the recruitment range of most jobs. Thus, when the black and white curves are superimposed as

Figure 8.2, the *average* IQ of blacks above the cutoff point for most jobs is lower than the white average above that point, as blacks cluster nearer the point itself. Since IQ correlates with income within as well as between occupations, whites outearn blacks when occupation is fixed.

A more refined model (Brown and Reynolds 1975) premises that each IQ increment on a job is worth more as the minimum IQ for the job rises. This model predicts smaller within-occupation variance for black wages than for white wages, since black IQs cluster near the cutoff point while white IQs are dispersed more widely, an effect Brown and Reynolds document (1975: 1003, also 1005–1006).

The intuitive point is that brilliant college students do better than less brilliant college students and brilliant policemen do better than less brilliant policemen. Among blacks and whites able to handle the demands of college or a job, proportionately more blacks are *barely* able, so mean white performance exceeds mean black performance. The higher the minimum IQ for a task, the greater the black/white attainment discrepancy within it. That is a simple explanation of why proportionately fewer black scientists conceive research projects good enough to win grants in blind competition.

By the same token, the average black SAT scores and grades in any college class should fall somewhat below those of whites, since qualifying blacks will cluster nearer the qualification cutoff than will whites. However, this does *not* however explain the overprediction of black performance, or, contrary to Fischer et al. (1996), the black/white credential gap in college classes. Computational experiments in Fischer et al. (1996: 246–7) show a mean 85-point SAT gap under unbiased admissions, and a mean 103-point gap if blacks are given a 50-point advantage. Since the gap between the mean SAT scores of blacks and whites currently admitted to college is about 200 points, blacks clearly benefit extensively from racial preference. We may go further. By Fischer et al.'s own reckoning, a boost of 50 SAT points knocks out 1% of better-qualified whites. As the relation of the boost given blacks to the percentage of whites disqualified is almost certainly non-linear, we may suppose that more than 4% of better-qualified whites have to be displaced to create the mean 200-point SAT gap between blacks and whites currently admitted to college. Since roughly 3,000,000 students apply to college each year, this means that, each year, at least 120,000 whites are displaced. Discrimination on this scale against blacks (even discrimination against 1% of black college applicants) would be considered a national scandal.

8.10. MOTIVATION AND RACE DIFFERENCES IN ATTAINMENT

The overprediction of black performance by standardized tests leads from cognitive to temperamental race differences.

Those who have speculated as to why blacks do less well than whites with identical objective test scores offer one of two explanations. Miller (1992) considers it a byproduct of test imprecision and regression to the mean. Since the mean mental ability of blacks is lower than that of whites, an outstanding black performance on an ability test is more likely to have resulted from lucky guesses than the same performance from a white. Defining an individual's "true" test score as the average of a series of retests, a black's true score is apt to be lower than that of a white earning the same nominal score. Retesting 100 blacks and 100 whites who score (say) 115 on a single test will tend to yield lower scores for the blacks. When IQ is used as a predictor, the blacks' criterion performances will be more in line with their true score, hence lower than that predicted by the 115. This hypothesis implies that overprediction of black performance should become more pronounced for test scores higher above the black mean, although I know of no relevant data.

Alternatively, overprediction may be due at least in part to motivational factors of the sort discussed in chapter 3—work habits, self-discipline, impulse control. I am aware of no study resembling Gottfredson's of the cross-racial effect of time preference on achievement, but certain personality traits are reported to explain some of the variance in academic achievement that remains when IQ is controlled for (Jensen 1980: 242). Socioeconomic status correlates less strongly with IQ for blacks than whites (Jensen and Reynolds 1982: 429), consistent with race/personality interaction. Occam's razor again requires that some of the residual attainment gap is explained by personality traits on which the races are known to differ. Certainly many writers, including defenders of affirmative action, cite impulsivity and lack of self-discipline as contributors to black failure. Banfield (1974) argues that the primary cause of black poverty is that "The lower-class person lives from moment to moment, [and] is either unable or unwilling to take account of the future or to control his impulses" (54). Black teenagers are distracted by "the manifold opportunities that the Negro district offers for 'action' " (114).[31] Boxill, an advocate of compensatory preferences, cites "chronic tardiness" (1978: 164) as a paradigm negative black work habit. Herrnstein and Wilson report:

In the slums, some persons may attach not simply a low value to reinforcers available from outside their neighborhood but a negative value. "Straight jobs" are for "suckers," outsiders who are robbed "deserve what they get," and having an arrest record is a badge of honor and a measure of toughness. (1985, 304)

Referring to studies of the poor conducted by others, they continue:

Every boy interviewed had been employed at one time, but the turnover was very high. When asked why they left a job, they typically answered that they found it monotonous or low paying. . . . [B]eing able to "make it" while avoiding the "work game" is a strong, pervasive, and consistent goal. (335)

A 1991 study by W. J. Wilson (reported in Whitman 1991, and eventually appearing in W. J. Wilson 1996) found that 40% of employers familiar with blacks found them "apathetic" or "arrogant," and that 75% of black fathers expressed the view that people have a right to "receive public aid without working."

Repeatedly, black respondents told ethnographers that their unemployed friends were "lazy" and questioned the diligence of those who quit jobs or had been fired. "Michael," for example, said that most of his friends don't want jobs. . . . He mentioned how several jobless friends failed to even apply for temporary slots at a downtown foo festival after he arranged to have them hired. "Clive" summed up a typical criticism when he claimed that "many black males don't want to work, and when I say don't want to work, I say don't want to work hard. They want a real easy job, making big bucks." Generally, ghetto males felt little or no obligation to marry the mothers of their children, preferring to chase other women and hang out with buddies. Several men candidly talk about "Mother's day," when men temporarily cozy up to ex-girlfriends after their monthly welfare check arrives. (Whitman: 1991)

W. J. Wilson (1996) blames the difficulties of the "new urban poor" on the "disappearance" of jobs (see n. 26), but on his own evidence the problem seems to lie with the traits of the unemployed. The attitudes he describes obviously impede attainment.

Boxill offers the only account of these unproductive habits open to a compensation theorist, that they are further disadvantaging effects of racism.

In order to survive and retain their sanity and equilibrium in impossibly unjust situations, people may have to resort to patterns of behavior and consequently develop habits or cultural traits which are debilitating and unproductive in a more humane environment. I see no reason why these cultural traits—which may be deeply ingrained and extremely difficult to eradicate—should not be classed as unjust injuries. (1978: 164)

This hypothesis faces by now familiar objections. First, young blacks currently disinclined to work were born after the passing of the circumstances Boxill has in mind. Second, not all groups respond to unpopularity by disinvesting in work: Jews in alien cultures have been so ambitious that "What Makes Sammy [Glick] Run" has become a cliché, and the industriousness of nineteenth century Chinese immigrants vexed white laborers. Such differential responses to similar environments indicate nonenvironmental differences in response readiness. Third, there are presocial race differences in temperament, and the within-group heritability of many personality traits relevant to attainment, such as cautiousness and achievement orientation, is high enough—near .5—to suggest a genetic element in between-group differences. Environmental factors reinforcing temperament may themselves be partly genetic in origin. Finally, the attitudes of American blacks toward time and the support of offspring resemble those reported in Africa and the Caribbean, so are unlikely to be caused by uniquely American factors.

On balance, unproductive black personality traits are probably not "unjust injuries."

8.11. WHY GENES? THE BURDEN OF PROOF AGAIN

Attentive readers will have noticed that I have used differences in *phenotypic* intelligence and temperament, not genes, to explain the attainment gap, an explanation consistent with these phenotypes being caused by environmental factors for which whites are not responsible. It follows that hereditarianism, while sufficient to exonerate whites, is not necessary. So why raise the contentious genetic issue at all?

It must be raised because it is widely, and reasonably, assumed that, among environmental factors, only oppression can produce an attainment gap as large as the one between the races. Boxill construes the phenotypic race difference in punctuality as an indirect effect of racism, and one may be sure that poor diets for black babies, should that prove to be an immediate contributor to the IQ discrepancy, would also be attributed to racism past or present. This construction is natural, even inevitable, once biology is ruled out. Adventitious factors should distribute randomly across races, canceling each other and leading blacks to excel whites in some respects, to fall behind in others, and to perform equally well in the rest. Yet blacks almost always do worse, and what (besides biology) can explain systematic failure except intentional imposition of disadvantage? Once environmentalism is accepted, the compensation argument returns at one remove: superior ability may give whites an advantage but, the cause of the superiority was a wrong, a wrong that must be annulled. This is why a thorough sifting of the compensation argument always finds the genetic issue.

I do not wish to beat the point into the ground, but it is of fundamental importance. Imagine a supporter of preferences confronted with hard phenotype data like Gottfredson's and Herrnstein and Murray's, indicating that blacks achieve less because of lower mean intelligence. He could reply, and the phenotype data give him no reason not to, that low black intelligence is an "unjust injury" caused by the "impossibly unjust situations" whites have created. The excessive cost of preferring blacks in competitive situations, he might insist, merely shows that reparation may have to take some other form. That is why Herrnstein and Murray could not be more wrong in saying that "The *existence* of the [race] difference has many intersections with policy issues. The *source* of the difference has none that we can think of, at least in the short term" (1994: 313).[32]

It cannot be proven beyond all doubt that intelligence is a valid construct measured by IQ, that the races differ with respect to it, that this difference explains race differences in outcome, and that this difference is due significantly to genes. But certainty on these points is not required. All that is required to rebut the compensation argument is that these propositions, taken together, offer an account of the attainment gap at least as plausible as the racism hypothesis.

Claims of damage must be *sustained*, a burden carried by the plaintiff. Jones cannot just hobble into court, accuse Smith of breaking his leg, and expect to collect damages; he must show that the condition of his leg is due to an action of Smith's. Compensation theorists must likewise show that whites damaged blacks. As the claim is one of tortious liability, the showing need meet only the relatively undemanding civil standard of the preponderance of the evidence: it must be more likely than not that white misdeeds opened the attainment gap. But show this the compensation theorist must—which means that a rebuttal need not prove categorically that white misdeeds did not open the gap, merely that it is at least as probable as not that the attainment gap was created by some innocent factor, such as genes. The hereditarian analysis need not be demonstrative, only *as likely* as "racism"—a conclusion that will be drawn, I believe, by any impartial student of the evidence.

Arguably a weaker defense suffices. Given the scope of the demands based

charges against whites, and the extraordinary harshness with which these
arges are brought, claims against whites might well be asked to meet the
gher standards of a criminal trial. The racism theory would then have to be
oven beyond a reasonable doubt, and it would be a sufficient defense of
hites that the hereditarian analysis cannot reasonably be rejected. The verdict
der that standard seems obvious.

Lest placement of the burden of proof seem legalistic quibbling in its own
ght, the reader should reflect that over the past three decades a great many
hites have watched jobs and resources go to less qualified blacks because of
e "racism" theory. It is only fair to ask that this theory be shown to be more
ausible than its rivals before more sacrifices from whites are demanded in its
me.

12. FINAL REFLECTIONS ON THE COMPENSATION ARGUMENT

Suppose whites *are* more successful than blacks because of past misdeeds,
d that the racial nexus makes exploitation of this advantage wrong. It might
ll be permissible for whites to retain this advantage, for two reasons. One
ncerns a white action that may have already repaid much or all of whatever
bt whites owe blacks, and the other concerns actions of blacks that may
ncel most or all of the debt remaining.

The white action was the Civil War, fought, as the popular idea has it, to
ee the slaves. Its immediate cause was the secession of the Confederacy,
stified as an assertion of states' rights, but the Confederacy seceded because
the federal ban on slavery in the new territories and fear of an eventual ban
slavery in the South itself. The suffering in this war of hundreds of
ousands of white Northerners for the benefit of blacks—one of the rare
amples in history of one group sacrificing itself for another unrelated by
nship or nationality—would seem to have discharged part of whatever
ligation whites as a group may have to reduce their net advantage over
acks. The suffering of non-slave-owning Southern whites—the vast
ajority—must also be included. Imagine it is 1870. Over here is an ex-slave,
orse off than he should have been, but also better off than he would otherwise
because of the Union soldier over there, who lost a leg at Shiloh. Surely the
st to the soldier and his family of helping bring the slave closer to where he
ould have been has reduced whatever illicit competitive advantage the soldier
d his descendants might once have had over the slave and the slave's
escendants. It is amazing that discussions of the duty of whites to even the
aying field by worsening their own prospects rarely mention the role of the
orth in the Civil War. If those sacrifices are too negligible to consider, one
onders how much preferential treatment will be deemed enough.

Suppose now that whites retain an illicit advantage after the cost of the
ivil War is deducted. An example makes plain why they may still be entitled
keep this advantage.

As a result of Jones' father having injured Smith's father, making it harder
r Smith's father to educate his son, Jones now excels Smith in ways he
ould not have. He got the college education Smith should have gotten. The
usal nexus is appropriate and the illicit advantage principle is ready to kick

in to demand recompense by Jones, but for one thing: Smith has recen[t] robbed Jones and vandalized his house. Smith did not do this to exa[ct] vengeance or rectification, but he has harmed Jones anyway. Common sen[se] would deduct Smith's damage to Jones' house from Jones' debt to Smith. [If] the education Smith should have gotten and Jones did get is worth $500,000 [in] lifetime earnings, but Smith stole or destroyed $500,000 of Smith's propert[y] they are quits. In theory two separate actions might be required: Jones mig[ht] have to pay Smith $500,000 in redress, and then Smith might have to hand [it] right back. But one way or another, their debts cancel. Any advantage Jon[es] still enjoys over Smith is Jones' to keep.

By that precedent the damage done to whites by black crime, and t[he] destruction by blacks of property created by whites, reduces the illi[cit] advantage of whites, hence the obligation of whites not to use the advanta[ge] they do have. Black-on-white robbery and murder rates are several times t[he] white-on-white rates.[33] Since half of all black robberies victimize whites a[nd] the black robbery rate is many times the white, whites experience greater loss[es] than they would if they had no contact with blacks. There is a steady flow [of] stolen resources from whites to blacks. In addition to its monetary losse[s] black crime imposes the psychic cost of anxiety. Most whites living in lar[ge] cities are more fearful in their everyday existence than they would be if t[he] black crime rate were as low as the white. The negative effect of black crime [on] noncriminal blacks does not reduce its effect on the white debt; Smith[s] robbing Jones reduces Jones' obligation to compensate Smith even if Smi[th] steals from his own family as well. In addition, the inability of blacks [to] maintain housing stock, apparent in any black neighborhood original[ly] occupied by whites, amounts to some, probably significant, portion of t[he] accumulated advantage in income of whites over blacks.

The power of race differences to explain ostensible damage to blac[ks] remains the most forceful reply to the compensation argument. But when t[he] Civil War, black crime and black destruction of property are added in, black[s] may turn out to owe whites.

8.13. DISTRIBUTIVE JUSTICE

A forward-looking perspective, although irrelevant to compensation, is n[ot] entirely misguided. Perhaps distributive justice or utility demand efforts [to] equalize the phenotypic intelligence of the races, or in some other way to ann[ul] the effects of the genetic difference. Opening this issue assumes that the blac[k] and white intelligence gap *can* be significantly narrowed, which now seem[s] unlikely in view of the failure of intervention programs to date. But to engag[e] the moral question I will suppose that effective intervention is possible. Wou[ld] it be obligatory?

The supposition that intervention is feasible opens more possibilities than [is] usually envisioned in statements of the interaction argument. Layzer (197[2:] 200), McGuire and Hirsch (1977: 91), Block and Dworkin (1976a) and Bloc[k] (1995) all speculate about the existence of some *one* environment in whic[h] white and black genotypes express themselves similarly. Each of these autho[rs] displays a hypothetical graph of the reaction ranges of black and white polygenes for IQ—that is, of black and white IQ phenotypes against environments—in which the phenotype lines meet. The E-coordinate of the intersection is an (hypothetical) environment in which the mean IQs of blacks and whites would be equal, so that creating this environment for both races would close the race gap. But (see section 4.2^{34}) the race gap might be reducible even if there is no one environment which equalizes black and white IQs. Distinct environments E and E', such that the expression of the mean black genotype in E is identical to the expression of the mean white genotype in E', would do; IQs could be equalized by creating E for blacks and E' for whites. This latter option, separate environments for whites and blacks, is the one more likely to present itself, and to bring with it problems of equity.

Conflating for now the different types of intervention, the first question is whether distributive justice requires any intervention at all.

Finding distributive justice an obscure notion, I will follow Benn (1967: IV–298) in taking it to mean "what is obligatory in the initial distribution of goods." As for what specifically that is, many contemporary writers take the inertial distributive state to be equality, any deviation from which needs justifying. Weak as this precept is—it merely requires treating two individuals alike when they are alike in relevant respects, a condition that merely shifts the issue to that of specifying "relevant respect"—it is plausible only when some agent is doling out the goods. There seems nothing prima facie wrong with two identical individuals at opposite ends of creation happening to end up unequally well off. A stronger case for equality is needed.

Egalitarianism that opposes inequalities on the ground that they symptomatize prior wrongdoing assumes that inequalities cannot be explained by natural differences. Jo. Baker for instance writes in *Arguing for Equality* (1988: 88) of the virtual impossibility "of accumulating wealth by consent. Anyone who thinks that's how capitalism got going in the first place should read Marx's description in *Capital*. Capitalism came into the world 'dripping from head to toe, from every pore, with blood and dirt.' " (He goes on [119–121] to question innate nonanatomical sex differences, and is confident that they cannot explain sex roles if they do exist.) On this theory, equality is good because as it restores things to where they would have been but for wrongdoing—leaving egalitarianism a compensatory claim of the sort already treated at length.

Any form of egalitarianism worth the name must consider equality good in itself, not good merely because it signals or promotes some more fundamental value. Egalitarianism of this sort would have the further advantage of targeting any factors that contribute to variation in outcome, as race differences in intelligence do, without inquiring into their origin. Yet egalitarians of this sort are extremely rare, and, despite appearances, may be nonexistent. Some seeming–egalitarians (Temkin [1986] may be an example) take great pains to expound what equality entails without presenting any reason to think it is intrinsically desirable. Other egalitarians resemble Fischer et al. (1996)^{35} in expressing great shock that inequalities exist in society, but never explaining why they are so distressed. And to judge by their metaphors and other rhetorical devices, ostensibly "ground floor" egalitarians willing to defend their commitment rely on independent values that equality is tacitly, perhaps unconsciously, assumed to serve, and inequality to violate. Documenting this properly would take a book of its own, but consider this central passage from Eric Rakowski's *Equal Justice*:

People at the brink of death, dragged to the precipice despite every effort to remain safe and hale, seem entitled to assistance from those who were luckier if they did not bring misery on themselves and cannot live without help. . . . [I]t appears self-evident that a little girl suffering from polio or a collapsed lung should be given medical assistance, even if her parents cannot or will not agree to cover the costs. No one should be condemned to anguished poverty or worse as a result of chance events against which she could not defend herself. From the standpoint of politics, people's lives matter equally, and the only way to accord them equal consideration is to grant each an equal bundle of resources. (1991: 94–95)

Insofar as Rakowski opposes condemning the innocent and pushing them over precipices, he is protesting *aggression*. Most people would agree that little girls who have done nothing should not be dragged anywhere, but—since polio viruses don't literally drag anyone—this has nought to do with distributive justice. Taken literally, Rakowski is saying that the involuntarily poor and sick have an enforcible right ("politics") to other people's money, which is hardly "self-evident."

Or consider Walzer's view (1973; also see 1983) that it is "a kind of tyranny" that people with more money should be able to get better medical care: "it is tyrannical of . . . the rich to gather to themselves social goods that have nothing to do with their personal qualities" (1973: 402). Again, it is not literally tyrannical for a rich man (whose qualities are more apt than Walzer supposes to be estimable, given the income/IQ correlation) to consult a pricey specialist. He does not forcibly and arbitrarily prevent poor men from consulting the same specialist, who presumably chooses his own patients, nor is he "unrestrained by law," the dictionary definition of "tyranny." Opposition to despots is welcome, but it too has nothing to do with inequality.

There is also the curious willingness of many egalitarians, noted by both their critics (Sesardic 1993a; Frankfurt 1986) and friends (Kerlie 1996), to relax the demand for equality so long as everyone is reasonably well off. The underlying value of egalitarians of this sort looks like improvement of the lot of the very badly-off, or reduction of the gap separating the very badly-off from the well-off, or a minimum level of well-being for all, but not equality per se. This impulse can be seen in the passage just cited from Rakowski, and Frankfurt (1986) offers a particularly acute analysis of Nagel (1979) along these lines. Self-styled egalitarians impelled by concern that everyone reach a minimum of well-being are not really egalitarians at all.

A recent attempt to state a "pure" form of distributive egalitarianism is Kerlie's (1996),[36] who bases equality of outcome on "a moral claim to an equal enjoyment of the good. . . . [A]n inequality is objectionable because it violates a claim of the worse-off person. That person has less than a fair share" (294–295). Kerlie says that this account is "teleologically" concerned with the value of equality rather than with the duty to distribute goods equally, but he recognizes its deontological sound. Moreover, he objects to inequality "wholly explained by natural causes" (see section 8.13) on the grounds that such inequality is "opposed to the relationship that should exist between different lives, [thereby] violating individual claims" (296). Surely, however, the language of "fair shares" and "claims" makes sense only for goods handed out by a human agent obliged to apportion them in a certain way. So Kerlie has not formulated, let alone defended, the view that equality per se is good; rather, he is thinking metaphorically, in terms of some agency somehow obligated to distribute equal shares. Who the distributor is and why he should distribute equal shares to actual persons are questions Kerlie does not reach.

A very common form of deviationism involves appeal to criteria. For instance, Nielsen (1997) repeatedly demands that "each person be treated with equal respect irrespective of desert" (209). So, does Nielsen think that, when A and B both need blood transfusions but there is only enough plasma for one, we should just flip a coin? Not at all. He assures the reader that equality

does not mean that in treating them [people] with respect you treat them in an identical way. In treating with equal respect a baby, a young person, or an enfeebled old man out of his mind on his death-bed, we do not treat them equally, that is identically uniformly, but with some kind of not very clearly defined proportional equality. (210) [37]

If A has been a more frequent blood donor, giving him the plasma "does not . . . run counter to justice as equality." If A is a young mother of three and B is a ninety-year-old, "more needs would be satisfied if A gets it than B." If A is a doctor and B an unemployable drunk, "we again quite rightly appeal to social utility" to justify favoring A. All equality demands is that we not "simply favor A because he was A" (216). (Presumably Nielsen, a university instructor, gives his better students higher grades. He would no doubt explain that this practice duly respects everyone's achievements, and that "inequality" would be, for instance, a huge fuss made over the good students and scorn for poor ones.) Nielsen's radical egalitarianism has degenerated down to the truism that individual differences in treatment must be warranted by differences between the individuals treated. There is nothing wrong with the warranting criteria Nielsen chooses, but his choices are ad hoc, and anyway people are patently unequal with respect to them.

We may cease classifying egalitarianisms, for opposing them all, at least as they bear on racial equality, is the common sense principle that the sheer possibility that a more favorable environment might make someone better off (or as well off as average) does not by itself oblige anyone to create it. Perhaps better training in his youth would have made clumsy Smith an average athlete, and some millionaire who knew young Smith did not hire him a coach. Surely the millionaire violated no canon of distributive justice. People have no right to be as athletic as other people, or as intelligent, so there is no correlative duty to make them so. Special circumstances create special obligations to create favorable environments, as parents are duty-bound to give their children the best education they can, but attractive possibilities, including possible environments that equalize groups, have no natural right to exist. As there is no general obligation to create more favorable environments just because they can be created,[38] whites are not obliged to enhance the mental development of blacks simply because it may be possible for them to do so.

Does past mistreatment of blacks create a special distributive obligation for whites? Not because of burdens it has placed on contemporary blacks, for then we are back to compensation. Perhaps a scheme *consciously contrived to allocate goods* should give any two individuals the same amount, other things being equal, and, by keeping slaves from learning to read and forcing their descendants to attend segregated schools, use inferior facilities and endure second-class citizenship, whites contrived an inegalitarian allocative system. So now they are specially obligated to raise black IQs if they can.

This argument too sails dangerously close to compensatory regions—for why does the fact (assuming it is a fact) that blacks in the past got less than their fair share impose an obligation to intervene in behalf of other blacks now, unless the situation of blacks now resulted from these past inequities in a way that illegitimately advantages contemporary whites? But rather than repeat objections to the compensation argument, two other points bear more directly on distributive justice.

First, slavery was endemic in black Africa, probably more so than in Europe or Asia (see Baker 1974: 364–365). African chiefs selling captives taken in tribal warfare were complicit in the overseas trade; European slavers apparently never had to use force directly. Hence, any obligation to contemporary American blacks imposed by the slave trade weighs less heavily on contemporary whites than on descendants of African kings. Since enslavement in America was not an obviously worse fate than capture in tribal warfare, blacks taken to the United States were not made grossly worse off than they would have been had they stayed in Africa.[39]

Second, the limits placed by whites on black education prior to 1954 did not, on the evidence, suppress an environment as favorable to mental growth as any that blacks would have found themselves in had they been left undisturbed in Africa. Black slaves on their own would almost certainly not have developed multistory buildings, written language, mechanical devices, the wheel, firearms, or a legal code, all features of the antebellum South. There blacks witnessed democracy and a tradition of free discussion, not found in their natural environment. Since no African society ever developed a written language, forbidding slaves to learn to read probably did not stifle a development that would otherwise have taken place. People should be allowed to learn skills useful in the environment in which they find themselves, so slaves should have been allowed to learn to read, but that is not to say blacks would have acquired literacy on their own.

Segregated schools having become an icon of injustice, it takes considerable imaginative effort to appreciate that no African society ever developed educational institutions remotely comparable in quality to the segregated schools of the Jim Crow era. Old, discarded white textbooks are better than no textbooks at all. Continuing to use as a baseline the environment that blacks attending segregated schools would have developed had they remained in Africa, those schools were abnormally stimulating, outside the range of environments in which blacks evolved. Relative to this same baseline, today's public schools in black neighborhoods are also abnormally stimulating, for they too would evidently not have been available to blacks had whites neve contacted the ancestors of the black children in attendance. (Many of these schools are dilapidated, but since the dilapidation is caused mostly by black students, it is not an external impediment.)

Whites have invested more resources in the education of blacks than blacks could have done themselves. In 1904, the heyday of segregation, 20% of the public education budget in Southern states was reserved for black schools, which served the 33% of the population that was black.[40] Each black child thus received $.60 in schooling for every $1 received by a white child. This would have violated the principle that other things being equal an allocative scheme should give everyone the same amount of the good being allocated, but other things were not equal. The mean per capita wealth of blacks in 1904 was $34, that of southern whites $885 ($1320 nationally). As public schools were and are funded primarily through property taxes—even today Federal spending accounts for only 6% of spending on schools—Southern whites may be assumed to have spent 885/34 = 26 times more for schools, per capita, than Southern blacks. In other words, each black child received .6 × 26 = $15.60 in schooling from taxes paid by whites for every $1 he received in schooling from taxes paid by blacks. Southern whites had a reason and (one would think) a right to spend more on white than black children; virtually all the money spent on education was their own. In 1984 white per capita wealth was still 4.7 times that of blacks (Jaynes and Williams 1989: 292); black children thus continue to have access to buildings, labs, books, computers, and other facilities that the black population left to itself gives no sign of being able to produce.

It might be replied that distributive racial equity does not depend on what would have been available where American blacks might have been, but what is available where they are, contemporary America. Blacks got (and get) more in the United States than they would have gotten in Africa, but they got (and get) less in the United States than American whites do, and that is the inequity. Now, the holdings of one's contemporaries may be an appropriate baseline for the distribution of resources in a state of nature, but schools, housing, and the other goods blacks are said to have too little of are products of human effort, and it seems a fixed point of everyday morality that the work of human hands should go to the hands doing the work. This is really the heart of the matter: the distribution of wealth should reflect the contributions of its creators, and whites create more wealth per capita than blacks. When Kozol (1992) complains (contrary to fact) of unequal funding for black and white schools, he writes as if whites were greedily monopolizing wealth that appeared out of nowhere; by ordinary standards, the level of funding for schools attended by blacks shows white generosity.

3.14. NATURAL INEQUALITIES

There is an ancient argument[41] for divorcing an individual's share in wealth from his contribution to it, recently revived by John Rawls (1971: 101–104).

Productivity is due to talent and effort. But talent and the capacity to exert effort, Rawls insists, are largely genetic in origin, bestowed by the "natural genetic lottery" rather than earned. The natural lottery is not wrongful—the lucky possessors of talent have not misappropriated anything, nor were the unlucky cheated—but it is arbitrary. From this Rawls concludes that natural gifts and their fruits belong to everyone, and that a productive individual may justly keep only as much of the net gain from his gifts as is needed as an incentive to remain active. Conjoining Rawls' conclusion to hereditarianism, whites just lucked into the genes that make them more intelligent and persistent than blacks, so do not deserve the advantages those genes bestow. This racial consequence has not to my knowledge been explicitly drawn by Rawls or his followers, but its obviousness may contribute to the prestige Rawls (1971) enjoys.

At one level this deduction of socialism from determinism collapses without a push from empirical data, for it confuses entitlement with entitlement to entitlement. An example clarifies this distinction (also see Levin 1979: 257; Levin 1992b: 12; Levin 1994). Suppose you find a gold nugget by glancing into a stream. It was fortuitous that you rather than someone else looked where and when you did; you had no right to be the one to find the nugget. You did nothing to deserve finding it. Yet, having found it, it is yours. You are entitled to keep it. Perhaps only those who cultivate alert habits deserve to be nugget-finders, but all you need do to deserve the *nugget* is to *be* a nugget-finder. You don't have to deserve to have found it.

The Rawlsian question, "Why give Einstein credit, when he didn't choose his brains?" falsely assumes that Einstein is admired for choosing his brains, or for something that requires that he chose them. In fact, Einstein is admired for discovering the theory of relativity, not for making himself clever enough to do so. He didn't deserve to be clever, or to make that discovery, but, having made it, he deserves whatever credit is owed discoverers.

"Ah," replies the Rawlsian, "the question is whether Einstein does deserve any credit (and whether you deserve the nugget). He was lucky to be born into a propitious environment with a superb brain and the will to use it. Why honor a triple dose of luck?" But here the Rawlsian oversteps himself, for in pushing the argument this far he pushes it beyond *genetic* causation. To reman consistent, he would have to deny people title to the product of their talents even if they did somehow choose their genes, so long as they did not chose the factors that cause them to chose the genes they do. Those who choose good genes would still just be lucky to have been caused to choose good genes by those unchosen factors, hence still deserve no credit for or profit from their genes' accomplishments. This same verdict would follow if people chose not only their genes but the factors that cause their gene choices *and* cause those choices as well, so long as at some point choice ends, as it presumably must with the natural causal lottery resuming control. Rawls' reason for denying people a right to the fruits of their genetically controlled labors is an equally good reason for denying them a right to the fruits of any labors, any trait whose ultimate cause lies outside their own wills, whether that ultimate cause is genetic, environmental or some third thing. The Rawlsian position is, at bottom, that no one deserves the fruits of extravolitional factors, from which it follows that, if all behavior is caused by extravolitional factors, no one deserves anything. It also follows that no one deserves the fruit of an uncaused behavior, as nothing could be more fortuitous than a random even

Only completely self-determined choices merit reward or profit. Despite perennial attractiveness to philosophers (see Campbell 1957; Chisholm 1964; Thorpe 1980; Nozick 1982: 294–362), a self-determining will is inherently obscure. It invites but cannot answer the question of what makes such a will determine itself to choose one way rather than another. Not external factors, for then the will is not *self*-determining. Not nothing, for a will acting by chance is not self-*determining*. Not a prior act of the will to choose its choice, for that launches an infinite regress. The alternatives exhausted, a self-determining will is impossible, and should such a will be necessary for desert, no one deserves anything at all.

That many people find this conclusion compelling rather than an occasion to rethink their premises shows that Rawls' argument touches a nerve. Why should some people, or groups, come into the world cleverer, more able than others? One can imagine Hamlet brooding over this. But the question, as posed, is unanswerable. Evolution cannot be shown to be fair. It just happened. However, unless it is already assumed that all persons should be equally endowed, evolution cannot be shown to be *un*fair, either. To repeat, it just happened. There is no positive reason to let its work stand, nor any positive reason to seek to undo it.

Rawls sometimes shows awareness of this. "The natural distribution is neither just nor unjust," he says; the positions people are born into "are simply natural facts" (1971: 102). Yet to argue for annulling the effects of the natural lottery because it is not just erroneously equates "not just" with *un*just. A situation may be "not just" in not being *required* by justice, without being *forbidden* by justice, that is, unjust. And, precisely as Rawls says, the natural genetic lottery is neither just *nor unjust*. That whites never earned their superior intelligence means that their title to it is not required by justice, not that their having superior intelligence is wrong. There is nothing right or wrong about it.

But the role of genes in behavior undermines the natural lottery argument in more concrete ways. Rawls derives the "difference principle," the duty of the well-endowed to use their talents "for the good of the least fortunate" (1971: 102), from the supposed fact that social contractors ignorant of their eventual positions in a society they are designing will reason as if they were going to find themselves in the worst possible position.[42] But a case can be made (see Gauthier 1984: 249)[43] that this derivation also assumes the factors shaping individuals to be entirely social. Certainly, if you think other people helped Jones develop his talents, you will be more inclined to feel he owes those others some of what his talents produce than you will be if you think Jones came by his talents some other way, for instance genetically. Genetic ability is not a gift from society, so does not demand reciprocation. Hence Caucasoid advantages seem less like resources available to others when conceived as significantly genetic.

In any event, the difference principle is not egalitarian: allowing only those advantages for the better off that help the worst off gives the worst off a veto denied other strata. Assuming blacks among the worst off in America, the difference principle gives priority to helping them. But here the argument takes a final turn. On its face, priority for helping blacks assigns the slums first claim on, say, the next education dollar, and commentators influenced by

Rawls (e.g., D. Richards 1973, 1977) do consider spending more on slow learners obligatory and spending more on the gifted unfair.[44] Yet together with relevant facts, a contrary conclusion may follow. Technical innovations, whose origin is the high tail of the bell curve, tend to help the worst off more than anyone else. In previous centuries, a peasant was unlikely ever to stray more than a few miles from his birthplace; the invention of the airplane has enabled everyone to travel great distances. Vacuum cleaners most benefit women unable to afford domestic help (see Fulda 1993, 1–3). The computer, that latest world-transforming product of Caucasoid ingenuity, most dramatically boosts the productivity and with it the earning power of workers whose talents confine them to routine clerical tasks. Giving that next dollar to the almost entirely white and Asian gifted might well benefit blacks more than giving it to blacks, in which case the difference principle awards it to the gifted.

8.15. UTILITY

What needs to be fed into the difference principle to decide where education money should go is whether intervention programs help blacks more than investment of the cost of those programs in the mostly nonblack gifted. Utilitarians, who weigh all policies on straight bang-for-the-buck calculations of overall effect, need similar data about the comparative educability of black and nonblack children. (The Herrnstein-Murray argument against intervention is wholly utilitarian.) Although common sense is not utilitarian, these issues deserve examination.

Utilitarian arguments for intervention often presume a threshold effect: so many more blacks than whites are extremely poor that raising the mean black environment gives hope of raising the IQs of black children dramatically, whereas there is no comparable intervention for whites. But the evidence does not support threshold effects. Concentration of blacks below a threshold across which the expression of the intelligence polygene changes markedly would mean lower heritability for IQ among blacks than whites. The deprived environment in which blacks live would swamp individual genetic differences between them (cf. the starving Extremes in section 4.2). Yet the within-race heritability of IQ for blacks appears to as great as that for whites (Jensen 1973: 175–188; Osborne et al. 1978). Black children are apparently not exposed to deprivation severe enough to affect mental development.

Absent threshold effects, environmental intervention that raised the IQs of black children would probably raise white and Asian IQs comparably. Therefore, reducing the IQ gap would require an extraordinary sort of affirmative action, the restriction of IQ raising measures to blacks. Let us continue to assume that some measure, perhaps training in verbalization during problem solving (see Carlson 1985), or training in "mental self-management" (see section 4.13), or enhanced nutrition in infancy, boosts IQ. For any such measure to narrow the race gap, black children would at the very least have to have more access to it than white. Black children could be assured of greater access without race preferences were black parents to seek the regimen more avidly than white parents, but that scenario is unlikely. Black parents now are less apt than white to expose to their children to written matter, despite the highly publicized benefits of reading—an outcome predicted by parental interest in education being an environmental correlate of genes. So allowing the market to allocate IQ-raising resources would probably widen the IQ gap, necessitating legal restrictions on white access to IQ-boosting techniques, through government programs available only to blacks, or bans on private purveyors (who would be preponderantly white) serving whites.

The "special treatment" problem is apt to run deeper. As we have seen, despite optimistic talk of gene/environment interaction[45] there are no known environments in which black and white IQs converge, or grow appreciably closer than 1 SD. Closing the race gap would therefore almost certainly require (and even this might not suffice) an *inversion* of environments, with black infants and children raised in especially rich ones and white infants and children raised in deprived ones. Goldberger and Manski tiptoe up to this proposal (1995: 764–765):

[A]n individual's observed IQ test score Y is the sum of her "genotype" Z and her "environment" U, so $Y = Z + U$. Imagine that Z and U are uncorrelated, so the variance of Y equals the variance of of Z plus the variance of U: $V(Y) + V(U)$ . . . [Herrnstein and Murray's] thought experiment called for equalizing environments, making $V(U) = 0$. Suppose instead that we preserve $V(U)$ at its current value, but make U perfectly negatively correlated with Z by introducing an extreme compensatory policy. Then IQ variance would fall from $V(Y)(h^2 + e^2)$ to $V(Y)(h^2 + e^2 - 2he) = V(Y)(h - e)^2$. So with $h^2 = 0.6$ and $e^2 = 0.4$, this intervention would reduce IQ variance to $(\sqrt{.6} - \sqrt{.4}) = 2$ percent of its current value $V(Y)$.[46]

The suggestion here, couched in opaque jargon and unconventional notation, is to reject the equalization of environments in favor of placing high genotypic intelligence into poor environments and vice-versa.

What would this intervention look like? Many commentators explain the achievement gap by pointing to the more stimulating character of white households. Would equalization then mean giving black children to white parents, and white children to black parents? Letting white parents keep their children but limiting the amount of printed matter kept in the home? It is hard to imagine anyone, white or black, agreeing to such steps, although the white majority in the US tolerates affirmative action, so the possibility cannot be ruled out. Coercion on a large scale would certainly be required. What is inconceivable is that turning society upside-down in this way, with the attendant destruction of individual liberty, could yield a net advantage.

The utilitarian case for raising everyone's IQs must be distinguished from a utilitarian case for raising the IQs of blacks only, or raising the IQs of blacks while lowering everyone else's. The presumed increase in black productivity and law-abidingness attendant on a higher mean black IQ is also a reason to raise the IQs of whites and Asians, whose productivity and law-abidingness would also presumably increase. On utilitarian grounds, low-IQ children of all races would merit treatment. The benefits of raising black IQ, insofar as they are benefits of raising IQ period, do not justify special efforts aimed at closing the race gap.

There is also no reason to think the gains for blacks per resource unit invested would exceed those for whites or Asians. One might anticipate an initially greater marginal payoff for black children because they have so far to go, but the difficulties of black children in coping with the elementary school curriculum and the failure of enrichment programs to date suggest that black children would be more resistant to IQ boosts. Chapter 4 noted the finding of Currie and Thomas (1995) that Head Start benefits in standardized test scores and grade retention are *more* lasting for white than black children. If all unit increases in children's IQs are equally desirable, but the resources needed to boost the IQ of a black child 1 point could boost the IQ of a white child 1.1 points and that of an Asian child 1.2 points, utility dictates that resources flow toward whites and Asians. Currie and Thomas calculate that Head Start costs $3,500 per child, and remark that

> when viewed strictly in terms of lasting benefits provided to children, Head Start programs serving African-american children are not cost-effective. . . . In contrast, the results for white children suggest that the potential gains are much larger than costs, since even a small decline in the high-school dropout rate has the potential to pay for itself in terms of future wage gains. (361) [47]

Other studies do not permit cross-racial cost/benefit comparisons. Using such indicators as employment, average salary, and avoidance of arrest, the net benefit to one individual of one year of the Perry Program was calculated to be $2515 (Spitz 1986: 197); the net benefit of two years of the program was –$1180, since there were no gains on the criteria after the first year. In the Milwaukee program, the average cost of raising one child's IQ one point was $23,000, implying an annual cost of $100 billion to raise to 100 the mean IQ of the approximately 400,000 black children in any one year cohort. Prototypes are often misleadingly expensive, but since the creation of stimulating environments requires physical facilities and personnel, it is not clear how massively replicating the Milwaukee pilot project could reduce unit cost to any significant extent. More important, these figures, calculated against black controls, shed no light on the opportunity cost of closing the race gap. Whatever one child-point of black IQ costs, the utilitarian also needs to know the cost of one child-point of white and Asian IQ. Again, the initial average gain of about 10 points in the experimental group over the control group in the Perry Program may be assumed to yield a mean payoff of $250 per IQ point per black child in the first year. Boosting the IQs of blacks for one year thus earns back its cost, but the utilitarian also needs to know the mean payoff of one point of white and Asian IQ. These comparisons would require more studies of white and Asian experimental groups with same-race and other-race controls.

Pursuit of equality together with awareness that few blacks can meet demanding academic standards has in recent decades led to lavish attention to inferior, disproportionately black students at the expense of abler predominantly white and Asians ones (Richardson 1993); what attention is paid to the gifted often takes the form of worries about finding the black gifted (Richardson 1993). The 8% of New York City public school students classified as disabled account for 23% of the school budget. Stuyvesant and the Bron

High School of Science, two New York City schools for the gifted, receive no more than other city high schools, many of whose students are barely capable of literacy. The Board of Education runs special tutorials closed to whites to help blacks and Puerto Ricans pass the entrance exams to these schools. Of the $8.6 billion spent directly on students by the federal government, 92% goes to the disadvantaged, and .1% to the gifted (Herrnstein and Murray 1994: 427–435). This disproportion is frequently defended with the demographic argument that, as blacks and Hispanics become a larger proportion of the population, sustaining the American standard of living requires that they be prepared to enter the technical professions. From the utilitarian point of view the reverse conclusion follows; as the population changes, it becomes more important to seek able individuals where the evidence shows they are to be found. The demographic defense of affirmative action is like arguing that when current oil wells begin to be exhausted, we should search more intensely where oil is unlikely to be found, and sink fewer new wells in the diminishing regions where it is likely to be more plentiful. Actually, sensible utilitarians disregard populations, and expect that educating each child in accordance with his objectively determined abilities will maximize general well-being.

The compensatory demand that a claimant be made better off makes clear who should finance the improvement and why: whoever harmed the claimant, because he did. Distributive justice cloaks the need to answer the same questions. Any intervention to equalize IQs, or raise black IQs, will be costly, and there is no one else to bear the cost except the white majority. Why should they? Not to annul illicit white advantages, since distributive justice was invoked precisely to find noncompensatory grounds for reducing the IQ gap. Not correction of the natural genetic lottery, which is not unjust (merely not just). Interventionists like Block and Lewontin urge that "we" (unidentified) keep experimenting with interventions, as if experiments consumed no resources that might be put to other uses. Repeated efforts to raise black IQ would consume white resources that whites might prefer to direct elsewhere, and there seems no reason why they should not do so. Compensatory indemnity at least has a chance of justifying the cost to whites, which is why all discussion of white debts sooner or later reverts to compensation, and why the failure of the compensation argument relieves whites of all burdens.

NOTES

1. About sex differences in parental involvement.
2. Preferences for women are discussed in Levin (1987, 1992).
3. The rationale for banding is that, since tests are imperfect predictors, there is some chance that a score of 1250 indicates as much ability as a score of 1350 (see Cascio et al. 1991). The width of the band can be adjusted to accord with a desired confidence level in prediction. However, while it is not certain that the true score corresponding to an observed 1350 exceeds the true score corresponding to an observed 1250, it is more probable, so candidates chosen by top-down are likely to be superior to those chosen by banding.
4. Specifically, the performance mean in 1990.
5. In this I follow Greenawalt (1983), who defines "discrimination" and "reverse discrimination" as, respectively, "a difference of treatment [that] is to the disadvantage

of members of some group," and "a difference in treatment that reverses the pattern of earlier discrimination" (16).

6. Rosenfeld's index lists "affirmative action" as defined on 335 of his text, where it is described as "remedy[ing] systematic deprivations of equality of opportunity for which the government can be held responsible." This unusually narrow definition is nonetheless compensatory.

7. Shaw also offers, as secondary arguments, that blacks who have not suffered discrimination themselves "have been affected by discrimination," and that "our whole history of racial and sexual discrimination" cannot be ignored. We will find that these appeals come down to compensation.

8. Chapter 10 argues that certain forms of private racial discrimination are not injurious, which implies that their consequences do not merit redress. However, nothing in the present chapter depends on that conclusion. For present purposes I allow "discrimination" to be a compensable wrong.

9. Also see the Lew Alcindor case in Pojman (1992: 185).

10. If a man wants both to smoke and to want to stop smoking, most people would say that the desire manifesting his autonomy and true self, and the one he should heed, is the second one (see section 9.11). To be sure, A's preferences about B's preferences cannot manifest A's autonomy, but they can serve another function of self-directed second-order preferences, namely evaluation. Just as a person's opinion of himself depends in part on his opinion of his own desires, his opinion of others depends in part on his opinion of their desires. Frankfurt (1971) makes a plausible case that man differs from the brutes in having second-order preferences. Being second-order is as much a virtue as a defect of desires.

11. He adds: "[T]he associational preference of a white law student for white classmates . . . may be said to be a personal preference. . . . But it is . . . parasitic upon external preferences: except in very rare cases a white student prefers the company of other whites because he has racist, social and political convictions [sic], or because he has contempt for blacks as a group" (236).

12. Endorsed, unenthusiastically, by Jencks (1992: 61).

13. Walter Block has pointed out to me in conversation that testers have strong psychological and financial interests in provoking differential treatment; under current law, discriminated-against testers, although never actually intending to accept a loan, a job, or housing have standing to sue. Hence black testers have an incentive to act uncreditworthy (cocky and rude), while white testers have an incentive to be respectable.

14. The classical justified-true-belief analysis of knowledge prevents a man who sees a mountain from knowing there is a mountain there unless he knows how vision works. Reliabilism lets a man ignorant of optics know about the mountain by sight.

15. When specified, these benefits often justify symmetrical preferences for whites. One common referent is consumer choice. Defending quotas in medical school admission, Herbert Nickens writes: "minority physicians [may] function in different ways compared with majority physicians. . . . Minority health care providers are likely to be more culturally sensitive to their populations" (1992: 2394–2395). A similar argument is made for preferring black teachers for black children. But white doctors may be more sensitive to whites, and white children may be more comfortable with white teachers.

16. Graglia (1988) expands on this circularity.

17. Biting this bullet, Rosenfeld rejects the tort law standard (1991: 81; "Deny men enforcement" should presumably read "enforce" in the sentence "Under such . . ."). If there are no women in a field that should have been half female, he asks, why isn't firing half the men as fair as firing none? Either way, 50% of some population is treated unjustly. One answer is that doing wrong is worse than letting wrong happen, especially when the doing is coercive, as all government action is. Also, given the

statistics on black crime, Rosenfeld's approach would justify extreme measures against blacks. Finally, the argument permits random redistribution of holdings in any imperfectly just society. If 20% of the average person's holdings are improper (or 20% of all holdings are), someone with n should only have $.8n$. Switching his n with someone else's m leaves total licit holdings at $.8(n + m)$. No injustice has been done!

18. So the shackled runner analogy was defective from the start. The runner's competitors could protest that they were not the ones who caused his lameness, which, relative to anything they did, could have resulted from an accident.

19. Thomson does not realize that the fecklessness she imputes may explain the circumstances of blacks.

20. "From the standpoint of the distributive system's efficiency, it might seem preferable to foreclose compensating victims of past education discrimination with jobs for which there are other persons who are more qualified. From the standpoint of the system's legitimacy, however, it may be inadequate to rely entirely on some other form of compensation, such as monetary damages. Indeed, the award of damages, even if coupled with better educational programs for subsequent generations, may relegate too many members of the discriminated-against group for too long to subordinate positions, and thus fail to ameliorate their sense of self-respect or to increase their confidence in the system. What is needed is a way to reintegrate the victims of past discrimination into the mainstream of society—which entails a share of the jobs allocated by society" (Rosenfeld 1991: 288).

21. I have occasionally asked black students benefiting from quotas how they feel about being selected over better-qualified competitors. Their reactions have ranged from "It's a job and I'm happy to have it" to anger at the suggestion that they do not merit special treatment. I have found no sense of stigmatization.

22. See for example Gross (1977a) and Sowell (1981). Rosenfeld argues that, rather than be answered, Sowell should be "excluded . . . from the dialogical process" because he is a successful black: successful blacks who oppose affirmative action do so because they wish to project the image that they are exceptions, so their opposition is "strategic rather than oriented toward communicative action" (1991: 312–313). Prof. Rosenfeld has informed me in conversation that he has not offered to resign his position in favor of a black. Is he then arguing in bad faith, and deserving of exclusion from the dialogue?

23. I do not see how to reconcile this sentence with the previous one.

24. Readers dissatisfied with the following material should consult Taylor (1992) and D'Souza (1995), who build similar cases more elaborately.

25. A five-minute tour of the bulletin boards near the author's office turned up announcements of the Mellon Minority Undergraduate Fellowships, the Ford Foundation Doctoral Fellowship for Minorities, the Ford Foundation Postdoctoral Fellowship for Minorities, the Gaius Charles Bolin Fellowship for Minority Graduate Students at Williams College, Project Focus for Minority College Freshmen and Sophomores of the American Society of Newspaper Editors, the General Motors Endowed Fellowship "to provide aid to minorities," the National Science Foundation Minority Graduate Fellowships, Graduate Fellowships in Business Administration for Minorities of the Consortium for Graduate Study in Management, the American Society for Microbiology Predoctoral Minority Fellowship Program, and the Wayne State University Post-Baccalaureate Program for Minority Students.

26. Even by the late 1960s Banfield could write "racial prejudice today is a different order of magnitude than it was prior to the Second War" (1974: 78). W. J. Wilson traces black difficulties to the disappearance of manufacturing jobs, without explaining why blacks have not adapted to a more service/information economy. The explanation suggested here is that the mean intelligence required for manufacturing tasks is closer to the black mean than that required for the jobs created by the new technologies.

27. Bureau of Labor Statistics, telephone interview, August 6, 1992.

28. There is often a threat in the telling. " 'Revenge and anger are the first two feelings I sometimes have [when a clerk follows me in a store] . . . ,' Dyami [the author's son] said. 'I'd like to yell and scream and curse that person out, maybe. But I know that doing that probably wouldn't solve anything, and, if anything, would make the situation worse.' The saleslady at the game store—like shop clerks and store owners throughout the city and the country—was lucky. Today." (Harper 1994)

29. "When a wrong has been as persistent and pervasive as racial discrimination and segregation in the public school systems . . . it seems beyond serious doubt that the wrong in question is the cause of a substantial reduction in the prospects of success of blacks in the employment market" (Rosenfeld 1991: 289). "Substantial" here can be taken to mean some proportion of the race difference, although Rosenfeld clearly intends "the entire race difference, which is substantial."

30. Since there is no reason to think blacks are on average more qualified than whites with respect to other job qualifications, IQ-based predictions of occupational representation are good proxies for predictions about bias effects.

31. "The lower-class forms of all problems are at bottom a single problem: the existence of an outlook and style of life which is radically present-oriented and which therefore attaches no value to work, sacrifice, self-improvement, or service to family, friends, or community" (235).

32. Brody too is amazingly short-sighted on this issue. He writes (1992: 310) that there are three "critical questions. (1) What are the reasons for the difference [in IQ]? (2) Can we eliminate it? (3) If we cannot eliminate [it], can we design an environment in which [its] effects are mitigated?. . . It is possible that the answer to the first question might enable us to eliminate the difference." He does not see a fourth question: Are whites to blame for the difference? or that the answer to the first question might answer this one as well.

33. To anticipate some data from chapter 9, blacks in 1992 murdered about 1500 whites, while whites murdered 5000 whites (Je. Adler et al. 1994: 26). As blacks are 12% of the population and whites 88%, blacks murdered whites at $1500/5000 \times .88/.12 = 2.2$ times the white rate.

34. Compare the distinction between (5), genetically diverse individuals being phenotypically identical in some common environments and (6), genetically diverse individuals differing phenotypically in all common environments, yet phenotypically identical when exposed to different environments.

35. Their title, *Inequality by Design*, is disingenuous. It makes them appear to be claiming, adventurously, that inequality is the *motive* of some powerful agency, but the authors later explain that all they mean is that inequality is a byproduct of certain intentional actions (usually government policies). That is like saying Columbus brought smallpox to the New World "by design" because the spread of the disease was a byproduct of his intentionally crossing the Atlantic.

36. One of Kerlie's main worries is egalitarian leveling down: harming the well-off without benefiting the badly-off simply to minimize the difference between them. He appears to settle for a "combined" strategy—other end-states as well as equality are good—which in some circumstances, however, prescribes leveling down.

37. He adds, "It is difficult to say what we mean here."

38. Kagan (1989) detects an almost unlimited obligation to help the worse off; see section 7.1, n. 2.

39. Weyl (1960) cites evidence that, as measured by birth, death, and infant mortality rates, the average health of slaves was better than that of manumitted freemen in the North.

40. These data from the entry *Negro* in the 11th edition of the *Encyclopaedia Brittanica* (1914).

41. To judge from the tone of Aristotle's discussion of this argument in *Nicomachean Ethics* III, it was familiar by his time. See Nagel (1987: 78–79) for another contemporary statement.

42. Rawls gives two reasons for not letting the contractors maximize the mean expected utility of positions (167–175). (a) Calculation of the relevant expectations requires each contractor to assign probability $1/n$ to his occupying each of the n possible positions in the society being formed, and the principle of indifference is irrational in a state of total ignorance. This restriction seems ad hoc, and anyway Rawls' contractors are stipulated to know a great deal of a general nature about society, for instance that they will have descendants. (b) Since no contractor knows his utilities, he cannot estimate his payoff in any position, hence cannot calculate average utilities for himself. This problem can be overcome by calculating the mean value of a position over all utility functions, and averaging those values over all positions. Let $\{p_i: i \le n\}$ be the set of positions in a given society and $\{u_j: j \le k\}$ the set of utility functions any contractor may have. The mean utility of positions is $\{\Sigma_j [\Sigma_i u_j (p_i)/k]\}/n$, the number maximized by contractors who are not pure risk avoiders.

43. Gauthier (1984: 221) comes close to invoking presocial genetic factors in individual productivity.

44. D. Richards (1973) argues that spending extra money on the disabled denies them equal opportunity by denying them an equal chance to "realize their good." Surely, bright students in boring classes geared to the less able are also denied an equal opportunity to realize their good.

45. In the technical sense: pairs of genotypes responding differently to the same environment, hence the difference in response of two genotypes varying with environment.

46. Goldberger and Manski immediately add "Of course, such calculations are fatuous." But they never explain why they think so, or disown the calculations, or withdraw them as a reason for declaring that heritability sets no limits on the effectiveness of environmental change.

47. They conclude: "If the factors preventing African-American children from maintaining the gains they achieve in Head Start could be removed, the program could probably be judged an incontrovertible success." One is reminded of the operation that succeeded but for the death of the patient.

9

Crime

Important as is the issue of justice, many people find the relation of race to crime of equal concern. This topic raises basic questions of risk avoidance, rights to self-defense, the use of race in judging individuals, the function of government, and, ultimately, free will.

9.1. BLACK CRIME

The impression that young black males are disproportionately likely to commit crimes against persons is supported by, and almost certainly results from intuitive awareness of, the relevant statistics. Every study of the subject finds a positive correlation between race and "serious victimful crime" (Ellis 1987b: 531–532). Blacks commit most of the FBI-indexed offenses, accounting for half of all arrests for assault and rape and 62% of arrests for robbery Rushton 1988a: 1016; Herrnstein and Wilson 1985: 461–466; Hacker 1992: 181; Gest 1995). Blacks commit more than 57% of all murders (Gest 1995; Hacker [1992] puts the figure at 55%), also the proportion derived from the FBI data for 1992. At 12% of the population, the average black is thus (.62/.12)/(.38/.88) = 12 times more likely than the average nonblack to have committed a robbery and (.55/.12)/(.45/.88) = 9 times more likely to have committed a murder. In Washington, D.C., Los Angeles, and Detroit the rate of homicide by firearms among 15- to 19-year-olds exceeds 220 per 100,000. According to Je. Adler et al. (1994), the national homicide rate for black males 15–24 is 160 per 100,000, as against 16 per 100,000 for white males. In addition to accounting for most violent crime, blacks are also disproportionately represented in all categories of felony except those requiring access to large amounts of money, such as stock fraud (Hacker 1992).

More than half the U.S. prison population is black (Klein, Petersilia, and Turner 1990; Walinsky 1995; Butterfield 1995; Gilliard and Beck 1995). Of black males over 18, 6.75% were held in federal, state or local jails in 1994 while .86% of white males were (Gilliard and Beck 1995: 7), a ratio of 7.8:1.

Some criminologists use the rule of thumb that a black male is ten times more likely than his white counterpart to be involved in homicide, rape, robbery, and aggravated assault (Berger 1987).

(Taylor notes that many of these comparisons understate the white/black difference by counting non-European Hispanics, whose crime rate exceeds that of European whites, as "white." A black is more than 9 times more likely than a non-Hispanic white to be in a federal prison.)

More than 30% of black males between the ages of 23 and 29 are incarcerated for the commission of a felony (Gest 1995; Mauer and Huling 1995: table 2), while the incarceration rate for white males in that age group is usually put below 3% (e.g., T. Moore 1988: 54; Walinsky 1995). The black incarceration rate increases with high black population density: at any time, 42% of the black male population of Washington, D.C. is in jail, on parole, on probation, or being sought by the police; in Baltimore the figure is 56%; between 45% and 55% of black males in Detroit, Los Angeles, Philadelphia, and Chicago spend time in jail, on probation, or being sought on arrest warrants (Raspberry 1992). In Little Rock, Arkansas, blacks, at 33% of the population, commit 83% of the homicides and 93% of the robberies, making a black 25 times more likely than a white to have committed a robbery. Since the black population is about 58,000,[1] and 7,500 blacks, virtually all young males, were arrested for violent crimes between May 1991 and April 1992 (Uytterbrouck 1993), about 25% of the black male population was arrested for felonies in that one year. It has been estimated that 85% of black males in the District of Columbia will be arrested once in their lives (Walinsky 1995).

There is a disparity between popular beliefs about black victimization and the facts. Georgia is thought to be populated by violent rednecks, threatening to blacks. But of the 2484 murders committed in Georgia between 1973 and 1979, 1676, or 67%, were perpetrated by blacks (Baldus, Woodworth and Pulaski 1992). At 26.5% of the population of Georgia in 1980,[2] blacks committed murder at $(67/26.5)/(33/73.5) = 5.6$ times the white rate.

The data do not sustain the idea that blacks are arrested more frequently, hence tried and convicted more frequently, because of police bias. Prevalence rates by race in arrests parallel prevalence rates by victim's reports, so do not represent bias in arrests (Hindelang 1978). In a California study, Klein, Petersilia, and Turner (1990) found no race differences in incarceration rates or sentencing when such variables as number of previous convictions and age of earliest conviction were fixed.[3]

When black population density is high, blacks commit literally scores of times more murders than whites. Murder rates for European countries run between 2 and 5 per 100,000 annually (Rushton 1994, Whitney 1995). In states like Utah whose residents are virtually all of European descent it fall below 1; in areas with mostly black populations, such as Washington, D.C. and Atlanta, Georgia, homicide victimization exceeds 60 per 100,000. The homicide rate among 15–18 year old black males reached 272 per 100,000 for Detroit in 1987 (Ropp et al. 1992: 2908). The correlation between murder rate and percentage of population black for the 50 states is .77 (Whitney 1995, rising to .82 when Washington, D.C., is added (Whitney 1990b). For the 17 most populous cities the correlation is .69 (Whitney 1995). In these same cities the correlation between percentage of population black and rape and robbery are .52 and .75, respectively (Kleck and Patterson 1993).

The major jump in black crime, along with increased racial divergence in rates for homicide and all arrests, occurred between 1966 and 1972, immediately after passage of the major civil rights legislation; see Murray (1984: 116, Figure 8.2), Herrnstein and Wilson (1985: 465, Figure 2), Jaynes and Williams (1989: 458–459, Figures 9-1 and 9-2), and Whitney (1995, Figure 3).

9.2. INTENSITY OF PREFERENCE FOR OTHER-RACE VICTIMS

Black crime concerns both blacks and whites, but three factors amplify white apprehensions.

One is the option enjoyed by whites but less available to blacks of fleeing black crime. Whites can in theory avoid contact with blacks, whereas in everyday life, kinship, if nothing else, prevents blacks from altogether avoiding young black males. Hence, it is more pertinent to ask whether whites are entitled to flee black crime than to ask this question of blacks.

The second is the character of the worst quintile of black criminal behavior. Like other race differences, differences in law-abidingness can be represented by two overlapping bell curves. That mean white conformity to laws against personal violence is ten times that of blacks amounts to a leftward shift of the black mean by as much as 2 SD,[4] a displacement inducing profound tail effects: the lowest black quintile should display viciousness almost unknown among whites. This expectation corresponds to a reality reported daily in the media—black males killing out of irritation (girlfriends' cranky babies), out of petty greed (for money to buy running shoes), out of anger (over a basketball call) (Lee, Schipp, and Tabor 1992). Whites are alarmed not only by the frequency of black crime, but by the extremes of depravity and indifference to human life it reaches. It is true that whites monopolize serial killing and have produced uniquely deranged specimens like Jeffrey Dahmer, but these instances are extremely rate. Casual brutality is far more common among blacks.

The third concern is the asymmetry in interracial crime. It is widely publicized that blacks are more likely than whites to be crime victims[5] and that their victimizers are more likely to be black than white: in 1989, for instance, 31.7% of aggravated assaults against blacks were committed by blacks, as opposed to 13.1% by whites.[6] What is less widely known is the preference of blacks for white victims. More than 97% of white crime is committed against whites, while one-half to two-thirds of black crime is also committed against whites. In 1987, 50.2% of simple assaults by blacks had white victims;[7] between 1979 and 1986, 2,416,696 of the 4,088,945 simple assaults committed annually by blacks were directed against whites (Whittaker 1990: Tables 1, 16), a cross-racial rate of 59%. Since blacks are 12% of the population, 88% of the victims of black (and white) crime would be white if victim choice were random. Thus, whites attacked blacks at about 1/4 the rate predicted by random victim choice, while blacks attack whites at about 2/3 the predicted rate. Taking the ratio of these fractions—.66/.25—as a measure of intensity of preference by race, blacks may be said to prefer white victims more than 2.6 times as intensely as whites prefer black victims. Informative as this ratio is, it seriously understates the racial asymmetry because of the absolutely greater crime rates of blacks. Since the average black is more than 2.5 times more likely to victimize a white than the average white is to victimize a black, and 10 times more likely to have committed a crime, the average black is about 25 times more likely to have victimized a white than the average white is to have victimized a black.

These measures indicate how much blacks prefer white victims as compared to how much whites prefer black victims, but not compared to how much blacks prefer black victims, and I have not found data that permit direct comparison of black and white preferences for black (or white) victims. Assuming these are the same is at the moment a convention. Whether black criminals prefer white to black victims also depends on questions of opportunity, but residential patterns do not present blacks with disproportionate contact with whites. Correction must also be made for assaults following prior personal contact (as in bar fights) as opposed to assaults with no prior contact (as in robbery), blacks predominating in the latter category (Katz 1989). Expected payoff is also a factor in robbery, with white victims likely to have more money than black, but, unlike opportunity, this factor would not mitigate the greater preference for white victims on the part of black criminals.

The intensity of preference ratio is robust. According to Ingrassia (1993), whites in 1992 murdered 5967 other whites and 392 blacks, while blacks murdered 6600 other blacks and 1216 whites. Hence, while the proportion of black victims of white victimizers was $392/5967 = .0616$, the proportion of white victims of black victimizers was $1216/7816 = .184$, yielding an intensity of preference ratio of $.184/.0616 = 3$. According to a survey described in the FBI's October 1993 *Uniform Crime Reports*, of 11,250 incidents of homicide in 1992 (47.3% of the 23,760 total committed that year), whites committed 4855, which included 291 black victims, while blacks committed 5984, which included 794 white victims. Again, assuming victim choice "ought" to be random, the intensity of black preference for white victims was 2.35. Proportionally, blacks killed 22 times as many whites as whites did blacks. Interracial rape statistics also show a pronounced preference on the part of black males for white victims when compared to white victimizations of blacks. The relevant figures are given in Table 9.1, from Tables 1 and 16 of Whittaker (1990). Black women are victimized by 3% of white rapes, or 3/12

Table 9.1 [8]

Interracial and Intraracial Rape

	Black Victimizer	White Victimizer
Black Victim	26,700	3,337
White Victim	25,032	105,130

Source: Whittaker (1990).

= .25 the expected rate if victim choice is random, while 48% of black rapes victimize white women, at 48/88 = .54 the expected rate, yielding an intensity of black preference for white victims of 2.26. Since blacks are reported as committing about 40% of all rapes, so that a black is almost 5 times as likely as a white to have committed rape, a black is about 11 times more likely to have raped a white woman than a white is to have raped a black woman. Whittaker cautions, incidentally, that the figure for white-on-black rape is based on fewer than 10 sample cases in the 125,000 households surveyed, a number too small to permit calculation of a standard error (1990: 10). Many people (e.g., Leslie 1990) use the belief that black men strongly desire white women, once prevalent in the American South, to illustrate the psychopathy of "stereotyping." In fact, this belief seems empirically correct. At the same time, the low incidence of white-on-black rape undermines the claim that one way whites oppress blacks is by treating black women as fair game.

The asymmetry in preference for other-race victims moderates but is far from reversed in the infamous American South. Using figures for Georgia from Baldus, Woodworth, and Pulaski (1992: 258, Table 13.1), 233 whites were murdered by blacks between 1979 and 1982, while 60 blacks were murdered by whites, meaning that a white was 233/60 = 3.9 times likelier to be murdered by a black than vice versa. However, as the ratio of blacks to whites in Georgia at that time was 26.5/73.5 = .36, random victim choice predicts that whites would be murdered by blacks 36% as frequently as blacks by whites; so a black was 3.9/.36 ≈ 10 times more likely to kill a white than vice versa.

Contrary to the impression given by extensive media coverage of a small number of white-on-black attacks, the great majority of serious interracial incidents involve blacks attacking whites; Baumeister, Smart, and Boden (1996) estimate that 80–90% of all interracial crime is black-on-white. Casual perusal of newspapers over a one year period turned up numerous instances of black-on-white mayhem [9] seemingly more heinous than the Rodney King case, the killing of Yusef Hawkins, or the death of Michael Griffith in Howard Beach.[10] There are other asymmetries in media coverage: the perpetrator's race is often omitted or mentioned fleetingly in accounts of black-on-white crime, stressed in descriptions of white-on-black crime. Black-on-white crime is not used to prove anti-white feeling on the part of blacks, as the King, Hawkins, and Griffith incidents continue to be used to demonstrate white racism.

9.3. ASSESSING RISK

The statistics in section 9.2 indicate that anyone of any race encountering a young black male in isolated circumstances is more warranted in considering himself a potential victim, in taking precautions, and in seeking to escape, than when encountering a white in like circumstances. One is justified, absolutely speaking, in believing himself in danger from a black, and whites are justified in believing themselves in more danger than they would be if they were black. Recall from chapter 2 the statistical rule that an hypothesis can be rejected only when its probability falls below .05. Since the odds now exceed 3 that any given black male is an offender, the possibility that a randomly encountered black male is an offender cannot be rejected.

This sort of reasoning is common at the informal level. If one in three Acme automobiles breaks down in traffic, you cannot ignore the possibility that the Acme you are about to step into will break down. Failure is not worth worrying about only if fewer than 5 Acmes per 100 break down. The failure rate for Acmes does not mean, nor need you believe, that the Acme you are thinking of taking on a test drive will conk out the instant you leave the showroom, but it does justify mistrust of a car that might fail at some inopportune time. While not always quantitative, common sense recognizes that an outcome need not be certain, or even more probable than not, to be taken seriously, so acting as if the Acme might break down—perhaps by preparing an alternate form of transportation home—does not require believing that it actually will. Acting as if a black might be a felon does not require believing he actually is one. It requires just the belief that the chance that he is one is great enough to take seriously.

Objectors to this reasoning might seize the concession in chapter 2 that .05 is not a magic number. Bayesian statisticians reject talk of "accepting" and "rejecting" hypotheses altogether, replacing it with talk of the *expected value* of *behavior*. To compute this quantity for an action A, first multiply the payoff of doing A given hypothesis h by h's probability: the sum of these weighted payoffs over all h is A's expected value, and the best act is the one whose expected valued is highest.[11] Here probabilities are used "neat," without reference to further epistemic assessment. Much scientific practice is Bayesian in spirit; scientists report the probability of data given hypotheses under test, or the level at which the data are significant, without presuming that any probability deserves or amounts to belief. Nothing important in the race/crime issue changes, however, under a shift from binary accept/reject mode to expectations; *comparative* acceptability judgments are invariant under choice of confidence level, or abandonment of confidence levels altogether. No matter what makes a judgment reasonable, it is more reasonable to act as if an Acme might break down than as if a Mercedes might, and more reasonable to be wary of blacks than whites: the expectation of wariness of blacks always exceeds that of wariness of whites. Nor is Bayesianism without its critics.[12] So nothing is altered by assuming a confidence level of .05.

(A useful generalization of expected value is "expected morality." Assume the world is in state W. The moral worth of an act A in W is the difference between the rights protected in W and the rights violated in W by doing A. Now multiply A's moral worth in W by the probability that the world actually is in W. The sum of these products for every world state is A's expected morality, and A is morally preferable to act B when A's expected morality is greater.)[13]

Suppose, then, that jogging alone after dark, you see a young black male ahead of you standing on the running track, not dressed in a jogging outfit and displaying no other information-bearing trait. Knowing nothing else about him, you must set the likelihood of his being a felon at .3. Since felons are not always on the prowl, the "conditional" probability that he will attack if he is felon is less than 1, so your chance of being attacked is under .3. Still, should the conditional probability that he will attack under the circumstances exceed .16, the all-in probability of danger (the product of the probability that he is felon with the conditional probability that he will attack if he is) still exceeds .05. On the other hand, it is rational to trust a white male under identical circumstances, since the probability of his being a felon is below .05, and, since whatever factors affect the probability of a felon attacking you—the isolation, your vulnerability, the chances that you carry money—are presumably invariant, the all-in probability of the white attacking likewise falls further when circumstances are considered. It remains rational to be more wary of the black.

The jogger case is not a contrivance to make a philosophical point. Many real-life situations resemble this one in requiring instant decisions. A woman is alone on an elevator, the door opens, a black male enters. Should she step out? A man alone in a subway car sees four black teenagers get on. Stay or go? Should a store owner keep a sharper eye on a black shopper about whom he knows nothing except his race? Should cab drivers be more reluctant to pick up black males, especially young ones? In these situations, race, sex and age are the only variables to go on. In each case it is rational to be more wary of blacks. In fact, the analogy between race and sex brought out by these cases is very strong. Females are known to be much more law-abiding than males, so you will see three teenage girls walking toward you late at night as less dangerous than three teenage boys. I daresay no one would deny that it is rational and sensible for a female hitchhiker to shun a ride with a male but accept one from another female. Anyone who denies the rationality and prudence of discriminating by race in the running track, elevator, shop, subway, and taxi cases must explain why it is worse than discriminating by sex in the hitchhiker case.

The probability of a black attacking you in any one running-track-like encounter might be less than .05. Still, since this probability is (much) greater than that of a white attacking in a similar situation, the number of such encounters one permits becomes important, since the number needed to raise above .05 the probability of being attacked eventually is smaller for blacks than whites; if blacks are n times more likely to be criminals than whites, the necessary number of encounters is $1/n$ that for whites. This inequality is of practical importance because rational conduct requires *policies*, general rules adopted in advance to cover all cases of a given sort. (While adherence to policies guarantees a few costly errors, the long-run cost of trying to decide each case on its merits is greater.) Even if the chance of attack in a single case is slight, the chance of attack sooner or later becomes ponderable more quickly if one avoids blacks no more frequently than one avoids whites. The extreme undesirability of even one attack thus warrants a *policy* of greater caution toward blacks.

It cannot be repeated too often that, although the odds are 7 out of 10 that the black on the running track is *not* a felon, you do not violate probabilities by acting as if he might be. The odds are 7 out of 10 that the proffered Acme will hold up in traffic, yet you do not violate probabilities by reserving other means of transportation. The presence of a raspy cough that indicates an operable tumor 30% of the time makes it more likely than not that you *don't* have lung cancer, yet in these circumstances you would quickly agree to a biopsy, since you are now much more likely to have cancer than if you did not cough at all. The odds of cancer have grown sufficiently high to initiate countermeasures. Expressions of statistical intuitions, like stereotypes, must be interpreted charitably. People may say "You can't trust Acmes" when they mean that *enough* Acmes (it need not be 50%) break down to bring all under suspicion. When (some) people say "You can't trust blacks," they probably mean, not that most or all blacks are criminals, but that blacks are likely *enough* to be criminals to warrant extra suspicion.

The black on the running track is definitely a felon or definitely not. Ideally, you would base your behavior on knowledge of which, just as, ideally, you would base a decision to buy the Acme in the showroom on determinate information about *its* reliability, and your decision to undergo a biopsy on information about the cause of your particular cough. But in none of these cases do you have this information. So far as you know, the black is a typical member of a class one third of whose members are felons, and the probability to be assigned to a typical member of that class being a felon is 1/3.

9.4. SOME TECHNICAL ISSUES

A number of technical objections might be brought against "running-track" reasoning.

Objection 1: This reasoning assumes indifference, the principle that, absent any consideration to the contrary, the probability of each of n mutually exclusive, jointly exhaustive outcomes is $1/n$.

Reply: The principle of indifference can indeed be abused, but it is legitimate here. The chief fallacy to which it lends itself is asymmetric partitioning of the reference class; the partition of throws of a die into "2" and "not 2," which yields a probability of .5 for a 2, is asymmetric. However, treating the black on the running track as typical does not asymmetrically partition the reference class of black males or humans generally, any more than treating a sight-unseen Acme as typical asymmetrically partitions the class of Acmes or automobiles generally.

Objection 2: The jogger is reasoning in ignorance, and according to John Rawls, the indifference principle should not be used under conditions of ignorance (see chapter 8, n. 42).

Reply: This Rawlsian premise is either untrue or inapplicable here. When reasoning in ignorance about the Acme, when all that is known about the Acme is that it comes from the Acme factory, common sense does treat it as chosen randomly. In any event Rawls' restriction on indifference applies at best to reasoning in *complete* ignorance, whereas we *do* know something about Acmes, or at least cars in general, for instance that they are mass produced. This background knowledge justifies treating any given one as typical. Likewise, we have enough background knowledge about people to justify appeal to indifference in everyday problems involving human behavior.

Objection 3: Mistrusting the black turns "objective" observed frequencie into "subjective" assessments.

Reply: So it does, but frequencies are ordinarily taken to support subjective assessments, whatever conceptual gulf may divide them. The frequency of defectiveness among all Acmes supports a credal assessment of this particular one.

Objection 4: Many contributors to the "epistemic decision" literature agree that an inquirer should seek more information when it is readily available, or when reliance on current information would be immoral. (A father should require more evidence than a jury before believing his son guilty of murder). It is possible to get more information about the black by approaching him, and arguably immoral not to try to do so.

Reply: With regard to the first rule (the second is discussed in 9.5), among the factors that may put information out of reach are cost of acquisition and time. Perhaps all Acmes that run for a week last ten years, but you may reasonably form a snap judgment if you need a car today, or if the salesman won't let you test drive one for a week unless you promise to buy expensive options. In general, extra information to guide a decision is too costly when the expected cost of acquiring the information exceeds the expected cost of deciding wrongly without it. It is silly to pay an investment counselor $1000 for advice on averting a $500 loss on the stock market. Consequently, no one need take steps to acquire data when the purpose to which the data will be put is that of helping to decide whether to take those very steps. No one buys an Acme to determine whether Acmes are worth buying. You want the data, presumably, because action taken without them is unacceptably risky, but action taken to acquire them must be done without their benefit—which is, by hypothesis, unacceptably risky. Yet this is just the situation on the running track. The problem is whether to continue toward the black; the only way to decide this is to find out more about him; the only to find out more about him is by continuing toward him. You know beforehand everything you need to know to make a decision about how to proceed.

Defending the obvious with complicated rejoinders to abstruse complaints, as is done in this and the previous section, invites ridicule. The reader may feel he is being led from New York to New Jersey by way of Alaska. Where race is concerned, however, people seem capable of doubting what they elsewhere find self-evident, so argumentative overkill is difficult to avoid.

9.5. PRIVATE MORALITY

What a private individual may do to avoid possible harm from blacks should be separated from what the state may do to protect citizens from harm, which raises additional problems.

Ordinary standards sanction private precautions against the anonymous black on the jogging track. Flight from perceived danger is always permitted so long as it harms no innocent bystander, and turning around harms no one. Indeed, the perception of danger is not required to be rational. An acrophobe who accidentally wanders onto the observation deck of the Empire State Building has a right to flee so long as he tramples no one on the way down. Be that as it may, the association between race and crime often makes a perception of danger in the presence of blacks perfectly rational. The jogger, the woman in the elevator and the man in the subway do nothing irrational (or injurious) in using race as a reason for flight.

What of the possibility mentioned in the last section, that flight is immoral because it risks offending an innocent black, or, in Armour's (1994) jargon-clotted words, that not waiting for the black "to clarify his intentions destroys what Professor Patricia Williams refers to as 'the fullness of African-Americans' public, participatory selves" (795)?[14] A perfectly adequate reply is that the jogger is not the one imposing the risk, or, should it occur, the harm of offense. Responsibility for a harm is borne by the author of the first wrong in the chain of events leading up to it; the jogger's flight was caused by fear of being victimized, caused in turn by the criminal acts of other blacks. The jogger may be said to initiate the offense since he flees voluntarily, but only in the sense in which (in Aristotle's example) a storm-beset captain voluntarily jettisons his cargo. Ideally, the black on the track will realize that *he* is not being judged, since the jogger knows nothing about him, but is merely a stand-in for a statistical judgment based on the behavior of others. The unwillingness of cab drivers to pick up blacks is one of the staple proofs of persistent racism, but understanding "racism" as *irrational* race-based aversion, this unwillingness is not racist. Cab drivers refuse blacks because they believe, rightly given the evidence, that black fares are more dangerous than others. Ideally, an innocent black unable to hail a cab will resent the individuals whose wrongful acts have caused his indignity, namely the blacks who have robbed so many taxis that cab drivers now fear all black males.

But let us grant that you the jogger are the one responsible for insulting the black you avoid. Your second exemption from blame is that the expected morality of flight exceeds that of continuing ahead. The moral cost of continuing is, at a first approximation, the odds that the black in front of you is a felon, namely .3, times the wrongness of assault, while the moral cost imposed by flight is .7 times the wrongness of insult. (Insulting criminals is not wrong.) If an attack is more than 7/3 times worse than an insult, as common sense surely agrees, fleeing minimizes expected moral cost. As I have already conceded, the expected moral cost of jogging ahead must be discounted by the odds that a felon will not attack, but the cost of flight must also be discounted by the odds that an innocent black will not notice, not care, or interpret your behavior charitably. Such calculations cannot be made entirely precise, but the right to avoid victimization by violence is so much stronger than the right not to be insulted that avoiding the risk of attack justifies inflicting a racial insult.

9.6. STATE ACTION

In asking whether the right to use race in seeking safety extends from the individual to the state, I am not proposing any specific race-based policy. In practice, the best responses to black crime may simply be the best race-neutral responses to crime, period. The question of race-based anticrime policies is of interest because there is no guarantee that the state can always ignore race in this regard and, more important, it raises general issues about race consciousness and equal treatment.

State use of nonracial information-bearing traits in suspect profiles, searches for probable cause, and Internal Revenue Service audits sets a precedent for the use of race for the same purpose and in the same manner. Customs agents may subject people carrying violins to special scrutiny if the suitability of violins for hiding contraband increases the chance that anyone possessing one is a smuggler. A glimpse of radios stacked in the back seat of a car entitles the highway patrol to search the car. The state "discriminates" on the basis of possession of violins and occupancy of cars with stacked radios in the sense that it calibrates its treatment of individuals by these traits. The IRS uses statistical data in deciding whom to audit. Once the ratio of deductions to income passes a certain (secret) threshold, the government "discriminatorily" calls the filer in. Indeed, officials charged with preventing and detecting crime are *obligated* to use relevant information in screening. Customs agents would be remiss in letting violin cases through airports unchecked in the circumstances described. But race bears information. Knowledge of race redistributes probabilities about past and potential crimes. So, absent countervailing considerations, the state seems entitled, indeed obligated, to use it in screening.

Reportedly, the New Jersey Highway Patrol at one time subjected young black males in expensive new cars to drug searches.[15] Given that the presence of a young black male in an expensive car makes involvement with drugs somewhat probable, and more probable than does the presence of a young white male, ordinary standards mandate searches of blacks while leaving optional, or precluding, searches of whites. If the probability of finding drugs in a vehicle varies with the driver's race, state agents may stop vehicles on the basis of driver's race, just as state agents may search violin owners if the probability of finding contraband varies with type of luggage. Gun-control advocates and the police agree that much of the violence in black neighborhoods is due to the number of young black males carrying guns. It therefore seems reasonable to allow the police to use race as a factor in deciding whom to frisk for weapons.

This conclusion seems to contradict a consensus in the literature on statistical proof in law (see, e.g., Tribe 1971). Much of this literature concerns the Blue Cab Co., which owns 60% of the taxicabs in a given city. When an accident involves an otherwise unidentified cab, there is, a priori, a 60% chance that the cab is a Blue. Yet—the consensus goes—use of this statistic as evidence of Blue's culpability is plainly improper.

The principle guiding judgment here, I believe, is that evidence sufficient to convict must not only raise the probability of guilt above some minimum, which statistical evidence might do, but must also be a *specific symptom* of guilt. That is, the evidence must be made more probable by the hypothesis of guilt. A paradigm specific symptom is a causal trace, like a chip of blue paint at the accident scene, which is more likely to be found if a Blue cab was involved than otherwise.[16] Statistical probabilities, on the other hand, are not specific symptoms; the hypothesis that a Blue was involved in the accident does not materially raise the probability that 60% of accidents involve Blues.[17] That seems to be why statistical probabilities do not convict.

However, the issues on which the Blue case turns are irrelevant to those raised by racial screening. Screening is not a determination of guilt or the imposition of punishment, so the evidence that warrants screening need not meet the specific symptom condition. In addition, stipulating that 60% of accidents involve Blues only because 60% of cabs are Blues is to assume that Blues, taken individually, are no more accident-prone than Greens, whereas blacks *are* more likely to commit crimes than whites.[18] Under the properly parallel assumption that Blues drive more recklessly than Greens, while it would be unreasonable to accuse Blues every time there is an accident, it would be reasonable for traffic officers to keep a sharper eye on them. (If the state is allowed to deter Blue heedlessness by imposing higher insurance premiums with expedited payments, expedited appeals procedures and more severe punishment might be warranted for convicted black offenders.)

Think of this test case. A policeman on his beat sees three adolescent black males enter store 1 and three adolescent Chinese males enter store 2. He wants to take preventive action against shoplifting by making his presence known, but has time to walk by just one store—walking by one will leave the other vulnerable. What should he do? It seems absurd to say that he should do nothing or flip a coin. Surely he should go where, using the facts available to him, a crime is more likely to occur, and that is store 1.

J. Adler (1994), a critic of race-conscious crime prevention, concedes that the policeman may walk past store 1 so long as he does not more actively interfere with the blacks in it. But the line Adler seeks to draw is inadequately motivated. Suppose the policeman sees one of the blacks ostentatiously keeping one hand in his pocket, either holding a gun or pretending to, to taunt the store owner. (The policeman has often seen black teens, but seldom boys of other races, taunt storekeepers in this way; in other words, he has found the provocativeness of a hand in a pocket to vary with the hand's color.) Is the policeman then permitted to enter store 1? To approach the boy with his hand in his pocket? May the boy be asked to empty his pockets, but not forced to if he refuses, or may the policeman make physical contact? Once the predictiveness of race is allowed into the policeman's initial deliberation, banning the use of race thereafter is arbitrary.

9.7. THE PURPOSE OF GOVERNMENT

The presumption favoring race consciousness created by suspect profiles can be rebutted by a relevant difference between race and other information-bearing traits, and we will presently examine a number of such proposed differences. Before that, however, a more systematic argument for racial screening is in order.

The argument begins from the premise that government exists primarily to ensure security. Hobbes and Locke, whose justifications of the state are the ones currently most favored, agree on this. They differ in that Locke sees the state as protecting antecedent natural rights, while for Hobbes the state is a device to end an inherently amoral war of all against all,[19] but both insist that the point of government is to keep each man safe from invasion by another. Furthermore, both are contractarians who derive the state from an agreement between individuals to transfer, to a single enforcer, their right (Locke) or "liberty" (Hobbes) to punish wrongdoers or invaders of "natural liberty." Both may grant that there never was a state of nature, and that the Prisoner's Dilemma would prevent rational egoists from ever leaving it; their point, rather, is that the only reason men in a state of nature would want to create civil society, the one benefit of a sovereign unavailable from voluntary association, is security.

One right each man transfers to the state is the use of information to gauge threats. Locke never explicitly mentions this prerogative, but traces of it remain in civil society, where anyone may preempt a clear and present danger to himself, as when in the absence of police a man may disarm a stalker advancing on him with a knife. Now, Locke accords everyone a natural "executive" right to enforce everyone's rights. On this theory, although my right to use information in civil society does not extend to my entering a car in which I see burglar's tools, I did have this right in the state of nature, since I had a right to act against perceived threats to anyone. The state acquires its right to preempt both nonspecific threats (a man carrying burglar's tools) and threats to unknown targets (a burglar concealing his destination) from these antecedent rights of individuals. A simpler derivation might trace the state's general right to preemption to each man's natural right to preempt threats to himself only, all of which have been transferred to the state; having everyone's right to self-defense, the state may move against nonspecific threats and threats to unknown targets because any threat endangers someone.[20] However derived, there is a police right to protect property by preemptively stopping cars with burglar tools.

The right to protection is not conceived as a correlative duty of *the state.* Individuals do not transfer their rights to the state by contracting with it—it is an artifact—but by contracting with each other to form a state. The effect is much the same, however. The state acquires each citizen's right to self-defense, with the understanding, between the citizens, that it will enforce those rights. Individuals thereby acquire a right to the enforcement of their rights by the state. (An automobile dealer's promise that the car he is selling will work gives the purchaser a right to a reliable car, although the promise binds the salesman, not the car.) Moreover, the state needs agents to perform its duties, and these agents—the police—quite literally promise to enforce the rights of other members of society. They are our deputies, to whom we have transferred our right (Locke) or liberty (Hobbes) to defend ourselves against attack. However the agreement to be a peace-keeper is understood, in making it each policeman obliges himself to protect everyone.[21] Agents of the state are thus obliged to provide security, and, since what is obligatory is permissible, the state as embodied in its agents may use otherwise permissible means to achieve this end. Given the relation between race and crime, the state's primary protective function permits and probably demands that it attend to race.

8. STRICT JUDICIAL SCRUTINY

This argument may be pursued by way of the U.S. Supreme Court's distinction between two rationales for government classification of individuals, corresponding to two "levels of scrutiny"—"intermediate" and "strict"—to which a classification may be held. (There is also a less demanding "rational basis" test, but not since *Plessy* has this test been deemed sufficient to warrant racial classifications.) A classification is "benign" when intended "to serve important governmental objectives, substantially related to the achievement of those objectives [and] not intended to burden any individual or group on the basis of race" (*Metro Broadcasting v. FCC* 1990). The Supreme Court upheld "benign" racial classifications in *Metro*, where preferences for blacks competing for broadcast licenses were permitted as serving the state's interest in "diversity."[22] A Lockean or Hobbist acquainted with this holding would insist that, if the state may identify individuals by race to enhance diversity, it may identify individuals by race to control violence. Controlling violence is a more important governmental objective than diversity, the race-crime correlation shows classifying by race to be substantially related to this goal, and this classification is not intended to burden blacks simply for being black.

Yet the Court has also held that race is a "suspect" category, that racial classifications must be scrutinized "strictly," and that such classifications are permitted only when "necessary" for a "compelling state interest." (The court resolves this seeming inconsistency by construing as benign those racial classifications beneficial to blacks, and as suspect those burdensome to blacks; see below.) Insistence on this more stringent standard does not disrupt the argument from Locke and Hobbes. Protecting the innocent against aggression is a compelling state interest; it is why the state exists. Black homicide and robbery rates 10 times those of whites, indicating that aggression is disproportionately black aggression, suggest that race-consciousness of some form is necessary for its adequate control. Greater police readiness to frisk young black males that decreased the incidence of murder, robbery and assault might carry these practices past strict scrutiny.

The Court's main reason for striking down burdensome race-based classifications is that they may reflect "racial antagonism [rather than] pressing public necessity" (*Korematsu v. United States* 1944), a finding that presumes that any such classification expresses only racial antagonism. Behind this presumption is the idea that classifications burdening blacks almost certainly have no rational use. What crime statistics show, precisely, is that special attention to young blacks can be driven by more than antagonism. Stopping crime, which young black males are more likely to commit by an order of magnitude, is a "pressing public necessity" if anything is.

Federal court decisions, it may be said, are a good guide to actual current Caucasoid standards. The Court's resolve to forbid classifications burdensome to blacks, and to permit—sometimes mandate—classifications burdensome to whites, shows that the value underlying this resolve has become a fixed moral point. To rephrase this claim less tendentiously, the Supreme Court (and our morality) finds benign those racial classifications whose purpose is not to burden whites, but to help the disadvantaged or increase diversity. Yet whatever the purpose of a legal classification, racial or otherwise, some individuals are better off because of it, and some are worse off. Under the racial classification permitted in *Metro*, whites are worse off than they would have been had that classification been struck down. At the same time, the Court has not been as ready to permit classifications under which blacks would be worse off than otherwise. Indeed, since *Griggs v. Duke* (1971), the Court has repeatedly applied the "disparate impact" test to classifications burdening blacks, finding practices, however intended, to be discriminatory when they affect blacks adversely. It is this asymmetry I mean when I speak of the impermissibility of just those classifications that burden blacks. The point of contemporary morality allegedly fixed by recent Court decisions is that just those racial classifications burdening blacks should be scrutinized strictly.

Allowing that the courts have embraced asymmetric scrutiny, vast differences remain between this value and such moral fixtures as honesty. Honesty seemed as obligatory to the Greeks as it does to us, whereas strict scrutiny is only a few decades old. More time than this is needed to establish fixity. Indeed, the period of asymmetric strictness may turn out to be a transitional one, during which the Court saw blacks as needing protection from whites and discrimination as the cause of all race problems. To historians a century hence, the asymmetry may seem as time-bound as *Plessy*'s separate-but-equal doctrine seems to us now.

Second, the paradigm fixed points of morality, as a class, give hope of subsumption under a single general principle, such as maximization of happiness or the golden rule. One-sided strict scrutiny does not seem to instance any general rule. The same point differently expressed is that strict scrutiny is complex whereas the familiar fixed points of morality are simple. There seems no clear, principled way to distinguish classifications burdening blacks from those burdening whites. Of course, any reasonable rule recognizes exceptions and gray areas; the injunction to honesty permits little white lies. However, hard as it may be to draw precisely, the line between forbidden and permitted lies is itself principled, whereas classifications burdening whites differ from those burdening blacks only in the race of the burdened.

At this point I will be reminded of all the reasons I myself cited in chapter 8 for distinguishing discrimination against blacks from discrimination against whites. Ironically, these reasons generate the deepest difference between asymmetric scrutiny and paradigm fixed points. Unlike the rule of honesty, one-sided color-consciousness rests on the historically specific factual assumptions that blacks were seriously mistreated in the past, that classifications burdening blacks reflect harmful "stereotypes," and that these classifications can be motivated only by racial hostility. The factual assumptions behind the rule of honesty, by contrast, are universal: that lying undermines trust, that lying pays only when it is the exception, that lying is parasitic on truthfulness because the meaning of a word is determined by its standard use. I am not (at this juncture) challenging either the empirical assumptions behind one-sided scrutiny or their ability to sustain it; I am pointing out that these assumptions carry *too much* of the weight for one-sided scrutiny to be a fundamental value. The most striking proof of the importance of these assumptions is that when they are relaxed—as when it is observed that whites were never mistreated by a dominant group—color-consciousness is allowed. Unlike the value of honesty, the value of one-sided scrutiny is not inherent in the human condition.

But let us allow that strict scrutiny for classifications burdening blacks has become integral to morality and law. We recur to the point that racial screening and other forms of race consciousness might yet be permitted, for they might be found to survive strict scrutiny. At this writing, they almost certainly would not. In the few cases where screening has been tested before the courts, such as the one involving the New Jersey Highway Patrol, it has been struck down. Yet times change, and someday the Supreme Court could, while continuing to scrutinize racial categories strictly, find control of black violence a compelling state interest and some racial classifications necessary to serve it. Were that to happen, strict scrutiny and race consciousness would coexist.23

Private and, especially, state race consciousness for crime control draw many other objections, most directed against the analogy between racial screening and ordinary suspect profiles. I turn to these objections with the reminder that they should be viewed from two perspectives: as bearing on the specific issue of crime and as raising general issues about race consciousness.

9.9. "RACISM" AGAIN

The analysis of "racism" in chapter 5 sheds light on whether stopping blacks but not identically situated whites is "racist." Since so describing a practice implies that it is unjustified, justified screening isn't racist, so (as usual) to criticize racial screening on grounds of racism begs the question. Screening must be found wrong for some independent reason before the epithet can apply. Alternatively, should one decide to use "racism" of all forms of race consciousness good or bad, racial screening is certainly "racist," but "racism" in this sense has ceased to be a defect, so again cannot by itself support censure. Either way, asking whether screening is "racist" does not help in determining its propriety.

9.10. "RIGHTS PRECEDE UTILITY"

It may be objected that rights against racial screening and frisking override their potential benefits.

This objection rather boldly begs the question of whether there *are* rights against racial screening, but that is not its main defect. Its main defect is that the argument for both ordinary and racial screening is not utilitarian at all, but rights-maximizing. When push comes to shove, according to section 7.4, common sense tolerates local invasions of rights to minimize rights invasions overall. Rights-maximizers are prepared to admit that detention without specific evidence of wrongdoing violates the detainee's rights, but insist that it is acceptable when on net it strengthens everyone's rights. Detaining Typhoid Mary violates her rights, but keeps her from negligently endangering others. Racial screening pits (what we may temporarily grant is) a right against race-based screening against the right to be secure; nothing is pitted against utility.

Seeing this, the screening critic might decide to refight the battle between absolute and goal-directed rules, arguing that certain rights, immunity to racial screening among them, are never to be compromised. This argument too begs the question of rights against race-based detainment, but—once again—there is no need to press that point. It is more instructive to stress the depths of our "maximizing" intuitions about the role of the state, to which end a glance at Robert Nozick's seemingly absolutist justification of the state (1975: 78–84, 110–113) is illuminating.

Initially accepting a categorical duty not to aggress, Nozick insists that the state is legitimate only if it can emerge from anarchy in a way that does not involve coercion. He easily shows how a dominant protective association might form by free association, but he cannot get it to become an enforcement monopoly, the mark of the state, without letting it impose its rules on "independents." To warrant this imposition, Nozick cites the anxiety created for association members by independents living among them who conduct their affairs by their own rules. He deems this anxiety so acute that the the association may impose its rules on erstwhile independents, so long as it compensates for the imposition by protecting them as well. Nozick's derivation thus requires that there be levels of mere threat serious enough to warrant a bit of preemptive aggression by the proto-state—essentially the rationale of screening. What is more, the proviso about compensatory protection corresponds to the commonsense demand that the state protect all other rights of screening detainees.

While hardly an inadvertent proof of anything, Nozick's appeal to maximizing shows how deeply maximizing intuitions inform our ideas of permitted state action.

9.11. "SLIPPERY SLOPE"

Racial criteria are said to grease a slope that bottoms out amidst concentration camps.

This criticism can hardly be decisive, for slippery slope problems dog any legitimate policy susceptible of abuse, which in practice means almost any policy whatever. "Reasonable" search and seizure can be interpreted overbroadly, yet, despite this danger, wiretaps are tolerated. Three possible tests have already been suggested to keep race consciousness within acceptable bounds: one might require race-conscious measures to be substantially related to an important government objective, or necessary for a compelling state interest, or, moving beyond Constitutional interpretation, to maximize expected morality. Each test can be misapplied in its turn, but so can any rule for interpreting a rule.

The invocation of Hitler is particularly obnoxious in connection with crime. It is a fact, not a speculative possibility, that the murder rate in American cities—and this largely means the black murder rate—has increased tenfold in the last half-century. These are real, unhypothetical deaths, and it seems perverse to reject steps that stand a real chance of reducing them because of a possibility, in context, of unimaginable remoteness. When was a death camp last erected on American soil?

9.12. RACE IS CAUSALLY IRRELEVANT TO CRIME

This objection develops the "proof by statistics" issue discussed earlier. Unlike skin color, familiar probable causes such as concealed rifles and stacks of car radios are causally related to the commission of crime. A stack of car radios carries information because it results from a crime. It is a specific symptom: few crimes produce stacks of radios in the back seats of cars, but most such stacks are produced by crimes. The presence of a gun carries information because it is either a cause of crime, or, along with crime itself, an effect of criminal intent. But race is not a cause or an effect of crime, or—let us grant for now—a coeffect along with crime of some underlying cause. The felt "unfairness" of race-consciousness is due in part to this causal irrelevance.

Reply: the assumed causal independence of race and crime still leaves race its predictive value, and it is predictive value that warrants use of a trait in screening. Traits are listed in suspect profiles because they carry information; why they carry information is a separate question. Even though a rifle is informative because of its causal role as an instrument in crime, it is the information the rifle carries, not the reason it does so, that justifies a search when a rifle is discovered.

Probable causes strike us as more trustworthy than unexplained correlations because, absent knowledge of a causal mechanism, we suspect that any correlation between two traits may be coincidental. But there are grounds other than a known mechanism for trusting a correlation. One is sheer persistence—indeed the very persistence of a correlation is evidence of an underlying mechanism at work. So long as we are confident of the probabilities generated by a crime-relevant trait, those probabilities, not the reasons for confidence in them, justify attending to the trait. Suppose fedoras attain popularity among drug couriers but hardly anyone else because most couriers hail from the one city in the world where fedoras are back in style. Neither a cause nor an effect of drug trafficking, nor a co-effect along with drug trafficking of some underlying cause, fedoras at airports become a reliable sign of drugs. Although it is an accident that fedoras are correlated with drugs, basing drug searches on headgear would surely be justified. These searches would of course be irrational if the correlation were liable to break down at any moment, and it is hard to imagine smugglers not noticing sooner or later that their headgear was giving them away. Nonetheless, so long as the couriers remained oblivious, reliance on the correlation would be reasonable.

Having held for many decades and across nations, the correlation between race and crime appears secure enough to support reliance. (The causal relation of race to crime is examined in below and in chapter 10.)

9.13. "RACE IS INVOLUNTARY"

People should not be penalized for what they cannot help, it is said, and blacks can't help their race. This is another component of the feeling that screening is unfair.

Both premises, though true, are irrelevant, for screening and allied forms of race-consciousness are not punishments. Were an at-large murder suspect known to bear an ineradicable birthmark, it would be proper to watch for men with such birthmarks, even though birthmarks cannot be helped. Caught and convicted, the suspect will be punished for murder, not for having a birthmark that betrayed him. If race must be ignored because it is unchosen, it is also impermissible to identify suspects by race, a practice now permitted, by eye color, by height, or—should there be a genetic disposition to obesity—by weight, consequences I assume are absurd.

The involuntary trait of age (usually 16) is used as a criterion for licensing drivers, in the interest of protecting the public from recklessness. Denial of the right to drive is a state-imposed burden, and an age floor is unjust to mature 15-year-olds. The reason an age standard is accepted is obvious: testing everyone for maturity is impossible, and age is a reasonable proxy. Of course, the denial of justice to a mature 15-year-old is temporary, since he will eventually become 16, whereas blacks do not become white. Still, the justice denied mature 15-year-olds is real, and permanent for mature 15-year-olds who die before turning 16.

Perpetrators of actual crimes differ from merely potential criminals in point of culpability. But identification, as it is not punishment, does not require culpability. Still, while it is one thing to keep an eye on a man with a birthmark when an actual crime is known to have been committed by some man with a birthmark, it seems quite another to keep an eye on a man who merely might do something that hasn't happened. A thought experiment helps clarify matters: Assuming such a thing were discovered, would it be permissible to use a genetic predictor of homicidal aggression to track potential criminals from birth? Initially this question cuts against racial screening, since many readers will say "no" right off. Yet it is far from clear that reliable tracking would be widely rejected were it concretely available. A decision is made by officials of the justice system to ignore the marker because "nobody chooses his genes," and every year postconviction testing shows that 12,000 murderers, committing half the murders in the United States, had the crime gene. Does the reader think it likely he would continue to object to tracking after he saw the corpses of victims who might have been alive had potential murderers been tracked?[24] Suppose, not that half of all murderers turn out to carry the gene, but that half the carriers eventually commit murder. Suppose 90% of carriers do. Would the reader still object to tracking because "nobody chooses his genes?" I suspect that at some point the average person would endorse severe restrictions on carriers. If so, intuitions about genetic tracking support racial screening.

9.14. ROOT CAUSES I: STEREOTYPES, RACISM, SELF-ESTEEM

On the popular view that black crime is a direct or indirect effect of white misdeeds, racial screening becomes a classic case of victim-blaming, unfairly persecuting a group for its reaction to persecution. This view is also thought to explain why nonstigmatizing traits like birthmarks and fedoras differ from race: the association of these traits with crime was not caused by wrongs. Racial screening is worthy of the Potter of the *Rubaiyat* "who threatens he will toss to Hell the luckless Pots he marr'd in making." Having made criminals of blacks, this popular view concludes, white society should address the root causes of black crime instead of picking on its victims, black criminals.

The first difficulty here is that addressing root causes, however important that may be, does not prohibit more immediate race-conscious steps. The root cause of arson in a city may be a depressed real estate market, but the fire marshal may stay alert anyway for suspicious cans of gasoline while economists work to buoy property values. If poor education contributes to black crime, schools can be improved at the same time that victims of past miseducation are stopped from harming (other) innocents.

Second is the selectivity of appeal to root causes. One seldom hears of measures against crime associated with other groups—the Italian Mafia, Chinese gangs, white-collar fraud—being judged by their attention to causes. Understanding black crime as an effect of social forces invites a parallel understanding of, say, insider trading as an effect of a materialistic culture that leaves Porsche-less stockbrokers feeling deprived. Lynching had its causes, as does everything in nature, yet an FBI investigation of a lynch mob is not called blaming the victim. Why isn't violent dislike of blacks excused, on grounds of anger caused by black crime? Why isn't attention paid to the root causes of the Holocaust? If some criminals are victims of the past, why aren't all?

Special pleading about the causes of black crime suggests the belief that its causes partially justify or excuse it. And many people do think something(s) very like this: that, since black crime is the effect of past injustice, either the two wrongs cancel out, or else that past injuries done to blacks have left them *unable to refrain* from crime. The latter idea is particularly powerful; all the complaining in the world about double standards does not still the sense that black muggers differ in degree of culpability from white embezzlers—an intuition taken up in section 9.17.

These crosscurrents complicate the root-cause issue. I eventually conclude, in section 9.18, that the causes of crime of any sort are irrelevant except as they suggest means to reduce it. People have a right not to be victimized, and a right to expect the state to prevent victimization. They do not care why victimizers act as they do, and they expect the authorities not to care. At the same time, the belief is so widespread that black crime is a response to white misdeeds, *and* that this fact limits what may be done about it, that it must be considered in their own right.

There are four basic root-causes hypotheses.

Blacks Commit Crime Because Whites Expect Them To

This theory, and the replies to it, echo the discussion of chapter 3. Whites admittedly do believe blacks are relatively crime-prone; the issue is why. If the causal arrow runs from expectation to crime, the origin of the expectation itself becomes obscure (although the expectation, once established, may reinforce black criminal behavior). On the other hand, the expectation is explained most parsimoniously as an effect of observation of black crime. The self-fulfilling prophecy theory of black crime has the defects of the self-fulfilling prophecy theory of stereotypes.

But assuming black crime is a self-fulfilling prophecy would not undercut race-conscious measures to prevent it. These measures seem wrong, on the self-fulfilling prophecy hypothesis, because no one should be penalized for a wrong created by the penalty itself. When a crime is created by the punishment, it is the author of the punishment, not the nominal criminal, who is responsible. What if the beat cop's suspicions of a group of black teenagers are well-founded only because these suspicions, by "labeling" the teenagers, dispose them to crime? The cop can insure that the blacks behave properly simply by ceasing to expect them to misbehave. Preventive screening is a generalized form of entrapment.

This seemingly solid argument collapses when two very different types of self-fulfillment are distinguished. A particular expectation may be called *specifically* self-fulfilling when that very expectation causes its own fulfillment, whereas a particular expectation is merely *generically* self-fulfilling when it is fulfilled because of other expectations like it. A policeman's expectation that black teenagers on the corner will harass passers-by is specifically self-fulfilling if the teenagers harass passers-by in response to that very expectation, generically self-fulfilling if those blacks harass passers-by because the suspicions of other policemen have alienated them from society. Now, the rule against entrapment only forbids acting on specifically self-fulfilled prophecies; if the blacks' potential for criminality was caused by the expectations of other policemen, not the expectations of the policeman now dispersing them, he is not responsible for the criminality he anticipates—others may be, but he is not—so in dispersing them he does not burden them for an inclination he himself has created. Perhaps those earlier suspicions should not have been harbored, but they were harbored and have had their effects. The policeman on the scene is obliged to prevent those effects from erupting into crime.

Armour's (1994) treatment of these issues deserves mention in this connection. He considers a white woman (an "intelligent Bayesian," in the phrase Armour borrows from Walter Williams) who shoots a black she fears is about to mug her—an extreme case involving private aggression, unlike my paradigms—and asks whether a legal defense can be based on the reasonableness of this fear. Struggling mightily with the statistics,[25] Armour seems to allow, equivocally, reluctantly, and merely for the sake of argument, that fear of blacks is rational. Yet he concludes that the 14th Amendment (see below) prohibits this defense lest race-based statistics tap into jurors' prejudices, "subvert the rationality of the fact-finding process" (796, also 798) and "further entrench stereotypes about blacks as criminals" (794). Armour backs up this judgment by citing the literature on the irrationality of stereotypes. So in the end he fails to take the crime statistics seriously, a failure that prevents him from properly engaging the issue of state action.

Racism

The claim that white racism has contributed to black crime, like the claim that it has handicapped blacks competitively, must be distinguished from the claim that racism is the sole cause. The serious issue is proportion of variance explained. Measures ruled out because black crime is an effect of victimization by racism are ruled out only to the extent that black crime is an effect of racism. Finding that white misdeeds explain only 10%, say, of the race difference in crime rates might leave race-conscious measures legitimate. These measures would not burden victims of racism, as black criminals would then be victims of racism only to a negligible extent.

As I hope the reader is by now convinced, this explanation is highly implausible. Black crime rates in other countries, including all-black countries, are comparable to those in the United States (Rushton 1988a, 1988b, 1994, 1995; Herrnstein and Wilson 1985). It is possible although ad hoc to hold that in the United States alone racism is a sufficient condition of black crime, or a necessary part of a more complex sufficient condition, but then black crime should decrease as racism abates. Yet black crime has increased in recent decades as discrimination has decreased, and increased most sharply after the passage of the Civil Rights Act, affirmative action, and busing ordered by the judiciary to improve the education of black children. The incarceration rate for blacks is now higher than in 1904 (*Encyclopaedia Britannica* 1914). Most black criminals have been blacks raised in a society in which they enjoy preferential treatment. So it is doubtful that racism explains more than a tiny portion of black crime.

Poverty and Unemployment

Crime is often explained by poverty, and black crime by the greater poverty of blacks. In this century, however, changes in the crime rate have varied almost inversely with poverty. Murder and robbery tend to increase with the employment rate (rubinstein 1992). Prison admissions rose during the prosperous 1920s, fell steadily between the early 1930s and the end of World War II, and have risen sharply since the late 1960s, even though the unemployment rate has held steady or dropped slightly. Unemployment fell from 6.6% in 1961 to 3.4% in 1969, but the crime rate doubled, and the rate for robbery tripled. (At this writing it seems to have fallen slightly during the early 1990s.)

Many studies have measured the relation of crime to economic factors in more sophisticated ways; some do find a positive correlation, but none that explains as much as half the variation in the crime rate. One literature review concludes:

The bulk of the studies examined here show some connection between unemployment (and other labor market variables) and crime, but they fail to show a well-defined, clearly quantifiable linkage. . . . In studies that include measures of criminal sanctions and labor market factors, sanctions tend to have a greater impact on criminal behavior than do market factors. (Freeman 1983: 106)

Another concludes: "Unemployment may affect the crime rate, but even if it does, its general effect is too slight to be measured. Therefore, the proper inference is that the effect of unemployment on crime rates is minimal at best" (Orsagh 1980: 183). A standard textbook judges the findings "contradictory"

(Vold and Bernard 1986: 134).

The strongest general statement that can be made is that, at any given time, chance of arrest varies inversely with income, which is consistent with low income and high criminality being joint effects of underlying causes. The factors causing high individual time preference may manifest themselves as a desire to steal—Banfield's "persons who want small amounts *now*"—and as aversion to labor.

The weak relation between poverty and crime has held specifically for blacks. In 1904, when the black incarceration rate was lower than at present, the mean per capita wealth of blacks was about 2.5% that of whites, whereas it is now about 20% that of whites (Jaynes and Williams 1989: 292). Between 1930 and 1990 the proportion of the black population counted as living in poverty fell steadily from over 90% to about 33% (Jaynes and Williams 1989: 278, Figure 6-1), as did the ratio of blacks to whites living in poverty, while the proportion of black prison inmates rose steadily from just under 25% to just over 50% (Hacker 1992: 197; rubinstein 1994). I have been unable to locate data on race differences in criminal behavior when income is controlled for, but as the ratio of white to black mean hourly wages is 1.33 (Jaynes and Williams 1989: 295, Table 6-5) while the black crime rate is roughly ten times the white, crime is almost certainly more prevalent among blacks at most or all income levels. Moreover, black crime, like white crime, is committed primarily by young, unmarried men. According to a *Los Angeles Times* survey, looters during the 1992 Los Angeles race riot primarily targeted electronic gadgets and liquor; only 9% of the stores ransacked carried food (Whitman and Bowermaster 1993: 58). The idea of starving blacks stealing necessities does not correspond to reality.

The experience of turn-of-the-century Jewish immigrants fleeing European anti-Semitism partially confirms the poverty/oppression/crime nexus claimed for blacks. Jews lived in great poverty on the Lower East Side of Manhattan and displayed unusual levels of crime in the sense that Jewish crime of that period exceeded Jewish norms elsewhere. It was not, however, disproportional relative to the overall American population (Joselit 1983: 32), as black crime is. Moreover Jewish crime, unlike black crime, was directed primarily against property rather than persons (Joselit 1983: 33). Finally, the relative prevalence of Jewish crime subsided by World War II (Joselit 1983: 158–159), the period 1900–1939 being less than that between *Brown* and the present.

Low Self-Esteem

As discrimination and material deprivation have lost their plausibility as explanations of black crime, psychic deprivation has replaced them. On this theory, young black males steal and sometimes kill for flashy goods because they must do something to stand out in a society that disdains them.

This theory is undermined by the almost unanimous finding that black self-esteem is higher than that of whites (see chapter 3). In fact, Baumeister, Smart, and Boden (1996) present a powerful case for associating violent aggression with *high* self-esteem,[26] and address black crime in this connection. They note that the large preponderance of interracial crime (they estimate 80–90%) is now black on white, a reversal of the pattern at the beginning of the century which they tentatively attribute to the growth in black self-esteem relative to white. They describe juvenile delinquents, primarily black, at length:

Delinquent boys [are] more likely than control boys to be characterized as self-assertive, socially assertive, defiant and narcissistic . . . the thoughts and actions of juvenile delinquents [suggest] that they held quite favorable opinions of themselves. . . . [Jankowski] found gang members were violent toward people "whom they perceived to show a lack of respect or to challenge their honor." . . . [T]he code of the street centers about "respect," which gangs regard as an external quality involving being "granted the deference one deserves." . . . [G]ang members believe they deserve to be treated as superior beings, [respect enhanced by] "nerve," which is essentially a matter of acting as if one is above the rules that apply to others and as if one disregards the rights of others. . . . Katz noted that many youthful circles and street subcultures afford respect mainly to the "badass" sort of person who transcends the pressures to conform to societal norms, rationality, and ideals. This prized identity is cultivated in part by creating the impression of being unpredictably prone to chaos and irrational violence. . . . Researchers from different disciplines concur in depicting these young men as egotistical in several ways, and they concur emphatically on the apparent preoccupation with respect and self-assertion. (21–22)

It is hard to reconcile all the dimensions of black crime with low self-esteem. Internalized disdain for things black might dispose blacks to attack other blacks, but it would not explain black-on-white crime. Low self-esteem translated into resentment might make blacks attack whites, but would leave unexplained the high rate of black-on-black crime. Explanations of black-on-black crime in particular—that blacks are "attacking the image they have been taught to hate" or "adopting the ways of the oppressor"—tend to be particularly soft and metaphorical.

In addition, the low-self-esteem theory treats self-esteem as externally imposed, when, to the extent that feelings of worth are under environmental control, they would seem to reflect an individual's assessment of his own abilities, and, to perhaps a lesser degree, those of the group(s) with which he identifies. These assessments, moreover, are comparative; being good at something is being better at it than most members of a reference class. A high school student who is good at science is better at science than most high school students; he may not be good at science compared to the average Nobel Laureate in physics. Were black self-esteem a function of acceptance of blacks by whites, blacks in integrated schools should think better of themselves than blacks in segregated schools, whereas the reverse is the case (although even in integrated schools blacks are more confident than whites; see section 3.10). The comparative-assessment theory explains this: most blacks in integrated schools find themselves performing below average at tasks everyone is asked to master. Contact with whites highlights black/white ability discrepancies less visible in segregated settings.[27]

Perhaps the most compelling commonsense reason to doubt the self-esteem theory is that the criminal behavior of young black males just does not look like an expression of despair. In account after account, these individuals come across as full of themselves and unrepentant. Consider this description by a group of black teenagers of the escalation of violence in their subculture (Lee, Schipp, and Tabor 1992: 28; italicized words the commentators').

Everyone wants some kind of "rep" or reputation—and one way to earn one for being tough is to carry a gun, the teenagers said. No one wants to be a "herb" or wimp.

SEAN: If a guy steps on your foot, it's one of your peers says, "Oh, you gone let that happen to you? He stepped on your foot." You know, you have the right mind that's like "I'm gone ignore that." But your friend says, "You ain't gone do nothing about that?" That puts you to the limit. You be like "Ahh, what I'm gone do?" It gets you thinking and you saying something: "Yo, what the hell are you doing?" or something like that. You start, you know, to break.

NATHANIEL EDWARDS: Say I talking in the hallway and somebody bumped me and they meant to bump me. And now, me, I'm the herb. I will turn and I will walk away from them. Next day, he gone bump me again. Now it's a Friday, right? And he comes and bump me and he says, "Yo, what's up man? If you keep walking away, we'll keep bothering you." . . . You can fight right there, but a gun gone come in sooner or later.

Disputes can start over what the teenagers acknowledge to be "the stupidest things." Kareem Smith said he almost fought a friend who tried to grab his cheeseburger. Others have fought when someone stepped on their sneakers or looked at them the wrong way. Rumors—"he said, she said," as Kenaisha Warren put it—also lead to "beef."

KAREEM: You get beef on a basketball court over a call or something. Be ready to fight. . . . People got shot over basketball right there at Gershwin (High School). Got shot dead. Two times in the head. It wasn't in school but it was after they had played basketball.

One way to assure status and protection is to join a "posse." Larry said he was a member of V. I. P. (Vanderveer [Housing Project] International Posse) until two years ago and, among others things, they stole money from kids who earned tips bagging groceries.

Threats to establish within-group dominance and alliances for exploiting the weak are not efforts to attract the attention of an indifferent world. They resemble instead what Hobbes called quarreling over "glory." "Every man looketh that his companion should value him, at the same rate he sets upon himself," said Hobbes, making men apt to "use violence for trifles: as a word, a smile, a different opinion, and any other sign of undervalue." The social relations of young black males strikingly bring to life Hobbes' "war of all against all."

9.15. ROOT CAUSES II: INDIVIDUAL FACTORS

Racism, poverty, low self-esteem, and the circular "culture of violence" offer less plausible explanations of black crime than do individual crime-elevant traits more prevalent among blacks.

The principal such trait is intelligence. Jensen writes: "*Across various racial and social-class groups, the prevalence of delinquency is approximately the same at any given IQ level*" (1980: 359). Herrnstein and Wilson (1985: Chapter 6) estimate that the mean IQ of the prison population is in the low 90s, with violent offenders lower. Gordon (1975a, 1975b) found that in a cohort of black and white 18-year-olds in Philadelphia, 51% of black youths and 17% of white youths had appeared at least once before the juvenile courts. This difference is explained almost completely by the assumption that both black and white criminals come entirely from the IQ range below 85, a section of the bell curve that includes 16% of the white population and 50% of the black. The assumed distribution is somewhat artificial, as Gordon himself observes. Taking the mean white IQ to be 101.8 with an SD of 16.4, and the mean black IQ to be 83.4 with an SD 12.9—Gordon's figures—a more realistic model assumes that all black criminals are recruited from the IQ range 75.7 to 91.1, with a mean of 83.4, and all white criminals from the range 72.5 to 87.7,[28] nearly homologous bands containing 50% of the black population and 17% of the white. In this sense as well, most of the difference in rates of black and white juvenile crime disappear when IQ is held constant.

While 13% of the blacks and 2% of the whites had been incarcerated in the 12,600-person sample studied in Herrnstein and Murray (1994), only 5% of the blacks had been incarcerated when IQ was held constant at the white mean of 100. This indicates both that IQ explains a great deal of the between-race variation in criminality, and that other factors are involved.

The correlation between crime and IQ does not by itself establish causation, but it is natural to direct the causal arrow from IQ to crime. As Gordon remarks, low IQ explains "the impulsive irrational quality of many offenses" (1975b: 278). Criminals have difficulty conceptualizing remote consequences for themselves and the way their actions look to their victims, although this latter factor shades over into the nonintellectual factor of empathy. Gordon observes that the race difference in crime rates, like that in IQ, decreases but does not vanish when social status is controlled for, and suggests regression to the mean as an explanation.

Mention of empathy and impulsivity leads to the role of temperament in black crime. Herrnstein and Wilson note the higher black scores on the MMPI scales (1985: 469), typical of the U.S. prison population (1985: 189). "[P]risoners [in studies cited] deviated from the population at large in the typical way—in those traits associated with high values on the Psychopathy, Schizophrenia and Hypomania scales of the MMPI, namely, deficient attachment to others and to social norms, bizarre thinking and alienation, and unproductive hyperactivity" (189). The reader will recall from chapter 3 that black scores exceed white on just these scales, particularly Hypomania. Black males are also more mesomorphic than whites (1985: 89), and criminals tend to be more mesomorphic and less ectomorphic than the general population. In general, mesomorphs low in ectomorphy tend to "unrestrained, impulsive self-gratification" (Herrnstein and Wilson 1985: 469). The reader will also recall Hacker's plea that black children be allowed to move around in the classroom because they are "more attuned to their bodies." There is no need to repeat here the evidence that race differences in IQ and temperament, variables which significantly affect criminal behavior, are significantly genetic in origin, and that, in light of the Minnesota Transracial Adoption Study, race differences in these variables seem to be significantly genetic as well.

Genetic factors need mediation, and one mediating mechanism may be the race difference (mentioned in chapter 4) in serum testosterone, known to facilitate aggression. Much current speculation has also focused on the neurotransmitter serotonin, whose presence seems inversely correlated to impulsivity and aggressiveness. Investigating possible race differences in levels of serotonin might shed a good deal of light on black crime.

Because the temperamental differences relevant to crime go farther than rates of discounting future reprisals, a purely economic analysis of race differences in crime will probably prove incomplete. Most people who obey the law do so not from fear of punishment or abstract respect for legality, but from habit. Since most laws proscribe aggression and deception in its various forms, this habit is rooted in empathy, the emotional pole of kantianism—a tendency to see other people as centers of experience like oneself, rather than as obstacles or resources. A common demand of blacks is that they be treated with "respect," usually understood to include deference and fear. Yet the high rate of violent black crime—a show of contempt for one's victim if anything is—indicates an unwillingness on the part of black males to treat others as they wish to be treated. The black advantage in criminality seems best attributed to lower kantianism, facilitated by lower intelligence and greater impulsivity themselves probably biological in origin.

Thomas (1990) argues that the bloodier wars waged by whites show that, far from being less criminal than blacks, whites are "worse." Judgments of value aside, it is pertinent to ask how can blacks be more innately prone to violence when, collectively, Europeans and Asians have been more destructive. The answer is probably the superior technology of Eurasians made possible by their greater intelligence, and a superior capacity to organize into large armies, also an effect of greater intelligence and ability to subordinate the self. Although tribal conflict is endemic to Africa, Africans have been unable to invent weapons of mass destruction or assemble themselves into million-man forces. Greater black criminality requires only that *individual* blacks be on average more aggressive than *individual* whites. The very individual aggressiveness that yields a higher black crime rate may well obstruct the formation of effective black armies. As Africa has disintegrated politically at the end of the twentieth century, civil wars are not being fought by disciplined units but by small bands with few lines of command and few goals beyond spoils. The result is not the fragmentation of large political units into smaller ones, but anarchy.

Natural experiments indicate that blacks do kill with unsurpassed ferocity when Western weapons are available. King Mutesa of Buganda, on being given a rifle by a visiting explorer, immediately used it to shoot a woman (Baker 1974: 389). The accessibility of handguns in the United States has produced a homicide rate among young black males an order of magnitude greater than that among whites. In 1994, after the publication of Thomas (1990), Rwandan Hutus, using small arms and machetes, killed a million Tutsi in a few weeks, an episode matching in ferocity the great slaughters of history.

In the course of discussing race as a predictor of crime I allowed that race and crime are causally unrelated, and distinguished predictive validity per se from explanations of validity. In retrospect, this concession was unnecessary; race/crime correlations meet the more stringent causality requirement, since race, or, strictly speaking, the observable indicators of race, share a common cause with the criminal behavior they predict. The indicators of African ancestry and mean black levels of intelligence and aggressivity appear to have been joint adaptations to an African environment, so, while indicators like skin color do not cause the behavior with which they associate, the association is not accidental. If a predictor may be used only when causally related to its criterion, race remains an admissible predictor of crime.

9.16. FREE WILL

The conclusion that genetic factors contribute heavily to black crime raises some old problems in a sharp new form. Nobody is to blame for his genes, so the race difference in crime rates is no more the fault of blacks than it is of whites. Yet this very concession seems to deny black criminals and blacks in general the responsibility for their behavior necessary for human dignity.[29] The issue of free will, hovering in the background, now steps to the fore.

I approach this topic by first taking up the general question of free will in the present section, then applying the results to black criminality in section 9.17. Many of the ideas in this section will be familiar to philosophers (and, in my experience, new to nonphilosophers), but the eventual conclusions are less expected.

Like many philosophers[30] I am a "compatibilist." I hold that the causal determination of behavior, indeed universal determinism, is compatible with freedom of the will. To understand this doctrine, reflect that, as "free" is ordinarily used, a man is free when he does what he wants to.[31] A "free" afternoon is one which you can spend as you please. You enjoy freedom of speech when you can say what you want without fear of reprisal. You eat lunch freely when you eat because you feel like eating, rather than being force-fed, ordered to eat at gunpoint, or, as in prison, during a prescribed lunch hour. One's own wants are not coercive; it is silly to think of your desire to visit a museum as forcing you to purchase admission. Now, to say an action is caused by a desire implies nothing whatever about the origin of the desire itself. The desire causing your museum visit might have had causes of its own such as genes and environment, which had causes of their own, and so back to the Big Bang. Inasmuch as "free" actions are simply actions caused by desire, the notion of freedom is likewise noncommittal about the possible causes of free action. An action can thus be both free and caused by factors outside the agent's control. It is free as long as it is caused by desires, wherever those desires came from. That is why free will is consistent with determinism.

Common sense is compatibilist. A snacker freely chooses to munch pizza even though, as everyone knows, his hunger resulted from involuntary changes in body chemistry, which resulted from usually involuntary food deprivation. You are free to eat lunch when you can eat when you want to, even though you do not control when you want to eat. Our snacker is freer than prisoners in mess hall even though his appetite has physiological determinants. Universal determinism does nothing to erase the difference between the two sides of a prison wall.

Belief in human freedom amounts to the conviction that to some extent we control our lives, and are not mere playthings of external forces. Compatibilism endorses this belief, since the everyday contrast between what is within and what is beyond one's control matches that between what happens because one wants it to happen and what happens despite one's wants. Under normal circumstances a driver controls his car because it goes where he wants it to; a car in a skid is out of control because its path is uninfluenced by his efforts. A driver would misleadingly alarm his passengers by telling them the car they are in is out of control as a way of saying that his decision to maintain a steady 60 MPH is an effect of his genes and experience. Indeed, uncaused events, which many people imagine is necessary for freedom, would lie as far outside anyone's control as can be imagined.[32] The last thing anyone would call free are actions based on random choices, unconnected to preferences and unforeseeable by the agent himself.[33]

Free will is of compelling interest because it is needed for moral responsibility—no one can be blamed for what he could not help—and compatibilism deftly makes sense of ordinary ascriptions of responsibility. People are ordinarily blamed for and credited with actions they choose to do, not for what happens to them independently of their choices. No one is blamed for his eye color because no one chooses it. You can desire green eyes all you want and it will make no difference. The mugger on the other hand is caused to rob you by his desire to rob you, which is why he is not excused despite his larcenous desire having been caused by his genes and upbringing. His desire did not *force* him to rob in the way in which a muscle spasm might force him to stop swimming, or his genes forced him to have green eyes; an individual is *forced* to do something when he does it *whether he wants to or not*[34] whereas the robber did what he did precisely because he wanted to. People are not punished for what they do not choose for a closely related reason. The point of punishing an action is to reduce its frequency by inhibiting the desire to do it, which is accomplished by associating the action with pain. Punishing outcomes not caused by desire is pointless, since it cannot reduce the frequency of those outcomes. Fines for people with naturally blond hair may make blondes wish they were brunettes, but, as hair color is unresponsive to wishes, the incidence of blondness will not fall. So the acts common sense says are punishable and creditable, namely those caused by wants, are the very ones the compatibilist calls free and responsible. Compatibilism makes good sense of not punishing a man for what he did not do freely.

People may do what they want, replies the metaphysical libertarian,[35] but, their wants are determined, in the end they have to do what they do, so are not really free. In saying this, however, the libertarian follows the critic of sociobiology in confusing intermediate causes with epiphenomena. When a choice is causally inevitable given its antecedents *and* necessary for a subsequent action, the inevitability of the choice does not mean the choice's antecedents would have produced the action in the absence of the choice. Jones frequents the Museum of Art because he likes Picassos, and he likes Picassos because of a genetically determined taste for distortion, but under normal circumstances it is simply untrue that he would have frequented the museum anyway, even if (per impossibile) that same genotype had made him dislike distortion; had Jones not wanted to go he would not have gone. He didn't *have to* visit the museum, that is, he would not have ended up there even if he had not wanted to. A desire can be inevitable, in other words, yet still matter. The question "Could Jones have done otherwise than visit the museum, given his genes?" is to be answered "yes." He could have done otherwise because he would have done otherwise had he not liked Picasso. "It was causally necessary that Jones want to do what he did" is consistent with "Jones would have done otherwise had he had different wants" (= "He could have done otherwise").[36] Bertrand Russell once quipped that we can do as we please but not please as we please; I am free to have lunch whenever I get hungry, but I do not go hungry freely.[37]

These victories may seem purely verbal. Perhaps we are "free," "responsible" and occasionally "culpable" in their ordinary senses of those words, but do freedom, responsibly and culpability in the ordinary sense justify blame and punishment, given determinism? Is it right to punish criminals for acting on preferences they did not choose to acquire, whether or not so acting is "free?"

These qualms rest on a false contrast between people and their characters. People don't choose their characters, it is true, but character does not happen to anyone, either. A person is his character, not a featureless wraith to which drives, preferences, and personality attach like barnacles.[38] The murderer neither chooses his violent impulses, nor finds himself saddled with them; without those impulses, that personality, there is no him. A murderer is not and cannot be punished for having a personality *he* was unlucky enough to be saddled with, for the same reason it cannot be unfair to punish a murderer for a personality *he* did not choose. Had "he" had an different personality, "he" would not have existed.

Libertarians admit the practical value of deterrence, but consider it *wrong* to punish what cannot be helped. They do not suggest that we change our formal and informal punitive institutions, but they would like us to feel uneasy about them. Certainly it *is* wrong to punish what cannot be helped—what, according to the compatibilist, did not proceed from choice—but, observes the compatibilist, punishing those behaviors is not deterrent, either. Society penalizes disvalued behaviors resulting from choice to extinguish the choices that cause the behavior. Since painful associations do not weaken involuntary behavior, people should be punished only for what is, as we say, in their power. The thief can't help wanting to steal, but that is not what he is punished for; he is punished for stealing, so he and others will subsequently find stealing less attractive.[39]

To bring these reflections to bear on race and crime: that we are free when we do what we choose to do, although our choices are caused by unchosen genes, preserves the freedom of individuals whose unchosen genetic aggressiveness leads them to chose lawbreaking. The corollary that blacks in particular retain the freedom needed for dignity despite higher mean aggressiveness is reinforced by two ancillary points. The first is our old frien that *social* causes of behavior lie as far outside individual control as genetic causes, so environmental causation—indeed any causation short of impossible self-causation—must be allowed to diminish criminal responsibility as much or as little as genetic causation does. The balance of genetic and social factors in criminal behavior is irrelevant to responsibility.

Second, allowing a genetic proneness to aggression to diminish the responsibility of black criminals requires, in consistency, that a genetic proneness to xenophobia and racial animosity diminishes the responsibility of xenophobes and bigots. Cattell (1992) and Rushton (1989b, 1995) argue that a tendency to xenophobia may be innate, since preference for bearers of observable correlates of genetic likeness, such as skin color similar to one's own, is adaptive and likely to have been selected in. Yet critics of "racism" would almost certainly not take this conjecture, should it prove correct, to excuse bigotry. Xenophobes would be told to fight their genes, and criticized, should they succumb, for not struggling hard enough. Why should genetic factors relieve black criminals but not bigots of responsibility?

9.17. RACE DIFFERENCES IN FREE WILL

The previous section applied to black crime what for centuries compatibilists have been saying about everyone. A genetic explanation of race differences in crime does not threaten black dignity because, while no aggressors choose to be aggressive, all are responsible for their aggressive acts. However, a final refinement of compatibilism significantly weakens this reassuring conclusion.

On reflection, freedom must be more than simply doing what one wants. Animals and infants act on their wants yet evidently lack some capacity possessed by adult humans. Compulsives doing what they want—smoking or gambling—also seem less than fully free. According to several recent writers (Frankfurt 1971; Jeffrey 1974; Sen 1974; Levin 1979; Davis 1979; Slote 1980),[40] the missing ingredient is that fully free, autonomous action proceeds from *desires the agent finds acceptable*. Animals and infants are heteronomous because they are incapable of assessing their own desires. A lion doesn't ask himself whether his taste for gazelle might be hard on gazelles. Compulsives can and do assess their desires, but are unable to act on the desires they desire to act on, and end by acting on desires they find unacceptable. Their second-order desires are impotent. Although the compulsive smoker wishes his yearning for a cigarette would vanish, it remains strong enough to cause him to light up.[41] Non-compulsives, by contrast, can weaken their cravings. A non-addicted smoker who wishes he didn't want to smoke will munch candy until the craving passes. Contrary to Russell we *can* to some extent control how we are pleased, and the more we can the freer we are.[42]

It follows that free will is a matter of degree, varying with the agent's capacity to step back from and reflect on his desires. The depth of this capacity depends in turn on his intelligence and self-control. A person of limited mental ability, not given to worrying about the quality of his desires or the likely consequences of following them, is relatively less free. So are people who follow an impulse as soon as it enters their heads. Compared to self-restrained, thoughtful individuals, neither acts on desires he finds acceptable, because their desires are subject to comparatively little internal scrutiny. That is why children, lacking the necessary self-critical attitude, are not held fully responsible for what they do. Every postadolescent is considered to have crossed some threshold of autonomy, but adults too vary in reflectiveness.

Just as individuals differing in intelligence and self-restraint differ in degree of autonomy, *groups* differing in mean intelligence and self-restraint differ in mean degree of autonomy. Some group comparisons are uncontroversial; although there are mature ten-year-olds and immature adults, the average ten-year-old is less responsible than the average adult. By the same token, the white advantage in intelligence and self-restraint implies that, on average, whites are more autonomous and responsible for their actions than are blacks (and Asians more autonomous than whites.)

There is some empirical support for this conclusion. Tashakkori (1993) reports that (a) black children and adolescents more than their white counterparts see what happens to them as the result of "uncontrollable external constraints of the environment," yet (b) while this perception predicts lower self-esteem for whites, it does not affect the self-esteem of blacks. Tashakkori tentatively links this finding to "the smaller internal self-control of African American children/adolescents" (1993: 597). The link may be that blacks are less inclined than whites to see events in the world, including their own actions, as flowing from a self. Presumably, if blacks are less aware of themselves as causes of what happens in the world, they are less capable of scrutinizing the self and its choices.

Appalling as it may sound to say that blacks have less free will than whites, many individuals, including a number nominally sympathetic to blacks, appear to agree. Consider those commentators, mentioned earlier, who subordinate criticism of black violence to a search for root causes while never subordinating criticism of stock fraud or lynching or the Holocaust to a search for root causes. One factor in this inconsistency, no doubt, is the belief that the external causes of black crime, but not white crime, were wrongs. Since whites rather than black criminals are thought to be to blame for black crime, the issue of responsibility for black crime is deemed closed, with only the question of causes left open.

But another factor, I believe, is a sense that reflection plays a relatively small role, hence external causes a larger one, in black behavior. A white stock swindler is assumed capable of realizing that what he is doing is wrong and thereupon stopping himself, whereas a black mugger is assumed much less capable of such self-monitoring. What looks like special pleading is an inclination to treat black crime as akin to an amoral natural force. Consider the following comments from a leading newspaper on two recent incidents. The first concerned the unmotivated stabbing of a white woman on a city street by a black vagrant: "The killing of Alexis Ficks Welsh on June 8, 1991, came to symbolize the failure of the mental health and criminal-justice systems to deal with the mentally ill homeless, even when there have been repeated danger signs" (Perez-Peña 1992: B3). The second incident was the killing of five-month-old Jeffrey Harden by the mother's boyfriend (not the biological father who immersed the infant in boiling water when it refused to stop crying.

Even in New York City's troubled child protection system, this oversight was remarkable: None of the caseworkers who investigated allegations that Doris Harden [the mother] had mistreated her children discovered that she had previously served three years in prison for holding down a 7-year-old girl while another young woman sexually assaulted the child with a toilet plunger. The four workers missed other important details as well: Ignorant of the prior abuse conviction, they initially did not speak with the mother's parole officers, who knew she was a crack user who had refused treatment. The workers interviewed Ms. Harden, but never talked to her boyfriends, despite warnings that one of the men was prone to violence and drug dealing. . . . Whether the caseworkers in the city's Child Welfare Administration could have prevented the child's death will never be known, but the flaws in handling of Jeffrey Harden's case illustrate what experts and advocates for children have long called the Achilles' heel of the child protection system in New York and the nation: the frequently poor quality of casework. (Dugger 1992: A1)

In both instances, responsibility and blame are reserved for the institutions charged with controlling violence. For its victims black violence is a "tragedy" (Perez-Peña)—an dreadful intervention of fate. This attitude, which extends to noncriminal matters,[43] betrays the belief that blacks are not fully responsible for their behavior. I venture to guess that this belief is also why conservative exhortations for blacks to take more control of their own fates seem perfunctory and pro forma, and are seldom taken seriously.

Implicit reluctance to treat blacks as moral agents is not confined to the Left. *The Turner Diaries,* a self-published book popular on the far Right (whose author's name, "Andrew Macdonald," is apparently a pseudonym), envisions a war waged by white Christians against what are represented as alien elements, particularly blacks. Yet the target of the author's special venom is not blacks themselves, but white women involved with black men and like white collaborators: long passages describe the hanging of these "race traitors," as well as detonation of an atomic bomb in front of a government building in Washington, D.C. The author, without perhaps quite realizing it himself, treats blacks as akin to germs that must be eradicated, holding whites alone up for moral condemnation.

18. APPROPRIATE RESPONSES TO DIMINISHED RESPONSIBILITY

Black crime provokes anger in some, sympathy for black criminals in others who sense, I conjecture, that blacks are not fully autonomous), responses that do not coexist happily in the same society. Putting emotion aside, how do lower levels of responsibility constrain efforts to control crime?

Contrary to what many people think, diminished responsibility does not imply greater leniency, only a lower degree of what might be called *punishability.* Since punishment is by definition administered for wrongful acts, it is less appropriate for the less responsible, who do not act in the full sense of the word. A child does not deserve to be punished for defacing property the way a normal adult vandal does. The subject of punishment must understand *why* he is made to suffer. Compatibilists of course agree: there is less point in threatening to discipline individuals whose behavior is less apt to changed by the threat. But lesser punishability is consistent with incapacitation, and with the infliction of aversive stimuli for other reasons. A naughty child may not deserve punishment, but he must be curbed; a disease carrier like Typhoid Mary, not responsible for being a vector and unable to see the threat she poses to others, must still be restrained. All that morality and logic require is that such measures not be called "punishment."

In fact, lower levels of self-control often warrant prompt infliction of unusually intense aversive stimuli. One often must be harder on a more difficult child to build especially firm associations between forbidden behavior and negative consequences. Less-deterrable robbers and killers must be stopped, even if deterring them requires especially harsh methods. Lower levels of self-control do not of course require harsh treatment if rewards for good behavior prove more effective, and there is no point in "punishing" wholly undeterrable madmen who can only be restrained or destroyed. Just what methods effectively deter at what levels of self-control is an empirical question; the philosophical point is that intermediate levels of responsibility permit the application of any otherwise appropriate measure. Just don't call it "punishment."

Equating responsibility with deterrability gives explanatory power to the hypothesis that blacks are less responsible than whites. I have remarked several times that the major environmental change coincident with the rise in black crime during the past thirty years has been the expansion of black rights and social mobility, the treatment of black and white offenders by identical rules, and an overall relaxation of the sanctions against crime. (One must not omit Aid to Families with Dependent Children and other forms of welfare, which have made once reckless behavior cost-free and allowed most black males to be reared without fathers.) Operationally, these changes have meant that a black who commits a violent crime, particularly against a white, is far less likely to suffer than previously. A black found guilty of a minor felony in the American South in 1920 would quickly find himself on a chain gang.[44] The same acts today are punished less severely, less swiftly, and less surely. With so many inhibitors relaxed, more crime is to be expected from genetically impulsive individuals. (I stress again that the environmental changes that triggered change in black criminal behavior did not trigger a like change in white criminal behavior, indicating, in light of the overall convergence of black and white noncorrelated environments, a genetic group difference.)

Higher rates of black crime may be accepted as the price of the current array of legal rights, but should that price seem too high, the following ideas might be worth thinking about.

(1) One intangible disinhibitor is current public opinion, which reflects constant reiteration of the statement that black crime is an understandable excusable response to mistreatment. "Through crime," says Hacker in a book that received dozens of adulatory reviews, "blacks are paying back whites" (1992: 187). Violent assaults on whites are "retributions for . . . injustice" (187) and the rape of white women is displaced rage against "the real center of power" (186–187), attitudes, Hacker implies, that should not be criticized (219). After a group of black boys raped and savagely beat a jogger in Central Park, her employer, Salomon Bros., donated several hundred thousand dollars to build a community center in the boys' neighborhood (Walls 1989

Midnight basketball leagues have been introduced as an antidote to black crime, on the theory that black youths commit crime from boredom and feelings of rejection. Blacks can hardly fail to notice that their violent behavior is being condoned and rewarded. These rewards should end—and presumably would end once people ceased attributing black crime to racism.

(2) Race differences in maturation rates, evidence for which was mentioned in chapter 4, might warrant treating blacks as adult offenders at an earlier age than whites. The justification once again is need for what statisticians call a "powerful" test of deterrability, one that erroneously treats some offenders too harshly to avoid catastrophic failure to treat corrigible serious offenders harshly enough. Since not every offender can be examined individually for corrigibility, the state uses age as a proxy. As the regression of corrigibility on age for whites may underpredict corrigibility for blacks, the law might use age differently for the two groups.

(3) The more rapid decay of reinforcement for blacks means that the same punishment will on average deter blacks less effectively than it does whites. Punishment schedules for given infractions might therefore be adjusted for race. The idea that punishment be person-independent is ostensibly codified in the equal protection clause of the 14th Amendment, but exceptions are recognized even now. Courts enjoy discretion in setting bail, and sentencing guidelines set ranges so that punishment can be varied from individual to individual, as a function, in part, of its probable effect. The principle that everyone who breaks a law should receive the same penalty may rest on the assumption that individual differences in conditionability are too small and uncertain to inform sentencing decisions. Were individual and group differences to prove significant and reliable, this assumption would have to be abandoned.

(4) The more rapid decay of reinforcement might also suggest the swifter administration of punishment to blacks, along with stricter limits on appeals. Once again, the idea that everyone is entitled to identical procedural justice may rest in part on the assumption that individual and group differences in response to delay are too small and uncertain to matter, another assumption that may prove false.

A propos (2)–(5), the norm of equal treatment, on its face requiring identical treatment for identical infractions, might be interpreted to mean *equally aversive* treatment for identical infractions. The two interpretations coincide when the same stimuli affect everyone alike, but diverge when responses differ. It seems sheer common sense that electrocution would be inappropriate for a convicted murderer somehow impervious to electricity, so, in consistency, prison is inappropriate for offenders who do not mind imprisonment. Reportedly, many young blacks view a jail term as a rite of passage into manhood, so the problem of equitable punishment across races is real. Use of reliable predictors of aversiveness to adjust punishment for all offenders would be race-neutral, but these predictors would presumably interact with race.

(5) Incapacitation. New technologies may make it possible to detect potential criminals early in life, track them, and reduce their aggressive impulses, possibly by chemical means. I will not speculate further, except to note that sufficiently reliable predictors of criminality would allow such tracking to be race blind. One dubious idea currently in the air is that, once the

neurological basis of crime is discovered, the brains of criminals might be altered. An unhappy side effect of this practice would be elimination of any deterrent against a first crime. Anyone who wished to kill an enemy need only do so, then turn himself in for a brain treatment. After the treatment has pacified him he might regret deeply what he had done, but before treatment it would seem like a painless way to get away with murder.

Suggestions (2)–(4) particularly will be opposed on the grounds that they treat people on the basis of their race, not as "individuals." Before taking up the overall issue of "individualism" in the next chapter, however, we must consider the possibility that everything said in sections 14 and 15 is wrong. What if black crime *is* caused by racism?

9.19. BEYOND CAUSES

Common sense takes the right to avoid and defend oneself against unprovoked attack to be independent of the cause of attack. The police protection I deserve from a burglar breaking into my home does not vary with why he is breaking in. I do not deserve less protection if my grandfather harmed his grandfather or the burglar himself. How could the cause of someone's seeking to violate my rights compromise my right to defend those rights? Crime, like everything else in nature, has causes, some of which were injustices; the presence of a wrong in the chain of causes leading to a crime does not normally constrain police action. It doesn't matter that Smith plans to steal Jones' car because a swindle in 1950 impoverished Smith's family. It doesn't matter that the state acquitted the swindler in a fixed trial. Just conceivably there is warrant for contralegal acts to retaliate against or rectify a specific past wrong (and, just conceivably, the police should turn a blind eye). That is why defenders of black crime call it pay-back. But the reprisal must be immediate, and closely linked to the wrong. Jones may run Smith down in hot pursuit to retrieve a watch Smith has just stolen, but he may not demand at gunpoint the return of a watch that Smith's grandfather stole from Jones' grandfather fifty years earlier. Past wrongs as far removed from the present as slavery cannot be relevant to state action.

It follows that race-conscious safety seeking is permissible whether or not black crime is due to white misdeeds. That the black male on the running track is more likely to attack the jogger because of discrimination or the jogger's suspicions of blacks does not diminish the jogger's right to escape danger.

The jogger might be said to be obliged to "break the vicious cycle of anger and fear," but his obligation to do so is no greater than his contribution to the cycle, weighted by the probability that self-endangering actions will actuall weaken it. In everyday cases, the likelihood that this good can be achieved b continuing ahead is smaller than the risk entailed. Arguably the fleeing jogge worsens the prospects for future victims of black crime by failing to diminis black crime as much as he can right now, but even then common sense permi flight. I have every right to lock my door even if so doing increases the chance that burglars will break into someone else's house; a woman has the right avoid a lonely stretch of road even if, by frustrating a lurking rapist, sł increases the risk of rape to other women. The potential victim has the right to increase the risk of harm to others because that risk is not the potential victim's fault. He is not himself a would-be burglar or rapist; he is merely defending himself against those who are.

Unlike a private individual's flight or a locked door, the state's protective steps, such as heightened attention to young black males, are invasive. Police searches interfere with individual liberty. However, as we have already seen, common sense permits a certain level of preemptive coercion against possible threats. Preemption is permitted even when the threat is caused by suspicion itself, so long as the suspicion is generic. The sheer wrongness of the suspicions that alienate blacks forbids preventive action only if, absurdly, preventive action may never be taken to avert wrongs caused by earlier wrongs. Nor does it matter whether the past wrongs that caused black crime were race-based. For one thing, race-conscious screening is generically *dis*similar to the race-conscious measures said to have caused contemporary black crime. Frisking black males is not enslavement, lynching, or restriction to separate public drinking fountains. So long as a crime is not caused by the specific measures taken to thwart it, its actual cause is irrelevant to how it should be treated.

The intuition that the racial character of wrongs creates relevance is rooted in the idea that race-consciousness is what started all the trouble. The countervailing intuition is: so what? Suppose the Holocaust turned many Jews to auto theft. A policeman sees a Semitic-looking man with a Mogen David and a number stenciled on his arm eying a car. Should the policeman disregard the man's ostensible background in deciding whether to enhance surveillance, because the crime perhaps being contemplated is an effect of anti-Semitism? Consider the question from the point of view of the car owner who expects the police to protect his property. For that matter, consider race-consciousness from the point of view of the social contractor in the street attacked by a black male. Imagine a policeman arriving just too late to intervene, who confides: "I saw him approach you and I suspected he might attack. But I didn't intercede even to the minimal extent of showing myself to discourage him because my belief that he might attack was race-based—I would not have been suspicious of a white male approaching you. I shouldn't act on thoughts I shouldn't think, and shouldn't think that way. My thoughts were bad because your attacker's life of crime was a result of his great-grandfather's enslavement, his father's inferior education, and his own limited opportunities, all the result of racial thinking. Doing anything to him because of his race is just the sort of thinking that caused him to attack you in the first place." Were you to retain your composure, you would insist that, whatever may have caused slavery and segregation, the policeman's suspicion was well founded. In signing the social contract, you created the office of policeman; when he agreed to assume that office he obligated himself to protect you. His belief that race predicts crime was untainted, innocently acquired. Why could he not act on it?

I have already alluded to the two reasons, seldom explicitly stated but often early at work, for thinking racism and discrimination make black crime special. One is that oppression has driven blacks mad. Recall Boxill's reference to the threat to "sanity and equilibrium" imposed by the "impossibly unjust situation" under which blacks have lived—a view which accepts the diminished responsibility for blacks but blames it on whites instead of biology. I doubt that proponents of this idea have thought through its implications. For if, for whatever reason, blacks are too unstable to be held fully responsible for their behavior, they are too unstable to be granted those privileges that depend on full rationality, including such aspects of citizenship as jury service and the vote. It would not be the fault of blacks that they are unsuited to vote or serve on juries, any more than it is the fault of 12-year-olds that they are too immature or the fault of a soldier driven mad by combat that he must be institutionalized, but deny them the franchise and jury service we do.

The other exemption is bolder: white misdeeds reduce or cancel the obligation of blacks to obey the law. Bruce Wright (1987) speaks of blacks breaking "a social contract that was not of their making." Elaine Brown, a former Black Panther, writes of extorting whites: "I don't know that I have to be moral with an immoral person. I don't know that I have to apply moral responses to an immoral situation" (quoted in Bray 1993: 68). Hacker concludes *Two Nations* with a question clearly inviting a negative answer:

A century and a quarter after slavery, white America continues to ask of its black citizens an extra patience and perseverance that whites have never required of themselves. So the question for white Americans is essentially moral: is it right to impose on members of an entire race a lesser start in life, and then to expect from them a degree of resolution that has never been demanded from your own race? (1992: 219)

Judging by the evidence, Hacker rests his question on false premises. Access to public assistance and an educational system better than any developed in black societies give blacks a better start in life than they would have gotten on their own, and, with affirmative action, better than the one whites receive. But suppose blacks' woes are an imposition of whites, and expecting blacks to commit no more crimes than whites asks more of blacks than whites ask of themselves. By ordinary standards the expectation remains right and reasonable, just as it would have been right and reasonable to ask any Jews still living in postwar Germany to refrain from robbing German citizens at random. The reason is simple: You cannot run a society in which one group feels less bound than the rest by rules against theft and violence. Let black crime be a cry for help, a call for attention, a demand for respect; some cries for help, some calls for attention, some demands for respect may not be uttered. Raping, stealing cars and packing guns may be ways to feel important, but they are not ways that can be allowed by a society hoping to survive. Should black males be dissatisfied with the respect they can get while conforming to the laws of Caucasoid societies, as is possible, they must settle for less respect than they wish or go elsewhere. Indeed, if it asks too much of blacks that they conform to law to the extent whites do, may it not be asking too much of whites that they show restraint when provoked by black crime? "War," according to Hobbes, "consisteth not in battle only, . . . but a tract of time wherein the will to contend by battle is sufficiently known." Without perhaps meaning to, those who accord blacks greater freedom than whites from the restraint of law are in effect endorsing race war.

That blacks have no right to break the law is not to say that blacks can conform themselves to law to the extent whites do. While it may—perhaps— prove possible to reduce by some extent the current racial disparity in law-abidingness, it seems highly likely that any multiracial society will find blacks less law-abiding than whites. Correlatively, any measures to curb crime will, of statistical necessity, abrade blacks more harshly than whites. These biological constraints on crime reduction will just have to be lived with, like the weather.

NOTES

1. U.S. Bureau of the Census, telephone interview, June 14, 1993.

2. U.S. Bureau of the Census, telephone interview, Aug. 6, 1995.

3. These variables did not mask race; for example, the typical black robber had no more prior convictions than the typical white robber.

4. Whitney (1990a) estimates a mean difference in "criminal liability" of 1 SD, based on the surprisingly high figure of .14 he cites for the probability that a white urban male will be arrested for an FBI-indexed crime in his lifetime.

5. Between 1979 and 1986, 44.3 out of 1000 blacks, as opposed to 34.5 out of 1000 whites, were victims of violent crime (Whittaker 1990: Table 1).

6. National Crime Survey, Department of Justice, telephone interview, June 28, 1990.

7. National Crime Survey, Department of Justice, telephone interview, June 28, 1990.

8. Numbers for rape of black and white women are given in Table 1 of Whittaker (1990) but not disaggregated into single- and multiple-offender victimizations. Table 16 of Whittaker (1990) disaggregates inter- and intraracial rape rates. Table 9.1 results from applying the single-offender rates from Table 16 to the numbers from Table 1. Whittaker's figure of 40% for the proportion of all rapists black is below the 50+% cited by some authorities but close to Hacker's (1992) 43%. The Whittaker study was based on a survey of households; other figures are based on other methods of reporting.

9. Highly publicized incidents include the murder of a rabbinical student by blacks after a black child was killed in a hit-and-run accident in Crown Heights, Brooklyn; the murder of a white guard and six white inmates by black inmates in an Ohio prison; Colin Ferguson killing six whites on a commuter train; a black teenager shooting a white schoolteacher to death to steal his bicycle in Brooklyn's Prospect Park; O. J. Simpson's probable killing of his white wife and her white friend; Darnell Hilton coldbloodedly killing three whites during a robbery. Other incidents buried in the back pages include: three black teenagers killing a white New York City school principal while shooting at each other (Lubasch 1992); a gang of blacks mugging old white women in 12 separate incidents, the leader of the gang reporting: "That's the way we planned it. It was a pact we made—only white people" (McConnell 1994); a 17-year-old black in Chicago raping a female gas company meter-reader, then killing her to prevent her talking; a deranged black vagrant stabbing a white women to death for no reason (Perez-Peña 1992); a black arsonist setting a fire that killed six white firemen; two blacks killing a white male and raping his sister (AP, Feb. 23, 1994); three black teenagers, one 14, arrested for the murder of a white grocery store owner (C. Wolff 1994); a black male prostitute stabbing a white attorney to death (Friefeld 1994); a black dental student hijacking a car, killing the driver, and wounding his companion Hanley 1994); a black burglar raping a white women in Queens (Rayman 1994); a teenager named Jamal Jefferson arrested for wounding Detective Joseph Vigiano

(Treaster 1994) (I take "Jamal" to be a black name and "Vigiano" Italian); a black arrested in the 13 year old murder of a white doctor (McFadden 1994); a Jewish landlord slain while collecting rent in Williamsburg, a black section of Brooklyn (Perez-Peña 1994: B2); a black executed for stabbing a pregnant white woman to death because "the best way to get back at whites was to attack white women" (AP 1995); a black sentenced to life imprisonment for the murder of a German tourist in Miami (Reuters 1995).

10. Rodney King was drunk and speeding; Michael Griffith initiated the dispute with the whites who chased him by spitting at them; Yusef Hawkins was mistaken for a black thought to be dating a white girl.

11. Thus, where $V(A/h_i)$ is the value of A given hypothesis h_i, the expected value $E(A)$ of A is $\Sigma_i V(A/h_i)p(h_i)$. A_i is to be preferred to A_k just when $E(A_i) > E(A_k)$.

12. G. Harman (1986: 22–26) stresses the unavailability of epistemic resources for the assignment of precise probabilities; Goldman (1986: 90, 324-343) cites widespread de facto violations of the probability axioms, and discusses neurological models for transforming information into binary yes/no decisions.

13. In the definition in n. 11, let $V(A/W_i)$ be conceived as the moral value of A in W_i, and h_i as the hypothesis that the world is in state W_i.

14. Armour's discussion is marred by a tendency to knock down straw men, such as the "involuntary negrophobe" traumatized into irrationality by prior encounters with blacks. Still, his essay is a useful statement of views quite contrary to those expressed here.

15. Those who oppose drug laws may use another example.

16. This is another inverse inference. Let e be some evidence, and G the hypothesis that Blue is guilty. $P(G/e) = [P(e/G) \times P(G)]/[(P(e/G) \times P(G)) + (P(e/\sim G) \times P(\sim G))]$, so $P(G/e)$ varies inversely with $P(e/G)$.

17. Single cases do significantly affect frequencies in small samples.

18. Presumably, the literature on statistics in the law restricts itself with this assumption to raise the question of "statistical guilt" in pure form.

19. On the uniqueness of security, see Levin (1982b); also Pennock (1984c), Levin (1984), Morris (1988), Schmidtz (1989), Levin (1989a), and Kavka (1986).

20. The state "existentially generalizes" on the disjunction of individual preemption rights.

21. For Hobbes, covenants are not contracts prior to the formation of the state, so an agreement to be a state enforcer cannot itself create a duty. Hobbes might insist that his third "Rule of Reason" and "Law of Nature" nonetheless tells policemen to keep their word, for the willingness of all to keep covenants is necessary for the creation of the state.

22. The Court held in *Adarand v. Peña* (1995) that federal affirmative action plans must be strictly scrutinized, without ruling that those plans failed the test, or commenting on the consistency of this holding with *Metro*.

23. While this book was in press, a number of Supreme Court decisions have modified government insistence on affirmative action. Racial districting is under a cloud. A California referendum has banned state affirmative action. By the time the reader sees these words, other changes in public policy may require modification of some of the comments about racial preferences in this and the previous chapter. The fundamental issues remain unaltered, however.

24. See Wertheimer (1971) on the effect visibly developing fetuses might have on the abortion debate.

25. "Although biases in the criminal justice system exaggerate the differences in rates of violent crime by race, it may, tragically, still be true that blacks commit a disproportionate number of crimes. Given that the blight of institutional racism continues to disproportionately limit the life chances of African-Americans, and tha

desperate circumstances increase the likelihood that individuals caught in this web may turn to desperate undertakings, such a disparity, if it exists, should sadden but not surprise us. . . . Further, for purposes of this analysis, I shall assume, perhaps counterfactually, that the rate of robberies is 'significantly' higher for blacks than for nonblacks. I shall also assume for the sake of argument that . . . greater fear of blacks results entirely from [the] analysis of crime statistics" (Armour 1994: 792–793).

26. They conjecture that violent aggression is triggered by challenges that produce uncertainty about the subject's self-appraisal, which are apt to be frequent when the appraisal is unrealistically high (12). This does not mean that the violent response "really" expresses a weak ego, they point out, since it begins with a sincere belief in one's superiority.

27. That the self-esteem of blacks in integrated schools still exceeds that of whites may indicate that blacks do not value academic success as much as whites do; see section 7.6; Hare (1985: 39).

28. Exactly half of a population lies within $.6z$ on either side of its mean; $83.4 - (12.9 \times .6 = 7.74) = 75.7$, and similarly for $+.6z$. The mode of the assumed IQ range for whites is also the black mean, but is $-1.12z$ relative to the white distribution. One must move $.26z$ to the right and $-.98z$ to the left to catch 17% of the white population.

29. On the relation of freedom and blameworthiness to dignity, see Strawson (1974).

30. Classic statements of compatibilism are Hume's essay "Of Liberty and Necessity," Chapter 23 of Mill's *An Examination of Sir William Hamilton's Philosophy*, and Hobart (1934). Aristotle's definition of the voluntary as "that whose cause lies within the agent" may anticipate compatibilism, although agency theorists (see n. 33 below) also claim him.

31. Here I ignore (a) differences between wanting, preferring, electing, and other synonyms for choice; (b) whether being caused by a want is necessary, sufficient, or necessary and sufficient for an act to be free (see Levin 1979: 252–255); and (c) constraints on how a want may cause a free act (a problem raised by benevolent telepathic Martians who bring about whatever I conceive a desire for).

32. A driver can at least try to escape a skid by manipulating distal causes like the angle of the tires; a man can at least try to influence his own caused desires by a program of conditioning (see below). But by hypothesis nothing can influence a causeless choice.

33. Libertarians such as Thomas Reid, and in modern times Chisholm (1964), reply that actions are not caused by the onset of wants or the occurrence of decisions at all, but by a *self*. These thinkers allow that natural events are caused by other events, and allow that a natural event uncaused by another event would be random, but (they insist) actions are caused by substances—selves, or persons. Standard usage seems to support this theory: the breaking of a window is caused by its being hit by a brick [an event], but *I* [a substance] order lunch. This support is quite weak, however. We commonly shorten "The impact of the brick broke the window" to "The brick broke it," so "I ordered lunch" may abbreviate "The onset of hunger caused my ordering." Everyday talk about agents thus need not be taken to imply agent causation. Anyway, the agent-causation theory must say why, in any specific case, an agent acts as he does. What caused me—I?—to order lunch? If nothing did, we are back to randomness. If something such as the onset of hunger did, that something had a cause, which had a cause, . . . , and we are back to determinism.

34. Spinoza describes a falling leaf that tells itself "Now I'll go this way, now that," and he asks why we, whose behavior is also necessitated, are not similarly deluded. The answer is that the leaf is *wrong* to think that it would take a different course if it wanted to, whereas the corresponding counterfactual beliefs humans hold about themselves are typically correct. A man who believes he is driving at 60 MPH because he wants to is usually *right* to think he would change speed if he so decides.

35. For example, Walker (1991: 455) and Chisholm (1964). Ethical libertarians, who condemn coercion, need not be metaphysical libertarians. One can believe it is wrong to thwart causally determined wants.

36. TV weathermen say the hurricane might have hit the coast without implying that the hurricane's path was undetermined. What they mean is that the hurricane would have hit the coast had conditions been somewhat different.

37. Chisholm (1964) says, of a murderer who would not have fired the fatal shot had he chosen not to: "if he could *not* have chosen *not* to shoot, then he could not have done anything other than just what it was that he did do." Not so. That the murderer could not have *chosen* other than he did does not mean that he could not have *done* other than he did; he could have done other than he did if he would have done otherwise had he chosen to. (To be sure, Chisholm would reject the analysis of "could have" under which the murderer could have acted but not chosen otherwise.)

38. Or (see section 5.2, n. 3) a person is the neurological basis of his drives and preferences.

39. Deterrence theory admits proportionality. The severity of punishment should match the seriousness of the crime, it says, because (a) deterring more serious offenses is more urgent and (b) maximal sanctions against less than the worst crimes leave perpetrators of these crimes no reason to refrain from worse ones. Capitalizing armed robbery would encourage armed robbers to kill policeman trying to apprehend them. Deterrence theory can also explain the intuition that punishment is owed, as if entailed by an agreement to which the wrongdoer is party. Hegel, following Kant, saw the evil visited on a wrongdoer as a consequence of the universalization of his own choice, tacitly willed in making that choice. Having implicitly agreed that it is all right to take life, the murderer can't complain when his life is taken. The murderer, as we say, asked for it. Deterrence is not meant to be retributive, but deterrent punishments tend to give wrongdoers what they "agreed" to accept.

40. Locke's discussion of "power" in the *Enquiry* anticipates the contemporary analysis.

41. On this analysis a smoker who doesn't mind smoking, who never regrets trading life expectancy for the pleasures of tobacco, counts as free, a conclusion concordant with common sense.

42. Ainslie (1992) views freedom as the ability to keep rapidly rising discount curves from overwhelming more slowly rising curves with higher maxima; see Appendix A. Ainslie's theory is a version of the present one when the desires one finds acceptable are identified with higher-maximum curves.

43. Milton Kotelchuck (1994) found that, in a large national sample, black mothers were twice as likely as white mothers to smoke, drink, and take drugs during pregnancy. He also found a slight difference, 36% versus 29%, in the proportion of black and white mothers told not to smoke. When Dr. Kotelchuk was asked in an interview to explain the heedlessness of black mothers, he said: "Is it because of implicit racism? That's probably too strong. But it is possible that people treat their clients differently on the basis of their social class" (AP 1994c, A16). Despite the ubiquity of warnings against smoking, drinking and drug-taking during pregnancy, Dr. Kotelchuk does not consider holding black mothers themselves responsible.

44. Alabama reinstituted the chain gang in 1995, eliminating it again in 1996.

10

Individualism and Discrimination

Objecting to racial screening on the grounds that it "ignores people's individuality" is one of many routes to the topic of individualism, which the reader has probably wished to raise for some time.

Let us call the rule that everyone should be judged and treated as an individual, not as a member of a group, the *principle of individualism*. This principle does not deny statistical race differences, but it bans appeal to them in practical contexts (especially legal contexts), and emphatically bans the application of statistical differences to individuals. It commands you to base your treatment of a person on *his* traits, not traits he has or is likely to have in virtue of a group to which he belongs.

The principle of individualism seems to forbid all race-based distinctions, and seems just as surely to be part of contemporary Western morality. Parents are expected to treat their children differently, teachers to respond to each pupil's particular strengths and weaknesses, and employers to judge an employee by his record. Yet, despite appearances, the principle disintegrates as soon as one tries to grasp it, or, to vary the metaphor, it shrivels in the light of analysis. What little of it remains is consistent with race-consciousness.

10.1. INDIVIDUALISM AND RACIAL PREFERENCES

However confused the principle of individualism may prove to be, it is clear at the outset that supporters of affirmative action must reject it. Preference benefits blacks on the basis of their race, without regard to the specifically demonstrated claim of any black beneficiary or the specifically demonstrated liability of any white male. On any of the rationales reviewed in chapter 8, preference must categorize by race.

Consider first the need to remedy injuries suffered by blacks, whether this is thought to require whites to restore a competitive advantage they themselves took, or to forego an advantage innocently acquired but still illicit. This rationale for choosing a particular black over a particular white for a position assumes that the black has been injured, but does not base this premise on specific knowledge about him. The premise rests, rather, on a generalization about his race, namely that, since discrimination has harmed a large but unspecified portion of blacks, this black is likely enough to have been harmed to merit an advantage now. Such an inference about an individual from group statistics is identical in form to the estimates of danger which prompt whites to be wary of blacks. Nickel (1975), who regards the correlation between race and injury as high enough to justify the "administrative convenience" of presuming all blacks to have been wronged, must and did agree that an equally high correlation between being black and untrustworthiness would justify discrimination against blacks.

Classification by race continues to play a role in the remedy rationale even when all blacks are assumed to have been injured to the benefit of all whites, for this posit is a generalization about groups from which the right to preference of each particular black is deduced. Absent a direct showing of injury by the white he is being preferred to, a black gains his advantage from membership in a group all of whose members are thought to deserve one. In fact, the compensation argument would retain a probabilistic element even if all blacks were known to have been injured by the whites they are preferred to, so long as the magnitude of each injury were uncertain. Once grant that no one knows whether the very blacks preferred ahead of Brain Weber would have been senior to him in a racially fair world, preferring them is compensatorily just only if they would *probably* have been his senior in that fair world. Since preferences so justified do not rest on a direct showing that *any particular black* would have been better qualified than any particular white, they must rely on an inference from the races to which each belongs that that black *probably* suffered an injury that accounts for the shortfall in his qualifications.[1]

The compensatory argument for role models obviously inherits this reliance. Role models intended to promote some other value also require grouping by race, although this may be a bit harder to see. Making it clear requires that we recur to the distinction in chapter 8 between the value allegedly promoted by role models being diversity itself, versus it being some independent end to which diversity is a means.

On one hand, role models intended simply to increase the number of blacks in some sphere make race one, in most cases the unique, qualification for entry into that sphere. So the promotion of diversity patently classifies by race. On the other hand, enlarging the black contingent to promote some further end requires that blacks possess traits relevant to that end either exclusively or disproportionately. At first glance appeal to exclusively black traits does not involve probability, so I will put off considering this alternative for a moment. Appeal to traits disproportionately displayed by blacks clearly *does* involve probability, for the justification of racial preference now becomes that blacks are *likelier* to have some contributory trait. For instance, anyone who argues that blacks should be preferred for judgeships because black judges are more likely to respect the rights of the accused must believe that blacks on average are more sensitive to the underdog—a belief that classifies by race.

The argument that preference for blacks helps the needy is statistical in the same way. To say that relieving need is a good, achieved more often if blacks receive the lion's share of help, is to treat neediness as a disproportionately black trait. Preferring a particular black without a specific showing that he is disadvantaged rests on a presumed correlation between race and need. When in *Metro Broadcasting v. FCC* the Supreme Court cited the need criterion as one reason to allow preferences for blacks competing to buy broadcast licenses, the Court did not demand a showing of disadvantage by particular black purchasers, nor did the FCC policy upheld by the Court extend preference to disadvantaged white purchasers. The thinking here is race-statistical.

Appeal to traits displayed only by blacks also involves probabilities, albeit indirectly, a link most salient in the argument that racial quotas weaken stereotypes. Hiring blacks alone can have this effect, it is said, because only conspicuous blacks can change minds about the abilities of blacks in general. (One assumes there should also be preferential hiring of incompetent whites, whose inadequacies will also change minds about relative white competence.) This argument plainly classifies individuals by their suitability for undermining stereotypes, an odd-sounding category but a category nonetheless, and one taken to contain none but blacks. The general point is one made repeatedly below: all judgments whatever classify in some way. This abstract formulation aside, the race-consciousness in using quotas to attack stereotypes is clear at the concrete level: more blacks are to be hired to raise everyone's estimate of blacks *as a whole*, and each white's expectations about the next black he meets. In addition, as the precise effect of preferring a particular black cannot be known in advance, hiring him is to be justified by its *likely* effect—resting the goal of destroying stereotypes on race-based probabilities in a second way.

That affirmative action distributes benefits and burdens by race, contrary to the principle of individualism, is not a deep insight. But it is enough to keep preference advocates from deploying the principle against racial screening, race-based frisking, or other race-conscious proposals. If categorizing by race is always wrong, categorizing by race because of past discrimination is wrong. If categorizing by race is permitted when race correlates with victimization, categorizing by race is sometimes permitted, the principle of individualism has been abandoned, and other race-conscious policies too must be judged on their individual merits. Indeed, preference compromises the vague principle of *Brown* that "separate is inherently unequal," for if (say) racially differentiated standards on a civil service exam can be considered "equal" because they cancel the effects of past discrimination, segregated schools with the same facilities might be "equal" because they let black and white children each learn up to their capacities.

At the same time, one can consistently defend racial screening yet oppose preferences, since classifying by race for some purposes does not compel classifying by race for all. By ordinary standards violence is so much worse than uncompensated discrimination (assuming quotas do compensate) that the expected benefits of screening far exceed those of preference—so much so, it may be held, that screening lies above, and preference below, the threshold of permissibility. The expected morality of screening is the value of all the assaults prevented, multiplied by the probability that screening will prevent them, less the injustice of detaining innocent blacks, multiplied by the probability that innocent blacks will be detained.[2] Likewise, the expected morality of preference is the value of compensating each act of discrimination, weighted by the probability that preference will compensate it, less the injustice of passing over innocent whites weighted by the probability that those whites will be passed over. Preventing assault, as I have urged, is ordinarily thought more urgent than compensating discrimination; the average person would rather be denied a job because of his race than be assaulted. Furthermore, submitting to a search seems less onerous than being denied a job. Finally, it is known that a large minority of black males are felons, but not known that a comparable minority of whites have benefited from discrimination—indeed, given the conclusions of chapter 8, virtually none have. Hence the expected morality of screening exceeds that of affirmative action. Moreover, any measure that reduced black crime would reduce crime against the innocent blacks the measure might incidentally penalize, while quotas offer no offsetting benefit for innocent whites. Whites cannot console themselves that their sacrifice serves justice (as Rosenfeld 1991: 310 appears to recommend), since, if they are innocent, their sacrifice is an injustice.

Of course, a screening advocate *can* also embrace affirmative action. In defiance of chapter 8 he can insist that discriminatory injuries done to blacks are extensive and serious enough to raise the expected morality of preference as high as is necessary. The principle of individualism can be subordinated to both security and restitution. But the logical point remains: even on assumptions favorable to affirmative action, one cannot consistently support affirmative action while opposing other race-based measures on the basis of individualism.

Nor can one consistently support race-conscious screening while opposing affirmative action on the basis of individualism. But a supporter of screening can oppose affirmative action for other reasons. The complaint against the compensatory argument in chapter 8, after all, was not that it classifies by race, but that it classifies by race *wrongly*, incorrectly presuming that most or all blacks have been injured. Screening advocates can allow that, given this presumption, the typical white probably does enjoy an unfair advantage over the typical black against whom he is competing, and, given a probability high enough for an injury severe enough, race is a reasonable proxy for compensability. The objection is that the presumption is false. Race-based compensation is uncalled-for because the white competitive advantage is not a result of white sins. Classifying by race per se is not an issue.

10.2. TREATING PEOPLE AS INDIVIDUALS: OUTLINE OF AN ANALYSIS

The argument of section 10.1 was tu quoque. That the principle of individualism is unavailable to preference advocates, despite their readiness to deploy it in other contexts, does not refute the principle itself. Still, the ease with which self-proclaimed champions of individualism abandon the principle does suggest there is less to it than meets the eye. In this instance, the instincts of individualists are sounder than their pronouncements: people are and must be judged by the categories they belong to, the traits they share with others. "Judging" someone consists in anticipating what he is like and what he will do, or assessing him on the basis of his past deeds. Anticipation is always based on correlations between traits a person is known to have and traits he may then reasonably be presumed to have. Retrospective assessment is general insofar as we judge a man by the *sorts* of things he has done. Jones who cheats at cards is contemptible because cheaters are contemptible.

Judging people is no more a conscious process than is slowing down in heavy traffic. At the conscious level we "size up" situations (Matson 1976), but sizing up, of traffic jams or people, is the upshot of associations formed within one's own experience and learned from others. A driver's awareness of what often happens to speeders makes him slow down, just as a person's past contact with and general knowledge of clergymen causes attitudinal adjustments when he meets one in a social setting. These adjustments are largely involuntary; whether our guard goes up or down, whether we feel excited or bored, depends on the resemblance of the present situation to others we have known.[3] "Individual" traits are simply those with proven predictive validity, traits on which it is reasonable (whether or not moral) to rely. You expect the clergyman you have just met to frown on smutty jokes, and you are right to expect this no matter how he may go on to conduct himself. Predictive traits are not always acquired voluntarily, so we will have to disentangle a trait's *deserving* to influence judgment from its voluntariness, but to a first approximation "treating someone as an individual" means treating him on the basis of his predictive general traits. Since race is predictive, classifying by race is consistent with individualism.

10.3. UNSATISFACTORY DEFINITIONS OF INDIVIDUALISM

There are other interpretations of "individualism" that capture some at least of what people have, rather confusedly, in mind, but these alternatives render the principle of individualism absurd, impossible to heed, or reduce it to predictiveness.

Perhaps the first explication that comes to mind of "treating people as individuals" is: (*a*) *Assessing each person and assigning him benefits and burdens on the basis of traits he alone has.*

This explication makes nonsense of the principle of individualism, since few of the traits that make people unique constitute reasonable grounds for judgment. One's full name and fingerprints are unique, but no one is or should be hired or befriended because of his name or fingerprints. These traits do not (as we say) make someone the person he is. Unique traits can be unimportant.

Perhaps, then, treating someone as an individual means (*b*) *Attending to his* important *unique traits, the traits that make him "who he is."*

The trouble with (*b*) is that at first glance—and in the end as well—important traits may not be unique. The treatment a person does and should receive depends, for instance, on his intelligence, sense of humor, kindness, and attractiveness, none of which are anyone's exclusive property. Using the terminology of chapter 7, our actual criteria of personal value are qualities people share.

At the same time, the precise degree to which someone exhibits important qualities may be unique, and should be responded to as such. So perhaps individualism means, or should mean, (c) *Attending to teach person's unique constellation of important traits.*

The relatively superficial problem facing (c) is that in some contexts a person's constellation of important traits is not unique. Everyone may differ from everyone else in some significant way if you look carefully enough, but there may be no time to look. A college admissions officer finds that the two top candidates for a scholarship have identical high school grades and SAT scores. He may break the tie by using further criteria, such as extracurricular activities, but there is no reason to think that the eventual winner's grades, SATs and clubs distinguish him from all students everywhere.

But even supposing each individual's unique mix of important traits can always be identified, what remains to be explained, without which (c) and in retrospect (b) get nowhere, is what makes a trait important in the first place, what distinguishes it from unimportant ones. No doubt a person should be judged by the traits that make him who he is, but the original question was how some traits but not others make people who they are. "By being important" just renames the problem, leaving us where we began.

All right—it may be replied—individualism cannot be defined via individuating or important traits. The principle of individualism is really meant to capture the completely different idea that (d) *People should be judged only by traits they* choose. *That is why race, an involuntary, immutable trait, is an improper basis for judgment.*

Voluntariness looms large in public discourse. The chief argument for the 1991 Americans with Disabilities Act, outlawing public or private practices which disparately affect the disabled, was that disabilities are unchosen.[4] The voluntariness criterion is often couched in the language of individuality, its advocates explain, because a person reveals himself, indeed creates himself, by his choices. It is for this reason that judging someone by an adventitious, immutable trait like race ignores his "individuality."

Yet, while choice is undoubtedly linked to the self—the link was insisted on in the treatment of free will in chapter 9—everyday judgments are frequently based on unchosen traits, by ordinary standards a perfectly acceptable practice. It is reasonable and (at this writing still) permissible for an employer to discharge an employee with a quick temper, even though no one chooses to be quick-tempered. Reflex speed is unchosen, yet a baseball manager is justified in rejecting rookie prospects who react slowly. (Nor do we think this policy more justified if reflexes are determined by choice of diet during a critical growth period.) The manager's decision is also probabilistic; he can't be sure that a boy who boots a grounder in practice will be equally inept in game play, but this has been his experience, and common sense allows him to base his decision on it.

Involuntary traits also play a large and blameless role in the spontaneous sizings up that determine personal relations. It is all right to avoid irascible people. Mates are chosen on the basis of appearance, intelligence, and personality, traits immutably determined by genes and environment, beyond the reach of the will. The law, too, classifies by unchosen traits, as when it requires nearsighted drivers wear glasses and denies licenses to the congenitally and accidentally blind[5] because of the (merely) statistical connection between highway safety and performance on eye tests. The ability to read a wall chart is a highly sensitive but nonetheless imperfect indicator of the capacity to drive safely. Tests which a handful of capable drivers might fail are justified because a test passed by every capable driver might also qualify a few incompetents, with calamitous results. Conversely, many voluntary traits are irrelevant to personal or institutional judgment. A college graduate's laboriously assembled collection of beer cans of the world will not influence the admissions committee of a law school, and will affect his appeal to women only slightly, and negatively. Voluntariness is neither sufficient nor necessary for relevance.

It is in any case disingenuous to call race an extrinsic characteristic. People take their race very seriously. Blacks particularly, when asked to conform to white norms, are quick to mention their blackness as part of "who we are" (see section 6.8.) Efforts to make race inessential just when so doing works in the interests of blacks smacks of special pleading.

One source of the perceived link between voluntariness and relevance has been the role of this link in civil rights advocacy. The original argument for banning race-based judgment was that everyone deserves to be "free of discrimination" whatever his empirical characteristics. But judgments based on race were also said to be irrational, because race carries no information about psychological or moral capacities. Any race differences, and consequently any information conveyed by race, were due to the very discrimination that civil rights laws would end. But when race differences persisted after the Civil Rights Act, the cart began to drift ahead of the horse. Attention to race, or certain types of attention, were still to be banned, but a new reason was needed for banning them, a new attribute of race that made it "irrelevant." Since race is patently involuntary (as are sex and disability,[6] further traits to be brought behind the civil rights aegis), involuntariness was the attribute chosen.

A more excusable error linking relevance to voluntariness is the notion that all goods, including jobs and housing, are rewards, and all evils punishments. Since only what is voluntary may be rewarded or punished, it appears to follow that the allocation of any good, including employment, housing—even self-esteem—should be based solely on traits that are directly or indirectly chosen. Hence race is a forbidden basis for allocation.

In reality, relatively few goods are bestowed as rewards. This is clearest for emotional goods like affection. A man does not propose to the woman he thinks he ought to, or the woman he thinks has done the most to earn his love, but the one he happens to want to marry. In the job market, the traits relevant to hiring decisions are determined by goals or tasks, and there is no reason to expect every trait that contributes to a goal to be under voluntary control. A manager is forced to classify prospective shortstops by reflex speed because, given where they are positioned, shortstops must be able to react quickly. Choice of the quickest prospect is not a reward for neural wiring; the goal of fielding a good team thrusts it on management, who in their turn wish to field a winning team not to reward skill, but to attract customers. Nor do baseball fans regard their attendance as a reward for winning baseball; they simply prefer teams that win to those that lose. (Nor, finally, do fans *decide* to like winning baseball.) Nowhere in this chain of justification do the criteria for hiring a shortstop include mutable, chosen traits. To be sure, other things being equal, managers also prefer players who try all-out and stay sober on the road, which are matters of will. Not all job criteria are involuntary. But many are.

Confusion between goods and rewards is abetted by a tendency to view competition solely from the competitors' perspective. Aspiring rookies see a place in the line-up as *merited* by outstanding performance; from their perspective, blind as it is to the task-relativity of hiring, retention feels like a reward and rejection like punishment. Since rewards and punishments are bestowed only on what people are responsible for, allocating jobs on the basis of involuntary traits appears, from this point of view, unfair and irrational. The best corrective is viewing the situation from the perspective of the team's owners, whose hiring decisions are dictated by the amoral, goal-directed criterion of winning baseball games.

10.4. WHY SOME TRAITS, INCLUDING RACE, MATTER

Once uniqueness and voluntariness are dismissed as reasons for deeming traits important, little besides predictiveness remains. The traits our thought-experiments have found unsuitable for judging people—fingerprints, scope of beer can collection—lack predictive power, while suitable traits—temperament, reflexes—possess it. So we may tentatively accept our initial conjecture, that treating someone as an "individual" consists in judging him by those of his traits which are known to predict, and, so far as is feasible, looking for more such traits.[7] Obviously, not all traits can be important simply because they predict other traits, which predict other traits . . . ; some qualities must be desired in themselves, particularly those involved in personal relations, while other predictive chains terminate at aptness for extrinsically specified tasks. But recognizing this does not force a retreat to uniqueness or voluntariness, for, as we saw, qualities valued in themselves are typically generic and unchosen.[8]

Proponents and opponents of race-based classification alike would agree in contrasting a student's high school grades with his race, as, respectively, "individualistic" and "group" criteria for college admission. But surely the use of grades is a paradigm of "judging a student on his own merits" because grades are a good predictor of academic success. Both before and after affirmative action was introduced, college admissions officers have favored white students with high grades because such applicants tend to do well. Students, like hopeful rookies, may see admission based on grades as "deserved," but admissions officers do not. It strengthens rather than weakens this hypothesis that grades reflect effort and choice as well as unbidden native intelligence, for admissions policies do not partition applicants' grades into a voluntary effort and an involuntary intelligence component. Wholly voluntary activities like club membership are considered, but again because they are taken to predict contribution to the college community. The sovereignty of predictiveness is tacitly admitted by the defense of quotas in admissions that merely mediocre grades from blacks indicate "promise" because of the hardships black students face. This argument in effect contends that low grades for black students *predict* what high grades for whites do. (The contention is erroneous, since grades predict college performance as well for blacks as whites, indeed—as admissions officers are doubtless aware—overpredict black performance.)

We have reached the central syllogism connecting race to individuality. Attention to any of a person's predictively valid traits (whether specific or generic, voluntary or involuntary) treats him as an individual; an individual's race is a valid, useful predictor of his intelligence, temperament and social behavior; therefore, the principle of individualism permits attention to race.

10.5. PROXIES

It might be objected (see Armour 1994: 792) that race predicts only as a proxy for the factors that actually produce behavior. Whenever race predicts a further trait such as intelligence or impulsiveness through being a coeffect, along with it, of some underlying causes (the objection continues), it is these causes that should be attended to, not race. Race per se is never a cause, so attention to it is always irrational.

It must be admitted that, ideally, we would ignore race in favor of true causes. The trouble is, we have relatively little information about what they are. At present the underlying mechanisms of behavior are either unknown or identifiable only by cumbersome, inexact techniques that cannot guide the quick, often informal judgments needed in everyday life. Intelligence and impulsiveness cannot be immediately discerned, and their neurological bases not at all, so observable surrogates must be used.

More deeply, discovery of the deeper causes of behavior would probably not decrease race-consciousness, and might well increase it. Evolution has made us efficient consumers of information, alert to observable predictors even when more recondite clues are available. (Recall the speculation about genetic factors in race-consciousness itself. Since aiding genetically related organisms enhances one's own fitness, alertness to observable predictors of genetic similarity, such as skin color, may have been selected for.) Indeed, discovering the causes of behavior can only strengthen the association of race with the criteria it now predicts; since race does correlate with these criteria, it must correlate with their causes. A superficial example that makes this point involves the ostensibly race-neutral variable of deportment. It has been urged (Thomas 1992: 38) that "bopping," the " 'bad' black walk" described by many writers (Thomas 1992: 38; Kochman 1981: 110; Wolfe 1987; P. Harrison 1972: 73) rather than race itself be used as a sign of aggressive intent. But the very fact that young black males are unusually aggressive means either that this walk is more common among blacks than whites, or more predictive of aggression for blacks, or both. The same is true for the oppositional/defiant mode of self-presentation that adolescent black males adopt more readily than white. Anti-"stereotypers" who set themselves to take race-neutral swaggering as a danger signal will inevitably notice that black males swagger more or that a black swagger has more meaning, and fall back into "stereotyping."

A thought-experiment may help the reader decide whether he himself favors using race as a basis for judgment. Suppose—never mind why—your young son is about to begin school in a neighborhood with many blacks, who regularly gang up on children. He will be in some danger starting tomorrow, and you want to advise him about whom to avoid. Telling him to "watch out for kids who look like trouble" is uselessly vague. Telling him to "stay alert" puts the onus on him, suggesting as it does that victimization would be his fault. The advice that one should avoid kids who sprinkle their conversation with "motherfuckah" is impractical and your son will sooner or later associate the phrase with blacks anyway. But suppose "Beware of black kids" would be an easy rule to follow. The warning becomes muddled once you start differentiating between kinds of blacks, cautioning that not all black boys are bad, or explaining that this generalization is an exception to the generalization that one should never generalize about people, but leaving it at "Watch out for blacks" maximizes you son's chances of escaping unpleasantness. What do you tell him?

10.6. CARICATURING THE USE OF RACE

Some critics of racial generalizations admit, overtly or tacitly, that race is a legitimate basis of judgment insofar as it predicts. I refer again to James Nickel, who disputes the right of "racists" to prefer whites on the empirical ground that blacks are equally trustworthy, yet consistently accepts the principle that "one important way of distinguishing justifiable from unjustifiable uses of racial classifications is in terms of the soundness of the alleged correlation between race and a relevant characteristic" (1975). Flynn (1980) is also clear on the essentially empirical character of "racism."

More typically, however, critics of race-consciousness attack straw men, thereby allowing themselves to avoid the hard questions. Recall Dworkin's failure to provide any motive for discriminating against blacks less opaque than "racism in political theory," and his confidence that "except in very rare cases a white student prefers the company of other whites because he has racist social and political convictions, or because he has contempt for blacks." Rosenfeld (1991) comes up with only two reasons for a "racist" wanting to discriminate: an otherwise unexplained belief that the white race is "superior, and that it is therefore of paramount importance to preserve racial purity" (149), and religious lunacy, "the claim of a religious fanatic who believes it to be his divine mission to establish a racially segregated society to preserve racial purity, to force others to convert to his religion and to give up the right to make moral decisions for themselves" (254). Rosenfeld does not consider even the slightly less demented fanatic who prefers a segregated society but has no desire to create one by force. Naturally, Rosenfeld has no trouble showing that "racist" views, so construed, "must simply be left out."

Another caricature touched on in chapter 5 is that of describing any negative attitude toward blacks as "race hatred," a word used of phenomena as disparate as lynch mobs and apprehension about falling property values when blacks enter a neighborhood. Even when not entirely off the mark, "hatred" may obscure the attitudes it is applied to. Many whites do seethe when required by law to live near blacks or send their children to schools with blacks, but this animus may be directed as much against the government coercing them as their new neighbors. "Hatred" almost by definition is excessive and impervious to reason; dismissing all negative responses to blacks as "hatred" begs (or ducks) the question of whether some of those responses might be legitimate and defensible.

An equally common caricature of race-consciousness attributes to whites the disparagement of blacks *because* of their race, with race often equated with literal skin color and skin color itself the imputed criterion of personal worth. One writer speaks of the United States as a society that accepts "a person's being black [as] right and proper grounds for denying that person full membership in the community" (Thomson 1977). Another defines a racist as one who "derives happiness" simply "from living in a racially segregated society" for no other reason than that there are two races (Rosenfeld 1991: 112). The most extreme version of this caricature is a widely cited essay by the legal philosopher Richard Wasserstrom (1985a). For Wasserstrom, "to be non-white—especially to be black—is to be treated and seen to be a member of a group that is different from and inferior to . . . adult white males." This pronouncement can be fully understood only in tandem with Wasserstrom's own ideal, "assimilationism":

a non-racist society would be one in which the race of an individual would be the functional equivalent of the eye color of individuals in our society today. . . . And for reasons we could fairly readily state we could explain why it would be wrong to permit anything but the mildest, most trivial aesthetic preference to turn on eye color. The reasons would concern the irrelevance of eye color for any political or social institution, practice or arrangement. According to the assimilationist ideal, a nonracist society would be one in which an individual's race was of no more significance in any of these areas than is eye color today.

To be sure, it is quite as absurd to treat people differently *just* because they differ in skin color as to treat people differently *just* because they differ in eye color. But hardly anyone thinks skin color in and of itself is more important than eye color; what some people think is that skin color carries information about other traits important by common consent. That paradigm racist, the Southern white, did not wish to deny blacks the vote or access to white drinking fountains because of the reflectance of their skin, but because he believed their judgment unstable and their personal hygiene deficient. Segregationists did not want to exclude blacks from white schools because they are black, but because the presence of blacks was believed to weaken academic standards. These beliefs may have been false, dogmatically held, and insufficient to justify the practices erected upon them, but they amounted to more than the belief that dark skin is inherently bad. Eye color differs from race—whose principal but not sole observable indicator is skin color—in lacking correlates. Anyone who endorses race consciousness believes that indicators of African descent do predict traits important to everyone, including, one may suppose, Wasserstrom. If blue eyes predicted high intelligence and low criminality, Wasserstrom would care very much about eye color.

(Eye color is an inapt analog for race on its own grounds, since it is indiscernible from more than a few feet, while race is salient from a distance. For purely perceptual reasons, eye color is bound to figure less prominently in human affairs.)

At one point Wasserstrom backhandedly concedes the dependence of the "assimilationist ideal" on predictiveness: "There do not appear to be any characteristics that are part of this natural [physiological] concept of race and that are in any plausible way even relevant to the appropriate distribution of any political, institutional, or interpersonal concerns in the good society" (23). As he realizes, this immediately invites the question, What if there were? and he admits in a footnote, "certain people believe that race is linked with characteristics that are prima facie relevant." His reply is another caricature: "even if it were true that such a linkage existed," he says, "none of the characteristics suggested would require that political or social institutions, or interpersonal relations, would have to be structured in a certain way." True, a "linkage" may not *mandate* any arrangements, but it might nonetheless make certain arrangements inevitable or permissible, or explain why people opt for them. That, essentially, was the point of chapters 8 and 9: linkages between race, intelligence, and temperament, while not mandating racial stratification, show how stratification can come about without wrongdoing and in this sense justify it; the linkage between race and crime justifies heightened suspicion of blacks. However reluctantly, Wasserstrom recognizes that the propriety of race-consciousness turns on the hinge of empirical correlation.

A final and more subtle caricature, playing directly on the principle of individualism, is conflation of racial awareness with awareness of nothing but race. There may well be people who see others exclusively as whites or blacks, and treating every black alike regardless of age, sex, deportment or other available cues certainly "denies individuality," just as it "denies individuality" to treat every white alike, or every woman, or every 17-year-old. Sensible people know that there are revealing characteristics beside race. However, the irrationality of attending to race to the exclusion of all else does not make it irrational to take race into account. It is absurd (but rhetorically convenient) to conflate a bank officer who uses an applicant's race as one factor among many in deciding whether to grant a loan with one who refuses all blacks loans no matter what.

At the same time, there are circumstances in which attending to race alone, or virtually alone, *is* appropriate. The jogger by himself, the woman in the elevator, a taxi driver scouting for fares, a policeman debating whether trouble threatens, have only moments to decide what to do when strangers present themselves, and the only data they have to go on is age, sex, and race. Race is decisive in these case because it is a good enough predictor to be used by itself when snap decisions are called for.

Do not think I am defending race-consciousness by turning it into something else—crime-consciousness or safety-consciousness. "Race-consciousness" is not only broader than the belief that race matters in and of itself, it need not include this belief at all. To observe that race matters because of its correlates is to explain *why* people notice it, not to deny that they do. The lonely jogger notices race out of fear of being hassled, not out of a lunatic concern with purity, but it is still race he notices. Nobody would care about race if black and whites were alike in every way except skin color. But they aren't, and that is why race is noticed.

10.7. TWO KINDS OF DISCRIMINATION

Once it is agreed that race can be overriding when time leaves it the only usable predictor, it must be agreed that the use of race is reasonable when other factors, such as numbers, leave race the only predictor. Lloyd Humphreys raises this issue in connection with school and housing integration:

Reactions to the relative proportions of blacks and whites in integrated schools reflect more than racism. Achievement levels in a previously all white school will decrease after a large influx of unselected black students, even if there were no effect on white achievement. Achievement on standardized tests is a principal criterion for judging the effectiveness of schools by parents and many others. Furthermore, unless great care is exercised and some differential treatment by race is exercised and some differential treatment by race is accepted, the standards for curriculum and achievement will drop. Teachers automatically adjust what they do in the classroom including the assignment of grades, to the students in their classes. The same large influx of black students, again on an actuarial basis, will result in an increase in interpersonal violence in the school, students against students and students against teachers. . . . [L]arge-scale housing integration poses a dilemma for which there is no easy answer. All resistance is not racist. Placement of subsidized public housing in a largely white neighborhood does increase the probability that white residents will become victims. . . . Evaluating each person in terms of individual worth as a neighbor . . . is not possible when large numbers of blacks move into a neighborhood. The actuarial information available to the average citizen [poses] a dilemma. . . . All resistance is not racist. (1991: 346–350)

Is resistance permissible? It is an "actuarial fact" that allowing 15-year-olds to drive would increase the number of traffic accidents. Fifteen-year-olds cannot be tested for maturity singly, and law must be uniform; either all otherwise qualified 15-year-olds may drive, or none. Since not all 15-year-olds can be allowed to drive, the decision is obvious. The state permissibly classifies by a single (involuntary) predictor, age, to prevent a statistically certain disaster, so individuals would seem entitled to use the statistical, involuntary predictor of race as a basis for classifying desirable neighbors, for instance by agreeing not to sell or rent to blacks, to avert what they consider disaster.

We thus reach the issue of private discrimination. The predictive validity of race justifies awareness of race, whether alone or as one factor among others; what sorts of race-conscious *action* is justified? There are now no laws, and I suspect that few people would want laws, against leaving an elevator when a black enters.[9] But the Fair Housing Act of 1968 forbids exclusionary agreements, and most of my other examples of the rational use of racial information are also illegal at present. Taxi drivers must accept black fares. Employers cannot base hiring decisions on race (except to prefer blacks over whites). Banks cannot consider race in extending credit (although consent decrees in "discrimination" cases involving banks and insurers have required banks and insurers to subsidize loans and insurance policies for blacks; see elnick 1996: 329–360). Yet I believe ordinary standards permit such actions.

Declining to live or deal with members of any race is an exercise of freedom of association permitted by the golden rule. It is easy to imagine a world in which everyone associates and deals with just those individuals he wishes to deal with and who wish to deal with him. The reader probably does not want undesired associations forced on him, so as a kantian he should have second thoughts about forcing associations on others. Actually, the golden rule plays a double role in the argument: it is the principle that permits whites to decline to deal with blacks, and the factor—the race difference in kantianism—that explains the wish to decline. Let us take up these issues in order.

"Racial discrimination" carries a bad aroma, in part, because of failure to distinguish its two different forms. There is aggressive, positive discrimination, or persecution, and nonaggressive, noninvasive, negative discrimination. Aggressive discriminators seek to harm individuals because of their race. Beating a man because he is black and mugging a woman because she is white discriminate aggressively. By ordinary standards such actions are wrong, some so wrong that others may prevent them by force. (There are no enforcible rights against more trivial forms of aggressive discrimination—maliciously breaking appointments with blacks to inconvenience them, for instance.) However, it is not necessary to reach their discriminatory character to criticize or criminalize them: that they are acts of aggression suffices. It is wrong to assault or break promises to blacks because assault and deceit are wrong. (It might be thought that racial motives make an assault worse because they imply a willingness to repeat; but the badness of a series of acts derives from the badness of the acts in the series, leaving us again with individual race-based assaults, in themselves no worse than any other assault.)

Negative discrimination, on the other hand, is the race-based refusal to bestow benefits. Refusing to hire a black, sell him a house, or cooperate with him in any other way because of his race discriminates negatively. The difference is that positive discriminators generally initiate interaction and leave their victim worse off than he was—they harm him—whereas negative discriminators are virtually always responding to an invitation to deal, and, more important, leave their "victim" no worse off than he was before interaction. You cannot refuse to hire a black or anyone else unless he volunteers to be your employee, and a black refused a job because of his race is no worse off than he was before he presented himself.[10] He lacked a job initially, and he lacks a job at the end. The whole notion of "victims" of negative discrimination is confused. It is not as if anyone minding his own business can be zapped out of the blue by it; you cannot be rebuffed because of your race unless you first ask a landlord, firm, or bank for an apartment, job, or loan, and you can hardly claim to have been victimized if your subsequent position is unchanged.

Negative discrimination is consistent with the golden rule, which lets people do what they want so long as their doing it prevents no one else from doing likewise. Universalizable liberties such as free association can be exercised capriciously, but negative racial discrimination need not be capricious. In chapter 7 the race difference in temperament was epitomized as a difference in kantianism, among whites the central criterion of personal worth. In nonracial contexts, likelihood of low kantianism is an excellent reason for refusal to deal, as when a landlord declines to rent to someone he believes will damage his property. Also, in nonracial contexts, inferences about kantianism drawn from statistics are also reasonable. A landlord who knows only that a would-be lessor belongs to a motorcycle gang may refuse to rent, or ask a higher rent, because of what he knows of the character of typical motorcycle gang members. In consistency, permitting negative discrimination against some groups membership in which predicts low kantianism permits negative discrimination against any such group.

The passage of civil rights legislation might seem to show that contemporary common sense subordinates negative liberty to other values, such as racial amity or the dignity of blacks, but this appearance may be misleading. The few decades civil rights legislation has been in effect is too short a time to confirm a major shift in values. Its passage and current acceptance may rest on empirical errors—the reader need only ask himself whether racial amity has improved—and conceptual confusions.

I have already mentioned the most significant confusion: that between harm and failure to help. No matter what A does, he does not harm B unless B is worse off than he would have been had A not so acted.[11] A's refusal to hire unemployed B, for whatever reason (including B's color) leaves B exactly as he was, jobless, before encountering A. B has lost nothing previously his. One reason for this confusion, I suspect, is the din of incessant talk of "race hatred." But another reason this distinction is ill grasped where race is concerned may be the failure of blacks to produce on their own the sorts of goods common in white society. This inability of blacks to acquire Caucasoid goods without Caucasoid cooperation makes white refusal to deal with blacks appear to be a *barrier*—and a barrier is indeed harmful, since those on the wrong side must expend more effort than previously to cross it, and are thus worse off. This perception is not unfounded: blacks would indeed have no television sets, automobiles, computers, or other products of Western technology without the compliance of whites. But refusal to offset an inability is not a barrier. A's unwillingness to sell to B something B cannot produce on his own does not take anything from B or leave B less productive than he was. Africans centuries ago unaware of Europe were not harmed by the sheer existence of unavailable European goods. Nor would Africans have been harmed had they known about and wanted to trade for those goods, but were unable to inform Europe of their desire. Those Africans would not have lost anything simply because Europe had more. But it follows that Europeans would not have harmed Africans had they learned of but spurned an offer to trade, for, once again, Africans would have ended up no worse off than they were before lines of communication opened. It also follows, from the fact that European refusal to deal with Africans would not have harmed them, that Africans would not have been harmed had the Europeans who refused to deal lived on the same continent, or shared the same territory, as do blacks and whites in the United States. The proximity of someone with goods you do not have may sharpen your desire for them, but his refusal to slake your desire does not make you worse off.[12]

I have also mentioned various empirical beliefs that have sustained civil rights laws, for instance that these laws would enhance racial harmony, and that, absent discrimination, blacks will come to act like whites. But the most important belief in civil rights theory has undoubtedly been that a wish to avoid blacks must stem from irrational motives. Americans value individual liberty for itself, but they grow suspicious of liberty they think is being misused.[13] Once it is accepted that the only possible motives for negative discrimination are vicious or insane—as are the motives customarily ascribed to "racists"—the freedom to avoid blacks is seen as a questionable indulgence. When the Civil Rights Act was before Congress, it was faulted mainly for "legislating morality," a silly criticism—morality is legislated all the time; murder is illegal because the right not to be killed is considered enforceable—which also conceded the immorality of negative discrimination. It is hard to work up much enthusiasm for a right to do what is wrong. But the persistent poor performance and disruptiveness of black schoolchildren, the higher black crime rate, the preference of black offenders for white victims, the black self-presentational style, and the overall lower kantianism of blacks are all comprehensible reasons to discriminate negatively. Some motives prompting white avoidance of blacks—negative discrimination—are by ordinary standards entirely reasonable, and have nothing to do with "hatred" or an urge to harm blacks.

A final conjecture: when association was purely voluntary, blacks could secure the compliance of whites needed for access to white wealth only by behaving in ways whites found acceptable.[14] Requiring whites to associate with blacks does away with the discipline mutual freedom enforces. That blacks no longer have to worry about pleasing whites is, perhaps, one of the significant environmental changes which, working together with genes, has affected black behavior over the past half century. Although I do not see this as a major argument against civil rights legislation, it may have been another unintended consequence of abolishing freedom of association.

An honest appraisal of black attitudes might find surprisingly little divergence. There is little doubt that blacks prize the amenities of advanced technological societies of the sort whites create, but it is less clear that blacks have any particular desire to associate with whites as an end in itself. Blacks and whites eat apart in every high school and college cafeteria in the country, and there is no evidence that black students want to socialize with whites any more than white students want to socialize with blacks. This voluntary self-segregation suggests that blacks wish to avoid whites for reasons having nothing to do with "hatred." Inclusive-fitness and kin-altruism theory predicts that most people prefer associating with those who bear observable indicators (like skin color) of genetic similarity.

It is received wisdom, endlessly repeated, that whites have no reason to wish to avoid blacks, and indeed that thinking that they might is evil. To give the thought a fair hearing, the reader should first ask himself whether the evidence concerning race and intelligence, academic performance, obedience to law, and personal deportment establish that differences exist. He should then ask himself whether they constitute a rational basis for negative discrimination, and if he thinks they do not to explain why. The reader prepared to grant the rationality of negative discrimination should now ask himself how third parties can rightly intervene against it.

10.8. RACE DIFFERENCES AND PUBLIC ASSISTANCE

A quite different example of the relevance of statistical race differences concerns public assistance. Despite extensive discussion by sociologists and economists of the effects of welfare incentives on blacks and the population as a whole, little thought has been given to race differences in response to these incentives. Herrnstein and Murray observe that major race differences in welfare dependency and illegitimacy remain after IQ is controlled for (1994: 331, 333), but say no more thereafter.

The argument for public assistance is familiar. Some people do not survive in the free market; illness strikes, accidents happen, breadwinners die, charity is unavailable. As letting the helpless perish seems inhuman, the more prosperous are enforcibly obliged to provide a "safety net" until another job is found, the children grow up, or health returns.[15] The conflict between public assistance and the negative liberty of taxpayers to keep their earnings may mark a confusion between negative liberty and the positive liberty of welfare recipients to get things they want or a fundamental shift in values.

"Relief," begun modestly during the New Deal as temporary aid for widows and abandoned wives with children, expanded dramatically in the 1960s to cover illegitimate children. Since that time there have been many debates and legislative shifts as to whether public support should be disbursed by the central government or localities, but public support itself has remained a fixture of American life. Yet from the beginning it has presented what economists call moral hazard. There is no reason to work when others will support you, or when the difference between relief and earned income is not worth the effort needed to procure it.[16] There is no reason to refrain from sex when others will pay for its possible consequences. The moral hazard problem did not become troublesome until the 1960s; most people continued to work and virtually all children were born and raised within marriage. In 1950 the white illegitimacy rate was 1.7% and the black rate 16% (Hacker 1992; Jaynes and Williams 1989). Since the extension of welfare benefits in the mid-1960s to mothers of illegitimate children, and the overall expansion of public support, the safety net has worked otherwise among blacks. At this writing more than 2/3 of black children are illegitimate, the majority of black female-headed households are on welfare, and altogether 21% of black women (as opposed to 4% of white)[17] were receiving Aid to Families with Dependent Children when it was (nominally) abolished in 1996. No longer temporary, welfare has become a multigenerational way of life among the black underclass, the major source of income in Harlem, Detroit, Camden, East St. Louis and other black areas.[18]

The important biosocial point is that, although welfare has been equally available to blacks and whites, these two populations have responded to it differently. In theory, white women could bear illegitimate children at the black rate and raise them on welfare until society went bankrupt, yet this has not happened. Despite the rise in white illegitimacy over the last half-century, most white children are still born to married parents, most white women still insist that the fathers of their children work to support them, most white men regard unemployment as shameful, and most whites of both sexes despise men who live off women. Differential rates of teenage pregnancy and sexually transmitted diseases indicate that promiscuity is more prevalent among black adolescent females than white. Whether whites think of their aversion to welfare as "pride" or the desire to live a better life than welfare provides, from a behavioral point of view blacks find welfare more attractive; welfare incentives affect the races differently. Current black illegitimacy rates may be "unnatural," since they reflect subsidies not found in African ancestral environments, but, since this environmental factor has been the same for blacks and whites, the race difference in response to the introduction of this factor indicates genetic variation. The hypothesis that blacks are more ready to devote effort to reproduction than child-care (see chapter 4) predicts a more rapid decline in pair-bonding among blacks than whites when others offer to support their children. Here, apparently, is more genotype/environment interaction: black and white mating and work behaviors are more similar in environments requiring self-support than under welfare.

Welfare is too new for its long-term effects on any group to be gauged precisely. Perhaps all welfare states eventually collapse as too much of their population exploits the public treasury. But for now, by ordinary prudential standards, a safety net for whites appears feasible. Sufficiently many whites continue to work and socialize their children in families for welfare to maintain itself. The case appears otherwise for blacks, who are more inclined than whites to regard public assistance as a legitimate means of support, and may be disposed to exploit welfare to a destabilizing extent. One implication is that welfare would be unstable in an all-black population, and another may be that welfare for blacks, at 12% of the population, will eventually bankrupt an otherwise prosperous white society. It thus may be imprudent to offer blacks the same safety net that whites make available to each other.

Like racial screening, racial distinctions in welfare may be said to violate norms of equal treatment, the subject of the next section. Denial of welfare is also said to punish innocent babies, and, since it "punishes" blacks disproportionately, is sometimes called genocide. These charges simply abuse language. A woman who remains childless because others will not support her has not been prevented from having a child, nor has her (non-existent) child been killed. The black population would probably grow more slowly were welfare eliminated, but demographic changes that are byproducts of individual choice rather than coercion[19] are acceptable by ordinary standards. Ceasing to force taxpayers to support illegitimate children does not coerce the mothers no longer supported. A heedless couple who produce a child they cannot support—or the mother cannot support when the father absconds—are the ones responsible for the child, and the ones who harm it by not providing for its support. Strangers not responsible for its existence do it no harm by refusing to help it. If through his own shortsightedness Smith is starving and Jones will not to feed him, Jones has not caused Smith's hunger, and only by fantastic hyperbole can Robinson's refusal to force Jones to feed Smith be called aggression on Robinson's part.

A last issue inviting differential treatment is jury nullification. There is evidence that black jurors are much more inclined than white to acquit clearly guilty black defendants (Weiss and Zinsmeister 1996): rates of felony conviction in areas with heavily black juror pools (and much black crime) are significantly below the national average. In several well-publicized trials, majority-black juries have acquitted with—to judge by their haste—no attention to the evidence. Obviously, a willingness by blacks to ignore the rules of evidence out of racial partiality would undermine the jury system, necessary to preserve order within a framework of rights acceptable to whites. Thought might therefore be given to limiting black admission to juries.

Welfare illustrates a general if incidental point. Conservative critics of government who point to the failure of the welfare system, or the decline of the public education, as proof that these institutions cannot work may be moving too hastily. The "failure" of public schools—falling test scores, illiterate graduates, attacks on teachers, chaotic classrooms, physical decline—has only become a matter of great concern since the start of integration (and has occurred as expenditures have risen). As just noted, welfare became problematic only after becoming widely available to blacks. Perhaps certain institutions, like public education, are viable in a white population, but not in a black or mixed one, and conservatives have mistaken the unworkability of these institutions in mixed-race populations for inherent flaws.

10.9. INDIVIDUALISM, EQUITY, EQUAL PROTECTION, AND THE GOLDEN RULE

Judgment based on group membership is said to violate the notion of equal treatment embedded in the 14th Amendment of the U.S. Constitution, indeed the very idea of equality embodied by the golden rule I have made so much of.

The 14th Amendment requirement of "equal protection under the laws" has drawn so much commentary that almost everything said about it will be considered controversial, but one point seems clear: it cannot require that the law treat everyone alike. Any law discriminates, if only between those who obey it and those who flout it. Laws against jaywalking burden jaywalkers. Laws setting conditions for military service single out those who meet these conditions. A legal system that "treats everyone alike" would be vacuous.

The minimal "procedural" reading of equal protection takes it to require only that the laws, whatever they are, apply to everyone in the same way. It is up to Congress, the states, and the municipalities whether and how to punish jaywalking, but under equal protection all jaywalkers must be punished alike. Where crossing midblock carries a $50 fine, all jaywalkers, black and white, tall and short, must pay $50. So understood, the 14th Amendment puts no limits on classification by race. It allows black jaywalkers to be $50 and white jaywalkers $100, so long as all black jaywalkers pay $50 and all white jaywalkers pay $100. It permits school segregation, so long as no member of one race is admitted to a school reserved for another. So understood, equal protection is irrelevant to public race discrimination.

This purely procedural interpretation, while logically the most defensible, is ahistorical, and violates the precept that legislatures do nothing in vain. The post–Civil War Congress could not have proposed an amendment placing no constraints on itself or the states with regard to race. The usual more expansive reading (see section 9.8) takes equal protection to forbid "arbitrary" legal classifications, defined as those unrelated to any proper government function.

While this reading endows the equal protection clause with some content, its bearing on race is still limited. A ban on arbitrary classifications does not by itself determine which similarities are arbitrary, so cannot by itself exclude racial classifications. Race is engaged via the empirical assumption that no racial classification[20] can serve a legitimate government purpose, an assumption, as we have seen, very much open to doubt. Race is an immutable trait at best statistically related to government functions, but the 14th Amendment is taken to permit other classifications based on immutable traits related only statistically to a proper governmental functions. Recall that the state classifies by visual ability, denying the blind permission to drive, although eye problems are usually congenital and not linked infallibly to highway safety. The difference between the blind and the normally sighted is relevant, although unchosen, immutable, and imperfectly related to state function, so discriminating between them is equitable.

It is natural to reply that equity is observed in the driving case because a blind man differs relevantly from the normally sighted in virtue of *his* eyesight, whereas a black need not differ from a white in respect to any of *his* traits. A black differs from all whites only because of the group he belongs to, and it is merely blacks as a group that differ from whites as a group. Yet the distinction between a person's own traits and those he inherits from group membership is untenable. A person's race is as much one of his properties as his eyesight. The difference between each black and all whites may be and on the evidence is relevant to such government activities as control of violence. True, each black's race per se is irrelevant to any governmental function, but each black—but not each white—has the property of belonging to a class 30% or more of whose members criminally aggress at some time.[21] "Belonging to a class 30% or more of whose members aggress" is a convoluted description, but the property it denotes is straightforward, and possessed by every black—just as "belongs to a group 99% of whose members cannot drive safely" is a property of everyone whose eyesight is 20×200. A less convoluted description of the same sort is: "being 30% likely to be a felon." A legally blind man, whether or not he personally can drive safely, belongs, in virtue of *his* eyesight, to a class 99% of whose members cannot drive safely, hence has the personal property of almost certainly being unable to drive. Statistical properties—properties an individual possesses in virtue of the statistical character of the classes he belongs to—are perfectly real. Should they be relevant to legitimate government objectives, legal classifications on their basis are nonarbitrary. Under the conventional assumption that most whites have profited from harm to blacks, *all* whites and *no* blacks possess the property of *probably* having profited from harm to blacks. With the additional assumption that redressing discriminatory harm is a legitimate government function, differential treatment of all whites and blacks is consistent with equal treatment. This is, in fact, the very reasoning behind state-mandated affirmative action.

Statistical properties feel unreal because reality in itself is definite. A concrete individual is or is not going to commit a crime; to say that he might just expresses our ignorance. Omniscient beings have no need of probability But probabilistic judgments are forced on nescient humans who must navigate in a world about which their knowledge is fragmentary. A textbook way of putting the relativity of statistical statements to human knowledge is that they depend on choice of reference class. Suppose we know that Jones owns a bicycle and a television, that 78% of bicycle owners jog, and that 12% of television owners jog. Our estimate of the probability that Jones jogs varies with whether we think of him as a bicyclist (likely) or a TV viewer (unlikely); it makes no sense to ask how likely it is, absolutely speaking, that he jogs. Vexing as this indeterminacy is, sense can be made of absolute probability judgments, at least at any given time. Simply take the narrowest, most fully described class to which Jones is known to belong—in our case bicycling televsion owners—and ask the proportion of joggers in that class.[22] Relative to a time, the right reference class for an individual with respect to some property (such as being a felon) is constructed from everything we know about him at that time that is relevant to the property in question. To ascribe a 30% chance of being a felon to an otherwise unspecified black male is to say that 30% of the narrowest class to which he is known to belong at the time of the ascription consists of felons. When the properties that are subject to statistical attribution are also relevant to some government function, discriminations based on varying statistical attributions support legal classifications, and are consistent with equal treatment.

I have dwelt on situations in which common sense, defying individualism, allows the use of statistical group properties. A certain degree of race consciousness in everyday life, I have argued, is as permissible as make-consciousness when buying a car. But people are not cars, I will be reminded, and asked, perhaps angrily, how I would like to be judged by statistical lights. Would I like to be judged on the basis of generalizations about Jews? Armour (1994) constructs from the emotions behind such a question a novel reading of the 14th Amendment: one of its longstanding purposes, he says, "has been the elimination of racial stigmatization" (814). Therefore, any state action which so much as countenances a "racially stigmatizing practice" will "infringe on protections guaranteed by the Equal Protection Clause" (815). The courts must therefore reject a statistically justified fear of blacks as relevant even to a claim of self-defense, lest it "reinforce derogatory cultural stereotypes." (Positive state action premised on crime statistics would presumably be beyond the pale.) The moral analog of this position, one imagines, is that any race-consciousness that translates into the avoidance of blacks—what Armour, following Kovel, calls "aversive racism"—improperly stigmatizes.

Armour's psychosocial interpretation of the 14th Amendment is neither historically nor hermeneutically tenable. As for the moral outlook it expresses, and the hypothetical question about generalizations about Jews, the only reply is to turn the question back on the reader. How would he like it? To find out, I propose a final thought experiment. Imagine yourself born with an indelible strawberry birthmark, it being known that bearers of this birthmark tend to be less intelligent than those without it, and that 30% of its bearers commit at least one violent crime. (These generalizations are not consequences of social attitudes toward the birthmark.) Would you be outraged if the police and shopkeepers examined you with unusual care? Would you be offended by women, particularly women lacking the birthmark, clutching their pocketbooks when you passed? Would you consider it unreasonable for people to be dubious of your ability to do college level work? If you would take such behaviors and attitudes amiss, why? What mistake are people making?

NOTES

1. Recall Rosenfeld's explicitness here: "[I]t seems reasonable to presume that the measure of the reduction in the prospects of success of blacks is roughly equivalent to the difference between the ratio of blacks with desirable employment to the total number of blacks in the workforce and that of whites with such desirable jobs to the total number of whites in the workforce" (1991: 290).

2. By rearranging terms in n. 11 of chapter 9. If $P(A,i)$ represents the assaults prevented by A in W_i, and $H(A, i)$ the harm of detaining innocent blacks in W_i, we have $V(A/W_i) = P(A, i) - H(A, i)$, so $\Sigma_i V(A/W_i)p(W_i) = \Sigma_i[P(A, i) - H(A,i)]p(W_i) = \Sigma_i[p(W_i)P(A, i) - p(W_i)H(A, i)] = \Sigma_i p(W_i)P(A, i) - \Sigma_i p(W_i)H(A, i)$.

3. And possibly their fit with innate perceptual schemata, as in encounters between the sexes.

4. As drinking and drug use are voluntary, this rationale does not cover the Americans with Disabilities Act ban on firing alcoholics and drug addicts in rehabilitation programs.

5. This example is C. D. Ankney's, in correspondence.

6. But see n. 4.

7. Thus does Dworkin ridicule the "catch-phrase" of "individualism": "It is said that Bakke has a right to be judged as an 'individual.' . . . What can that mean? Any admissions procedure must rely on generalizations about groups that are justified only statistically. . . . Suppose that decision [to reject Bakke] had been based on the following administrative theory: it is . . . unlikely that any white doctor can do as much to counteract racial imbalance in the medical professions as a well-qualified and trained black doctor. That presumption is, as a matter of fact, more plausible than the . . . presumption [that] applicants whose grade-point average fall below the cutoff line [will fail]. If the latter presumption does not deny the alleged right of individuals to be judged as individuals in an admissions procedure, then neither can the former" (1977a: 143–144).

8. So importance is recursive. Traits valued in themselves are important, traits that predict important traits are important, and no other traits are important.

9. It is not entirely clear that such behavior is legal. Civil rights legislation has been interpreted to ban "harassment" in the workplace, and conceivably a black avoided by a female colleague in such an encounter would have standing to sue his employer. Whether such a suit would be successful cannot be predicted.

10. Displaying a "No blacks or Irish need apply" sign may cause blacks and Irish uninvited distress, but so do many exercises of free speech. A right implies a right to announce one's intent to exercise it, so a right to refuse an association implies a right to announce one's intent to refuse it.

11. B must stay above this baseline at all points during the transaction. A cannot drag B below the baseline, make immediate reparations, and claim not to have harmed him.

12. Arguably, A harms B by intentionally showing B something B will want but can get only from A, but American whites do not create attractive goods for the purpose of tantalizing blacks.

13. Kristol (1978) discusses this ambivalence.

14. Suggested by a conversation with Hans Hoppe.

15. The claim is also made that we are all in this together, or, in Rawls' words,

that "society is a cooperative venture for mutual advantage" (1971: 4). Since the currently unfortunate have indirectly helped the currently more prosperous, the argument continues, the latter should help the former. Unfortunately, this is more metaphor than argument, and a flawed metaphor at that, since not every taxpayer has literally been helped by every vagrant.

16. The incentives problem is intractable. Continuing benefits after recipients get jobs violates the point of welfare—helping those who can't help themselves—and produces the infuriating anomaly of those on welfare working less but earning more than those who never were. Workfare yields a return on tax dollars, but the jobs are often make-work. Training to prepare current recipients for work also makes the taxpayer subsidize someone else's education, and workfare with childcare for unmarried mothers makes the taxpayer subsidize someone else's children as well.

17. A figure that includes Hispanics.

18. Natural experiments purportedly showing that welfare does not remove disincentives to illegitimacy suffer from range restriction: correlations between illegitimacy and the value of welfare will be indiscernible when the annualized reduction in welfare is small. A well-designed test would end stipends, medical care, housing, food stamps, and all other benefits. Absent this ideal experiment one must rely on common sense, which says that people are more industrious and prudent when their survival depends on their own actions rather than guarantees by the state.

19. Less coercion, since the taxpayer will no longer have to pay.

20. I remind the reader that, since 1965, this has meant "no racial classification burdening blacks."

21. Strictly speaking, belongs to a natural, "projectible" class fitting this description. Each white belongs to many contrived classes most of whose members are criminals, such as the class consisting of himself, Charles Manson, and Jeffrey Dahmer, but which do not correspond to genuine properties. Why "being myself or Charles Manson or Jeffrey Dahmer" is not projectible is an open philosophical problem, but that does not make the distinction baseless.

22. Some descriptions may be redundant. Where a is an individual, F the criterion, and $F_1, \ldots, F_n, \ldots$ the known properties of a at time t, F_i is redundant when $P(Fa/F_1a \,\&\, \ldots \,\&\, F_ia \,\&\, \ldots \,\&\, F_na \ldots) = P(Fa/F_1a \,\&\ldots\&\, F_{i+1}a \,\&\, F_{i+1}a \,\&\, \ldots F_na \,\&\, \ldots)$. Where i ranges over indices of irredundant properties, the probability of Fa at t is $P(Fa/\&_i F_ia)$.

Afterword

A Hypothetical Address by the President of the United States of America to a Joint Session of Congress and the American People

My fellow Americans: I wish to speak to you tonight more frankly than is usual about the topic of race.

You may wonder what remains to be said on this subject. Everyone knows there is a "race problem," which has been discussed constantly for the past half century. However, these discussions, hobbled by taboos, have been misguided. The first step in tackling any problem is to define it precisely, and most versions of the "race problem" rest on assumptions completely at variance with reality.

The problem is not "racism." Common sense tells us that systematic adverse treatment of blacks ended decades ago. And, despite endless repetition, it has never been shown that negative attitudes toward blacks have the devastating effects they are supposed to.

Nor is the problem racial inequality. Inequalities in and of themselves are not bad. Most people see nothing wrong with one athlete being better than another and gaining greater rewards from his skill, or with some groups outperforming other groups. Russians are stronger than Japanese and make superior weight-lifters. Everyone accepts these facts of nature, and anyway nothing can be done about them, certainly not with our present knowledge.

It has been denied again and again that racial inequalities are natural. But a review of the evidence, which has become increasingly accessible, has convinced me that they are. On average, blacks are less intelligent than whites and more impulsive, for largely biological reasons. This at bottom is why blacks don't do as well as whites in most of the endeavors our society rewards. I realize you may be shocked to hear me say this, but I assure you it is so. The cause of racial inequality is not malice or defective institutions. This is nonsense we must stop telling each other.

My advisors have urged me to enter disclaimers about race differences as soon as I mention them: to point out that they do not justify judging individuals solely by their race, that these differences are merely statistical, that some blacks are more intelligent than most whites. Such qualifications are correct, but what is the hurry to roll them out? Dwelling on them defensively suggests that the facts about race need to be softened, that mentioning them is something to apologize for—when more candor, not less, is needed. Most people realize that average group differences permit overlap. Nobody finds it necessary to say there are tall Japanese. Why, then, does what goes without saying elsewhere have to be underlined about race? Rushing to say what race differences are not, I fear, is a way of avoiding what they are.

The "race problem," then, is the friction produced when two populations differing in intelligence, emotional intensity, and concern for the future occupy the same geographical territory. What can be done about this friction? What would count as doing something?

I will put some possibilities before you in a few moments. But let me emphasize one absolutely vital, indispensable condition for making any progress at all: we must accept the facts as they are. Convincing you of this is the most important thing I can do tonight. Let me cite the words of the physicist Enrico Fermi: "Whatever Nature has in store for mankind, however unpleasant, men must accept, for ignorance is never better than knowledge." I am convinced that once we recognize the reality of race differences, fully and openly, sound ideas will follow. But should we continue to pursue the illusion that our troubles will be ended by more anti-discrimination laws and more hand-wringing about racism, America will become increasingly Balkanized, and perhaps cease to exist.

Many people say the idea of racial differences is a rationalization for the status quo. This charge gets everything backward. If the races do differ, the reasons for the present state of affairs are not "rationalizations," and the state of affairs itself needs no excuses. Suppose one day at the beach you announce that you are going to swim past the breakers. Someone, himself a good swimmer, warns you that you are not strong enough to fight the undertow. Do you instantly attack his motives, and assume he is rationalizing the status quo of his being thought the better athlete? Do you accuse him of trying to talk you into resigning yourself to an inferior status? No. The first thing you would want to know is whether he might be right. If he is, if you aren't able to swim past the breakers safely, there is nothing wrong with the status quo, his advice has probably been offered in good faith, and heeding it may save your life. We must stop attacking unwelcome news about race on the grounds that the messengers have bad motives, and then attack the motives of the messengers on the grounds that what they tell us are lies.

When claims of race differences are said to discourage blacks, I ask in reply: if the races are truly not equal in ability, what is gained by pretending they are? What is gained by saying blacks would be as successful and prosperous as whites if the rules were not slanted against them? Which message, the truth or the flattering lie, is more likely to inflame blacks to anger over constant failure? Which one requires so much backing and filling that its dishonesty must sooner or later become obvious to everyone?

"All right," say my advisors, who in this respect are like most Americans, "there are race differences. Now we know what the race problem is. What is the solution?" My friends, if a "solution" is supposed to be some magic formula, some way to make these differences go away or keep them from counting—

make the United States as harmonious as a racially uniform society—there may not be a solution. Looking for that kind of measure may be like trying to build a perpetual motion machine. We had better stop looking for "solutions" and start trying to do the best we can in the situation facing us.

And isn't it a little facile to expect that we can just announce that thinking about race has been misguided for many decades, and then move briskly ahead? New truths, like old ones rediscovered, take time to sink in. When the Earth was found to revolve around the Sun, refuting the age-old notion that man is the center of the universe, it would have been absurd to ask the very next day, "Now what? How should religion, philosophy and morality adjust to this new development?" Major errors must be unlearned. We have to retreat along all the false paths we have taken before we can start up the true one. This process, of accustoming ourselves intellectually and emotionally to thinking about race in new ways, cannot be hurried. I believe the impulse to acknowledge race differences as quickly as possible, and then demand "solutions," is yet another effort to avoid the facts themselves.

At some point in the future we will have to take concrete steps, and it is my duty as president to suggest some directions in which we might move, as well as some of the obstacles we may encounter. In doing so I must speak in general terms because, as I have said, neither I nor anyone else has all the answers now. A few voices, with new ideas, are beginning to be heard; perhaps this talk tonight will encourage still others. But I must warn you that any useful idea is likely to be radical. As Claudius says in *Hamlet*, "diseases desperate grown / By desperate appliance are relieved, or not at all."

It is obvious, to begin with, that policies based on error must end. One such policy is affirmative action, conceived as compensating blacks for the harm done them by whites. No damages are owed when no damage has been done, and the difficulty blacks have in competing in a white world are not the legacy of past wrongs—however regrettable those wrongs may have been—but a result of biology for which whites are not to blame. Affirmative action is an injustice to whites that whites legitimately resent. Ending it will help calm racial tempers.

As for more systematic approaches, there are three that can be taken: minimizing race differences, controlling their negative aspects, and laissez-faire. These strategies cut across the conventional classifications of Right and Left, and they are not mutually exclusive. Nonetheless, they represent different basic orientations.

"Minimizing," as the name suggests, means large-scale social engineering efforts to reduce the race gap as much as possible. The government would undertake job training and job creation for blacks, daycare for the children of trainees and workers, childhood enrichment programs, programs to convince black girls not to have illegitimate children, programs for antisocial youths, and preferences for blacks in employment and the academy—all openly intended to create equality, not right past wrongs. At the international level, reducing race differences would mean support of Africa and Haiti by the United States and prosperous European and Asian countries.

One drawback to this approach is white resentment. The richer white majority would have to fund the programs I have mentioned, and, once race differences are widely understood, whites may not wish to make sacrifices to end inequalities they did not cause. As I have said, equality as such is not a value many people find urgent. Advocates of minimizing the race gap might be forced to fall back on the rhetoric of "fighting racism" to defend policies whose aim is egalitarian, leading to hypocrisy and eventual exposure.

A second drawback of the minimizing approach is the uncertainty, indeed the unlikelihood, of success. One of my predecessors thought to combat black crime with basketball leagues, on the theory that young black males turn to crime because of the dreariness of their lives. This theory is very dubious, and treatments based on misdiagnoses seldom work. Training programs geared to individual ability are more promising, but they too may be ineffective, for reasons that require a brief foray into science. Genes help shape human behavior—that can no longer be denied—but exactly what genes accomplish depends on their environment. This interplay leads optimists to hope that the input of genes can always be canceled by manipulating the environment. The truth is that genes often limit what environmental manipulation can accomplish. Some effects of genes cannot be overcome at all, others can be reduced but not eliminated—that is why I have spoken of minimizing—and others still can be reduced only at prohibitive cost. There is no guarantee that the race gap in intelligence, school achievement, productivity, or impulsiveness can be reduced to any significant extent. Certainly, intervention efforts to date give no grounds for optimism. We must be prepared to discover that no amount of training and childhood enrichment can shrink the race gap. We must be prepared to recognize, at some point, that further experiments in social engineering would be throwing good money after bad. We must be prepared, in other words, to find minimizing a dead end.

Programs to help disadvantaged individuals of all races would avoid some of these problems. However, if training benefits whites more than blacks, another possibility that cannot be ruled out, race-blind social engineering will actually increase the race gap.

A longer-range intervention is use of incentives, including cash payments, to encourage people with desirable traits to have children, and those with undesirable traits to abstain. Some have talked of licensing parents. Consider the fact that college-educated black women have proportionally fewer children than college-educated white women, while less-educated black women have proportionally more children than less-educated white women. Encouraging the more intelligent members of all races to reproduce, it is said, would shrink the race gap in intelligence, as well as benefit society in other ways.

I oppose these programs. I don't think anyone wants to see the government deciding who should and should not have children. Even using public funds as an incentive to influence private reproductive decisions is highly problematic. And whatever their pros and cons, eugenic measures do not show results for several generations, which in human terms is close to a century. We don't have that much time.

Rather than trying to reduce race differences, we might try to control their disruptiveness. Since it is black crime, illegitimacy, and unemployment that deviate from the norms of the dominant society, a control approach would aim to instill self-discipline, respect for law, and the work ethic in blacks. Among the measures that might encourage these traits are the easing of regulatory burdens on black businesses, a more efficient criminal justice system, and the restriction of welfare to those employed on public works, or those with legitimate children, or unmarried women who agree to contraception. Perhaps welfare would be ended altogether. Black churches might be informally enlisted.

I have no doubt that curtailing public assistance would necessitate more responsible black behavior. The problem with this course of action is that, for better or worse, a great many people have come to depend on public assistance. Hundreds of thousands of women—of all races, but disproportionately black—have illegitimate children, no husband, and no marketable skills. The problem is especially acute in cities, where "living off the land" is not possible. A little later I will give you my ideas on this subject.

Crime control would emphasize discipline, swifter justice, and more effective punishment—possibly corporal punishment for those undeterred by imprisonment. Capital punishment would be a basic element of this approach. One can envisage new technologies such as brain scans and DNA testing that permit monitoring of potential criminals, although the most effective step might simply be a return to now-discredited practices, like the chain gang. Measures like these, though they would affect blacks disproportionately, would not be inherently racial. Other anti-crime measures might well be. Race, for instance, might become a legitimate "probable cause" for police intervention. The police would be allowed to search all black males for weapons, confiscating any they found.

The downside of many of these steps is their conflict with traditional civil liberties, potential overbroadness, and the risk of provoking disorder. At the same time, many blacks might be willing to surrender some civil liberties to restore the level of order their communities knew in earlier eras. It would be useful to poll blacks on this matter.

Both the minimizing and control approaches call for government intervention, although they disagree about what sort. Opposed to both is classical laissez-faire liberalism, the policy, or nonpolicy, of leaving almost everything to the free market. The laissez-faire approach would also end all forms of public assistance, on the moral ground that people should not be forced to support each other as well as the practical ground that it encourages sloth. At the same time, all laws against private discrimination would be repealed, allowing individuals to live, work, and trade with whomever they please. Private affirmative action would be allowed; those who continue to feel that blacks deserve compensation, or that preferring blacks is desireable for some other reason, would be free to do so—as would those who prefer whites, and those who wish to ignore race. One can imagine various sorts of bargaining going on, as both black and white workers sought employment with discriminators by offering, for instance, to accept lower wages. Firms that refer whites, those that prefer blacks, and those that prefer merit would compete against each other, in a real-world test of which policy is most efficient.

This prospect may be called "turning back the clock," but if many people have a strong wish to associate with others of their own race, fighting this natural impulse by insisting that they live, work, and attend school with members of other races only heightens animosity.

A laissez-faire government, aiming at efficiency, would be largely race neutral. Merit hiring and contracting with lowest bidders are cheaper than quotas and set-asides. However, as classical liberals regard control of violence as a key government function, they might permit some race-consciousness in the area of crime. As they have traditionally been civil libertarians as well, a balance would have to be struck on measures like race-based surveillance.

The free market accommodates race differences more easily than the other two approaches, since it does not aim at a preset outcome that biology may have put out of reach. It imposes discipline. A woman who knows that she alone will be responsible for herself and a baby she bears will be more careful about having babies. In the long run, letting the racial chips fall where they may is an attractive prospect.

But in the short run, as I have said, ending welfare will leave many in the lurch. Various temporary measures might soften these consequences. Announcing on a certain date that all forms of public assistance will end in five years, with support to be gradually reduced each year in the meanwhile, would give everyone time to seek work, help from private sources, or new skills from temporary public training programs. Present recipients of welfare with very young children could be grandfathered in.

A basic problem during this transition and in a postwelfare world—again I will be blunt—is that the labor of many blacks is not valuable to most people, and will become less valuable as brain-power becomes more important in an increasingly high-tech world. Consequently, I urge the repeal of the minimum wage. A black whose skills cannot command $5 an hour may yet find someone willing to pay him $1 an hour, and it must be possible for him to accept employment at that rate. Many people recoil at such a prospect as if it were a return to feudalism, but it is surely better that a laborer with limited skills receive a low salary for work that snobs consider demeaning than that he find no work at all. Other legislation and regulation that increases the cost of hiring low-skilled labor should be repealed. A commonplace of contemporary cant is worry that blacks can only get "dead-end" jobs, which forgets that it is the individual who is obliged to make himself useful to others.

I cannot promise that that segment of the black underclass grown dependent on public funds will find the end of welfare painless. And similar dilemmas are developing at the international level. The African population is outstripping its capacity to grow or buy food, and is being ravaged by AIDS Interventionists will want the United States to take collective action; others will insist that American taxpayers not be forced to support anyone else. But above all, *we must not assume that help is possible.* Perhaps the cycle of overpopulation and starvation in black Africa is beyond human control Triage, the denial of scarce resources to those past help, may be unavoidable These are hard sayings, but I remind you again that not every problem has a agreeable solution: we must learn to live with reality as it is, not as we wish to be.

Cutting across the three approaches I have outlined is the divide between race-consciousness and race-blindness. Right now public policy is schizophrenic jumble of both. For example, attorneys may not consider race in peremptorily challenging jurors, yet the racial makeup of juries is carefully monitored for enough blacks. Consistency and coherence should be restored.

Now you might imagine that if race differences exist, it automatically follows that they should always be taken into account. This is not so. Social policy can acknowledge race differences while reducing the relevance of race far below its present level.

The approach I have in mind might be called *realistic race blindness*: Racial classifications are never to be used, but disparate impact is accepted as a natural consequence of race differences. Of any course of action, the race-blind realist asks: Is it a good idea apart from its racial impact? Would we follow it if races did not exist? If the answer is "yes" we should follow it, with eyes open to the likelihood that it will affect the races differently. Disparate impact is anticipated but discounted, never allowed to exercise a veto.

Let me illustrate how this rule would treat an issue facing educators, the grouping of students by ability. The main argument in favor of "tracking," as it is also called, is that it allows bright students to be challenged without being held back by slow learners, and slow learners to move at their own pace without falling behind. Realistic race-blindness would accept tracking if the only argument against it were that it places disproportionately many blacks in the slower groups. Decide what is best on educational grounds, according to this attitude, and accept racial stratification as a by-product of differences in intelligence.

This rule would sometimes be difficult to apply. Reasons for balking are always ready to hand when disparate impact looms. I have heard educators say, to my mind preposterously, that tracking harms the bright student, when what really bothers them, I suspect, is white and Asian dominance of classes for the gifted. Race-blindness is impossible without recognition that differences in ability will assert themselves in almost every facet of life.

I have already mentioned that race might be used as one among many predictors of crime. Yet most crime-control measures could be race blind. The basic equation of social order is that, other things being equal, crime falls when its cost rises, so crime can be reduced by making punishment swifter, surer, and more disagreeable. Limits on appeals, corporal punishment, and the application of new technologies are, as I pointed out, inherently race neutral. Realistic race blindness would have us evaluate such steps on their own merits, while recognizing that any measure to curb crime will affect blacks more than whites. It is hard to balance a tolerable level of crime against the measures necessary to attain it, but race-blind realism keeps the prospect of disparate impact off the scales.

Some issues, though, present a different picture. One is public assistance. A purely race-blind assessment of welfare would focus on just three points: the morality of taxing Peter to support Paul, the incentives created by welfare, and the consequences of ending it—all the while recognizing that, since black families are many times more likely than white families to receive public assistance, any cutback will affect blacks disproportionately. Yet black and white responses to welfare seem to differ in ways that must be taken into account. For many years before 1960, when public assistance as we know it did not exist, the black illegitimacy was a flat 16%. This was far above the white illegitimacy of that period, but still below the 70% it is approaching. White illegitimacy has also risen, but not nearly as much, and, while both trends are disturbing, the steeper black increase seems tied to a greater willingness on the part of blacks to rely on public assistance. If reducing black illegitimacy is important, welfare policy may reasonably take this race difference into account.

Explicit consideration of race, or *realistic race consciousness*, would be triggered by a two-part test: a policy's racial impact is to be taken into account when individual reactions to the policy cannot be predicted, but it is known that blacks and whites will on average react to it in significantly different ways. Welfare reform meets this test because, while no one can tell how any particular teenage girl is going to be influenced by the availability of public subsidies for illegitimate children, teenage blacks are on average more likely than teenage whites to take advantage of it. Hence the two-part test permits race-consciousness about welfare. Likewise, it is impossibile to measure the maturity of every youthful offender, but if blacks mature more quickly than whites on average, which scientists tell us they may well do, black offenders might be treated as adults at an earlier age than white offenders. If blacks are on average less deterred than whites by the punishments currently attached to crime, and deterrence is an important goal, the two-part tests says that race can be taken into account in sentencing. It may be that, while the objectivity of particular jurors cannot be measured in advance, blacks tend on average to side with black victims and defendants regardless of the evidence in interracial cases. If so, the two-part test permits the exclusion of black jurors from such cases. It may be that blacks tend on average to be less objective than whites about black defendants no matter what the victim's race, in which case the two-part test sanctions limits on the number of blacks in juries.

I present the option of realistic race consciousness to make clear that there is nothing wrong in principle with taking race into account. Nonetheless, I expect that in practice realistic race blindness will usually prove superior. Equity and efficiency seldom requires explicit attention to race.

I would also encourage people to use realistic race blind standards for personal behavior. Should you wish to write a letter to the editor criticizing rap music on aesthetic grounds, and all that stops you is fear of "insulting the taste of blacks," write the letter. If your children and their friends want to form a science club, but you fear it will "exclude blacks" because blacks are less interested in science, help them form the club anyway if you think it is otherwise a good idea. Private decisions that are sound apart from their racial side-effects are sound, period.

Every course of action I have described requires complete candor. We canno try to close the race gap unless we recognize it for what it is and asses: progress, or its lack, accordingly. We cannot control black crime unless we ar honest about its causes. We cannot have a free market unless we are prepare for racial stratification. Classroom presentations of "cultural diversity" tha evade race differences are deceitful. "Speech codes" and "sensitivity training in universities should end. Centuries of warfare between enlightenment an obscurantism have made it plain that enlightenment is always better.

I will give you two examples of what we cannot have. In 1974, a journal of social thought published an article[1] whose authors, while disavowing the use of force against hereditarians, accepted "responsibility" for the chance that their readers might misunderstand them and take violent action. During the same period, an article in *Scientific American* concluded:

In the present racial climate of the US, studies on racial differences in IQ, however well intentioned, could easily be misinterpreted as a form of racism and lead to an unnecessary accentuation of racial tensions. Since we believe that, for the present at least, no good case can be made for such studies on either scientific or practical grounds, we do not see any point in particularly encouraging the use of public funds for their support.[2]

There is a case for limiting public funds for science to research that bears on national security. But defunding should not be used as a threat to silence unpopular opinions.

I am not alone in deploring censorship, but I believe racial censors are often misunderstood. I am convinced they do not think of themselves as seeking to suppress harmful truths, but as protecting the public, and particularly blacks, from harmful untruths. Were they to admit the reality of race differences, to themselves as well as others, they would no longer see merit in censorship. The issue underlying censorship is, once again, whether the races do in fact differ, and that, I reassure you, can no longer be denied.

I understand, and regret, that talk of genetic differences will upset many blacks. Nobody wishes to hear his group called unintelligent. I don't like to think that my words may distress or seem to demean my black friends and associates, and I wish I did not have to say what I am saying tonight. Let me emphasize, in this connection, that race differences are no excuse for personal unpleasantness. Members of each race should continue to treat members of every other race with the same courtesy they expect to receive. No-one should adopt an attitude of superiority in individual encounters. The fact remains, though, that certain distressing truths about group characteristics need to be said, and everyone, black and white, must come to terms with them.

It seems cruel to speak in what seems a negative way of a racial group because people don't choose the race they belong to. At the same time, most people acquire their religious convictions early in life, and have no more choice in being offended by perceived insults to their faith than by slights against their race, yet we do not for that reason refrain from criticizing religion or otherwise saying things that give offense. Many religious persons are disturbed by Darwinism, but few people oppose open discussion of evolution. Loss of cherished illusions and abandonment of dreams is often the price of wisdom. The impossibility of our hopes is seen first as a crisis, then a chronic problem, then, finally, accepted as part of the human condition. So it will be with race.

And, while the feelings of blacks deserve respect, so do those of whites. Not only have whites, particularly white males, been made to pay reparations for what they did not do, they have been vilified and ridiculed in public discourse to an extent unthinkable for other groups. They must be permitted to defend themselves, and if calling attention to unwelcome facts about nonwhites is part of their defense, that too must be allowed.

When Niels Bohr realized that atomic fission could be harnessed in a weapon, his assessment was terse: "We are in a completely new situation." So is America at the end of the 20th century. There has never been a society as racially heterogeneous as ours, and we are just now realizing that almost everything that happens between the races has a biological component. Race matters. The rethinking forced on us by this great fact will not be easy. Let it begin, bravely and honestly.

Thank you.

NOTES

1. Block and Dworkin (1976a).
2. Bodmer and Cavalli-Sforza (1970): 27.

Appendices

Appendix A

Time Preferences

Time discounting is often treated as exponential. Suppose you won't trade one unit of good G today unless assured of n Gs one unit of time (say, a day) hence. Then you discount time at the rate of $1/n = p$. Therefore, you won't trade today's one G without assurance of n^t Gs t days hence, so one G in t days is worth $p^t G$ to you now. The smaller p, the higher your time preference. (So, if black time preferences exceed white, a smaller p characterizes blacks.) Time discounting is exponential in Axelrod's (1984) theory of the evolution of cooperation.

It has been suggested more recently (Herrnstein 1990a, 1990b; Herrnstein and Prelec 1991, n.d.; Mazur 1987; Ainslie 1992) that time discounting is hyperbolic. On this analysis, the value of a unit G to you t days hence is proportional to $G/(Z + It)$, where the "coefficient of impulsivity" I measures the rate at which time-delay reduces perceived value and Z is a constant that keeps the value of G finite at the moment of consumption; see Ainslie (1992: 70–71). A natural hypothesis is that the mean black coefficient of impulsivity exceeds the white.

One widely noted difference between the two analyses may bear on race in an unexpected way. If at any time the perceived value of a good G exceeds that of another good G', then, on the exponential theory, the perceived value of G exceeds that of G' at any other time. Suppose $G' > G$, that G's consumption will be t days hence, G''s consumption will be t' days hence, $t' > t$, and that, at t, the discounted value of G exceeds that of G', that is, $p^t G > p^{t + [-(t - t')]} G' = p^{t'} G'$. This situation is represented in Figure A.1, where the x axis is the delay, so at the origin the consumption of G at 0 is t days away, and the delay for G' is $t + [-(t - t')] = t'$. Then clearly $G > p^{(t' - t)} G'$. Now pick any time t'' between t and 0. The perceived value of G at t'' is $p^{t - (t - t'')} G = p^{t''} G$, and that of G' at t'' is $p^{-(t - t')} + (t - (t - t'')) G''$, that is, $p^{t - t' + t''} G'$. But $p^{t''} G > p^{t - t' + t''} G'$ just in case $G > p^{(t' - t)} G'$, which we had by hypothesis; again see Figure A.1.

Hyperbolic curves may cross, however. Suppose for instance that $I = 2$,

Figure A.1
Exponential Discounting

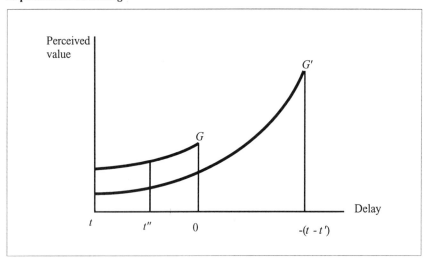

$Z = 1$, $G = 10$, and $G' = 50$, that attainment of G is 10 days off and attainment of G' is 20 days off. This situation is represented in Figure A.2, where the x axis is again the delay—so at the origin G is 10 days away and G' is 20—curve G is the discounted value of G as consummation approaches point, and curve G' is the

Figure A.2
$I = 2$

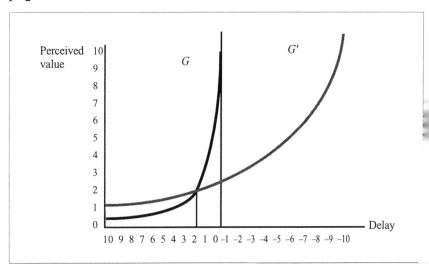

Figure A.3
$I = 10$

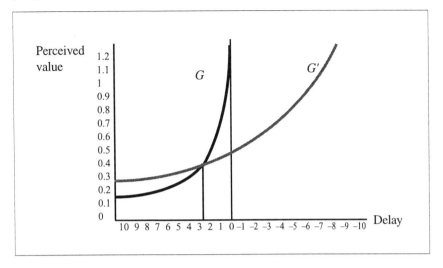

discounted value of G'. For the first 8 days, the discounted value of G' exceeds that of G, but for the 2 days before the consummation G, G is preferred. The puzzling experience of weakness of will, a knowing choice of the worse, may be construed as the temporary crossing of two hyperbolic valuation curves. The option chosen is objectively "worse" even at the moment of choice in that the chooser will regret it later and may at the time know he will regret it later. If the subject chooses G, he will feel regret at every moment after the G' curve exceeds 10, that is, after day 18. Moreover, since—before the curves cross—we want preference G' to prevail, we identify with G' (Ainslie [1992] describes psychological mechanisms by which we do so), and succumbing to G is experienced as something happening to us.

Race differences come into play with the observation that, as the impulsivity coefficient I increases, the discount curve for the dispreferred good tends to overtake the curve for the preferred good more quickly. Thus, if with the same parameters as before except for an I of 10 rather than 2, the G curve reaches the G' curve in 7.6 days (see Figure A.3). Calling the interval between the point at which $G = G'$ and the consummation of G the "period of weakness," periods of weakness lengthen with increasing impulsivity. If black coefficients of impulsivity are larger than white, blacks will more tend to behave in ways they subsequently regret, and which are perceived as happening to them rather than as actions of their own. This adds empirical substance to the idea broached in chapter 9 that blacks are less free than whites or Asians.

Appendix B

A Formal Treatment of Genetic Causation

Let G be a genotype expressed by phenotypic variable P, and let G's home organism O display phenotypic value P_i in environment E_i. The proposed analysis takes G as a many-one function $G(E)$ from the E_i to the P_i. $G(E)$ can also be more explicitly written as G_O or $G(E, O)$. The apparent circularity in defining G in terms of a phenotype expressing G can be avoided by taking G as the set $\{ <E_i, P_i>: O \text{ displays } P_i \text{ in } E_i \}$. Corresponding to (1)–(10) in section 2 are the following:

(1) *G expresses itself differently in different environments* when $(\exists E, E')G(E) \neq G(E')$. *Interaction occurs* when $(\exists G, G', E')G(E) \neq G'(E')$. The expressed phenotypic difference between genotypes *varies with environment* when $(\exists G, G', E, E')\, G(E) - G'(E) \neq G(E') - G'(E')$.

(2) *O and O' have the same gene for P* when $G(O) = G(O')$, that is, when $(\forall E)G(E, O) = G(E, O')$. Hence

(3) *O and O' have different genes for P* when $(\exists E)G(E, O) \neq G(E, O')$.

(4) A phenotypic difference between O and O' is *genetic in origin* in environment E when $G_O(E) \neq G_{O'}(E)$.

(5) *Genetic diversity, phenotypic identity*: We may have $G_O \neq G_{O'}$ yet $(\exists E)G(E, O) = G(E, O')$, which follows trivially from many-oneness.

(6) *Genetic diversity in all environments, phenotypic identity*: We may have $G_{O'} \neq G_O$ *and* $(\forall E)G(E, O) \neq G(E, O')$, yet $(\exists E, E')[E \neq E' \text{ and } G(E, O) = G(E', O')]$.

(7) *Phenotypic identity does not imply genotypic identity*: $(\exists E, E')G(O, E) = G(O', E')$ does not imply $(\forall E)G(E, O) = G(E, O')$—another consequence of many-oneness.

(8) *Genotypic identity does not imply phenotypic identity*: We may have $G(O) = G(O')$, yet $(\exists E, E')\{E \neq E' \ \& \ G(E, O) \ [= G(E, O')] \neq G(E', O')\}$.

(9a) Let P be considered a set of values, and $R(f)$ the range of function f. Then P is *in O's genes* when $(\exists G)(\forall E, R(G_O) \in P$. Phenotypic value P_i is in O's genes when $(\exists G)(\forall E)G(O, E) = P_j$.

(9b) Call the environments in which O's ancestors evolved O-*ancestral* environments. P is *natural* when $(\exists G)(\forall E)(E$ is O-ancestral $\supset R(G_O) \in P)$.

(10a) Where P is a metric trait expressing G, O is *innately more* P than O' if $(\forall E)G(O, E) > G(O', E)$.

(10b) Where A and B are humans, A is innately more P than B *for all practical purposes* if $(\forall E)$(Human society is possible in $E \supset G(A, E) > G(B, E))$.

(10c) O is *naturally more* P *than* O' if $(\forall E)(E$ is O- and O'-ancestral $\supset G(O, E) > G(O', E)$.

Let E be the domain of G, O a set of organisms, and P a set of values. G and E are *correlated* if for some i, $R(G \upharpoonright \{E_i \times O\})$ is smaller than P. If P expresses G, let $P(t)$ and $E(t)$ be the values of P and E at time t. Then a G/E correlation is active-reactive just in case P is recursive in G. For clearly $P(t) = G(E(t))$. But there is also a function G' mapping phenotypes to environments at just-later times, so that $E(t) = G'(P(t-1))$. If we let $G\# = G(G')$, $P(t+1) = G\#(P(t))$. The correlation is positive when $G\#$ is monotone increasing, negative when $G\#$ is monotone decreasing. In general P is not relative recursive, since G' is not always a function.

Appendix C

A Logical Difficulty

Naturalism says that no act a is obligatory, forbidden or permitted. Yet the conjunction of

1. $(\forall a)\sim(a$ is obligatory)
2. $(\forall a)\sim(a$ is forbidden)
3. $(\forall a)\sim(a$ is permitted)

appears inconsistent. Since by definition an act is permitted if and only if it is not forbidden (or neither forbidden nor obligatory), 3 seems to imply $\sim$2 (or 3&2 seems to imply $\sim$1). An analogous problem arises, incidentally, for relational theories of spacetime, which reject absolute velocity and absolute rest. A relational theory asserts both

4. $(\forall x)(\forall n > 0)\sim(x$ is moving at n MPH)

and

5. $(\forall x)\sim(x$ is at rest),

yet 4 seems to imply $\sim$5.

Both problems can be solved by distinguishing sentential negation, written here as "$\sim$," from predicate complementation, expressible as "$\neg$." "$\neg Px$" holds when x satisfies the complement of P, while "$\sim[Px]$" holds when x fails to satisfy P. $\sim[Px]$ usually implies $\neg Px$—if x does not satisfy P, x usually satisfies not-P—but when P is ill-defined we have both $\sim[Px]$ and $\sim[\neg Px]$. Hence the implication from "not not Px" to "Px" can be blocked by reading the first "not" as negation and the second "not" as complementation.

One can also now deny that an object x is moving at 0 MPH without affirming that x moves at n MPH for some $n > 0$. As x moves at 0 mph $\equiv$ ($\forall n > 0)(\neg x$ is moving at n mph), 5 may be rewritten as

5'. $(\forall x)\sim(\neg x$ is moving at n MPH),

which is consistent with 4. Similarly, 3 rewritten as

3'. $(\forall a)\sim(\neg a$ is forbidden),

which is consistent with 2.

References

Adams, H. 1988. *Melanin: Biophysics of Melanin and Consciousness*. Videocassette. New York: KM-WR Science Consortium.

Adarand Construction, Inc. v. Peña. 1995. 115 S.Ct. 2097.

Adler, J. 1993. "Crime Rates by Race and Causal Relevance." *Journal of Social Philosophy* 24: 176–184.

———. 1994. "More on Race and Crime: Levin's Reply." *Journal of Social Philosophy* 25: 105-114.

Adler, Je., McCormick, J., Washington, F., Carroll, G., Liu, M., Brant, M., Shenitz, B., Joseph, J., and Gill, J. 1994. "A Week in the Death of America." *Newsweek*, Aug. 15: 24–43.

Aiello, L. 1993. "The Fossil Evidence for Modern Human Origins in Africa: A Revised View." *American Anthropologist* 95: 98.

Ainslie, G. 1992. *Picoeconomics*. Cambridge: Cambridge University Press.

Alexander, L. 1992. "What Makes Wrongful Discrimination Wrong? Biases, Preferences, Stereotypes and Proxies." *University of Pennsylvania Law Review*, Nov.: 149–219.

Allport, G. 1954. *The Nature of Prejudice*. New York: Addison-Wesley.

Ama, P., Simonau, J., Boulay, M., Serresse, O., Theriault, G., and Bouchard, C. 1986. "Skeletal Muscle Characteristics in Sedentary Black and Caucasian Males." *Journal of Applied Physiology* 61: 1758–1761.

American Association of University Women. 1991. *Shortchanging Girls, Shortchanging America*. Washington, D.C.: Author.

Anderson, E. 1994. "The Code of the Streets." *Atlantic Monthly* 273, 5: 81–94.

Andreasen, N., Flaum, M., Swayze, V., O'Leary, D., Alliger, R., Cohen, G., Ehrhardt, J., and Yuh, T. 1993. "Intelligence and Brain Structure in Normal Individuals." *American Journal of Psychiatry* 150: 130–134.

Anson, Robert Sam. 1988. *Best Intentions: The Education and Killing of Edmund Perry*. New York: Vintage.

AP. 1993. "U.S. Judge, Citing Racism, Gives Black Defendant Lesser Sentence." *New York Times*, Feb. 20: A8.

———. 1994a. "Trade Schools Win Reprieve on Aid Cutoff." *New York Times*, June 29: A17.

———. 1994b. "Killing and Rape in a Florida National Forest." *New York Times*,

Feb. 23: A10.

———. 1994c. "Study Finds Racial Disparity in Warnings to the Pregnant." *New York Times*, Jan. 20: A16.

———. 1995. "Texas Executes Man Convicted in 1977 Killing." Feb. 22: A11.

Applebome, P. 1992. "Mississippi Hearings in Evers Slaying Pits Trial Rights against Civil Rights." *New York Times*, Oct. 18: A18.

———. 1995. "Equal Admission Standards Leave Mississippi's Black Universities Wary." *New York Times*, Apr. 24: 1996.

Argyle, M. 1988. *Bodily Communication*, 2nd ed. London: Methuen.

Armour, J. "Race Ipse Loquitur: Of Reasonable Racists, Intelligent Bayesians, and Involuntary Negrophobes." *Stanford Law Review* 46: 781–816. *1994*

Arvey, R., Abraham, L., Bouchard, T., and Segal, N. 1989. "Job Satisfaction: Environmental and Genetic Components." *Journal of Applied Psychology* 74, 2: 187–192.

Ashenfelter, O., and Mooney, J. 1968. "Graduate Education, Ability and Earnings." *Review of Economic Statistics* 50, 1: 78–86.

Asimov, I. 1979. *In Memory Yet Green*. New York: Doubleday

———. 1989. *Chronicle of Science and Discovery*. London: Grafton Books.

Asmin, S. 1995. "Metaphors of Race." *American Philosophical Quarterly* 32: 13–29.

Axelrod, R. 1984. *The Evolution of Cooperation*. New York: Basic Books.

Axelrod, R., and Hamilton, W. 1981. "The Evolution of Cooperation." *Science* 211, 27: 1390–1396.

Ayer, A. 1982. *Philosophy in the Twentieth Century*. New York: Random House.

Baier, K. 1958. *The Moral Point of View*. Ithaca, N.Y.: Cornell University Press.

Baird, C. 1996. "Rights, Freedom, and Rivalry." *The Freeman* 46: 667–669.

Baker, J. 1974. *Race*. Oxford: Oxford University Press.

Baker, Jo. 1988. *Arguing for Equality*. London: Routledge Chapman & Hall.

Baldus, D., Woodworth, G., and Pulaski, A. 1992. "Law and Statistics in Conflict: *McClesky v. Kemp*." In Kagehiro and Laufer, eds.: 251–271.

Banfield, E. 1974. *The Unheavenly City Revisited*. Boston: Little, Brown.

Barbanel, J. 1993. "School Financing Not Less for the Poor, Study Says." *New York Times*, Oct. 5: B3.

Baughman, E. B., and Dahlstrom, W. G. 1968. *Negro and White Children: A Psychological Study in the Rural South*. New York: Academic Press.

Baumeister, R., Smart, L., and Boden, J. 1996. "Relation of Threatened Egotism to Violence and Aggression: The Dark Side of High Self-Esteem." *Psychological Review* 103, 1: 5–33.

Baumrind, D. 1991. "To Nurture Nature." *Behavioral and Brain Sciences* 14: 386–387.

Bayley, N. 1965. "Comparisons of Mental and Motor Test Scores for Ages 1–15 Months by Sex, Birth Order, Race Geographical Location, and Education of Parents." *Child Development* 36: 379–477.

Beals, K., Smith, C., and Dodd, S. 1984. "Brain Size, Cranial Morphology and Time Machines." *Current Anthropology* 25: 301–315.

Beardsley, T. 1995. "For Whom the Bell Curve Really Tolls." *Scientific American*, Jan.: 14–17.

Bedini, S. 1972. *The Life of Benjamin Banneker*. New York: Scribner's.

Beer, W. 1987. "Resolute Ignorance; Social Science and Affirmative Action." *Society*, May/June: 63–69.

———. 1988. "Sociology and the Effects of Affirmative Action: A Case of Neglect." *The American Sociologist* 19: 218–231.

Bell, W. 1983. "Bias, Probability, and Prison Populations: A Future for Affirmative Actions?" *Futurics* 7: 19.

Benbow, C., and Lubinski, D., ed. 1992. *From Psychometrics to Giftedness: Essays in*

Honor of Julian C. Stanley.

Benbow, C., and Stanley, J. 1984. "Sex Differences in Mathematical Reasoning: More Facts." *Science* 222: 1029–1031.

Benjamin, J., Li, L., Patterson, C., Greenberg, B., Murphy, D., and Hamer, D. 1996. "Population and Familial Association between the D$_4$ Dopamine Receptor Gene and Measures of Novelty Seeking." *Nature Genetics*, Jan. 12: 81-84.

Benn, S. I. 1967. "Justice." *Encyclopedia of Philosophy*. New York: Macmillan.

Berger, J. 1987. *New York Times News Service*, June 19.

Bergmann, B. 1996. *In Defense of Affirmative Action*. New York: Basic Books.

Berlin, I. 1939. "Verifiability in Principle." *Proceedings of the Aristotelian Society* 39: 225–248.

————. 1969. *Four Essays on Liberty*. Oxford: Oxford University Press.

Bernal, M. 1991. *Black Athena: The Afroasiatic Roots of Classical Civilization*. New Brunswick, N.J.: Rutgers University Press.

Berofsky, B. 1971. *Determinism*. Princeton, N.J.: Princeton University Press.

Berry, J., and Dasen, R., eds. 1974. *Culture and Cognition*. London: Methuen.

Blake, A., and Bray R. 1993. "A Stolen Bicycle, An Innocence Lost." *New York Times*, Sept. 12: 13, 29.

Blinkov, S., and Glezer, I. 1968. *The Human Brain in Figures and Tables*. Trans. Basil Haigh. New York: Basic Books.

Block, N. 1995. "How Heritability Misleads about Race." *Cognition* 56: 99–128.

Block, N., and Dworkin, G. 1976a. "IQ, Heritability and Inequality." In Block and Dworkin, eds.: 410–540.

————, eds. 1976b. *The IQ Controversy*. New York: Pantheon.

Block, W. 1976. *Defending the Undefendable*. New York: Fleet Press.

————. 1992. "Discrimination: An Interdisciplinary Analysis." *Journal of Business Ethics* 11: 241–254.

————. 1993. "The Economics of Discrimination: Some Further Thoughts." Unpublished.

Bloom, A. 1988. *The Closing of the American Mind*. New York: Simon and Schuster.

Bluestone, N. 1987. *Women and the Ideal Society*. Amherst: University of Massachusetts Press.

Blum, K., Noble, E., Sheridan, P., Finley, O., Montgomery, A., Ritchie, T., Ozkaragoz, Fitch, R., Sadlack, F., Sheffield, D., Dahlmann, T., Halbardier, S., and Nogami, H. 1991. "Association of the A1 Allele of the D$_2$ Dopamine Receptor Gene with Severe Alcoholism." *Alcohol* 8: 409–416.

Bock, G., and Goode, J., eds. 1996. *Genetics of Criminal and Antisocial Behavior*. Chichester: John Wiley & Sons.

Bodmer, W., and Cavalli-Sforza, L. 1970. "Intelligence and Race." *Scientific American* 223: 19–27.

Bolick, C. 1996. *Affirmative Action Fraud*. Washington, D.C.: Cato Institute.

Boorse, C. 1975. "On the Distinction between Disease and Illness." *Philosophy and Public Affairs* 5: 49–68.

Bouchard, T. 1983. "Do Environmental Similarities Explain the Similarity of Identical Twins Reared Apart?" *Intelligence* 7: 175–184.

————. 1984. "Twins Reared Together and Apart: What They Tell Us About Human Diversity." In Fox, ed.: 147–184.

————. 1987. "The Hereditarian Research Program: Triumphs and Tribulations." In Modgil and Modgil, eds.: 55–75.

————. 1990. "The Genetic Architecture of Human Intelligence." In Vernon, ed.: 33–93.

————. 1994. "Genes, Environment, and Personality." *Science* 264: 1700–1703.

————. 1995. "IQ Similarity in Twins Reared Apart: Findings and Responses to

Critics." In Sternberg and Grigorenko, eds., in press.

Bouchard, T., Lykken, D., McGue, M., Segal, N., and Tellegen, A. 1990. "Sources of Human Psychological Differences: The Minnesota Study of Twins Reared Apart." *Science* 250: 223–228.

Bouchard, T., Lykken, D., Tellegen, A., and McGue, M. 1992. "Genes, Drives, Environment and Experience: EPD Theory-Revised." In Benbow and Lubinski, eds., in press.

Bouchard, T., and McGue, M. 1981. "Familial Studies of Intelligence: A Review." *Science* 212: 1055–1059.

Boxill, B. 1978. "The Morality of Preferential Hiring." In Wasserstrom, ed.: 158–174.

———. 1984. *Blacks and Social Justice*. Totowa, N.J.: Rowan Allendheld.

Boyd, R. 1988. "How to Be a Moral Realist." In Sayre-McCord, ed: 181–228.

Boykin, A. 1986. "The Triple Quandary and the Schooling of Afro-American Children." In Neisser, ed.: 57–92.

Braithwaite, R. 1964. *Theory of Games as a Tool for the Moral Philosopher*. Cambridge: Cambridge University Press.

Brand, C. 1987. "The Importance of General Intelligence." In Modgil and Modgil, eds.: 251–267.

———. 1996. *The g Factor*. Chichester: John Wiley & Sons.

Bray, R. 1993. "A Black Panther's Long Journey." *New York Times Magazine*, Jan. 31: 21–76.

Breland, H. 1979. *Population Validity and College Entrance Measures*. New York: College Board.

Brimelow, P., and Spencer, L. 1993a. "The Hidden Clue." *Forbes,* Jan. 4: 48.

———.1993b. "When Quotas Replace Merit, Everybody Suffers." *Forbes,* Feb. 15: 80–102.

Brink, D. 1989. *Moral Realism and the Foundations of Ethics*. Cambridge: Cambridge University Press.

Brody, N. 1992. *Intelligence*, 2nd. ed. New York: Academic Press.

Brody, N., and Zuckerman, M. 1988. "Oysters, Rabbits and People: A Critique of 'Race Differences in Behavior' by J. P. Rushton." *Personality and Individual Differences* 9: 1025–1033.

———. 1987. "Jensen, Gottfredson, and the Black/White Difference in Intelligence Test Scores." *Behavioral and Brain Sciences* 10: 507–508.

Broman, S., Brody, N., Nichols, P., Shaughnessy, P., and Kennedy, W. 1987. *Retardation in Young Children*. Hillsdale, N.J.: Erlbaum.

Brooks-Gunn, J., Klebanov, P., and Duncan, G. 1996. "Ethnic Differences in Children's Intelligence Test Scores: Role of Economic Deprivation, Home Environment and Maternal Characteristics." *Child Development 67:* 396–408.

Brown, W., and Reynolds, M. 1975. "A Model of IQ, Occupation, and Earnings." *American Economic Review* 65 1002–1007.

Burd, S. 1993. "Racial Imbalance at NIH." *Chronicle of Higher Education*, Nov. 10: A28.

Burnham, S. 1993. *America's Bimodal Crisis,* 3rd. ed. Athens, Ga.: Foundation for Human Understanding.

Burt, C. 1966. "The Genetic Determination of Differences in Intelligence: A Study of Monozygotic Twins Reared Together and Apart." *British Journal of Psychology* 57: 137–153.

Butterfield, F. 1995. "More in U.S. Are in Prisons, Report Says." *New York Times*, Aug. 10: A14.

Caccone, A., and Powell, J. 1989. "DNA Divergence Among Hominoids." *Evolution* 43: 925–942.

Calhoun, C. 1988. "Justice, Care, Gender Bias." *Journal of Philosophy* 85, 9:

451–463.

Campbell, C. A. 1957. *On Selfhood and Godhood*. London: Unwin.

Carey, G., et al. 1982. "Genetics and Personality Inventories: The Limits of Replication with Twin Data." *Behavior Genetics* 8: 299–313.

Carrey, J. 1993. "The Great White Dopes." *San Jose Mercury News*, June 15.

Carlson, J. 1985. "The Issue of *g*; Some Relevant Questions." *The Behavioral and Brain Sciences* 8: 225.

Carnap, R. 1936. "Testability and Meaning." *Philosophy of Science* 3: 419–471.

Carroll, J. 1991. "Cognitive Psychology's Psychometric Lawgiver." *Contemporary Psychology* 36: 557–559.

———. 1993. *Human Cognitive Abilities*. Cambridge: Cambridge U. P.

Cartwright, N. 1983. *How the Laws of Physics Lie*. Oxford: Oxford University Press.

Cascio, W. F., Outtz, J., Zedeck, S., and Goldstein, I. 1991. "Statistical Implications of Six Methods of Test Score Use in Personnel Selection." *Human Performance* 4: 233–264.

Cattell, R. 1950. *Personality*. New York: McGraw-Hill.

———. 1992. "Virtue in 'Racism.' " *Mankind Quarterly* 32: 281–284.

Cavalli-Sforza, L., Menozzi, P., and Piazza, A. 1993. "Demic Expansions and Human Evolution." *Science* 259: 639–646.

———. 1994. *The History and Geography of Human Genes*. Princeton, N.J.: Princeton University Press.

Celis, W., 1992. "Push for Higher Admission Standards." *New York Times*, Dec. 4: C8

Chafetz, Z. 1990. *Devil's Night, and Other True Tales of Detroit*. New York: Random House.

Chakraborty, R., Kamboh, M., Nwanko, M., and Ferrell, R. 1992. "Caucasian Genes in American Blacks: New Data." *American Journal of Human Genetics* 50: 145–155.

Chipeur, H., Rovine, M., and Plomin, R. 1990. "LISREL Modelling: Genetic and Environmental Influences on IQ Revisited." *Intelligence* 14: 11–29.

Chisholm, R. 1964. *Human Freedom and the Self*. Topeka: University of Kansas Press.

Chomsky, N. 1976. "The Fallacy of Richard Herrnstein's IQ." In Block and Dworkin, eds.: 285–298.

Churchland, P. 1981. "Eliminative Materialism and Propositional Attitudes." *Journal of Philosophy* 78, 2: 67–90.

Clark, L., and Halford, G. 1983. "Does Cognitive Style Account for Cultural Differences in Scholastic Achievement?" *Journal of Cross-Cultural Studies* 14: 279–296.

Clarke, C., and Sheppard, R. 1966. "A Local Survey of the Distribution of Industrial Melanic Forms in the Moth *Biston betularia* and Estimates of the Selective Values of them in an Industrial Environment." *Proceedings of the Royal Society of Biology* 165: 424–439.

Coate, S., and Loury, G. 1992. *Will Affirmative Action Policies Eliminate Negative Stereotypes?* Boston: Ruth Pollack Series of Working Papers on Economics, Department of Economics, Boston University.

Cohen, J. 1983. *Statistical Power Tests*, 2nd ed. New York: Academic Press.

Cohen, M., Nagel, T., and Scanlon, T., eds. 1977. *Equality and Preferential Treatment*. Princeton, N.J.: Princeton University Press.

Cole, N. 1973. "Bias in Selection." *Journal of Educational Measurement* 10: 237–255.

Cole, N., and Moss, P. 1989. "Bias in Test Use." In Linn, ed.: 201–219.

Coleman, J. S., Campbell, E. Q., Hobson, C. J., McPartland, J., Mood, A. M., Weinfeld, F. D., and York, R. L. 1966. *Equality of Educational Opportunity*. Washington, D.C.: U.S. Department of Health, Education, and Welfare.

Commons, M., Mazur, J., Nevin, J., and Rachlin, H., eds. 1987. *Quantitative Analyses of Behavior, V: The Effect of Delay and of Intervening Events on Reinforcement Value*. Hillside, N.J.: Erlbaum.

Copp, D., and Zimmerman, D., eds. 1984. *Morality, Reason and Truth*. Totowa, N.J.: Rowman and Allenheld.

Coren, S. 1994. *The Intelligence of Dogs: Canine Consciousness & Capabilities*. New York: The Free Press.

Countryman, J. 1994. "The Other Side of the Wall." *New York Times Book Review*, Jan. 23: 2.

Crocker, J., and Major, B. 1989. "Social Stigma and Self-Esteem: The Self-Protective Properties of Stigma." *Psychological Review* 96: 608–630.

Cronbach, L., and Drenth, P., eds. 1981. *Mental Tests and Cultural Adapation*. The Hague: Mouton.

Crow, J. F. 1969. "Genetic Theories and Influences: Comments on the Value of Diversity." *Harvard Educational Review* 39: 301–309.

Crow, J., and Kimura, D. 1970. *An Introduction to Population Genetics Theory*. New York: Harper and Row.

Culotta, E. 1992. "Two Generations of Struggle: A Special Report on Minorities in Science." *Science* 258: 1176–1237.

Currie, J., and Thomas, D. 1995. "Does Head Start Make a Difference?" *American Economic Review* June: 341–364.

Dahlstrom, W. 1986. "Ethnic Status and Personality Measurement." In Dahlstrom, Lachar and Dahlstrom.: 3–23.

Dahlstrom, W., Lachar, D., and Dahlstrom, L. 1986. *MMPI Patterns of American Minorities*. Minneapolis, Minn.: University of Minnesota Press.

Daly, M. 1996. "Evolutionary Adaptationism: Another Biological Approach to Criminal and Antisocial Behavior." In Bock and Goode, eds.: 183–191.

Daly, M., and Wilson, M. 1988. "Evolutionary Social Psychology and Family Homicide." *Science* 242: 519–524.

Daniels, N. 1979. "Wide Reflective Equilibrium and Theory Acceptance in Ethics." *Journal of Philosophy* 76: 256–282.

Darnton, J. 1994a. " 'Lost Decade' Drains Africa's Vitality." *New York Times*, June 19: A1–A10.

———. 1994b. "In Poor, Decolonized Africa, Bankers are New Overlords." *New York Times*, June 20: A1–A9.

Davidson, D. 1970. "Mental Events." In Swanson and Foster, eds.: 79–102.

Davis, L. 1979. *Theory of Action*. Englewood Cliffs, N.J.: Prentice-Hall.

Dawkins, R. 1976. *The Selfish Gene*. Oxford: Oxford University Press.

———. 1986. *The Blind Watchmaker*. New York: Norton.

DeBerry, K. 1991. "Modeling Ecological Competence in African American Transracial Adoptees." Unpublished Doctoral Dissertation. Charlottesville VA: University of Virginia.

DeFries, J. 1972. "Quantitative Aspects of Genetics and Environmental Determination of Behavior." In Ehrman, L., Omenn, G., and Caspari, E., eds.: 6–16.

DeFries, J., Loehlin, J., and Plomin, R. 1979. "Genotype-Environment Interaction and Correlation in the Analysis of Human Behavior." *Psychological Bulletin* 34: 309–322.

DeFries, J., Vandenberg, S., and McClearn, G. 1976. "Genetics of Specific Cognitive Abilities." *Annual Review of Genetics* 10: 179–207.

DeMott, B. 1996. "Sure, We're All Just One Big Happy Family." *New York Times*, Jan. 7: 2-1–2-31.

De Palma, A. 1992. "Drop in Black Ph.D.'s Brings Debate on Aid for Foreigners." *New York Times*, Apr. 21: A1–A18.

Deparle, J. 1992. "The Civil Rights Battle Was Easy Next to the Problems of the Ghetto." *New York Times*, May 17: 4, 1.

Derr, R. 1989. "Insights on the Nature of Intelligence from Ordinary Discourse." *Intelligence* 13: 113–118.

De Vos, G., Wetherall, W., and Stearman, K. 1983. *Japan's Minorities: Burakumin, Koreans, Ainu and Okinawans*. New York: Minority Rights Group.

Diamond, J. 1994. "Race without Color." *Discover*, Nov.: 83–89.

Dietrichson, P. 1967. "What Does Kant Mean by 'Acting from Duty'?" In Wolff, ed.: 314–330.

Diop, C. 1974. *The African Origin of Civilization: Myth or Reality?* New York: L. Hill.

Donnellan, K. 1966. "Reference and Definite Descriptions." *The Philosophical Review* 75: 218–304.

Draper, P. 1989. "African Marriage Systems: Perspectives from Evolutionary Ecology." *Ethology and Sociobiology* 10: 145–169.

Dreger, R., and Miller, K. 1960. "Comparative Psychological Studies of Negroes and Whites in the United States." *Psychological Bulletin* 57: 361–402.

———. 1968. "Comparative Psychological Studies of Negroes and Whites in the United States: 1959–1965. *Psychological Bulletin Monograph Supplement*, 70.

Dretske, F. 1971. "Conclusive Reasons." *Australasian Journal of Philosophy* 49: 1–22.

———. 1981. *Knowledge and the Flow of Information*. Cambridge, Mass.: MIT Press.

D'Souza, D. 1995. *The End of Racism*. New York: The Free Press.

Dugger, C. 1992. "Litany of Signals Overlooked in Child's Death." *New York Times* Dec. 24: A1-B2.

———. 1993. "Young, Impressionable and Accused of Murder." *New York Times*: May 15: A1.

Dunham, W. 1990. *Journey Through Genius*. New York: Penguin.

Dworkin, R. 1977a. "Why Bakke Has No Case." In Wasserstrom, ed.: 138–147.

———. 1977b. *Taking Rights Seriously*. Cambridge, Mass.: Harvard University Press.

Ebstein, R., Novick, O, Umansky, R., Priel, B., Osher, Y., Blaine, D., Bennett, E., Nemanov, L., Katz, M., and Belmaker, R. 1996. "Dopamine D4 Receptor (D4R4) Exon III Polymorphism Associated with the Human Personality Trait of Novelty Seeking." *Nature Genetics*, Jan. 12: 78–80.

The Economist. 1993. "Great Expectations." June 26: 21.

Egan, V., Chiswick, A., Santosh, C., Naidu, K., Rimmington, J., and Best, J. 1994. "Size Isn't Everything: A Study of Brain Volume, Intelligence and Auditory Evoked Potentials." *Personality and Individual Differences* 17, 3: 357–367.

Ehrman, L., Omenn, G., and Caspari, E., eds. 1972. *Genetics, Environment, Behavior*. New York: Academic Press.

Elion, V., and Megargee, E. 1975. "Validity of the MMPI *Pd* Scale among Black Males." *Journal of Consulting and Clinical Psychology* 43: 166–172.

Ellis, L. 1987a. "Evolution and the Nonlegal Equivalent of Aggressive Behavior." *Aggressive Behavior* 12: 57–71.

———. 1987b. "The Victimful-Victimless Crime Distinction, and Seven Universal Demographic Correlates of Victimful Criminal Behavior." *Personality and Individual Differences* 9: 525-548.

———. 1991. "A Biosocial Theory of Social Stratification Derived from the Concepts of Pro/Antisociality and r/K Selection." *Politics and the Life Sciences* 10: 5–17.

Ellis, L., and Hoffman, H., eds. 1990. *Crime in Biological, Social and Moral Contexts*. New York: Praeger.

Ellis, L., and Nyborg, H. 1992. "Racial/Ethnic Variations in Male Testosterone Levels: A Probable Contributor to Group Differences in Health." *Steroids* 57: 72–75.

Ely, J. 1983. "Professor Dworkin's External/Personal Preference Distinction." *Duke Law Journal* 1983: 959–986.

Encyclopedia Britannica, 11th ed. 1914. London: Encyclopedia Britannica Corporation.

Englehardt, T. 1986. *Foundations of Bioethics*. Oxford: Oxford University Press.

Epstein, R. 1992. *Forbidden Grounds*. New York: Cambridge University Press.

Eysenck, H. A. 1993. "The Biological Basis of Intelligence." In Vernon, ed.: 1–32.

Ezorsky, G. 1991. *Racism and Justice: The Case for Affirmative Action*. Ithaca, New York: Cornell University Press.

Falconer, D. S. 1989. *Introduction to Quantitative Genetics*, 3rd ed. New York: John Wiley & Sons.

FBI 1993. *Uniform Crime Reports of the FBI for 1992*. Washington, D.C.: Department of Justice.

Feinberg, J. 1989a. "Psychological Egoism." In Feinberg, ed.: 489–500.

Feinberg, J., ed. 1989b. *Reason and Responsibility*, 7th ed. Belmont, Calif.: Wadsworth.

Feldman, M., and Lewontin, R. 1975. "The Heritability 'Hang-Up.'" *Science* 190: 1168.

Fischer, C., Hout, M., Jankowski, M., Lucas, S., Swidler, A., and Voss, K. 1996. *Inequality by Design*. Princeton, N.J.: Princeton University Press.

Fleming, E., and Anttonen, R. 1971. "Teacher Expectancy as Related to the Academic and Person Growth of Primary-Age Children." *Monographs of the Society for Research in Child Development* 36.

Fletcher, R. 1991. *Science, Ideology and the Media: The Cyril Burt Scandal*. New Brunswick, N.J.: Transaction.

Floderus-Myrhed, B., et al. 1980. "Assessment of Heritability of Personality Based on a Short Form of the Eysenck Personality Inventory: A Study of 12,898 Twin Pairs." *Behavior Genetics* 10: 153–162.

Flynn, J. 1980. *Race, IQ and Jensen*. London: Routledge and Kegan Paul.

———. 1984. "The Mean IQ of Americans: Massive Gains 1932 to 1978." *Psychological Bulletin* 95: 29–51.

———. 1987a. "Massive IQ Gains in 14 Nations: What IQ Tests Really Measure." *Psychological Bulletin* 101: 171–191.

———. 1987b. "Race and IQ: Jensen's Case Refuted." In Modgil and Modgil, eds.: 221–232.

Fodor, J. 1974. "Special Sciences, or the Disunity of Science as a Working Hypothesis." *Synthèse* 28: 77–115.

Fogelin, R., and Sinnott-Armstrong, W. 1997. *Understanding Arguments*, 5th ed. New York: Harcourt Brace Jovanovich.

Forrester, J. 1989. *Why You Should: The Pragmatics of Deontic Speech*. Hanover, N.H.: Brown University Press.

Fox, R., ed. 1975. *Biosocial Anthropology*. New York: John Wiley and Sons.

Fox, S., ed. 1984. *Individuality and Determinism*. New York: Plenum.

Frank, B. 1992. "Race and Crime: Let's Talk Sense," *New York Times* Jan. 13: A15.

Frankfurt, H. 1971. "Freedom of the Will and the Concept of a Person." *Journal of Philosophy* 68: 5–20.

———. 1987. "Equality as a Moral Ideal." *Ethics* 98: 21–43.

Freeman, R. 1983. "Crime and Unemployment." In Wilson, J., ed. *Crime and Public Policy*. Rutgers, N.J.: Transaction.

Freifeld, K. 1994. "Jury: Killer Didn't 'Snap'." *Newsday*, June 10: A28.

French, H. 1994. "As War Factions Shatter, Liberia Falls into Chaos." *New York Times*, Oct. 22: A12

———. 1995. "Mobutu, Zaire's 'Guide,' Leads Nation into Chaos." *New York Times*, June 10: A1–A5.

Frey, D. 1994. *Last Shot: City Streets, Basketball Dreams*. Boston: Houghton Mifflin.

Friedman, D. 1989. *The Machinery of Freedom*, 2nd ed. La Salle, Ill.: Open Court.

Fulda, J. 1993. *Are There Too Many Lawyers? and Other Vexatious Questions.* Irvington-on-Hudson: Foundation for Economic Education.

Fullenwider, R. 1980. *The Reverse Discrimination Controversy*. Totowa, N.J.: Littlefield, Adams.

Garber, H. 1988. *The Milwaukee Project: Preventing Mental Retardation in Children at Risk*. Washington, D.C.: American Association on Mental Retardation.

Gardner, H. 1983. *Frames of Mind*. New York: Basic Books.

Garner, W., and Wigdor, A. 1982. *Ability Testing I & II*. Washington, D.C.: National Academy Press.

Garrett Press. N. d. *Directory of Special Programs for Minority Group Members*. Garrett Park, Md.: Garrett Park Press.

Gates, H. 1989. "TV's Black World Turns—but Stays Unreal." *New York Times*, Nov. 12: H1.

Gauthier, D. 1967. "Morality and Advantage." *Philosophical Review* 76: 460–475.

———. 1984. *Morals by Agreement*. Oxford: Oxford University Press.

Genovese, E. 1967. *The Political Economy of Slavery*. New York: Pantheon.

George, N. 1992. *Elevating the Game*. New York: HarperCollins.

Gert, B. 1970. *The Moral Rules*. New York: Harper & Row.

Gest, T. 1995. "A Shocking Look at Blacks and Crime." *U.S. News & World Report*, Oct. 16: 53–54.

Gewirth, A. 1978. *Reason and Morality*. Chicago: University of Chicago Press.

Gibbard, A. 1982. "Human Evolution and the Sense of Justice." *Midwest Studies in Philosophy* 7: 31–46.

Gibbons, A. 1990. "Our Chimp Cousins Get That Much Closer." *Science* 250: 376.

———. 1995. "Out of Africa—At Last?" *Science* 267: 1272–1273.

Gibbs, W. 1995. "For Biological Studies, Minorities Need Not Apply." *Scientific American*, Mar.: 106.

Gibson, M., and Ogbu, J., eds. 1991. *Minority Status and Schooling: A Comparative Study of Immigrants and Involuntary Minorities*. New York: Garland.

Gilliard, D., and Beck, A. 1995. "State and Federal Prisons Report Record Growth During Last 12 Months." Washington, D.C.: Bureau of Justice Statistics, Office of Justice Programs, U.S. Department of Justice.

Gilligan, C. 1983. *In a Different Voice*. Cambridge, Mass.: Harvard University Press.

Goldberg, C. 1995. "Joker, 12, Dies in a Rage of Bullets and a Youth, 16, is Held." *New York Times*, July 20: B1-B5.

Goldberg, S. 1977. *The Inevitability of Patriarchy*, 2nd ed. London: Temple Smith.

———. 1992. *When Wish Replaces Thought*. Buffalo, NY: Prometheus.

Goldberger, A., and Manski, C. 1995. "Review Article: *The Bell Curve* by Herrnstein and Murray." *Journal of Economic Literature* 33: 762–776.

Goldin, D. 1995. "Increasingly, Those Paying Full Tuition are Supporting Their Poorer Peers." *New York Times*, Mar. 22: B7.

Goldman, A. 1967. "A Causal Theory of Knowing." *Journal of Philosophy* 64: 357–372.

———. 1986. *Epistemology and Cognition*. Cambridge, Mass.: Harvard University Press.

Goleman, D. 1990. "Anger Over Racism Seen as Cause of Blacks' High Blood Pressure," *New York Times*, Apr. 14: C3.

Goodman, N. 1966. *Fact Fiction and Forecast*, 3rd ed. Indianapolis, Ind.: Bobbs-Merrill.

Gordon, R. 1975a. "Crime and Cognition: An Evolutionary Perspective." *Proceedings of the 11th International Symposium on Criminology:* 7–55. Sao Paulo:

International Centre for Biological and Medico-Forensic Criminology.
———. 1975b. "Prevalence: the Rare Datum in Delinquency Measurement and Its Implications for the Theory of Delinquency." In Klein, M., ed.: 201–284.

Gottfredson, L. 1986. "Societal Consequences of the *g* Factor in Employment." *Journal of Vocational Behavior* 29: 379–410.

———. 1987. "The Practical Significance of Black-White Differences in Intelligence." *The Behavioral and Brain Sciences* 10: 510–512.

———. 1988. "Reconsidering Fairness: A Matter of Social and Ethical Priorities." *Journal of Vocational Behavior* 33: 293–319.

Gould, J. 1982. *Ethology*. New York: Harper and Row.

Gould, S. 1978. "Morton's Ranking of Races by Cranial Capacity." *Science* 200: 503–509.

———. 1981. *The Mismeasure of Man*. New York: Norton.

———. 1984. "Human Equality is a Contingent Fact of History." *Natural History*, Nov.: 26–33.

———. 1994. "Curveball." *The New Yorker*. Nov. 28: 139–149.

———. 1995. "Age-Old Fallacies of Thinking and Stinking." *Natural History*, June: 6–13.

Graglia, L. 1988. "The 'Remedy' Rationale for Requiring or Permitting Otherwise Prohibited Discrimination: How the Court Overcame the Constitution and the 1964 Civil Rights Act." *Suffolk Law Review* 22: 569–621.

Green, M., and Winkler, D. 1980. "Brain Death and Personal Identity." *Philosophy and Public Affairs* 9, 2: 105–133.

Greenawalt, K. 1983. *Discrimination and Reverse Discrimination*. New York: Knopf.

Greenwood, J. 1994. *Realism, Identity, and Emotion*. Beverly Hills, Calif.: Sage.

Grice, H. P. 1957. "Meaning." *Philosophical Review* 66: 377–388.

Griggs v. Duke Power Co. 1971. 401 U.S. 424.

Gross, B. 1977a. "Is Turnabout Fair Play?" In Gross, ed.: 379–388.

———, ed. 1977b. *Reverse Discrimination*. Buffalo, N.Y.: Prometheus.

Gutmann, A. 1980. *Liberal Equality*. Cambridge: Cambridge University Press.

Gynther, M. 1972. "White Norms and Black MMPIs: A Prescription for Discrimination?" *Psychological Bulletin* 78: 386–402.

Hacker, A. 1992. *Two Nations*. New York: Scribner's.

Hacking, I. 1995. "Pull The Other One." *London Review of Books* 26 January: 3–5.

Hagen, R. 1979. *The Biosexual Factor*. New York: Doubleday.

Hahn, M., Hewitt, J., Henderson, N., and Benno, R., eds. 1990. *Developmental Behavior Genetics*. New York: Oxford University Press.

Haier, R. 1993. "Cerebral Glucose Metabolism and Intelligence." In Vernon, ed.: 317–332.

Haier, R., Siegel, B., MacLachlan, A., Soderling, E., Lottenberg, S., and Buchsbaum, M. 1992. "Regional Glucose Metabolic Changes After Learning a Complex Visuospatial/Motor Task: A Positron Emission Tomographic Study." *Brain Research* 570: 134–143.

Haier, R., Siegel, B., Nuechterlein, K., Hazlett, E., Wu, J., Paek, J., Browning, H., and Buchsbaum, M. 1988. "Cortical Glucose Metabolic Rate Correlates of Abstract Reasoning and Attention Studied with Positron Emission Tomography." *Intelligence* 12: 199–217.

Haier, R., Siegel, B., Tang, C., Lennart, A., and Buchsbaum, M. 1992. "Intelligence and Changes in Regional Cerebral Glucose Metabolic Rate Following Learning." *Intelligence* 16: 415–426.

Halpern, D. 1995. "The Skewed Logic of the Bell-Shaped Curve." *Skeptic* 3: 64–71.

Hamill, P. 1993. "How to Save the Homeless—and Ourselves." *New York Magazine* Sept. 22: 34–39.

Hamilton, W. 1963. "The Evolution of Altruistic Behavior." *The American Naturalist* 97: 354–356.

———. 1964. "The Genetical Theory of Social Behavior, I, II" *Journal of Theoretical Biology* 7: 1–52.

———. 1975. "Innate Social Aptitudes of Man: An Approach from Evolution Genetics." In Fox, ed.: 133–155.

Hanley, R. 1994. "Carjack Suspect Confessed, Rockland Prosecutor Says." *New York Times*, Jan. 11: B3.

Hanushek, E. 1986. "The Economics of Schooling: Production and Efficiency in Public Schools." *Journal of Economic Literature* 24, 3: 1141–1177.

Hare, B. 1985. "Stability and Change in Self-Perception and Achievement Among Black Adolescents: A Longitudinal Study." *The Journal of Black Psychology* 11, 2: 29–42.

Hare, R. 1952. *The Language of Morals*. Oxford: Oxford University Press.

Harman, G. 1977. *The Nature of Morality*. Oxford: Oxford University Press.

———. 1986a. "Moral Explanations of Natural Facts." *The Southern Journal of Philosophy* 24: 57–68.

———. 1986b. *Change of View*. Cambridge, Mass.: MIT Press.

Harman, H. 1976. *Modern Factor Analysis*. Chicago: University of Chicago Press.

Harper, P. 1994. "Treating My Son Like a Thief." *Newsday*, June 27: A22–A24.

Harrison, E. 1993. "Richmond Will Stop 'Clustering' White Students." *Los Angeles Times*, Feb. 25: A1.

Harrison, G., Weiner, J., Ranner, J., and Barnicott, W. 1964. *Human Biology*. London: Oxford University Press.

Harrison, P. 1972. *The Drama of Nonmo*. New York: Grove.

Hart, M. 1992. *The 100: A Ranking of the Most Influential Persons in History*. New York: Citadel.

Hartigan, J., and Wigdor, A. 1989. *Fairness in Employment Testing*. Washington, D.C.: National Academy Press.

Harwood, S. 1991. "Abstract: Fullenwider on Affirmative Action as Compensation." *Proceedings of the American Philosophical Association* 64: 68.

———. 1993. "The Justice of Affirmative Action." In Hudson and Peden, eds.: 77–89.

Helmreich, W. 1982. *The Things They Say Behind Your Back*. New York: Doubleday.

Hempel, C. 1952. "Problems and Changes in the Empiricist Criterion of Meaning." *Revue Internationale de Philosophie* 11: 41–63.

———. 1965a. "The Theoretician's Dilemma." In Hempel, 1965b: 173–226.

———. 1965b. *Aspects of Scientific Explanation*. New York: The Free Press.

Henderson, N. 1982. "Human Behavior Genetics." *Annual Review of Psychology* 33: 403–440.

Hendrick, P. 1984. *Population Biology*. Boston: Jones and Bartlett.

Henry, W. 1994. *In Defense of Elitism*. New York: Doubleday.

Herrnstein, R. 1990a. "Behavior, Reinforcement and Utility." *Psychological Science* 1: 217–223.

———. 1990b. "Rational Choice Theory: Necessary but Not Sufficient." *American Psychologist*, Mar. 356–367.

Herrnstein, R., and Murray, C. 1994. *The Bell Curve*. New York: The Free Press.

Herrnstein, R., and Prelec, D. 1991. "Preferences or Principles: Alternative Guidelines for Choice." In Zeckhauser, ed.: 319–339.

———. N.d. "Melioration." Freiburg, Germany: European Study Group for Evolutionary Economics.

Herrnstein, R., and Wilson, J. 1985. *Crime and Human Nature*. New York: Simon and Schuster.

Herszenhorn, D. 1995. "S.A.T. With Familiar Anxiety but New (Higher) Scoring." *New*

York Times, Apr. 2: A22.

Hettinger, E. 1997. "What is Wrong with Reverse Discrimination?" In Mappes and Zembaty, eds.: 304–314.

Hindelang, M. 1978. "Race and Involvement in Common Personal Crime." *American Sociological Review* 43: 93–109.

Hirsch, E. 1989. *Cultural Literacy*. Boston: Houghton Mifflin

Hirsch, J. 1970. "Behavior-Genetic Analysis and Its Biosocial Consequences." *Seminars in Psychiatry* 2: 89–105.

Hirschi, T., and Hindelang, M. 1977. "Intelligence and Delinquency: A Revisionist View." *American Sociological Review* 42: 571–587.

Hitchcock, D. 1976. *Intellectual Development and School Achievement of Youths 12–17 Years: Demographic and Socioeconomic Factors*. Washington, D.C.: Department of Health, Education and Welfare.

Ho, K-C., Roessmann, U., Straumfjord, J., and Monroe, G. 1980. "Analysis of Brain Weight: I. Adult Brain Weight in Relation to Sex. Race, and Age." *Archives of Pathology and Laboratory Medicine* 104: 635–639.

Hobart, R. 1934. "Free Will as Involving Determinism and Inconceivable Without It." *Mind* 43: 1–27.

Hocutt, M. 1977. "Skinner on the Word 'Good': A Naturalistic Semantics for Ethics." *Ethics* 87: 319–338.

Hoffman, P. 1994. "The Science of Race." *Discover*, Nov.: 4.

Hogg, R., and Craig, A. 1970. *Introduction to Mathematical Statistics*, 3rd ed. London: Macmillan.

Hohfeld, W. 1923. *Fundamental Legal Conceptions*. New Haven: Yale University Press.

Holly, E. 1996. "Waiting for a Black Heathcliff." *New York Times*. Mar. 26: A22.

Holt, J. 1994. "Anti-Social Science." *New York Times*. Oct. 19: A23.

Horai, S., Hayasaka, K., Kondo, R., Tsugane, K., and Takahata, N. 1995. "African Origin of Modern Humans Revealed by Complete Sequences of Hominoid Mitochondrial DNAs." *Proceedings of the National Academy of Sciences* 92, 2: 532–536.

Horgan, J. 1993. "Eugenics Revisited." *Scientific American* 268, 6: 122–127.

Horwich, P. 1991. *Truth*. Cambridge, Mass.: Basil Blackwell.

Howison, C. and Urbach, P. 1985. *Scientific Reasoning*. La Salle, Ill.: Open Court.

Hudson, Y., and Peden, C., eds. 1993. *The Bill of Rights: Bicentennial Reflections*. Lewiston, N.Y.: Edward Mellen.

Humphries, L. 1991. "Limited Vision in the Social Sciences." *American Journal of Psychology* 104: 333–353.

Humphries, L., Fleischman, A., and Lin, P. C. 1977. "Causes of Racial and Socioeconomic Differences in Cognitive Tests." *Intelligence* 11: 191–208.

Hunter, J. 1986. "Cognitive Ability, Cognitive Aptitudes, Job Knowledge, and Job Performance." *Journal of Vocational Behavior* 29: 340–362.

Hyland, W. 1991. "Downgrade Foreign Policy." *New York Times*, May 20: A15.

Ingrassia, M. 1993. "Three Generations of Single Mothers." *Newsweek*, Aug. 30: 25.

Irvine, S. and Berry, J., eds. 1988. *Human Abilities in Cultural Context*. Cambridge: Cambridge University Press.

Itzkoff, S. 1994. *The Decline of Intelligence in America*. Westport, Conn: Praeger.

Jacoby, T. 1992. "A Portrait of Black and White." *Wall Street Journal,* March 24 A14.

James, G. 1952. *Stolen Legacy*. Trenton, N.J.: African World Press.

Jaynes, G., and Williams, R., eds. 1989. *A Common Destiny: Blacks in America. Society*. Washington D.C.: National Academy Press.

Jeffrey, R. 1974. "Preference Among Preferences." *Journal of Philosophy* 71: 377–391.

Jencks, C. 1972. *Inequality*. New York: Basic Books.
———. 1992. *Rethinking Social Policy*. Cambridge, Mass.: Harvard University Press.
Jensen, A. 1969. "How Much Can We Boost IQ and Scholastic Achievement?" *Harvard Educational Review* 39: 1–123.
———. 1972. *Genetics and Education*. New York: Harper and Row.
———. 1973. *Educability and Group Differences*. New York: Harper and Row.
———. 1980. *Bias in Mental Testing*. New York: The Free Press.
———. 1981. *Straight Talk About Mental Tests*. New York: The Free Press.
———. 1985a. "Compensatory Education and the Theory of Intelligence." *Phi Delta Kappan*, April: 554–558.
———. 1985b. "The Nature of the Black-White Difference on Various Psychometric Tests: Spearman's Hypothesis." *The Behavioral and Brain Sciences* 8: 192–218.
———. 1986. "*g*: Artifact or Reality?" *Journal of Vocational Behavior* 29: 301–331.
———. 1987. "Further Evidence for Spearman's Hypothesis Concerning Black-White Differences on Psychometric Tests." *Behavioral and Brain Sciences* 10, 3: 512–519.
———. 1989. "Raising IQ Without Increasing *g*? A Review of Garber, *The Milwaukee Project*." *Developmental Review* 9: 234–258.
———. 1991a. "Spearman's *g* and the Problem of Educational Equality." *Oxford Review of Education* 17: 169–187.
———. 1991b. "Understanding *g* in Terms of Information Processing." *Educational Psychology Review* 4: 271–308.
———. 1993. "Spearman's Hypothesis Tested with Chronometric Information-Processing Tasks." *Intelligence* 17: 47–77.
———. 1994. "Psychometric *g* Related to Differences in Head Size." *Personality and Individual Differences* 17, 5: 597–605.
Jensen, A., and Figueroa, 1975. "Forward and Backward Digit-Span Interaction with Race and IQ: Predictions from Jensen's Theory." *Journal of Educational Psychology* 67: 882–893.
Jensen, A., and Inouye, A. 1980. "Level I and Level II Abilities in Asian, White and Black Children." *Intelligence* 4: 41–49.
Jensen, A., and Johnson, F. 1994. "Race and Sex Differences in Head Size and IQ." *Intelligence* 18: 309–333.
Jensen, A., and Naglieri, J. 1987. "Comparison of Black-White Differences on the WISC-R and the K-ABC: Spearman's Hypothesis." *Intelligence* 11: 21–43.
Jensen, A., and Reynolds, C. 1982. "Race, Social Class and Ability Patterns on the WISC-R." *Personality and Individual Differences* 3: 423–438.
Jensen, A., and Sinha, S. 1993. "Physical Correlates of Human Intelligence." In Vernon, ed.: 139–242.
Jerison, H. 1973. *Evolution of the Brain and Intelligence*. New York: Academic Press.
Jones, J., and Hochner, A. 1973. "Racial Differences in Sports Activities." *Journal of Personality and Social Psychology* 27: 86–95.
Jordan, J. 1969. "Physiological and Anthropometric Comparisons of Negroes and Whites." *Journal of Health, Physical Education, and Recreation* 40: 93–99.
Joselit, J. 1983. *Our Gang; New York Crime and the New York Jewish Community, 1900–1940*. Bloomington: University of Indiana Press.
Joynson, R. 1989. *The Burt Affair*. London: Routledge.
Judson, G. 1993. "Goal: Integration by Consensus." *New York Times*, Nov. 11: B1–B29.
Juel-Nielsen, N. 1965. "Individual and Environment: A Psychiatric-psychological Investigation of Monozygotic Twins Reared Apart." *Acta Psychiatrica Scandinavia* Supplementum 183.

Kagan, S. 1989. *The Limits of Morality*. Oxford: Oxford University Press.

Kagehiro, D., and Laufer, S., eds. 1992. *Handbook of Psychology and Law*. New York: Springer-Verlag.

Kamin, L. 1974. *The Science and Politics of IQ*. Hillsdale, N.J.: Erlbaum.

———. 1995. "Behind the Curve." *Scientific American*, February: 99–103.

Kaniel, S., and Fisherman, S. 1991. "Level of Performance and Distribution of Errors in the Progressive Matrices Test: A Comparison of Ethiopian Immigrant and Native Israeli Adolescents." *International Jouranl of Psychology* 26: 25–33.

Kaplan, R. 1992. "Continental Drift." *The New Republic*, Dec. 28: 15–20.

Katz, J. 1989. "Statement to the Senate Committee on the Judiciary Concerning the Relationship between Race and the Death Penalty." Oct. 2. Unpublished.

Kavka, G. 1986. *Hobbesian Moral and Political Philosophy*. Princeton, N.J.: Princeton University Press.

Keller, L., Arvey, R., Bouchard, T., and Segal, N. 1992. "Work Values: Genetic and Environmental Influences." *Journal of Applied Psychology* 77, 1: 79–88.

Kemper, T. 1990. *Social Structure and Testosterone*. New Brunswick, N.J.: Rutgers University Press.

Kendall, M. 1960. *The Advanced Theory of Statistics*, 3rd ed. New York: Halfner.

Kendall, M., Verster, M., and Von Mollendorf, J. 1988. "Test Performance of Blacks in South Africa." In Irvine and Berry, eds.: 299–339.

Kennedy, W., et al., 1963. *A Normative Sample of Intelligence and Achievement of Negro Elementary School Children in the Southeastern United States. Monographs for the Society for Research on Child Development* 28: 6.

Kerlie, D. 1996. "Equality." *Ethics* 106: 274–294.

Kim, J. 1990. "Supervenience as a Philosophical Concept." *Metaphilosophy* 221: 1–27.

———. 1993. *Supervenience and the Mind*. New York: Cambridge University Press.

Kimble, G. 1956. *Principles of General Psycholog*. New York: Ronald.

Kleck, G. 1991. *Point Blank: Guns and Violence in America*. New York: Aldine de Gruyten.

———. 1995. Personal Communication: Aug. 17.

Kleck, G., and Patterson, E. 1993. "The Impact of Gun Control and Gun Ownership Levels on Violence Rates." *Journal of Quantitative Criminology* 9, 3: 249–287.

Klein, M., ed. 1975. *The Juvenile Justice System*. Beverly Hills, Calif.: Sage.

Klein, S., Petersilia, J., and Turner, S. 1990. "Race and Imprisonment Decisions in California." *Science* 247: 812–816.

Klich, L. 1988. "Aboriginal Cognition and Psychological Nescience." In Irvine and Berry, eds.: 427–452.

Klineberg, O. 1935a. *Negro Intelligence and Selective Migration*. New York: Columbia University Press.

———. 1935b. *Race Differences*. New York: Harper and Bros.

Kochman, T. 1983. *Black and White: Styles in Conflict*. Chicago: University of Chicago Press.

Kohlberg, L. 1981. *The Philosophy of Moral Development*. New York: Harper and Row.

Kolata, G. 1987. "Early Signs of School Age IQ." *Science* 236: 774–775.

Kolbert, E. 1993. "TV Viewing and Selling by Race." *New York Times*, Apr. 5: D3.

Korner, S., ed. 1974. *Practical Reason*. New Haven, Conn.: Yale University Press.

Korematsu v. United States. 1944. 323 U.S. 214, 65 S. Ct. 193, 89 L.Ed. 194.

Kotelchuk, M. 1994. "The Adequacy of Prenatal Care Utilization Index: Its U.S Distribution and Association with Low Birthweight." *American Journal of Public Health* 84, 9: 1486–1489.

Kozol, J. 1992. *Savage Inequalities*. New York: HarperCollins.

Knowlton, G. 1995. "Questions About Remedial Education in a Time of Budget Cuts." *New York Times*, June 7: B11.

Krauthammer, C. 1994. " 'Recentered' Scores Just Another Step Toward Mediocrity." *Chicago Tribune*, June 17: A18.

Kripke, S. 1973. *Naming and Necessity*. Cambridge, Mass.: Harvard University Press.

———. 1980. *Wittgenstein on Private Language and Following a Rule.* Cambridge, Mass.: Harvard University Press.

Kristol, I. 1978. *Two Cheers for Capitalism*. New York: Basic Books.

Ladd, J. 1973. *Ethical Relativism.* Belmont, CA: Wadsworth.

Larry P. v. Wilson Riles 1979. 48 U.S.L.W. 2298.

Layzer, D. 1976. "Science or Superstition? A Physical Scientist Looks at the IQ Controversy." In Block and Dworkin, eds.: 194–241.

Lee, F., Schipp, E., and Tabor, M. 1992. "Life at 'Jeff': Tough Students Wonder Where Childhood Went." *New York Times* Mar. 7: 28.

Lee, M.-W. 1992. " 'Programming' Minorities for Medicine." *Journal of the American Medical Association* 267, 17: 2391–2394.

Lee, Y. 1991. "Koreans in Japan and the United States." In Gibson and Ogbu, eds.: 131–167.

Lefcourt, H. 1965. "Risk-Taking in Negro and White Adults." *Journal of Personality and Social Psychology* 2: 765–770.

Lefkowitz, M. 1996. *Not Out of Africa.* New York: Basic Books.

Lemann, N. 1991. *The Promised Land.* New York: Knopf.

Leonard, J. 1983. "Anti-discrimination or Reverse Discrimination: The Impact of Changing Demographics, Title VII and Affirmative Action on Productivity." Working Paper 1240. Boston: National Bureau of Economic Research.

Leplin, J., ed. 1984. *Scientific Realism*. Berkeley: University of California Press.

Lerner, R. 1996. "Acquittal Rates by Race for State Felonies." In Reynolds, ed.: 85–93.

Leslie, C. 1990. "Scientific Racism: Reflections on Peer Review, Science and Ideology." *Social Science and Medicine* 31, 8: 891–912.

Levay, S. 1991. "A Difference in Hypothalamic Structure Between Heterosexual and Homosexual men." *Science* 253: 1034–1037

Levin, M. 1974. "Kant's Derivation of the Formula of Universal Law as an Ontological Argument." *Kant-Studien* 65, 1: 50–66.

———. 1979. *Metaphysics and the Mind-Body Problem.* Oxford: Oxford University Press.

———. 1980. "Reverse Discrimination, Shackled Runners and Personal Identity." *Philosophical Studies* 37: 13–149.

———. 1981. "Equality of Opportunity." *Philosophical Quarterly* 31: 110–125.

———. 1982a. "Review of S. Gould, *The Mismeasure of Man* and A. Jensen, *Straight Talk about Mental Tests*."*Policy Review* 21: 173–178.

———. 1982b. "A Hobbesian Minimal State." *Philosophy and Public Affairs* 11: 338–353.

———. 1984a. "Negative Liberty." *Philosophy and Social Policy* 2: 84–100.

———. 1984b. "Why Homosexuality is Abnormal." *The Monist* 67: 251–283.

———. 1984c. "Reply to Pennock." *Philosophy and Public Affairs* 13: 263–267.

———. 1984d. "What Kind of Explanation is Truth?" In Leplin, ed.: 124–139.

———. 1987a."Rigid Designators: Two Applications." *Philosophy of Science* 54: 283–294.

———. 1987b. *Feminism and Freedom.* New Brunswick, N.J.: Transaction.

———. 1989a. "To the Lighthouse." *Philosophia* 19: 461–470.

———. 1989b. "Conditional Rights." *Philosophical Studies* 55: 211–213.

———. 1989c. Review of "L. May, *The Morality of Groups*." *Constitutional*

Commentary 6, 2: 523–532.

———. 1990. "Realisms." *Synthése* 85: 115–138.

———. 1991. "The Realism-Reification Controversy in Marketing." *Journal of Macromarketing* 11, 1: 57–65.

———. 1992a. "The Reader Replies." *American Scholar* 61 479–480.

———. 1992b. "Women, Work, Biology and Justice." In Quest, ed.: 9–25.

———. 1992c. "Responses to Race Differences in Crime." *Journal of Social Philosophy* 23, 1: 5–29.

———. 1993a. "Stove on Gene Worship." *Philosophy* 68: 240–243.

———. 1993b. "Review of C. McGinn, *Mental Content.*" *Nous* 27, 1: 137–139.

———. 1993c. "Reliabilism and Induction." *Synthése* 88: 307–334.

———. 1994a. "Comment on the Minnesota Transracial Adoption Study." *Intelligence* 19: 13–20.

———. 1994b. "Race, Biology and Justice." *Public Affairs Quarterly* 1994, 8: 267–285.

———. 1995. "Tortuous Dualism." *The Journal of Philosophy* 97: 313–323.

———. 1996. "Homosexuality, Abnormality, and Civil Rights." *Public Affairs Quarterly* 10, 1: 31–48.

———. 1997. "Plantinga on Functions and the Theory of Evolution." *Australasian Journal of Philosophy* 76. To appear.

Levy, C. 1993. "Norwalk Police Find Focus Turns to Them After a Death." *New York Times*, July 23: B1–B4.

Lewis, B. 1990. *Race and Slavery in the Middle East.* Oxford: Oxford University Press.

Lewontin, R. 1976a. "Race and Intelligence." In Block and Dworkin, eds.: 78–92.

———. 1976b. "Further Remarks on Race and Intelligence." In Block and Dworkin, eds.: 107–112.

———. 1976c. "The Analysis of Variance and the Analysis of Cause." In Block and Dworkin, eds.: 179–193.

———. 1982. *Human Diversity.* New York: Scientific American Library.

Lewontin, R., Rose, S., and Kamin, L. 1984. *Not in Our Genes.* New York: Pantheon.

Liebow, E. 1967. *Talley's Corner.* Boston: Little, Brown.

Lindsay, T. 1996. "The Impact of Multiculturalism on Education in America." *The Journal of Social, Poltical and Economic Studies* 20, 3: 289–298.

Linn, R. 1973. "Fair Test Use in Selection." *Journal of Education Research* 43: 139–161.

———. 1982. "Ability Testing: Individual Differences, Prediction, and Differential Prediction." In Garner and Wigdor, eds.: 335–388.

———. 1989, ed. *Educational Measurement*, 3rd ed. New York: Macmillan.

Loehlin, J., Lindzey, G., and Spuhler, J. 1975. *Race Differences in Intelligence.* San Francisco: Freeman.

London, R. 1991. "Judge's Overruling of Crack Law Brings Turmoil." *New York Times*, Jan. 11: B5.

Loury, G. 1992. "Incentive Effects of Affirmative Action." *Annals of the AAPSS* 523: 19–29.

Lovejoy, C. 1981. "The Origin of Man." *Science* 211: 341–350.

Lubasch, A. 1992. "Murder Indictments for 3 Youths in Killing of Brooklyn Principal." *New York Times*, Dec. 24: B3.

Luce, R., and Raiffa, H. 1957. *Games and Decisions.* New York: John Wiley and Sons.

Ludlow, H. G. 1956. "Some Recent Research on the Davis-Eels Game." *School and Society* 84: 146–148.

Lycan, W. 1986. "Moral Facts and Moral Knowledge." *The Southern Journal of Philosophy* 24: 79–93.

Lynch, K. 1984. "Totem and Taboo in Sociology: The Politics of Affirmative Action Research." *Sociological Inquiry* 54: 124–141.
———. 1989. *Invisible Victims: White Males and the Crisis of Affirmative Action.* Westport, Conn: Greenwood Press.
Lynch, K., and Beer, W. 1990. " 'You Ain't the Right Color, Pal.' " *Policy Review.* Winter: 64–67.
Lynn, R. 1978. "Ethnic and Racial Differences in Intelligence: International Comparisons." In Osborne et al., eds.: 261–286.
———. 1987. "The Intelligence of the Mongoloids: A Psychometric, Evolutionary and Neurological Theory." *Personality and Individual Differences* 8: 813–844.
———. 1990a. "New Evidence on Brain Size and Intelligence: A Comment on Rushton and Cain and Vanderwolf." *Personality and Individual Differences* 11: 795–799.
———. 1990b. "Testosterone and Gonadotropin Levels and r/K Reproductive Strategies." *Psychological Reports* 67: 1203–1206.
———. 1991a. "Race Differences in Intelligence: A Global Perspective." *Mankind Quarterly* 31: 254–296.
———. 1991b. "The Evolution of Racial Differences in Intelligence." *Mankind Quarterly* 32: 99–121.
———. 1993. "Further Evidence for the Existence of Race and Sex Diferences in Cranial Capacity." *Social Behavior and Personality* 21, 2: 89–92.
———. 1994a. "Some Reinterpretations of the Minnesota Transracial Adoption Study." *Intelligence* 19: 21–27.
———. 1994b. "The Intelligence of Ethiopian Immigrant and Israeli Adolescents: A Comment on Kaniel and Fisherman." *International Journal of Psychology* 29: 55–56.
———. 1996. "Racial and Ethnic Differences in Intelligence in the United States on the Differential Ability Scale." *Personality and Individual Differences* 20: 271–273.
Lynn, R., and Holmshaw, M. 1990. "Black-White Differences in Reaction Times and Intelligence." *Social Behavior and Personality* 18: 299–308.
Macallum, G. 1967. "Negative and Positive Freedom." *Philosophical Review* 76: 312–334.
Mack, E. 1996. "What is Multiculturalism?" *The Free Market*, October: 684–687.
Maclean, C., Adams, M., Leyshon, W., Workman, P., Reed, T., Gershowitz, H., and Weitkamp, L. 1974. "Genetic Studies on Hybrid Populations. III. Blood Pressure in An American Black Community." *American Journal of Human Genetics* 26: 614–626.
Maclean, F. 1993. "When the Marines Went to Haiti." *Smithsonian* 23: 44–55.
Magnet, M. 1993. *The Dream and the Nightmare.* New York: William Morrow.
Majors, R., and Billson, J. 1993. *Cool Pose: The Dilemma of Black Manhood in America.* New York: Simon and Schuster.
Malina, R. 1969. "Growth and Physical Performance of American Negro and White Children." *Clinical Pediatrics* 8: 476–483.
———. 1973. "Biological Substrata." In Miller and Dreger, eds.: 54–123.
———. 1988. "Racial/Ethnic Variation in the Motor Development and Performance of American Children." *Canadian Journal of Sports Science* 13: 136–143.
Mappes, T., and Zembaty, J., eds. 1997. *Social Ethics*, 5th ed. New York: McGraw-Hill.
Marinoff, L. 1990. "The Inapplicability of Evolutionarily Stable Strategy to the Prisoner's Dilemma." *British Journal for Philosophy of Science* 41: 461–472.
Marriott, M., Brant, B., and Boynton, T. 1995. "CyberSoul Not Found." *Newsweek*, July 31: 62–64.
Matarazzo, J. 1992. "Biological and Physiological Correlates of Intelligence."

Intelligence 16: 257–258.

Matson, W. 1976. *Sentience*. Berkeley, Calif.: University of California Press.

Mauer, M., and Huling, T. 1995. *Young Black Americans and the Criminal Justice System: Five Years Later*. Washington, D.C.: The Sentencing Project.

May, L. 1987. *The Morality of Groups*. Notre Dame, Ind.: University of Notre Dame Press.

Mazur, J. 1987. "An Adjusting Procedure for Studying Delayed Reinforcement." In Commons, et al.: 55–73.

McCall, N. 1994. *Makes Me Wanna Holler*. New York: Random House.

McClellen, D. 1976. "Testing for Competence Rather than 'Intelligence.' " In Block and Dworkin, eds.: 45–71.

McConnell, S. 1994. "The Conspiracy of Silence." *New York Post*, April 1: 19.

McFadden, R. 1994. "13 Years Later, an Arrest in a Manhattan Doctor's Slaying." *New York Times*, July 31: 33.

McGinn, C. 1989. *Mental Content*. London: Basil Blackwell.

McGue, M., Bouchard, T., Iacono, W., and Lykken, D. 1992. "Behavioral Genetics of Cognitive Ability: A Life-Span Perspective." Unpublished. Based in part on a paper presented to the Centennial Meeting of the American Pschological Association, August, 1992, Washington, D.C.

McGuire, T., and Hirsch, J. 1977. "General Intelligence (*g*) and Heritability (*H²*, *h²*)." In Uzgiris and Weizmann, eds.: 25–71.

Mead, L. 1990. *The New Politics of Poverty*. New York: Basic Books.

Mealy, L. 1995. "The Sociobiology of Sociopathy: An Integrated Evolutionary Model." *Behavioral and Brain Sciences* 18: 523–541.

Meehl, P. 1989. "Law and the Fireside Induction." *Behavioral Sciences and the Law* 7, 4: 521–550.

Melnyk, A. 1989. "Is There a Formal Argument Against Positive Rights?" *Philosophical Studies* 55: 205–209.

Mercer, J. 1984. "What is a Racially and Culturally Nondiscriminatory Test? A Sociological and Pluralistic Perspective." In Reynolds and Brown, eds.: 295–356.

Metro Broadcasting, Inc. v. FCC. 1990. 497 U.S. 547.

Michael, J. 1988. "A New Look at Morton's Craniological Research." *Current Anthropology* 29, 2: 349–354.

Miele, F. 1979. "Cultural Bias in the WISC." *Intelligence* 3: 149–164.

Miller, E. 1992. "Some Implications of Bayes' Theorem for Personnel Selection." To appear.

Miller, J., and Rose, R. 1982. "Familial Resemblance in Locus of Control: A Twin-Family Study of the Internal External Scale." *Journal of Personality and Social Psychology* 42, 3: 535–540.

Miller, K., and Dreger, R., eds. 1973. *Comparative Studies of Blacks and Whites in the United States*. New York: Seminar.

Mills, C. 1992. "Reflections on 'Recognition and Development of Academic Talent in Educationally Disadvantaged Students.' " *Exceptionality* 3: 189–192.

Mischel, W. 1958. "Preference for Delayed Reinforcement: An Experimental Study of a Cultural Observation." *Journal of Abnormal and Social Psychology* 56: 57–61.

———. 1961a. "Preference for Delayed Reinforcement and Social Responsibility." *Journal of Abnormal and Social Psychology* 62, 1: 1–7.

———. 1961b. "Father-Absence and Delay of Gratification: Cross-Cultural Comparisons." *Journal of Abnormal and Social Psychology* 63, 1: 116–124.

Modgil, S., and Modgil, C., eds. 1987. *Arthur Jensen: Consensus and Controversy*. London: Falmer International.

Montagu, A. 1972. *Statement on Race: An Annotated Elaboration and Exposition of* the Four Statements on Race Issued by the United Nations Educational, Scientific and Cultural Organization, 3rd ed. New York: Oxford University Press.

Montague, R. 1974. Formal Philosophy. New Haven: Yale University Press.

Montie, J., and Fagan, J. 1988. "Racial Differences in IQ: Item Analysis of the Stanford-Binet at 3 Years." Intelligence 12: 315–33.

Moore, T. 1988. "The Black-on-Black Crime Plague." US News and World Report, Aug. 22: 49–55.

Moore, D., and Erickson, P. 1985. "Age, Gender and Ethnic Differences in Sexual and Contraceptive Knowledge, Attitudes and Behavior." Family and Community Health 8: 38–51.

Moore, E. 1986. "Family Socialization and the IQ Test Performance of Traditionally and Transracially Adopted Black Children." Developmental Psychology 22, 3: 317–326.

Morgan, J., and Falkner, D. 1993. Joe Morgan: A Life in Baseball. New York: Norton.

Morris, C. 1988. "A Hobbesian Welfare State?" Dialogue 27: 653–663.

Mosteller, F., Rourke, R., and Thomas, G. 1970. Probability with Statistical Applications. Reading, Mass: Addison-Wesley.

Munnell, A., Tootell, G., Browne, L., and McEneaney, J. 1996. "Mortgage Lending in Boston: Interpreting HMDA Data." American Economic Review 86, 1: 25–53.

Murray, C. 1984. Losing Ground. New York: Basic Books.

Mussen, P., Harris, S., Rutherford, E., and Keasey, C. 1970. "Honesty and Altruism Among Pre-Adolescents." Developmental Psychology 3: 167-191.

Myrdal, G. 1944. An American Dilemma. New York: Carnegie Foundation.

Nagel, T. 1970. The Possibility of Altruism. Oxford: Oxford University Press.

———. 1986. The View from Nowhere. New York: Oxford Univesity Press.

———. 1987. What Does it All Mean? New York: Oxford.

———. 1979. Moral Questions. New York: Cambridge University Press.

Narveson, J. 1987. "Have We a Right to Non-discrimination?" In D. Poff and W. Waluchow, eds., Business Ethics in Canada. Scarborough, Ontario: Prentice-Hall.

NCAA. 1991. NCAA Academic Performance Study, Report 91-01. NCAA.

Neisser, U., ed. 1986. The School Achievement of Minority Children. Hillsdale, N.J.: Erlbaum.

Neisser, U., et al. 1996. "Intelligence: Knowns and Unknowns." American Psychologist 51: 77–101.

New York Times. 1989. "San Francisco Blacks Found Progressing Slowly." Aug. 23: II, 5.

———. 1991. "Quarter of Newborns in U.S. Were Born to Single Women." June 15: C21.

———. 1993a. "Judge Delays Move to 'Uncluster' White Pupils in Richmond School." Feb. 2: A14.

———. 1993b. "Racial Remark Stalls Job Seeker." July 7: A18.

———. 1993c. "Remark Ends a Job Candidacy." July 29: A21.

Newman, H., Freeman, F., and Holzinger, K. 1937. Twins: A Study of Heredity and Environment. Chicago: University of Chicago Press.

Nichols, R. 1978. "Twin Studies of Ability, Personality and Interests." Homo 29: 158–173.

Nickel, J. 1975. "Preferential Policies in Hiring and Admissions: A Jurisprudential Approach." Columbia Law Review 75: 534–555.

Nickens, H. 1992. "The Rationale for Minority-Targeted programs in Medicine in the 1990s." Journal of the American Medical Association 267, 17: 2390–2395.

Nielsen, K. 1985. Equality and Liberty. Totowa, N.J.: Rowman and Allenheld.

———. 1997. "Radical Welfare Egalitarianism." In Pojman and Westmoreland, eds.: 204–217.

Niemann, Y., and Secord, P. 1995. "Social Ecology of Stereotyping." *Journal for the Theory of Social Behavior* 25: 1–13.

Noble, C. 1978. "Age, Race, and Sex in the Learning and Performance of Psychomotor Skills." In Osborne et al., eds.: 287–378.

Noble, K. 1991. "Once a Colonial Jewel, a City Hurtles Backward." *New York Times*, Nov. 15: A4.

———. 1994. "Zaire's Rich Mines Are Abandoned to Scavengers." *New York Times*, Feb. 21: A3.

Nowell-Smith, P. 1954. *Ethics*. London: Penguin.

Nozick, R. 1975. *Anarchy State and Utopia*. New York: Basic Books.

———. 1982. *Philosophical Explanations*. Cambridge: Harvard University Press.

———. 1989. *The Examined Life*. New York: Simon and Schuster.

Office of the Assistant Secretary of Defense (Manpower, Reserve Affairs and Logistics). 1982. *Profile of American Youth*. Washington, D.C.: Department of Defense.

Ogbu, J. "Variability in Minority School Performance: A Problem in Search of an Explanation." 1987. *Anthropology and Education Quarterly* 18: 312–334.

Orenstein, P. 1994. *School Girls, Young Women, Self-Esteem and the Confidence Gap*. New York: Doubleday.

Orsagh, T. 1980. "Unemployment and Crime." *Journal of Criminal Law and Criminology* 71, 2: 181–183.

Osborne, T. 1978. "Race and Sex Differences in Heritability of Mental Test Performance: A Study of Negroid and Caucasoid Twins." In Osborne et al., eds.: 137–169.

Osborne, T., Noble, C., and Weyl, N., eds. 1978. *Human Variation: The Biopsychology of Age, Race and Sex*. New York: Academic.

Owen, K. 1992. "The Suitability of Raven's Standard Progressive Matrices for Various Groups in South Africa." *Personality and Individual Differences* 13: 149–159.

Papenoe, D. 1992. "Review of S. Hewlett, *When the Bough Breaks* and *Childhood's Future*." *Family Affairs* 5: 16–18.

Papineau, D. 1993. *Philosophical Naturalism*. Oxford: Blackwell.

Parker, R., Nichter, Mi., Nichter, Ma., Vukovic, N., Sims, C., and Ritenbaugh, C. 1995. "Body Image and Weight Concerns among African American and White Adolescent Females: Differences that Make a Difference." *Human Organization* 54, 2: 103–114.

Parks, R., Crockett, D., Tuokko, H., Beattie, B., Ashford, J., Coburn, K., Zec, R., Becker, R., McGeer, P., and McGeer, E. 1989. "Neuropsychological 'Systems Efficiency' and Positron Emission Tomography." *Journal of Neuropsychiatry* 1: 269–282.

Parks, R., Loewenstein, D., Dodrill, K., Barker, W., Yoshii, F., Chang, J., Emran, A., Apicella, A., Shermata, W., and Duara, R. 1988. "Cerebral Metabolic Effects of a Verbal Fluency Test: A PET Scan Study." *Journal of Clinical and Experimental Neuropsychology* 10: 565–575.

Pedersen, N., Plomin, R., Nesselroade, J., and McClearn, G. E. 1992. "A Quantitative Analysis of Cognitive Abilities During the Second Half of the Life Span." *Psychological Science* 3: 346–353.

Peoples, C., Fagan, J., and Drotar, D. 1995. "The Influence of Race on 3-year-old Children's Performance on the Stanford-Binet: Fourth Edition." *Intelligence* 21: 69–82.

Perez-Peña, R. 1992. "Ex-Rockette's Kin Find Relief in Ruling." *New York Times* Dec. 24: B3.

———. 1994. "Landlord Slain in a Robbery and Superintendant is Injured." *New Yor*

Times, Feb. 2: B2

Persaud, H., Ron, M., Baker, G., and Murray, R. 1994. "Volumetric MRI Measurements in Bipolars Compared with Schizophrenics and Healthy Controls." *Psychological Medicine* 24: 689–699.

Pettigrew, T. 1964. *A Profile of the American Negro*. Princeton, N.J.: Van Nostrand.

Pfennig, D., and Sherman, P. 1995. "Kin Recognition." *Scientific American* 272, 6: 98–103.

Pennock, J. 1984. "Correspondence." *Philosophy and Public Affairs* 13: 255–262.

Perry, R. 1926. *General Theory of Value*. Cambridge, Mass.: Harvard University Press.

Piaget, J. 1948. *The Moral Development of the Child*. New York: The Free Press.

Pierson, R. 1993. "Scofflaw's Mom: It's Not a Race Issue." *New York Post*, Jan. 10: 5.

Pieterse, J. 1992. *White on Black: Images of Africa and Blacks in Western Popular Culture*. New Haven, Conn.: Yale University Press.

Pitt, D. 1989. "Despite Revisions, Few Blacks Passed Police Sergeant Test." *New York Times*, Jan. 13: 1.

Plomin, R. 1990a. *Nature and Nurture*. Belmont, Calif.: Wadsworth.

———. 1990b. "The Role of Inheritance in Behavior." *Science* 248: 183–188.

Plomin, R., and Bergeman, C. 1991. "The Nature of Nurture: Genetic Influence on'Environmental' Measures." *Behavior and Brain Science* 14: 373–386.

Plomin, R., and Daniels, D. 1987. "Why Are Children in the Same Family So Different from One Another?" *Behavior and Brain Sciences* 10, 1: 1–16.

Plomin, R., and DeFries, J. 1980. "Genetics and Intelligence: Recent Data." *Intelligence* 4: 15–24.

Plomin, R., and Neiderhiser, J. 1992. "Genetics and Experience." *Current Directions in Psychological Science* 1, 5: 160–163.

Plomin, R., Owen, M., and McGuffin, P. 1994. "The Genetic Basis of Complex Human Behaviors." *Science* 264: 1733–1739.

Plous, S. 1994. "William James' *Other* Concern: Racial Injustice in America." *The General Psychologist* 30, 3: 80–88.

Pojman, L. 1991a. "A Critique of Contemporary Egalitarianism." *Faith and Philosophy* 8: 481–504.

———. 1991b. "Is Contemporary Moral Theory Founded on a Misunderstanding?" *Journal of Social Philosophy* 22, 2: 49–59.

———. 1992a. "The Moral Status of Affirmative Action." *Public Affairs Quarterly* 6: 181–206.

———. 1992b. "Do Animal Rights Entail Moral Nihilism?" *Public Affairs Quarterly* 7: 165–186.

———. 1992c. "Are Human Rights Based on Equal Human Worth?" *Philosophy and Phenomenological Research* 52: 605–622.

Pojman, L., ed. 1991. *Introduction to Philosophy*. Belmont, Calif.: Wadsworth.

Pojman, L., and Westmoreland, R., eds. 1997. *Equality*. New York: Oxford University Press.

Popenoe, D. Review of "S. Hewlett, *When the Bough Breaks*, and R. Louv, *Childhood's Future*." *Family Affairs* 5, 1–2: 16–17.

Porter, J. R., and Washington, R. W. 1979. "Black Identity and Self Esteem. A Review of Studies of Black Self-Concept and Identity." *Annual Review of Sociology* 5: 53–74.

Prior, E. 1982. "Review of J. Richards, *The Skeptical Feminist*." *Australasian Journal of Philosophy* 60, 1: 90–94.

Putnam, H. 1975. *Philosophical Papers I & II*. Cambridge: Cambridge University Press.

Quest, C., ed. 1992. *Equal Opportunities*. London: IEA Health and Welfare Unit.

Quine, W. 1951. *From a Logical Point of View.* Cambridge, Mass.: Harvard.
———. 1960. *Word and Object.* Cambridge, Mass.: MIT Press.
———. 1968. *Ontological Relativity.* New York: Columbia University Press.
———. 1981. *Theories and Things.* Cambridge, Mass.: Harvard University Press.
Rachels, J. 1986. *The Elements of Moral Philosophy.* New York: McGraw-Hill.
Railton, P. 1986. "Moral Realism." *Philosophical Review* 95, 163–207.
Rakowski, E. 1991. *Equal Justice.* New York: Oxford.
Raspberry, W. 1992. "The Making of Certified Criminals." *Washington Post*, Dec. 30: A19.
Rawls, J. 1971. *A Theory of Justice.* Cambridge, Mass.: Harvard University Press.
Rayman, G. 1994. "Queens Woman Raped, Assaulted." *Newsday*, May 26: A8.
Raz, N., Torres, I., Spencer, W., Millman, D., Baertschi, J., and Sarpel, G. 1993. "Neuroanatomical Correlates of Age-Sensitive and Age-Invariant Cognitive Abilities: An *in vivo* MRI Investigation." *Intelligence* 17, 407–421.
Rector, R. 1991. "Food Fight." *Policy Review* 58: 38–43.
———. 1992a. "Requiem for the Welfare State." *Policy Review* 61: 40-46.
———. 1992b. Telephone Interview, July 16.
———. 1992c. *How the Poor Really Live.* Washington, D.C.: The Heritage Foundation.
Rector, R., and Mclaughlin, M. n.d. "A Conservative's Guide to State Level Welfare Reform." Washington, D.C.: The Heritage Foundation.
Reed, T. 1969. "Caucasian Genes in American Negroes." *Science* 165: 762–768.
———. 1995. "Defining 'Race' from World Gene Distributions." Unpublished.
Regan, T. 1985. "The Case for Animal Rights." In Singer, ed.: 13–26.
Reuning, H. 1981. "Psychological Studies of Kalahari Bushment." In Cronbach and Drenth, eds.: 171–181.
———. 1988. "Testing Bushmen in the Central Kalahari." In Irvine and Berry, eds.: 453–486.
Reuters. 1995. "Life Sentences for 2 in Miami In '93 Murder of a Tourist." July 3: Miami.
Reynolds, C., and Brown, R., eds. 1984. *Perspectives in Bias in Mental Testing.* New York: Plenum.
Reynolds, G., ed. 1996. *Race and the Criminal Justice System.* Washington, D.C.: Center for Equal Opportunity.
Richards, D. 1973. "Equal Opportunity and School Financing: Toward a Moral Theory of Constitutional Adjudication." *University of Chicago Law Review* 41: 32–71.
———. 1977. *The Moral Criticism of Law.* Encino, Calif.: Dickinson.
Richards, J. 1980. *The Sceptical Feminist.* Boston: Routledge and Kegan Paul.
Richardson, L. 1993. "Public Shools are Failing Brightest Students, a Federal Study Says." *New York Times*, Nov. 23: A23.
Ridgely, J. 1992. "Toward Equal Access." *Academe*, Sept./Oct.: 13–18.
Rios v. Enterprise Association Steamfitters Local 698. 1984. 501 F 2d. 2nd cir.
Robin, D. 1977. "Comment on the Nature and Scope of Marketing." *Journal of Marketing* 41: 136–138.
Rockwell, Llewellyn. 1993. "Nader Aims at Banks—Unsafe at Any Rate?" *Washington Times*, Aug. 24: F2.
Root, Michael. 1993. *Philosophy of the Social Sciences.* Cambridge, Mass.: Blackwell.
Ropp, L., Visintainer, P., Uman, J., and Treloar, D. 1992. "Death in the City.' *Journal of the American Medical Association* 267, 21: 2905–2909.
Rosenfeld, M. 1991. *Affirmative Action and Justice.* New Haven, Conn.: Yale University Press.

Rosenthal, R., and Jacobson, L. 1968. *Pygmalion in the Classroom*. New York: Rinehart and Winston.

Ross, R., Bernstein, L., Judd, H., Hanisch, R., Pike, M., and Henderson, B. 1986. "Serum Testosterone Levels in Healthy Young Black and White Men." *Journal of the National Cancer Institute* 76: 45–48.

Rothman, S., and Powers, S. 1994. "Execution by Quotas?" *The Public Interest*, Summer: 3–18.

Rothman, S., and Snyderman, M. 1988. *The IQ Controversy*. New Brunswick, N.J.: Transaction.

Rubinstein, D. 1992. "Don't Blame Crime on Joblessness." *Wall Street Journal*, Nov. 9: A10.

———. 1993. "Capitalism, Social Mobility, and Distributive Justice." *Social Theory and Practice* 19: 183–204.

———. 1994. "Cut Cultural Root of Rising Crime." *Insight*, Aug. 8: 18–20.

Rushton, J. P. 1988a. "Race Differences in Behavior: A Review and Evolutionary Analysis." *Personality and Individual Differences* 9: 1009–1024.

———. 1988b. "The Reality of Racial Differences: A Rejoinder and New Evidence." *Personality and Individual Differences* 9: 1035–1040.

———. 1989a. "Japanese Inbreeding Depression Scores: Predictors of Cognitive Differences between Blacks and Whites." *Intelligence* 13: 43–51.

———. 1989b. "Genetic Similarity, Human Altruism, and Group Selection. *"Behavioral and Brain Sciences* 12: 503–519.

———. 1990a. "Race Differences and r/K Theory: A Reply to Silverman." *Ethology and Sociobiology* 11: 131–140.

———. 1990b. "Race And Crime: A Reply to Roberts and Gabor." *Canadian Journal of Criminology* 32: 315–334.

———. 1991a. "Racial Differences: A Reply to Zuckerman." *American Psychologist* 46: 983–987.

———. 1991b. "Mongoloid-Caucasoid Differences in Brain Size From Military Samples." *Intelligence* 15: 35–359.

———. 1991c. "Reply to Willerman on Mongoloid-Caucasoid Differences in Brain Size." *Intelligence* 15: 365–367.

———. 1991d. "Do r-K Strategies Underlie Human Race Differences? A Reply to Weizman et al." *Canadian Psychology* 32: 29–42.

———. 1992. "Cranial Capacity Related to Sex, Rank, and Race in a Stratified Random Sample of 6,325 U.S. Military Personnel." *Intelligence* 16: 401–413.

———. 1993. "Corrections to a Paper on Race and Sex Differences in Brain Size and Intelligence." *Personality and Individual Differences* 15: 229–231.

———. 1994. "Sex and Race Differences in Cranial Capacity from International Labour Office Data." *Intelligence* 19: 281–284.

———. 1995a. "Race and Crime: International Data for 1989–1990." *Psychological Reports* 76: 307–312.

———. 1995b. *Race, Evolution, and Behavior*. New Brunswick, N.J.: Transaction.

Rushton, J., and Bogaert, A. 1987. "Race Differences in Sexual Behavior: Testing an Evolutionary Hypothesis." *Journal of Research in Personality* 21: 529–551.

———. 1988. "Race versus Social Class Differences in Sexual Behavior: A Follow-up Test of the r/K Dimension." *Journal of Research in Personality* 22: 259–272.

———. 1989. "Population Differences in Susceptibility to AIDS: An Evolutionary Analysis." *Social Science and Medicine* 28: 1211–1220.

Rushton, J., Fulker, D., Neale, M., Nias, D., and Eysenck, H. 1986. "Altruism and Aggression: The Heritability of Individual Differences." *Journal of Personality and Social Psychology* 50: 1192–1198.

Russell, B. 1929. *Our Knowledge of the External World*. New York: Norton.

Rutenberg, J. 1993. "Civil Rights Lawyers Blast Village Crackdown." *The West Side Spirit*, March 17: 18.

Ruth, D. 1989. "CBS News Looks at Housing Woes of Chicago's Poor." *Chicago Sun-Times*, Aug. 31: 2-1.

Ryle, G. 1949. *The Concept of Mind*. London:Hutchinson's University Library.

———. 1974. "Intelligence and the Logic of the Nature-Nurture Issue." *Proceedings of the Philosophy of Education Society of Great Britain* 8: 52–60.

Sanoff, A., Minerbrook, S., Thornton, J., and Pezzullo, E. 1993. "Race on Campus." *US News & World Report,* Apr. 19: 52–64.

Sayre-McCord, G., ed. 1988. *Essays on Moral Realism*. Ithaca, N.Y.: Cornell University Press.

Scarr, S., ed. 1981. *Race, Social Class, and Individual Differences in IQ.* Hillsdale, N.J.: Erlbaum.

———. 1992. "Developmental Theories for the 1990s: Development and Individual Differences." *Child Development* 63: 1–19.

Scarr, S., Caparulo, B., Bernardo, M., Tower, R., and Caplan, J. 1983. "Developmental Status and School Achievements of Minority and Non-Minority Children from Birth to 18 Years in a British Midlands Town." *British Journal of Developmental Psychology* 1: 31–48.

Scarr, S., and McCartney, K. 1983. "How People Make Their Own Environments: A Theory of Gene-Environment Effects." *Child Development* 54: 424–435.

Scarr, S., and Weinberg, R. 1981. "IQ Test Performance of Black Children Adopted by White Families." In Scarr, ed.: 109–135.

Schlacter, G., and Goldstein, S. N.d. *Directory of Financial Aid for Minorities*. San Carlos, Calif.: Reference Service Press.

Schlesinger, A. 1992. *The Disuniting of America.* New York: Norton.

Schmidt, F. 1988. "The Problem of Group Differences in Ability Test Scores in Employment Selection." *Journal of Vocational Behavior* 33: 272–292.

Schmidtz, D. 1989. "Contractarianism Without Foundations." *Philosophia* 19: 461–470.

Schoendorf, K., Hogue, C., Kleinman, J., and Rowley, D. 1992. "Mortality Among Infants of Black as Compared with White College-Educated Parents." *The New England Journal of Medicine,* 326, 23: 1522–1526.

Schoenfield, C. 1988. "Blacks and Violent Crime: A Psychoanalytically Oriented Analysis." *Journal of Psychiatry and Law* 16: 269–301.

Schönemann, P. 1987. "Jensen's g: Outmoded Theories and Unconquered Frontiers." In Modgil and Modgil, eds.: 313–327.

Schwartz, J. 1991. Public Debate at Antioch College, Nov. 25.

Schwartz, R. 1986. "I'm Going to Make You a Star." *Midwest Studies in Philosophy* 11: 427–438.

Scott, R. 1993. *Education and Ethnicity: The U.S. Experiment in School Integration*. Washington, D.C.: CSES.

Seligman, D. 1987. "The Esteem Team." *Fortune*, November 14: 134.

———. 1992. *A Question of Intelligence*. New York: Birch Lane.

Sen, A. 1974. "Choice, Orderings and Morality." In Korner, ed.: 54–67.

Sesardic, N. 1993a. "Egalitarianism and Natural Lottery." *Public Affairs Quarterly* 7: 57–69.

———. 1993b. "Heritability and Causality." *Philosophy of Science* 60: 396–418.

———. 1996. "Recent Work on Human Evolution." *Ethics* 106: 128–157.

Seymour, R. 1988. "Why Plaintiffs' Counsel Challenge Tests, and How They Can Successfully Challenge the Theory of 'Validity Generalization.' " *Journal of Vocational Behavior* 33: 331–364.

Shaw, W. 1996. *Business Ethics*, 2nd ed. Belmont, Calif.: Wadsworth.

Sher, G. 1975. "Justifying Reserve Discrimination in Employment." *Philosophy and Public Affairs* 4: 159–170.

———. 1987. *Desert*. Princeton, N.J.: Princeton University Press.

Shields, J. 1962. *Monozygotic Twins Brought Up Apart and Brought Up Together.* London: Oxford University Press.

Shimahara, N. 1991. "Social Mobility and Education: Burakumin in Japan." In Gibson and Ogbu, eds.: 327–353.

Shuey, A. 1966. *The Testing of Negro Intelligence*, 2nd ed. New York: Social Science Press.

———. 1978. "Own-Race Preference and Self-Esteem in Young Negroid and Caucasoid Children." In Osborne, et al., eds.: 199–260.

Singer, P., ed. 1985. *In Defense of Animals*. Oxford: Basil Blackwell.

———. 1993. *Practical Ethics*, 2nd. ed. Oxford: Oxford University Press.

Slote, M. 1980. "Understanding Free Will." *Journal of Philosophy* 77, 3: 136–151.

Smith, B. 1988. *Contingencies of Value*. Cambridge, Mass.: Harvard University Press.

Sniderman, P., and Piazza, T. 1993. *The Scar of Race*. Cambridge, Mass.: Harvard University Press.

Sommers, C. 1984. "Ethics Without Virtue." *The American Scholar* 53: 381–389.

Sorensen, R. 1988. *Blindspots*. Oxford: Oxford University Press.

Sowell, T. 1981. *Markets and Minorities*. New York: Basic Books.

———. 1993. *Inside American Education*. New York: Basic Books.

———. 1995. "Ethnicity and IQ." *The American Spectator*, February: 32–36.

Spanier, J. 1962. *American Foreign Policy Since World War II*. New York: Praeger.

Spitz, H. 1986. *The Raising of Intelligence*. Hillsdale, N.J.: Erlbaum.

Statement on Race. 1951. New York: United Nations Educational Social and Cultural Organization.

Steele, C. 1992. "Race and the Schooling of Black Americans." *The Atlantic Monthly*, April: 68–78.

Steele, S. 1990. *The Content of Our Character*. New York: St. Martin's.

Sterelny, K. 1995. "Understanding Life: Recent Work in Philosophy of Biology." *British Journal for Philosophy of Science* 46: 155–183.

Sternberg, R. J. 1985. "The Black-White Differences and Spearman's *g*: Old Wine in New Bottles that Still Doesn't Taste Good." *The Behavioral and Brain Sciences* 8: 244.

———. 1988. *The Triarchic Mind*. New York: Viking.

———, ed. 1994. *Encyclopedia of Human Intelligence.* New York: Macmillan.

Sternberg, R., Conway, B., Ketron, J., and Bernstein, M. 1981. "People's Conception of Intelligence." *Journal of Personality and Social Psychology* 41, 1: 37–55.

Sternberg, R., and Grigorenko, C., eds. 1995. *Intelligence: Heredity and Environment*. New York: Cambridge University Press.

Stove, D. 1992a. "The Demons and Dr. Dawkins." *The American Scholar*, Winter: 67–78.

———. 1992b. "A New Religion." *Philosophy* 67: 233–240.

Strawson, P. 1974. *Freedom and Resentment and Other Essays*. London: Methuen.

Stringer, C. 1990. "The Emergence of Modern Humans." *Scientific American*, Dec.: 98–104.

Stringer, C., and Andrews, P. 1988. "Genetic and Fossil Evidence for the Origin of Modern Humans." *Science* 239: 1263–1268.

Sturgeon, N. 1984. "Moral Explanation." In Copp and Zimmerman, eds.: 49–78.

———. 1986. "Harman on Moral Explanations of Natural Facts." *The Southern Journal of Philosophy* 24: 69–79.

Swanson, J. and Foster, L., eds. 1970. *Experience and Theory.* Amherst, Mass.: University of Massachussett Press.

Symons, D. 1979. *The Evolution of Human Sexuality*. New York: Oxford University Press.

Szasz, T. 1970. *Ideology and Insanity*. New York: Anchor Doubleday.

Tashakkori, A. 1993. "Race, Gender and pre-Adolescent Self-Strucure: A Test of [sic] Construct-Specificity Hypothesis." *Personality and Individual Differences* 14: 591–598.

Tashakkori, A., and Thompson, V. 1991. "Race Differences in Self-Perception and Locus of Control during Adolescence and Early Adulthood." *Genetic, Social and General Psychological Monographs* 117: 135–152.

Taylor, J. 1992. *Paved with Good Intentions*. New York: Carol and Graf.

———. 1994. "Race, Crime and Gun Control." *National Review*, May 16: 44–45.

Tellegen, A., Lykken, D. et al. 1988. "Personality Similarity in Twins Reared Apart and Together." *Journal of Personality and Social Psychology* 54, 6: 1031–1039.

Temkin, L. 1986. "Inequality." *Philosophy and Public Affairs* 15: 99–121.

Terry, D. 1995. "Detroit Family in the Jaws of a Monster." *New York Times*, Dec. 4: A12.

Thoday, J. 1976. "Educability and Group Differences." In Block and Dworkin, eds.: 146–155.

Thomas, L. 1990. "Next Life, I'll Be White." *New York Times* Aug. 13: A15.

———. 1992. "Statistical Badness." *Journal of Social Philosophy* 23: 31–41.

Thompson, L, Detterman, D., and Plomin, R. Differences in Heritability Across Groups Differing in Ability Revisited." *Behavior Genetics* 23: 331–336.

Thomson, J. 1977. "Preferential Hiring." In Cohen, Nagel and Scanlon, eds.: 19–39.

Thorpe, J. 1980. *Free Will*. London: Routledge and Kegan Paul.

Tobias, P. 1970. "Brain-size, Grey Matter and Race—Fact or Fiction?" *American Journal of Physical Anthropology* 32: 3–36.

Tooby, J., and Cosmides, L. 1990. "On the Universality of Human Nature and the Uniqueness of the Individual: The Role of Genetics and Adaptation." *Journal of Personality* 58, 1: 17–67.

Travis, J. 1993. "Schools Stumble on an Afrocentric Science Essay." *Science*: 262: 1121–1122.

Treaster, J. 1992. "Teen-Age Murderers: Plentiful Guns, Easy Power." *New York Times*, May 24: 1–42.

———. 1994. "Brooklyn Youth, 17, Is Arrested in Hiding After Officer Is Shot." *New York Times*, Aug. 4: B2.

Tribe, L. 1971. "Trial by Mathematics." *Harvard Law Review* 84: 1329–1393.

Trivers, R. 1971. "The Evolution of Reciprocal Altruism." *Quarterly Review of Biology* 46: 35–57.

United Steelworkers of America v. Weber. 1979. 443, U.S. 193, 99 S. Ct. 2721, 61 L.Ed.2d 480.

University of California Regents v. Bakke. 1978. 438 U.S. 265, 98 S. Ct. 2733, 57 L.Ed.2d 750.

Urbach, P. 1974. "Progress and Degeneration in the IQ Debate, I & II." *British Journal for Philosophy of Science* 25: 99–135, 235–259.

Uytterbrouck, O. 1993. "Police Study Links Blacks to 80% of Violent Crime." *Arkansas Democrat-Gazette*, June 13: 16.

Uzgirix, I., and Weizman, F., eds. 1977. *The Structuring of Experience*. New York: Plenum.

Vacco, D., and Donzigen, S. 1995. "Should Prisoners Have Voting Rights?" *New York Daily News*, Dec. 21: 17.

Van Fraassen, B. 1980. *The Scientific Image*. Oxford: Oxford University Press.

Van Sertima, I. 1983. *Blacks in Science: Ancient and Modern*. New Brunswick, N.J.:

Transaction.

Van Valen, L. 1974. "Intelligence and Brain Size in Man." *American Journal of Physical Anthropology* 40: 417–424.

Vernon, P., ed. 1993. *Biological Approaches to the Study of Human Intelligence.* Norwood, N.J.: Ablex.

Vincent, K. 1991. "Black/White IQ Differences: Does Age Make the Difference?" *Journal of Clinical Psychology* 47, 2: 266–270.

Vining, D. 1983. "Mean IQ Differences in Japan and the United States." *Nature* 301: 738.

Vold, G., and Bernard, T. 1986. *Theoretical Criminology*, 3rd ed. New York: Oxford University Press.

Waldman, I., Weinberg, R., and Scarr, S. 1994. "Racial-Group Differences in IQ in the Minnesota Transracial Adoption Study: A Reply to Levin and Lynn." *Intelligence* 19: 29–44.

Walinsky, A. 1995. "The Crisis of Public Order." *The Atlantic.* June 1995: 39-54.

Walker, C. 1995. *Long Time Coming: A Black Athlete's Coming-of-Age in America.* New York: Grove.

Walker, L. 1991. "A Libertarian Defense of Moral Responsibility." In Pojman, ed.: 453–457.

Walls, J. 1989. "Salomon Jogging Park Memories." *New York Magazine*, Aug. 28: 11–12.

Walzer, M. 1973. "In Defense of Equality." *Dissent*, Fall: 399–408.

———. 1983. *Spheres of Justice.* New York: Basic.

Wards Cove Packing Co. v. Antonio. 1989. 490 U.S. 642.

Washington Post. 1992. "The Making of Certified Criminals." Dec. 30: A19.

Wasserstrom, R. 1985a. "On Racism and Sexism." In Wasserstrom, ed.: 1–29.

Wasserstrom, R., ed. 1985b. *Today's Moral Problems*, 3rd ed. New York: Macmillan.

Wattenberg, B. 1987. *The Birth Dearth.* New York: Pharos.

Wechsler, D. 1939. *The Measurement of Adult Intelligence.* Baltimore: Williams and Wilkins.

Weinberg, R. 1989. "Intelligence and IQ." *American Psychologist* 44: 98–104.

Weinberg, R., Scarr, S., and Waldman, I. 1992. "The Minnesota Transracial Adoption Study: A Follow-up of IQ Test Performance at Adolescence." *Intelligence* 16: 117–135.

Weinberg, W., Dietz, S., Penick, E., and McAlister, W. 1974. "Intelligence, Reading Achievement, Physical Size, and Social Class." *Journal of Pediatrics* 85, 4: 482–489.

Weiss, M., and Zinsmeister, K. 1996. "When Race Trumps Truth in the Courtroom." In Reynolds, ed.: 57–64.

Weiss, R. 1970. "The Effect of Education on the Earnings of Blacks and Whites." *Review of Economic Statistics* 52, 2: 150–159.

Weizmann, F., Wiener, N., Wiesenthal, D., and Ziegler, M. 1991. "Eggs, Eggplants and Eggheads: A Rejoinder to Rushton." *Canadian Psychology* 31: 1–13.

Wertheimer, R. 1971. "Understanding the Abortion Debate." *Philosophy and Public Affairs* 1, 1: 67–95.

West, C. 1997. "Nihilism in Black America." In Mappes and Zembaty, eds.: 334–343.

Weyl, N. 1960. *The Negro in american Civilization.* Washington, D.C.: Public Affairs Press.

Whitman, D. 1991. "What Keeps the Poor Poor?" *U.S. News and World Report* Oct. 21: 42–46.

Whitman, D., and Bowermaster, D. 1993. "A Potent Brew: Booze and Crime." *U.S. News World Report.* May 31: 57–59.

Whitney, G. 1990a. "On Possible Genetic Bases of Race Differences in Criminality." In

Ellis and Hoffman, eds.: 134 –149.

———. 1990b. "A Contextual History of Behavior Genetics." In Hahn, et al.: 7–24.

———. 1995. "Presidential Address to the Behavioral Genetics Association." Richmond, Va., unpublished.

Whittaker, K. 1990. *Black Victims*. Washington, D.C.: Bureau of Justice Statistics, Office of Justice Programs, U.S. Department of Justice.

Wicker, T. 1987. "The Greatest Tragedy." *New York Times*, June 17: A22.

Wickett, J., Vernon, P., and Lee, D. 1994. "*In Vivio* Brain Size, Head Perimeter, and Intelligence in a Sample of Healthy Adult Females." *Personality and Individual Differences* 16, 6: 831–837.

Wilkerson, Isabel. 1990. "Call for Black Militia Stuns Milwaukee." *New York Times*, Apr. 6: A12.

Willerman, L., Schultz, R., and Rutledge, J. 1991. "*In vivo* Brain Size and Intelligence." *Intelligence* 15: 223–228.

Williams, C. 1974. *The Destruction of Black Civilization; Great Issues of a Race From 500 B.C. to 2000 A.D.* Chicago: Third World Press.

Wilson, E. O. 1975. *Sociobiology*. Cambridge, Mass.: Harvard University Press.

———. 1978. *On Human Nature*. Cambridge, Mass.: Harvard University Press.

Wilson, W. J. 1987. *The Truly Disadvantaged*. Chicago: University of Chicago Press.

———. 1996. *When Work Disappears:The World of the New Urban Poor*. New York: Knopf.

Winch, P. 1958. *The Idea of a Social Science*. London: Routledge and Kegan Paul.

Wittgenstein, L. 1953. *Philosophical Investigations*. Trans. G. Anscombe. Oxford: Basil Blackwell.

Wober, M. 1974. "Towards an Understanding of the Kiganda Concept of Intelligence." In Berry and Dasen, eds.: 261–280.

Wolfe, T. 1987. *Bonfire of the Vanities*. New York: Farrar, Strauss Giroux.

Wolff, C. 1994. "Boy's Arrest Is Greeted With Sorrow, Not Surprise." *New York Times*. Feb. 7: B4.

Wolff, R., ed. 1967. *Kant*. Garden City, N.Y.: Doubleday.

Wong, D. 1984. *Moral Relativity*. Berkeley: University of California Press.

Woodfield, A. 1976. *Teleology*. Cambridge: Cambridge University Press.

Worthy, M., and Markle, A. 1970. "Racial Differences in Sports Activity." *Journal of Personality and Social Psychology* 16: 439–443.

Wright, B. 1987. *Black Robes, White Justice*. New York: Lyle Stuart.

Wright, L. 1973. "Functions." *Philosophical Review* 82: 139–168.

———. 1976. *Teleological Explanations*. Berkeley: University of California Press.

Wright, R. 1995. "The Biology of Violence." *The New Yorker*, March 13: 68–77.

Wycliff, D. 1990. "Science Careers are Attracting Few Blacks." *New York Times*, June 8: A1–A4.

Yates, S. 1991. *Beyond Affirmative Action*. Chicago: Heartland Institute.

———. 1992. "Multiculturalism and Epistemology." *Public Affairs Quarterly* 6: 435–456.

Yee, A. 1992. "Asians as Stereotypes and Students: Misperceptions that Persist." *Educational Psychology Review* 4: 95–132.

Yee, A., Fairchild, H., Weizmann, F., and Wyatt, G. 1993. "Addressing Psychology's Problem's with Race." *American Psychologist* 48: 1132–1140.

———. 1995. "Readdressing Psychology's Problems with Race." *American Psychologist* 50: 46–47.

Zeckhauser, R., ed. 1991. *Strategy and Choice*. Cambridge, Mass.: The MIT Press.

Zelnick, B. 1996. *Backfire*. Washington, D.C.: Regnery.

Zigler, E., and Berman, W. 1983. "Discerning the Future of Childhood Intervention." *American Psychologist* 38: 894–906.

Zindi, F. 1994. "Differences in Psychometric Performance." *The Psychologist* 7, Dec.: 549–552.

Index

About the Author

MICHAEL LEVIN is Professor of Philosophy at the City College of New York and the Graduate Center, City University of New York. Among his earlier publications are *Metaphysics and the Mind-Body Problem* and *Feminism and Freedom*.

ISBN 0-275-95789-6

HARDCOVER BAR CODE